Introductory Algebra
Equations and Graphs

Katherine Yoshiwara
Los Angeles Pierce College

Bruce Yoshiwara
Los Angeles Pierce College

THOMSON
™
BROOKS/COLE

Australia • Canada • Mexico • Singapore • Spain • United Kingdom • United States

THOMSON

BROOKS/COLE

Publisher: *Bob Pirtle*
Senior Editor: *Jennifer Huber*
Sr. Assistant Editor: *Rachael Sturgeon*
Editorial Assistant: *Carrie Dodson*
Technology Project Manager: *Star MacKenzie*
Sr. Marketing Manager: *Leah Thomson*
Marketing Assistant: *Jennifer Gee*
Advertising Project Manager: *Bryan Vann*
Project Manager, Editorial Production: *Mary Vezilich*

Print/Media Buyer: *Barbara Britton*
Production Service: *Matrix Productions*
Text Designer: *Carolyn Deacy*
Illustrator: *Asterisk Inc.*
Cover Designer: *Roy Neuhaus*
Cover Photo: *Shigeru Tanaka*
Cover Printer: *CTPS*
Compositor: *Better Graphics*
Printer: *CTPS*

Printed in China
4 5 6 7 8 09 08 07

For more information about our products, contact us at:
Thomson Learning Academic Resource Center
1-800-423-0563
For permission to use material from this text, contact us by:
Phone: 1-800-730-2214   **Fax:** 1-800-730-2215
Web: http://www.thomsonrights.com

Library of Congress Control Number: 2002106709

Student Edition: ISBN-13: 978-0-534-35824-2
ISBN-10: 0-534-35824-1
Instructor's Edition: ISBN-13: 978-0-534-39668-8
ISBN-10: 0-534-39668-2

Brooks/Cole–Thomson Learning
20 Davis Drive
Belmont, CA 94002
USA

Asia
Thomson Learning
5 Shenton Way #01-01
UIC Building
Singapore 068808

Australia/New Zealand
Thomson Learning
102 Dodds Street
Southbank, Victoria 3006
Australia

Canada
Nelson
1120 Birchmount Road
Toronto, Ontario M1K 5G4
Canada

Europe/Middle East/Africa
Thomson Learning
High Holborn House
50/51 Bedford Row
London WC1R 4LR
United Kingdom

Latin America
Thomson Learning
Seneca, 53
Colonia Polanco
11560 Mexico D.F.
Mexico

Spain/Portugal
Paraninfo
Calle Magallanes, 25
28015 Madrid, Spain

Contents

Preface

Introductory Algebra: Equations and Graphs was written with two goals in mind: to present the skills of algebra in the context of modeling and problem solving, and to engage students as active participants in the process of learning. To achieve these goals, we present ideas from verbal, numerical, algebraic, and graphical viewpoints. We use graphs extensively to illustrate algebraic technique and to help students visualize relationships between variables. Although functions are not explicitly part of an elementary algebra course, our book builds an intuitive framework for later study of functions.

Content

Throughout the book, new topics are studied in the context of applications. Chapter 1 begins with a study of tables and graphs, providing groundwork for the concept of variable. Chapter 2 develops algebraic skills for solving linear equations alongside the corresponding graphing skills, and concludes with a section on lines of best fit. In Chapter 3, graphing examples of direct variation motivate the definition of slope as a rate of change. In Chapter 4, students begin the study of quadratic equations by using trial and error with a table of values to find the price that results in a given revenue. They then use the table to graph the revenue function and to solve the equation graphically. Factoring skills and the quadratic formula are presented in conjunction with locating the intercepts of a parabola.

Chapter 5 continues the study of polynomials begun in Chapter 4 and includes a section on inverse variation. These functions have many applications suitable for beginning students, and their graphs, while simple in shape, display the most important features of more general rational functions. Chapters 5 and 6 include more challenging algebraic skills: "simplifying" algebraic fractions and expressions containing radicals. After examining texts from numerous fields that use mathematics, we have tailored the complexity of the calculations to a level that students may actually encounter in later courses, and tried to provide some context for their use.

Organization

Our goals have also shaped the style and delivery of the text. A number of design features contribute to the interactive nature of the textbook. Each section of the text has two parts, a Reading assignment and a Lesson. The Reading assignments are intended to help students learn to read a math text, and students can complete them outside of class. At first, the Reading assignments are quite short, but fairly soon increase in length and content. We have varied the format of these Reading sections, so that some present an application to motivate the material in the Lesson, and others discuss preliminary concepts and definitions.

The Reading assignments end with a set of Skills Review problems (with answers) for students to complete on their own. The Skills Review problems focus on an arithmetic or algebraic technique needed for the Lesson that follows. The Reading assignment and Skills Review may be discussed in class, or they can be the subject of a short quiz. (Some possible quiz questions for each section are provided in the Instructor's Manual.)

The Lessons are also designed to encourage student participation. Some Lessons consist of Activities for students to complete in groups or with guidance from the instructor; others can be presented in a lecture format. All the Lessons include Exercises for students, so that instructors who wish to do so can incorporate group work into class sessions. The Activities and Exercises include space for students to record their work and blank grids for graphing.

Each section concludes with a set of Homework problems to reinforce and extend the topics in the Lesson. We have tried to break up "skills" practice into smaller sets of exercises and to combine them with conceptual questions, graphing, and applications of various types. In addition to the Homework problems, each chapter includes Midchapter Review problems and a Chapter Summary and Review. Answers to odd-numbered Homework problems are provided, as well as a Glossary of mathematical terms. A Review of Arithmetic Skills, with Exercises, appears in Appendix A.

Graphing

Because choosing appropriate scales for the axes is a time-consuming task for beginning students, the text includes labeled grids for most of the graphing exercises. Ready-made grids allow students to consider a wider range of examples (with "harder" numbers) and to focus on the properties of the graphs, such as intercepts and slope, and on interpreting the information given by the graph.

A note on technology: Some instructors may wish to use graphing calculators with their classes, and we offer suggestions for calculator activities in the *Instructor's Manual*. However, beginning students also need experience drawing graphs by hand. The correspondence between a graph on paper and the calculator image is not always apparent to beginners, and (at least for the present generation of calculators) is not perfect. Students must learn to "see" the idealized graph behind the calculator's screen, and comparing a calculator graph with a hand-drawn graph is a good way to do this. In addition, the ability to sketch a graph quickly by hand is a skill we take for granted, but it must be learned through practice. Students are often reluctant to spend sufficient time with graphing, and we find that providing scaled grids encourages participation.

For the Instructor

- **Annotated Instructor's Edition** (0-534-39668-2) This special version of the complete student text contains a Resource Integration Guide and answers printed next to all respective exercises. Graphs, tables, and other answers appear in a special answer section in the back of the text.
- **Test Bank** (0-534-38644-X) The *Test Bank* includes 8 tests per chapter as well as 3 final exams. The tests are made up of a combination of multiple-choice, free-response, true/false, and fill-in-the-blank questions.
- **Complete Solutions Manual** (0-534-38647-4) The *Complete Solutions Manual* contains teaching notes, reading questions, and complete worked-out solutions to all of the problems in the text.

- **Online Tutoring with InfoTrac®** With Brooks/Cole Assessment's online tutoring, your students have dynamic, flexible online tutorial resources at their fingertips. By entering a PIN code packaged with their textbook, students gain access to *BCA Tutorial*, a text-specific tutorial with step-by-step explanations, exercises, and quizzes. They also have access to *vMentor* with live, one-on-one help from an experienced mathematics tutor. In addition to robust tutorial services, your students also receive anytime, anywhere access to Info Trac College Edition. This online library offers the full text of articles from almost 4000 scholarly and popular publications, updated daily and going back over two decades. Both adopters and their students receive unlimited access for four months.

- **BCA Instructor Version** (0-534-38649-0) With a balance of efficiency and high-performance, simplicity and versatility, *Brooks/Cole Assessment* gives you the power to transform the learning and teaching experience. *BCA Instructor Version* is made up of two components, *BCA Testing* and *BCA Tutorial. BCA Testing* is a revolutionary, Internet-ready, text-specific testing suite that allows instructors to customize exams and track student progress in an accessible, browser-based format. *BCA* offers full algorithmic generation of problems and free response mathematics. *BCA Tutorial* is a text-specific, interactive tutorial software program, that is delivered via the Web (at http://bca.brookscole.com) and is offered in both student and instructor versions. Like *BCA Testing*, it is browser-based, making it an intuitive mathematical guide even for students with little technological proficiency. So sophisticated, it's simple, *BCA Tutorial* allows students to work with real math notation in real time, providing instant analysis and feedback. The tracking program built into the instructor version of the software enables instructors to carefully monitor student progress. The complete integration of the testing, tutorial, and course management components simplifies your routine tasks. Results flow automatically to your gradebook and you can easily communicate to individuals, sections, or entire courses.

- **http://mathematics.brookscole.com** When you adopt a Thomson-Brooks/Cole mathematics text, you and your students will have access to a variety of teaching and learning resources. This Web site features everything from book-specific resources to newsgroups. It's a great way to make teaching and learning an interactive and intriguing experience.

- **Text-Specific Videotapes** (0-534-38648-2) This text-specific videotape set, available at no charge to qualified adopters of the text, features 10- to 20-minute problem-solving lessons that cover each section of every chapter.

- **MyCourse 2.0** (0-534-16641-5) Ask us about our new free online course builder! Whether you want only the easy-to-use tools to build it or the content to furnish it, Brooks/Cole offers you a simple solution for a custom course Web site that allows you to assign, track, and report on student progress; load your syllabus; and more. Visit us at http://mycourse.thomsonlearning.com

The above items are available to qualified adopters. Please consult your local sales representative for details.

For the Student

- **Student Solutions Manual** (0-534-38645-8) The *Student Solutions Manual* provides worked-out solutions to the odd-numbered problems in the text.

- **Student Resource Manual** (0-534-40110-4) The *Student Resource Manual* contains additional application problems and projects related to the text.

- **http://mathematics.brookscole.com** At this Web site, students will have access to a variety of learning resources. The Web site features everything from book-specific resources to newsgroups. It's a great way to make learning an interactive and intriguing experience.

- **Online Tutoring with InfoTrac**® Brooks/Cole Assessment's online tutoring gives students dynamic, flexible online tutorial resources at their fingertips. Students enter a PIN code packaged with their textbook to gain access to *BCA Tutorial*, a text-specific tutorial with step-by-step explanations, exercises, and quizzes. They also have access to *vMentor* with live, one-on-one help from an experienced mathematics tutor. In addition to robust tutorial services, students also receive anytime, any-where access to InfoTrac College Edition. This online library offers the full text of articles from almost 4000 scholarly and popular publica-tions, updated daily and going back as much as two decades. Access is unlimited for four months.

- **BCA Tutorial Student Version** (0-534-38643-1) This text-specific, interactive tutorial software is delivered via the Web (at http://bca.brookscole.com). It is browser-based, making it an intuitive mathematical guide, even for students with little technological profi-ciency. So sophisticated, it's simple, *BCA Tutorial* allows students to work with real math notation in real time, providing instant analysis and feedback. *BCA Tutorial Student Version* is also available on CD-ROM for those students who wish to use the tutorial locally on their computer and not via the Internet.

- **vMentor** is an ideal solution for homework help and tutoring. It is an efficient way to provide supplemental assistance that can substantially improve student performance, increase test scores, and enhance techni-cal aptitude. When *vMentor* is bundled with a Brooks/Cole mathematics text, students will have access, via the Web, to highly qualified tutors with thorough knowledge of the textbooks. If students get stuck on a particular problem or concept, they need only log on to *vMentor*, where they can talk to *vMentor* tutors who will skillfully guide them through the problem using the whiteboard for illustration. To take a live test-drive of *BCA* or *vMentor*, visit us online at: http://bca/brookscole.com

- **Interactive Video Skillbuilder CD** (0-534-38646-6) Think of it as portable office hours! The *Interactive Video Skillbuilder CD-ROM* con-tains more than eight hours of video instruction. The problem for each video lesson is shown next to the viewing screen so that students can try working the problem before watching the solution. To help students evaluate their progress, each section contains a 10-question Web Quiz (the results of which can be emailed to the instructor) and each chapter contains a chapter test, with answers to each problem on each test. The CD also includes *MathCue* tutorial and testing software. *MathCue* is keyed to the text and includes these components:

- *MathCue Skill Builder*—Presents problems to solve, evaluates answers and tutors students by displaying complete solutions with step-by-step explanations.

- *MathCue Quiz*—Allows students to generate large numbers of quiz problems keyed to problem types from each section of the book.

- *MathCue Chapter Test*—Also provides large numbers of problems keyed to problem types from each chapter.

- *MathCue Solution Finder*—This unique tool allows students to enter their own basic problems and receive step-by-step help as if they were working with a tutor.

- Score reports for any *MathCue* session can be printed and handed in for credit or extra-credit.

Acknowledgments

We would like to thank the following reviewers for their helpful comments and suggestions: *Phil Farmer*, Diablo Valley College; *Nancy Henry*, Indiana University; *Kathryn Hodge*, Midland College; *Bruce Hoelter*, Raritan Valley Community College; *Judy Jones*, Valencia Community College; *Martin MacDonald*, West Los Angeles College; *Jeff McGowan*, Central Connecticut State University; *Brian Peterman*, Century College; *Shirley Robertson*, High Point University; *Carol Rychly*, Augusta State University; *Debra Swedberg*, Casper College; *Betty Weissbecker*, J. Sargent Reynolds Community College.

Katherine Yoshiwara
Bruce Yoshiwara

Note to Students

This workbook contains tables and grids for the activities and problems in your textbook. The grids are already labeled with appropriate scales for the graphs you will draw. We hope that providing these grids will eliminate some of the time-consuming work involved in drawing a graph, and allow you to concentrate on the mathematics.

The lessons in your textbook include Exercises printed on a blue background. There are copies of these Exercises in the workbook, with space for you to show your work and record your answers. You should try these Exercises as you read the text, to see if you understand the material.

You should try to keep your workbook up to date as your course progresses. You will also need a spiral or loose-leaf notebook for class notes and the rest of your homework problems. Your workbook and notebook will be useful study aids when you are preparing for exams.

How to Be Successful in Your Math Class

The key to success in a math class (as in most endeavors) is persistence. You cannot learn mathematics in one great rush the night before the exam; but you can master it in small chunks a little at a time. You should plan to study math for at least one hour every night. Don't give up until you have a good grasp of the lesson and can work the problems on your own. If you get behind in a math class it is very difficult to catch up.

1. **Attend class every day.**
   Studies have shown that success in math classes is correlated strongly with attendance. If you must miss a class, find out beforehand what the class will cover. Read the lesson and complete the assignment anyway, just as if you had attended.

   a. Use class time wisely. This is your best opportunity to learn the material.

   b. Take notes. Learn to summarize what the instructor says, not just what he or she writes on the board.

   c. Don't be afraid to ask questions when you don't follow the lesson.

2. **Read the text book.**
   Reading a math book is not like reading a novel. You will need to read the material more than once to understand and retain it.

   a. Read the new material *before* it will be covered in class.

   b. Read with a pencil in hand so that you can make notes to yourself, underline important points, or put question marks in the margins.

   c. Read the section again after it has been covered in class.

3. **Look over your handouts and class notes.**
   The sooner you can review your notes after class, the better. People forget most of what they hear very quickly, and reviewing your notes will help you retain the new information.

   a. Look for points where your notes reinforce the material in the textbook.

   b. Try to fill in any steps or information you may have missed in class.

   c. Write a sentence or two summarizing the main points of the lesson.

4. Do the homework problems.

Most of your learning takes place when you work problems. If you do some of your work in a study group or tutoring center, you will have someone to consult as soon as you hit a snag.

  a. If you get stuck on a problem, refer to the textbook or your notes for help.

  b. Call a classmate on the phone and try to figure out together the problems you had trouble with.

  c. Mark any problems you can't get, but don't stop! Skip those problems for now, and continue on to the end of the assignment.

5. Get help right away.

Mathematics builds upon earlier material, so if you don't understand today's lesson you will have even more trouble tomorrow or the next day.

  a. Make a list of points you don't understand and problems you need help with.

  b. Ask your instructor or a tutor for help *today*—don't put it off!

  c. Fill in your notes with the answers to your questions, and make sure you can work all the problems that gave you trouble.

6. Prepare for exams.

In addition to keeping up with daily work, you must prepare specifically for exams. Always study 100% of the material the exam will cover. If you omit some topics, you won't be sure which problems you should work on during the exam!

  a. Begin studying for the exam a week ahead of time, so that you will have a chance to get help on any topics you are unsure about.

  b. Make a check-list or outline of the material the exam will cover, and review each topic until you have mastered it.

  c. Have a classmate or tutor make up a sample exam (or make one yourself), and practice working problems under exam conditions.

Variables and Equations

Algebra is the language of mathematics. Using algebra, you will be able to express the relationship between two quantities in a precise way and to organize information in useful forms. If we understand how one quantity depends on another, we can often predict how they will behave in the future and plan our actions accordingly. Professionals in almost every field use mathematics to formulate policy and strategies. Environmental scientists plan resource management, health officials try to control the spread of epidemics, and business investors seek to understand the forces that shape the market. All of these tasks require a solid understanding of graphs and equations.

1.1 Tables and Bar Graphs

Reading

Numerical information is often presented in tables. A table organizes data and makes it easier to draw conclusions from them. In this section, we use tables to investigate how life expectancy has changed during this century, how average income depends on education, and how long savings will last with different interest rates and annual withdrawals.

Skills Review

In this chapter, you will need skills in rounding decimal numbers and using percents. Try the following review exercises:

1. Round this number to hundredths: 387.256.

2. Round this number to three decimal places: 2.6498.

3. What is 30% of 42?

4. 16 is 40% of _____.

5. What percent of 36 is 5.76?

Answers: *1.* 387.26   *2.* 2.650   *3.* 12.6   *4.* 40   *5.* 16%

If you had trouble with the exercises, review the following sections of Appendix A before starting the lesson:

Rounding decimal numbers (Section A.6)
Using percents (Section A.9)

Lesson

Tables

Activity 1

Table 1.1 gives the number of complaints received, and the number of passengers carried, by 15 U.S. airlines in 1995.

Table 1.1

Airlines	Complaints	Passengers (in thousands)	Calculations and results
Alaska	53	10,084	
American	497	79,511	
American Eagle	35	11,900	
Continental	368	35,013	
Delta	504	86,909	
Hawaiian Airlines	31	4776	
Markair	263	990	
Nations Air Express	36	82	
Northwest	257	49,313	
Southwest	107	50,039	
Sun Jet International	142	486	
TWA	291	21,551	
United	597	78,664	
USAir	379	55,674	
ValuJet	83	5145	

Source: *Los Angeles Times*, June 2, 1996.

Use the table to help you answer these questions:

1. Which airline had the most complaints in 1995? Which had the fewest?

2. Which airline carried the most passengers in 1995? Which carried the fewest?

3. Locate the data for Nations Air Express and for American Eagle. Which of these two airlines had the worse complaint record in 1995? Why did you choose the one you did?

4. Would you say that the airline with the most complaints in 1995 had the worst complaint record? How can you use the data to compare the complaint records fairly? Please describe your method: What computations will you use?

5. Use the last column of Table 1.1 to carry out your method. (Round your results to three decimal places.) According to your calculations, which airline had the worst complaint record in 1995?

Activity 2

Table 1.2 shows life expectancy at birth and at age 50 for each decade of the 20th century. (Life expectancy at birth is the average age to which all people born in that year will live. How would you define life expectancy at age 50?) Use the table to help answer these questions:

1. What is the average life span for people born in 1930? What is the average life span for people who were 50 years old in 1930?

2. In which decade did life expectancy at birth increase the most?

3. Is there any decade in which either life expectancy at birth or life expectancy at age 50 decreased?

4. Over the century, what has happened to the gap between life expectancy at birth and life expectancy at age 50? Why do you think this has occurred?

5. Do you think that life expectancy will continue to increase indefinitely?

Table 1.2

Life expectancy		
Year	At birth	At age 50
1900	49	71
1910	51	71
1920	58	73
1930	59	72
1940	63	73
1950	68	74
1960	70	75
1970	71	76
1980	74	78
1990	75	79

Source: *Consumer Reports*, January 1992.

Bar Graphs

To answer questions such as 4 and 5 in Activity 2, it is helpful to look at a graph of the data. A graph can reveal trends in the data that might be difficult to see in a table. The graph in Figure 1.1 on page 4 is called a **bar graph**. The years listed in the first column of Table 1.2 are displayed along the graph's horizontal scale, or **axis**. The heights of the bars at each decade give the life expectancies at birth and at age 50 for that decade.

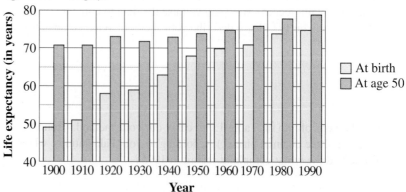

Figure 1.1 A bar graph

Activity 3

Use the graph to help you answer these questions:

1. Locate the bars corresponding to 1940 on the graph. Find the heights of the bars by comparing them to the vertical scale on the left side of the graph. Do the heights match the values for life expectancy given in Table 1.2?

2. Look at your answer to question 2 of Activity 2. Explain how to answer that question by looking at the graph instead of at the table.

3. In what year did life expectancy at birth catch up to life expectancy for 50-year-olds in 1900?

4. Look at the bars for 1900. In 1900, the gap between life expectancy at birth and life expectancy at age 50 was $71 - 49$, or 22 years less. In which decade did this gap first decrease to 10 years or less? Check your answer using Table 1.2.

5. Life expectancy has been increasing throughout the century. Is it increasing more rapidly or more slowly as time goes on? Does this change your answer to question 5 of Activity 2, or support it? Why? (Scientists believe that life expectancy will level off at 85 years, unless the fundamental rate of aging can be slowed.)

Making a Bar Graph

It is not hard to create a bar graph from a table of information. Table 1.3 shows the number of bicycle commuters in the United States from 1983 to 1992, in millions.

Table 1.3

Year	1983	1984	1985	1986	1987	1988	1989	1990	1991	1992
Number of commuters (in millions)	1.5	1.6	1.8	2.0	2.2	2.7	3.2	3.5	4.0	4.3

Activity 4

Follow these steps to make a bar graph for this information:

1. Look at the first row of Table 1.3, the years. These values will lie along the horizontal axis of your graph. How many values are there?

2. Draw a horizontal line for the year axis, and place the years from 1983 to 1992 between *equally spaced* tick marks from left to right. (Use the grid below to help you.)

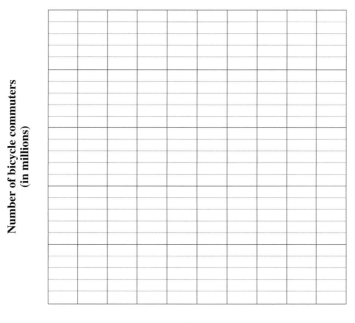

Year

3. Now consider the second row of the table, the numbers of bicycle commuters. These values will be measured against the vertical axis of your graph. You must answer two questions about these values in order to construct the vertical axis:

> What is the largest value shown in the table?
> What scale is appropriate to show all the values?

The largest value in the second row of the table is 4.3. The values are given in tenths, so you could mark off the vertical axis in tenths, but this would require at least 43 tick marks! Instead, use intervals of 0.5 on the vertical axis, and estimate the heights of the bars.

4. Draw a vertical line at the left end of your horizontal axis for the numbers of commuters. The tick marks on this vertical axis must also be *equally spaced*. Choose a convenient spacing, and mark off the vertical axis with tick marks from 0.5 to 4.5.

5. Above each year on the horizontal axis, draw a vertical bar whose height, measured against the vertical axis, is given by the value in the table.

6. Describe any trends in bicycle commuting that your bar graph reveals.

Histograms

Delbert conducted a survey of the cost of off-campus housing at his university. He polled 40 freshmen to determine how much they pay monthly. The bar graph in Figure 1.2 summarizes the results.

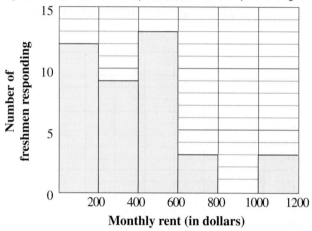

Figure 1.2 Results of a survey on the cost of off-campus housing

Figure 1.2 is an example of a **histogram**. The height of the bar over each response shows the frequency of that response in the poll. For example, the first bar has a height of 12, which means that 12 freshmen pay between $0 and $200 monthly for rent.

Activity 5

Use the histogram to help you answer the following questions:

1. Fill in the table below with the appropriate frequencies.

Monthly rent	$0–$200	$200–$400	$400–$600	$600–$800	$800–$1000	$1000–$1200
Number of freshmen						

2. Which of the given ranges of monthly rents is most common? How can you tell from the histogram?

3. Which rent range has more respondents: $600–$800 or $800–$1000? How can you tell from the histogram?

4. Verify that the total number of freshmen polled was 40.

HOMEWORK 1.1

1. The table shows average annual incomes for full-time workers in California, broken down by ethnic group and educational level.

   a. Does a higher educational level seem to pay off in higher income for all ethnic groups? Are there any instances when it does not?

   b. At each educational level, who has a higher income on average, Asians or African-Americans?

   c. At each educational level, calculate the average income for Latinos as a percent of the average income for Anglos. For example, for workers with no high school diploma,

$$\frac{\$16{,}487}{\$26{,}115} \approx 0.63$$

On average, Latinos with no high school diploma make 63% of what Anglos at the same educational level make. (See Appendix A.9 to review percents.)

Educational level	Average annual income			
	Anglo	Latino	African-American	Asian
No high school diploma	$26,115	$16,487	$21,678	$18,517
High school diploma	$27,376	$21,121	$22,040	$21,608
Bachelor's degree	$44,426	$33,817	$34,290	$33,758
Master's degree	$52,787	$41,431	$42,254	$45,550
Doctorate	$59,348	$46,873	$54,205	$53,792
Professional degree	$77,877	$41,029	$61,015	$59,603

Source: Los Angeles Times, January 1993.

   d. What increase in income will a higher level of education provide? First, calculate the actual increase in dollars, and then divide the increase by the lower income. For example, Anglos with a master's degree earn

$$\$52{,}787 - \$44{,}426 = \$8361$$

more than Anglos with a bachelor's degree. This is an increase of

$$\frac{\$8361}{\$44{,}426} \approx 0.188$$

or 18.8%. Use this method to calculate the percent increase in income from a high school diploma to a bachelor's degree for each ethnic group.

2. How long will a "nest egg" of $100,000 last you? It depends on how much you withdraw each year and on what interest rate your savings earn. The table shows what happens if you make withdrawals that start at the given level and increase by 4% each year to cover inflation.

   a. Choose a particular level of annual withdrawal, and look at the values in that row of the table. Does increasing the interest rate increase the length of time the money will last?

Annual withdrawals starting at	Years money will last at each interest rate				
	4%	6%	8%	10%	12%
$ 2000	50	151	forever	forever	forever
$ 3000	33	52	forever	forever	forever
$ 4000	25	34	76	forever	forever
$ 5000	20	25	38	forever	forever
$ 6000	17	20	25	43	forever
$ 8000	13	15	17	22	39
$ 9000	11	13	15	18	25
$10,000	10	11	12	14	17
$12,000	9	9	10	12	14
$15,000	7	8	8	9	10

Source: Newsweek, June 5, 1992.

   b. Choose a particular interest rate, and look at the values in that column of the table. Does increasing the annual withdrawal increase the amount of time the money will last?

   c. Explain how it is possible that $100,000 could last "forever."

   e. Does doubling the annual withdrawal cut in half the time the money will last? Support your answer with some specific values from the table.

   d. In general, does doubling the interest rate double the amount of time the money will last? Support your answer with some specific values from the table.

Nationality	Number in United States in 1980 (in thousands)	Increase from 1980 to 1990	Increase from 1980 to 1990	Number in United States in 1990 (in thousands)
Chinese	806	104.1%	$1.041 \times 806 \approx 839$	$806 + 839 = 1645$
Filipino	775	81.6%		
Japanese	701	20.9%		
Indian	361	125.6%		
Korean	353	126.3%		
Vietnamese	262	134.8%		

Source: Christian Science Monitor, July 27, 1993.

3. The table above shows the U.S. populations of people of various Asian nationalities in 1980 and the percent increase in those populations from 1980 to 1990.

  a. Fill in the last column of the table. To find the number in 1990, calculate the increase from 1980 to 1990, and then add the increase to the number in 1980. The first one is done for you.

  b. The bar graph at right shows the populations of the various Asian nationalities in 1980. On the same graph, sketch in bars to show these populations in 1990.

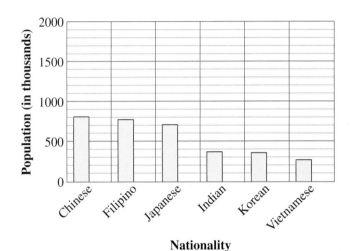

Nationality

U.S. population from selected Asian nations, 1980

4. The bar graph below shows the growth in the number of jobs in Southern California in the three largest sectors of the economy: manufacturing, trade, and services.

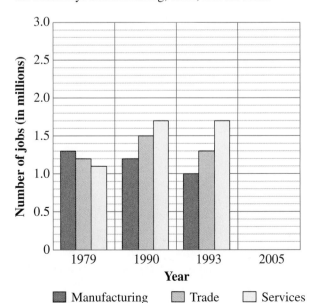

Job growth in Southern California
Source: Los Angeles Times.

  a. Which sector showed the greatest job growth from 1979 to 1993? Which sector showed a decline in number of jobs?

  b. Use the graph to fill in the values in the table.

	Number of jobs (in millions)			
Sector	1979	1990	1993	2005
Manufacturing				1.1
Trade				1.7
Services				2.6

  c. The table gives projections for the numbers of jobs in the three sectors in 2005. Add this information to the bar graph.

5. The table shows the number of calories and the number of grams of fat in 1 ounce of each of three popular brands of chips, the low-fat version of each brand, and each brand's new product that incorporates the artificial fat Olestra.

Follow the steps to make two bar graphs of this information: one showing calories on the vertical axis, and one showing grams of fat. The type of chip should appear on the horizontal axis in each graph. Use the grids to help you.

a. How many items will you display on the horizontal axes of your graphs? Draw the horizontal axis for the calories graph.

Type of chip	*Calories*	*(Fat in grams)*
Lay's Potato Chips	150	10
Baked Lay's	110	1.5
Lay's Max	75	0
Ruffles	150	10
Ruffles Reduced Fat	130	6
Ruffles Max	75	0
Tostitos Tortilla Chips	130	6
Baked Tostitos	110	1
Tostitos Max	90	1

Source: Consumer Reports, August 1996.

b. What is the largest number of calories you must accommodate on the vertical axis? Draw the vertical axis, and mark it off in intervals of 10 calories.

c. Use the table to complete the bar graph showing calories for each type of chip.

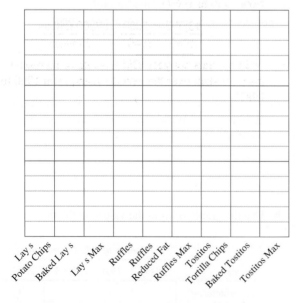

d. Now consider the graph for grams of fat. What is the largest number of grams of fat you must display on the vertical axis? Draw both axes, and mark off the vertical axis with appropriate intervals.

e. Use the table to complete the bar graph showing grams of fat.

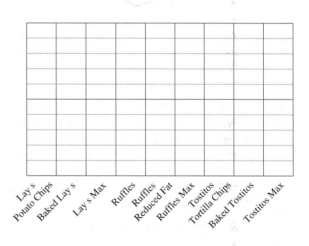

6. The table shows the average annual salary of NFL football players from 1940 to 1990. Make a bar graph of this information by following the steps below. Use the grid to help you.

a. The horizontal axis will show the year. What are the earliest and latest years included in the table? Mark off the horizontal axis in 5-year intervals to include all the values in the table.

b. Locate the years given in the table on your axis. (Be careful—the years are not evenly spaced!)

c. What is the largest salary value you must accommodate on the vertical axis? Mark off the vertical axis in $25,000 intervals.

d. Once you have set up the axes, use the information in the table to construct the bar graph.

Year	*Salary*	*Increase per year*
1940	$ 8,000	—
1960	$ 15,000	
1970	$ 23,200	
1975	$ 39,600	
1980	$ 78,700	
1985	$193,000	
1990	$355,000	

Source: Los Angeles Times, November 3, 1991.

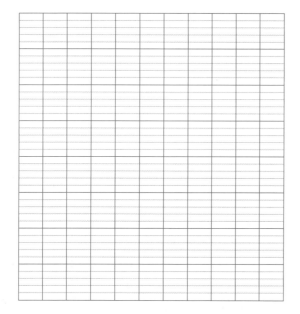

e. Did NFL salaries increase at a *constant* rate over the 50 years from 1940 to 1990? Fill in the last column of the table by computing the following ratio for each time period:

$$\frac{\text{change in salary}}{\text{number of years}}$$

For example, the increase per year in average salary from 1940 to 1960 was

$$\frac{15,000 - 8000}{1960 - 1940} = \frac{7000 \text{ dollars}}{20 \text{ years}}$$

or 350 dollars per year. Did this rate of increase remain constant over the next 30 years, from 1960 to 1990?

7. The histogram shows the results of a quiz in a biology class.

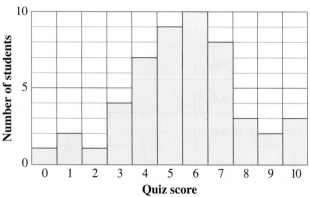

Biology quiz results

a. Fill in the table with the appropriate frequencies.

Quiz score	0	1	2	3	4	5	6	7	8	9	10
Number of students											

b. What is the mode of the quiz scores—that is, what is the most frequent quiz score? (See Appendix A.11 to review the terms *mode*, *median*, and *mean*.)

c. How many students took the quiz?

d. What is the median quiz score?

e. What is the total of the quiz scores of all the students? (Note that the score of 0 adds nothing to the total, but the combined scores of the three top students add 30 points to the total.)

f. What is the mean quiz score?

8. The histogram shows the results of a poll on the numbers of persons living in various households.

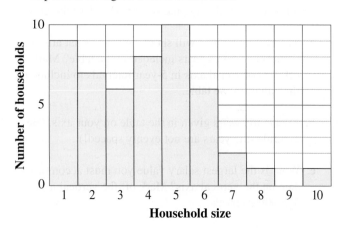

Results of poll on household size

a. Fill in the table with the appropriate frequencies.

Household size	1	2	3	4	5	6	7	8	9	10
Number of households										

b. What is the mode of the household sizes?

c. How many households were polled?

d. What is the median household size?

e. What is the total number of persons living in all the households polled?

f. What is the mean household size?

9. a. Find an example of a bar graph in a newspaper, a
magazine, or one of your textbooks.
  b. What quantities are displayed on the horizontal and
vertical axes?

  c. Describe one or two interesting features of the graph,
and explain what they tell you about the quantities it
represents.

10. a. Survey the students in your class to find out how
many units each student is taking this semester.
  b. Use graph paper to make a histogram of the data.
  c. What is the mode of the number of units being
taken?

  d. Find the median number of units.

  e. Calculate the mean number of units.

1.2 Line Graphs

Reading

In Section 1.1, we used bar graphs to analyze information given in tables. In
many cases, we can simplify the graph and still provide a good picture of the data.

You made a bar graph of average NFL salary for Problem 6 in Homework 1.1. The graph is shown in Figure 1.3. The average salary in each year is
given by the height of the vertical bar at that year. However, we do not have to
draw the entire bar in order to show its height. We can just place a dot at the top
of each bar, as shown in Figure 1.3(b). The heights of these dots, measured
against the vertical scale, give the players' average salary.

Because any trend in the data may be hard to see from a collection of dots, we
often connect the dots with line segments, as shown in Figure 1.4. This type of
graph is called a **line graph.** It gives the same information as a bar graph but is
easier to draw. In addition, the connecting line segments make it easier to see how
fast the data values (in this case, average salaries) are increasing or decreasing. In
fact, Problem 6e in Homework 1.1 shows that a faster *rate* of salary increase appears on the line graph as a steeper connecting line segment.

Figure 1.3 Bar graphs of average NFL salary

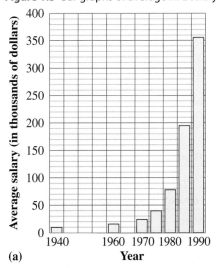

(a)

Figure 1.4 Line graph of average NFL salary

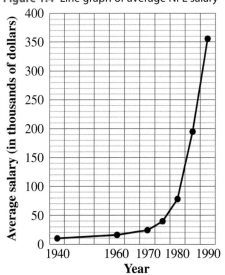

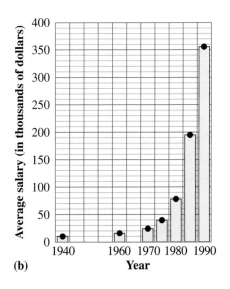

(b)

Skills Review

Recall that to multiply a fraction by a whole number, we write the whole number with a denominator of 1. For example,

$$6 \times \frac{2}{3} = \frac{6}{1} \times \frac{2}{3} = \frac{12}{3} = 4.$$

Find each product mentally (without using pencil, paper, or calculator):

1. $8 \times \dfrac{3}{4}$

2. $12 \times \dfrac{5}{6}$

3. $9 \times \dfrac{4}{3}$

4. $6 \times \dfrac{5}{2}$

5. $8 \times \dfrac{7}{8}$

6. $5 \times \dfrac{3}{5}$

7. $35 \times \dfrac{2}{5}$

8. $32 \times \dfrac{3}{4}$

Answers: *1.* 6   *2.* 10   *3.* 12   *4.* 15
5. 7   *6.* 3   *7.* 14   *8.* 24

Lesson

Making a Line Graph

Activity 1

Table 1.4 shows the wholesale price of natural gas in each month over a 2-year period.

    Follow these steps to make a line graph of the data in the table. Use the grid to help you.

1. Plot time on the horizontal axis and prices on the vertical axis. How many tick marks will you need on the horizontal axis?

2. Mark off the horizontal axis with equally spaced tick marks for the 24 months. Label them J, F, M, and so on.

Table 1.4

Month	Price (per 1000 cubic feet)	
	1989	**1990**
January	$1.99	$2.22
February	$1.81	$1.85
March	$1.89	$1.56
April	$1.56	$1.50
May	$1.61	$1.47
June	$1.65	$1.49
July	$1.65	$1.50
August	$1.61	$1.51
September	$1.55	$1.57
October	$1.58	$1.79
November	$1.66	$1.99
December	$1.92	$2.07

Source: Christian Science Monitor, September 24, 1991.

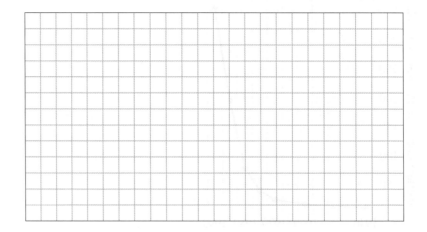

3. What are the highest and lowest prices that appear in Table 1.4? If you mark off the vertical axis in intervals of $0.10, starting at $0.00, how many tick marks will you need?

4. Mark off the vertical axis in equally spaced intervals of $0.10.

5. Above each month, place a dot to represent the price of natural gas in that month. (You will need to estimate the vertical position of the dot.)

6. Connect the dots on your graph with line segments.

7. Describe any trends in the price of natural gas that are illustrated by your graph. Do these trends agree with what you would predict on the basis of common sense?

8. Compare the monthly prices in 1989 to the monthly prices in 1990. What can you say in general about gas prices in those two years? Do you think this comparison is easier to make with the graph or with the table?

Smoothing Out a Line Graph

Imagine you are traveling by plane to another city. Table 1.5 shows the altitude of the plane (in thousands of feet) at 10-minute intervals after takeoff. The two quantities shown in the table—time and altitude—are called **variables.**

Table 1.5

Time (in minutes)	0	10	20	30	40	50	60	70	80	90	100	110
Altitude (in thousands of feet)	0	8	20	21	23	28	29	30	25	13	5	0

Activity 2

Make a line graph of this information, with time on the horizontal axis and altitude on the vertical axis.

1. First, note the largest and smallest values of the altitude in the table. Decide how to mark off the horizontal and vertical axes on the grid in Figure 1.5.

2. Label the axes, and plot the data points at the appropriate heights. Connect successive data points with line segments.

Figure 1.5

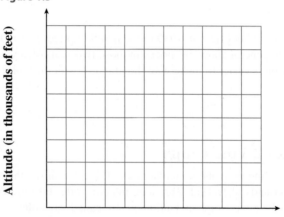

Time (minutes)

Figure 1.6

Altitude (in thousands of feet)

40

30

20

10

0   10 20 30 40 50 60 70 80 90 100 110

Time (minutes)

The line graph does not give a completely accurate picture of the airplane's motion. For example, we know that the altitude did not suddenly jump from 8000 feet to 20,000 feet at exactly 20 minutes into the flight. Instead, the altitude increased gradually as the plane ascended. It is also unlikely that the *rate* of ascent changed abruptly from very rapid to fairly gradual at 20 minutes, as the line segments suggest. Instead, the altitude probably increased smoothly during the first part of the flight. A more realistic graph of altitude over time is shown in Figure 1.6.

3. Draw a smooth curve through the data points on your graph so that it resembles the graph in Figure 1.6.

4. Use your new graph to estimate the plane's altitude 15 minutes into the flight.

5. Use your new graph to estimate when the plane first reached an altitude of 25,000 feet. Were there other times when the plane was flying at that altitude? What were those times?

6. How long did it take the plane to reach its maximum altitude? Over that time, what was its average rate of ascent, in feet per minute? Did the plane climb at that same rate during the entire ascent?

Many, but not all, line graphs increase or decrease smoothly from one data point to the next. In the preceding example, we were given the plane's altitude at 10-minute intervals. If we had more information, we could plot the altitude every minute or every second (or as often as we liked). With enough data points, the line graph would appear to be a smooth curve.

HOMEWORK 1.2

1. In Section 1.1, you made a bar graph showing the number of bicycle commuters in the United States from 1983 to 1992, as shown in the figure.

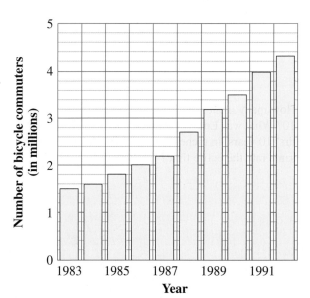

U.S. bicycle commuters (1983–1992)

a. What variable is measured on the horizontal axis? What variable is measured on the vertical axis?

b. Place a piece of thin paper over the graph and trace the axes and their scales.
c. Next, in order to convert the bar graph to a line graph, place a dot at the top of each bar.
d. Do you think the number of bicycle commuters increased smoothly over time? If so, connect your data points with a smooth curve. If not, use line segments to connect the points.

2. The table shows Emily's height on each birthday up to her tenth. Emily's age is given in years and her height in inches.

a. What are the two variables shown in the table? Fill in the third column of the table with Emily's growth each year.

b. Make a line graph with Emily's age on the horizontal axis and her height on the vertical axis. First, notice the largest and smallest values of Emily's height. Decide how you will mark off the vertical axis on the grid below.

c. Label your axes, and plot the data points at the appropriate heights.

d. Do you think you should connect the data points on your graph with line segments or with a smooth curve? Why?

Age	Height (in inches)	Growth this year
0	19	—
1	29.5	
2	33.5	
3	38.5	
4	41.25	
5	43.75	
6	46.5	
7	49	
8	52.5	
9	55	
10	58	

e. Use your graph to estimate when Emily's height reached 50 inches.

f. Use your graph to estimate Emily's height when she was 8 years and 6 months old.

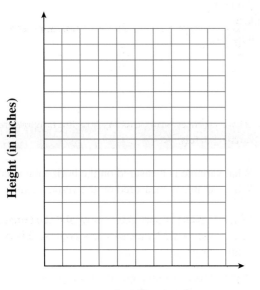

Age (in years)

g. How many inches did Emily grow from birth to age 10? What was Emily's average rate of growth over her first 10 years, in inches per year? Did she grow at this same rate the entire time?

3. Francine recorded the temperature in degrees Fahrenheit
   on her patio every 2 hours over an entire day, starting at
   noon. The results are shown in the table.

   a. Make a line graph with time on the horizontal axis and
      temperature on the vertical axis. How many tick
      marks will you need along the horizontal axis?

Time	Temperature
noon	68°F
2 P.M.	76°F
4 P.M.	80°F
6 P.M.	72°F
8 P.M.	57°F
10 P.M.	51°F
midnight	46°F
2 A.M.	43°F
4 A.M.	41°F
6 A.M.	40°F
8 A.M.	42°F
10 A.M.	50°F

   b. What are the highest and lowest temperatures
      recorded in the table? Mark off the vertical axis in
      intervals of 4°F.

   c. Label your axes, and plot the data points.

   d. Do you think you should connect the data points with
      line segments or with a smooth curve? Why?

   e. Use your graph to estimate the temperature at 1 P.M.
      and at 9 A.M.

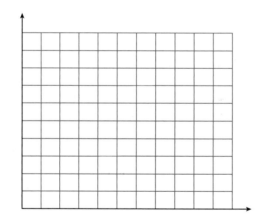

   f. At what times during the day was the temperature
      approximately 70°F?

   g. During which 2-hour time interval did the temperature
      change the most? How can you recognize this fact
      from your graph?

4. The table shows the number of immigrants, in thousands, entering the United States in each decade since 1900.

Decade	Number of immigrants (in thousands)	Cumulative total (in thousands)	Rounded total
1901–1910	8795	8795	
1911–1920	5736	8795 + 5736 = 14,531	
1921–1930	4107	14,531 + 4107 = 18,638	
1931–1940	528		
1941–1950	1035		
1951–1960	2515		
1961–1970	3322		
1971–1980	4493		
1981–1990	7338		

a. Complete the third column of the table, showing the total number of immigrants from 1900 to the given decade. The first few entries are done for you.

b. Round each cumulative total to the nearest thousand.

c. Label the axes for a line graph with decades on the horizontal axis and total number of immigrants on the vertical axis. Mark off your vertical axis in units of 2000; use the rounded numbers to plot the data. Connect the data points with line segments.

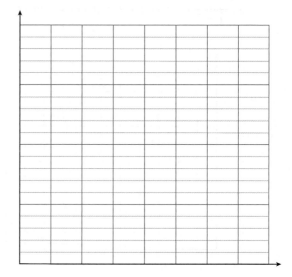

5. In Section 1.1, we considered a table of life expectancies given by decade. The table below gives more detailed information about life expectancies at birth from 1975 to 1986.

Year of birth	1975	1976	1977	1978	1979	1980
Life expectancy (in years)	72.6	72.9	73.3	73.9	73.9	73.7

Year of birth	1981	1982	1983	1984	1985	1986
Life expectancy (in years)	74.3	74.5	74.6	74.7	74.7	74.8

a. Make a line graph of these data. For this graph, mark off your vertical axis in tenths of a year, starting at 71 instead of zero. Do you think you should use a smooth curve or line segments to connect the data points?

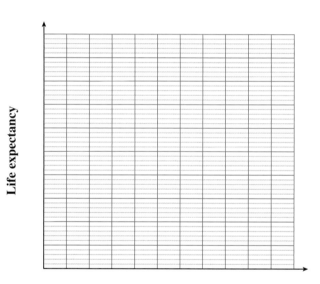

Year of birth

b. Did life expectancy at birth increase continuously from 1975 to 1986?

c. In which year (or years) did life expectancy increase the most?

d. Are there any time intervals over which life expectancy remained constant? How is this stability reflected in the graph?

6. What is the likelihood that at least two people in a group will have the same birthday? The probability depends on the size of the group, but it is probably greater than you think! The table lists the percent probabilities for groups of various sizes.

Size of group	Percent probability	Rounded probability value
10	11.7%	
20	41.1%	
25	56.9%	
30	70.6%	
40	89.1%	
50	97.0%	
60	99.4%	

a. Round each percent probability to the nearest whole number, and decide on an appropriate scale for the vertical axis.

b. Make a line graph of the data. Will you use a smooth curve or line segments to connect the data points?

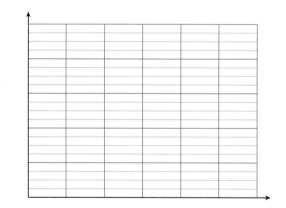

Size of group

c. Using your graph, estimate the number of people needed for the percent probability to exceed 75%.

d. Describe the shape of your graph in one or two sentences. Will the graph continue to increase forever, or will it reach a highest value? If so, when?

Use the line graphs provided to answer the questions in Problems 7–10. (You may have to estimate some of your answers.)

7. Suppose you invest $2000 in a retirement account that pays 8% interest compounded continuously. The graph shows the amount of money in your account each year.

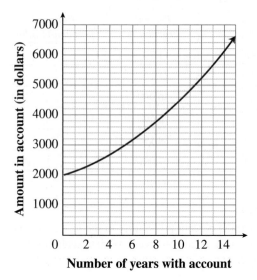

Number of years with account

a. What variable is displayed on the horizontal axis? On the vertical axis?

b. How much money will be in the account after 5 years?

c. When will the amount of money in the account exceed $6000?

d. By how much will the account grow between the second and third years?

e. By how much will the account grow between the 12th and 13th years?

8. A team of biologists released a small flock of an endangered species of quail on a remote island. They monitored the number of quail and recorded the data in the graph.

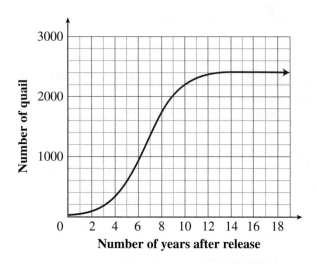

Number of years after release

a. What variable is displayed on the horizontal axis? On the vertical axis?

b. How many quail were on the island 3 years after the flock was released?

c. When did the population of quail reach 500?

d. When did the growth of the quail population begin to slow down?

e. It appears that the island can support only a certain maximum population of quail. What is that maximum population?

9. Wendy brings a Thermos of hot soup for her lunch while hiking in the mountains. When she pours the soup into a bowl, it begins to cool off. The graph shows the temperature of the soup after it is poured.

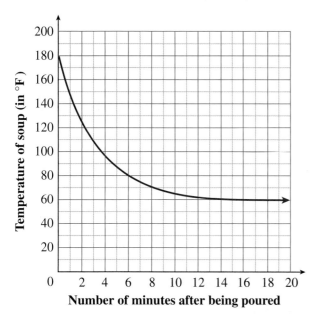

a. What variable is displayed on the horizontal axis? The vertical axis?

b. How long does it take the soup to cool to 100°F?

c. What is the temperature of the soup after 8 minutes?

d. By how many degrees does the soup cool during the first 2 minutes?

e. After a while, the temperature of the soup levels off and becomes the same as the outside temperature. What is the outside temperature?

10. Galileo discovered that all objects fall at the same rate when dropped from a height, regardless of their weights. Some students at Cal Tech dropped a watermelon from the top of a 200-foot building and used a video camera to record its descent. Their data are graphed below.

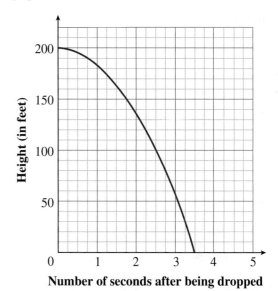

a. What variable is displayed on the horizontal axis? The vertical axis?

b. What was the height of the watermelon 2 seconds after it was dropped?

c. How long did the watermelon take to fall to a height of 150 feet?

d. How far did the watermelon fall during the first second of its descent? How far did it fall as the clock ticked from 2 to 3 seconds?

e. Approximately how long did the watermelon take to reach the ground?

For Problems 11–14:

a. Plot each pair of values on the grid. Connect the points with a smooth curve.

b. Which graphs appear to be straight lines?

11.

t	0	1	2	5	6	8
v	16	14	12	6	4	0

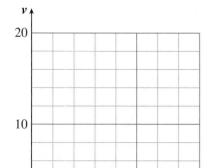

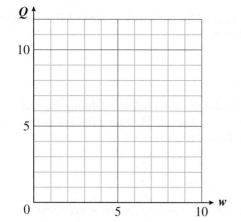

12.

m	0	1	2	4	6	8
z	2	2.25	3	6	11	18

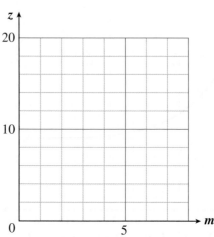

13.

w	0	1	2	4	6	10
Q	12	8	6	4	3	2

14.

u	0	2	3	6	9	10
H	6	$6\frac{2}{3}$	7	8	9	$9\frac{1}{3}$

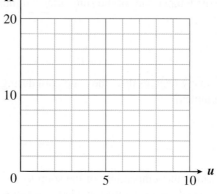

15. A cantaloupe is dropped from a tall building. Its speed increases until it hits the ground. Which of the four graphs best represents the speed of the cantaloupe? Explain why your choice is the best one.

16. The driver of a moving car slams on the brakes, and the car comes to a stop. Which of the four graphs that you examined for the previous question best represents the speed of the car? Explain why your choice is the best one.

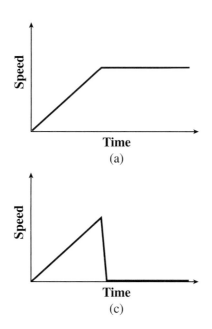

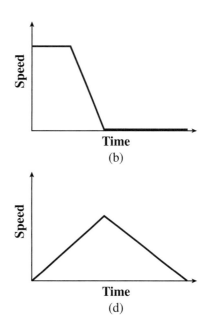

17. a. Match each of the following stories with the appropriate graph below.

   i. Delbert walks directly from home to school and stays there.

   ii. Delbert walks toward school but stops at a coffee shop for a cappuccino; then he continues on to school and stays there.

   iii. Delbert walks toward school but then decides to return home, where he stays.

b. Write your own story for the remaining graph.

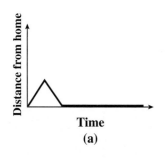

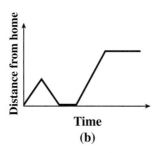

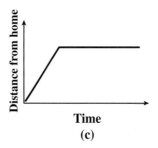

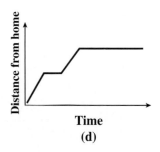

18. a. Match each of the following situations with the most appropriate graph below.

    i. A ball is dropped from a height and allowed to bounce.

    ii. Francine is on a swing, going higher on each pass.

    iii. Francine is on a spinning Ferris wheel.

b. Write your own story for the remaining graph.

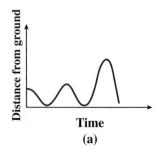

Time

(a)

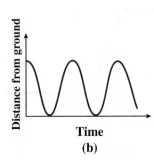

Time

(b)

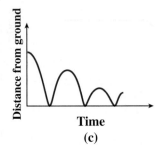

Time

(c)

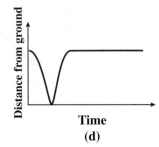

Time

(d)

19. a. Find an example of a line graph in a newspaper, a magazine, or one of your textbooks.

  b. What quantities are displayed on the horizontal and vertical axes?

  c. Describe one or two interesting features of the graph, and explain what they tell you about the quantities it represents.

20. Find an example of a graph that shows two line graphs on the same set of axes. What two quantities are being compared?

1.3 Variables

Reading

A **variable** is a numerical quantity that changes over time or in different situations. In Sections 1.1 and 1.2, we considered several examples of variables. Life expectancy is a variable that depends on the year in which you were born. The number of calories in a serving of chips depends on the brand of chips. The life span of a $100,000 nest egg depends on the interest rate being earned and on the amounts of any annual withdrawals. All of these quantities are variables.

By comparing the values of two variables in a table or a graph, we sometimes see trends in the data. For instance, we saw that life expectancy has been increasing slowly over time and that the average salary of NFL football players has been increasing ever more rapidly over the past 50 years. It would be even more useful if we could discover a formula or rule to help us *predict* the values of an interesting variable in the future or in different situations. The various life spans of the $100,000 nest egg were calculated using just such a rule: If we know the interest rate being earned and the amounts of any annual withdrawals, we can use a formula to calculate how long a nest egg will last.

In this section, you will learn how to write a formula relating two different variables. You will also learn how to use letters as a kind of shorthand to represent variable quantities. The formula relating the two variables is a type of **equation.**

EXAMPLE 1

Fernando plans to share an apartment with three other students next year. The table shows his share of the rent for apartments of various prices.

Rent per month	280	300	340	360	400	460	500
Fernando's share	70	75	85	90	100		

   a. Fill in the blanks in the table. Describe in words how we found the unknown values:

   Divide the rent by 4.

   b. Write a mathematical sentence that relates Fernando's share of the rent to the total rent:

   Fernando's share = Rent ÷ 4

   c. Let the letter R stand for the total rent, and let F stand for Fernando's share. Write an equation for Fernando's share of the rent:

   $F = R \div 4$

Skills Review

Choose the correct arithmetic operation (addition, subtraction, multiplication, or division) to answer each question.

1. An air conditioner keeps the inside temperature 16°F cooler than the outside temperature. If the outside temperature is 90°F, how can you find the inside temperature?

2. Tom's recipe for punch calls for three times as much fruit juice as soda. If he has half a gallon of soda, how can he find the amount of fruit juice he needs?

3. A clothesline should be 2 feet longer than the distance between the supporting poles so that it can be tied at each end. If the poles are 20 feet apart, how can you find the length of the clothesline?

4. The weight of a bridge is supported equally by eight pillars. If the bridge weighs 1 million tons, how can you find the weight each pillar must support?

5. Katrin has 4 hours to complete both her math and her geography homework. If her math assignment takes $2\frac{1}{2}$ hours, how can she calculate how much time she has for geography?

6. The cost of leasing a compact car is 63% of the cost of leasing a luxury car. If the lease on the luxury car is $500 per month, how can you find the cost of leasing the compact car?

Answers: *1.* Subtract 16 from the outside temperature.   *2.* Multiply the amount of soda by 3.   *3.* Add 2 feet to the distance between the poles. *4.* Divide the weight of the bridge by 8.   *5.* Subtract the time for math from 4 hours.   *6.* Multiply the cost of the luxury car by 0.63.

Lesson

Writing Mathematical Sentences

Activity 1

Barry lives with his aunt while he attends college. Every week he gives her $20 from his paycheck to help pay for groceries. Fill in the table:

Barry's paycheck	45	60	75	100	125	p
Calculation	$45 - 20$					
Amount Barry keeps	25					

1. Explain in words how to find the amount Barry keeps from his weekly paycheck.

2. Write your explanation as a mathematical sentence:

   Amount he keeps =

3. Let p stand for the amount of Barry's paycheck and k for the amount he keeps. Write an equation for k in terms of p.

4. Plot the data from the table and connect the data points with a smooth curve.

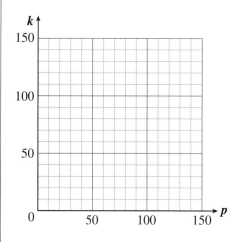

Activity 2

Liz makes $6 an hour as a tutor in the Math Lab. Her wages for the week depend on the number of hours she works. Fill in the table.

Hours worked	3	5	6	8	15	h
Calculation	6×3					
Wages	18					

1. Explain in words how to find Liz's wages for the week.

2. Write your explanation as a mathematical sentence:

   Wages =

3. Let h stand for the number of hours Liz works and w for her wages. Write an equation for w in terms of h.

4. Plot the values from the table and connect the data points with a smooth curve.

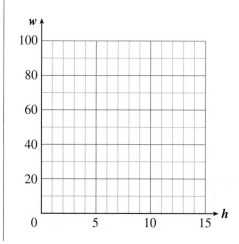

Activity 3

Together, Ralph and Wanda weigh 320 pounds. Fill in the table with Wanda's weight for various possible values of Ralph's weight.

Ralph's weight	150	165	180	195	210	R
Calculation	320 − 150					
Wanda's weight	170					

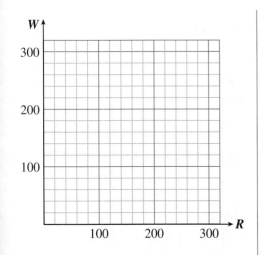

1. Explain in words how to find Wanda's weight.

2. Write your explanation as a mathematical sentence:

   Wanda's weight =

3. Let *R* stand for Ralph's weight and *W* for Wanda's weight. Write an equation for *W* in terms of *R*.

4. Plot the values from your table and connect the data points with a smooth curve.

Activity 4

Nutrition experts tell us that no more than 30% of the calories we consume should come from fat. Fill in the table with the number of fat calories allowed daily for various total calorie levels. (To review converting percents to decimals, see Appendix A.9.)

Total calories	1200	1500	2000	2400	2800	C
Calculation	0.30×1200					
Fat calories	360					

1. Explain in words how to find the number of fat calories allowed.

2. Write your explanation as a mathematical sentence:

   Fat calories =

3. Let *C* stand for the total number of calories and *F* for the number of fat calories. Write an equation for *F* in terms of *C*.

4. Plot the values from your table and connect the data points with a smooth curve.

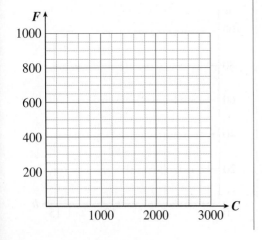

HOMEWORK 1.3

1. The temperature in Sunnyvale is usually 12°F hotter than it is in Ridgecrest, which is at a higher elevation.

a. Fill in the table.

Temperature in Ridgecrest	70	75	82	86	90	R
Calculation						
Temperature in Sunnyvale						

b. Explain in words how to find the temperature in Sunnyvale.

c. Write your explanation as a mathematical sentence:

Temperature in Sunnyvale =

d. Let R stand for the temperature in Ridgecrest and S for the temperature in Sunnyvale. Write an equation for S in terms of R.

e. Plot the values from the table and connect the data points with a smooth curve.

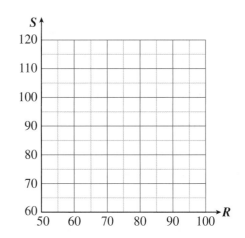

2. Delbert has a coupon for $12 off at Music City.

a. Fill in the table.

Regular price	18	26	54	76	90	r
Calculation						
Amount Delbert pays						

b. Explain in words how to find the amount Delbert pays.

c. Write your explanation as a mathematical sentence:

Amount Delbert pays =

d. Let r stand for the regular price and D for the amount Delbert pays. Write an equation for D in terms of r.

e. Plot the values from the table and connect the data points with a smooth curve.

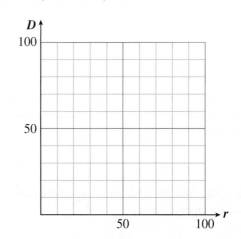

3. Jerome is driving from his home in White Falls to his parents' home in Castle Heights, 200 miles away.

a. Fill in the table.

Miles driven	40	60	95	120	145	170	d
Calculation							
Miles remaining							

b. Explain in words how to find the number of miles Jerome has left to drive.

c. Write your explanation as a mathematical sentence:

Miles remaining =

d. Let *d* stand for the number of miles Jerome has driven and *r* for the number of miles that remain. Write an equation for *r* in terms of *d*.

e. Plot the values from the table and connect the data points with a smooth curve.

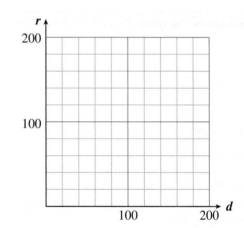

4. Tam is traveling on an express train at 70 miles per hour.

a. Fill in the table.

Hours traveled	2	3	5	8	10	h
Calculation						
Miles traveled						

b. Explain in words how to find the distance Tam has traveled.

c. Write your explanation as a mathematical sentence:

Miles traveled =

d. Let *h* stand for the number of hours Tam has traveled and *m* for the number of miles he has traveled. Write an equation for *m* in terms of *h*.

e. Plot the values from the table and connect the data points with a smooth curve.

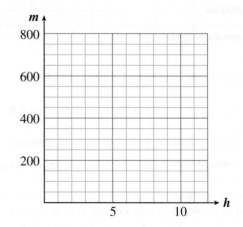

5. Milton goes to a restaurant with two friends, and they agree to split the bill equally.

   a. Fill in the table.

Total bill	24	30	33	45	54	81	b
Calculation							
Milton's share							

   b. Explain in words how to find Milton's share of the bill.

   c. Write your explanation as a mathematical sentence:

      Milton's share =

   d. Let b stand for the total bill and s for Milton's share. Write an equation for s in terms of b.

   e. Plot the values from the table and connect the data points with a smooth curve.

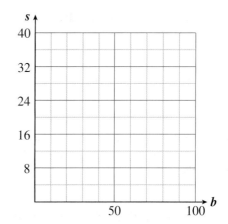

6. Milton and his friends from Problem 5 decide to tip 15% of the bill. (See Appendix A.9 to review percents.)

   a. Fill in the table.

Bill	20	24	36	44	50	55	b
Calculation							
Tip							

   b. Explain in words how to find the tip.

   c. Write your explanation as a mathematical sentence:

      Tip =

   d. Let b stand for the bill and t for the tip. Write an equation for t in terms of b.

   e. Plot the values from the table and connect the data points with a smooth curve.

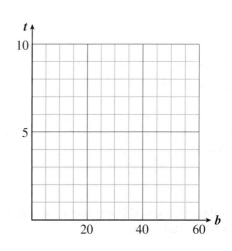

In Problems 7–14, find the pattern and fill in each table. Then write an equation for the second variable in terms of the first variable.

7.

m	2	3	5	10	12	16	18	m
g	5	6	8	13				

Equation: $g =$

8.

h	0	2	3	4	6	7	10	h
v	0	6	9	12				

Equation: $v =$

9.

t	0	2	4	5	6	10	12	t
w	20	18	16	15				

Equation: $w =$

10.

k	4	6	9	10	15	18	20	k
s	28	26	23	22				

Equation: $s =$

11.

b	0	2	4	5	6	8	9	b
x	0	1	2	2.5				

Equation: $x =$

12.

n	2	5	10	12	15	18	25	n
p	0.4	1	2	2.4				

Equation: $p =$

13.

z	3	6	8	12	15	18	20	z
r	2	4	$\dfrac{16}{3}$	8				

Equation: $r =$

14.

d	1	2	4	8	12	18	20	d
y	$\dfrac{3}{4}$	$\dfrac{3}{2}$	3	6				

Equation: $y =$

In Problems 15–18, make your own table for each rule. Choose any values for the first variable, and use the rule to find the values of the second variable.

15. $W = 1.2 \times n$

n					
W					

16. $t = 12 - a$

a					
t					

17. $M = \dfrac{3}{2} \times x$

x					
M					

18. $B = \dfrac{5}{3} \times t$

t					
B					

In Problems 19 and 20, use the graph to:
a. Fill in the table.
b. Find a formula expressing the second variable in terms of the first.

19.

x	0	10	30	40	60	70
y						

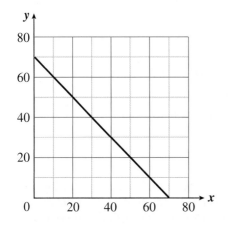

20.

x	4	6	10	12	15	20
y						

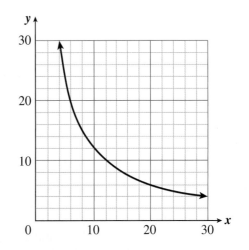

1.4 Algebraic Expressions

Reading

An **algebraic expression**, or simply an **expression**, is any meaningful combination of numbers, variables, and operation symbols. For example, these are algebraic expressions:

$$4 \times g \qquad 3 \times c + p \qquad \text{and} \qquad \frac{n - 7}{w}$$

An important part of algebra involves translating word phrases into algebraic expressions. You have already written some algebraic expressions in Section 1.3. In this section, you'll learn some terminology and notation used in algebra and become familiar with some useful algebraic expressions.

Sums

When we add two numbers a and b, the result is called the **sum** of a and b. We call the numbers a and b the **terms** of the sum.

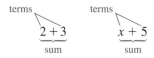

EXAMPLE 1

Write sums to represent the following word phrases:

    a. Six more than x **b.** Fifteen greater than r

Solution

    a. $6 + x$ or $x + 6$ **b.** $15 + r$ or $r + 15$

    In Example 1 the terms can be added in either order. This is a consequence of the commutative law for addition.

Commutative Law for Addition

If a and b are any numbers, then

$$a + b = b + a$$

Products

When we multiply two numbers a and b, the result is called the **product** of a and b. We call the numbers a and b the **factors** of the product. In arithmetic, we use the symbol $\times$ to denote multiplication. However, in algebra, that symbol may be confused with the variable x, which occurs frequently. Therefore, to express multiplication we use a dot centered between the numbers or parentheses around the numbers.

$$\text{factors} \diagdown \qquad \text{factors} \diagdown \qquad \text{factors} \diagdown$$

$$\underbrace{2 \cdot 3}_{\text{product}} \qquad \underbrace{2(3)}_{\text{product}} \qquad \underbrace{(2)(3)}_{\text{product}}$$

Products involving variables may be written in these same ways or with the symbols side by side. For example,

ab means   "the product of a times b"

$3x$ means   "the product of 3 times x"

EXAMPLE 2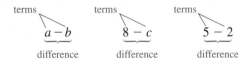

Write products to represent the following word phrases.

    a. Ten times z **b.** Two-thirds of y **c.** The product of x and y

Solution

    a. $10z$ **b.** $\dfrac{2}{3}y$ **c.** xy or yx

In Example 2a, $10z = 10 \cdot z = z \cdot 10$ because two numbers can be multiplied in either order to give the same answer. This is the commutative law for multiplication.

Commutative Law for Multiplication

If a and b are any numbers, then
$$a \cdot b = b \cdot a$$

However, in algebra it is customary to write products with the numeral first. Thus we write $10z$ for 10 times z.

Differences

When we subtract b from a, the result is called the **difference** of a and b. As with addition, a and b are called *terms*.

$$\text{terms} \diagdown \qquad \text{terms} \diagdown \qquad \text{terms} \diagdown$$

$$\underbrace{a - b}_{\text{difference}} \qquad \underbrace{8 - c}_{\text{difference}} \qquad \underbrace{5 - 2}_{\text{difference}}$$

CAUTION!

$5 - 2$ is *not* the same as $2 - 5$. When we interpret a phrase such as "*a* subtracted from *b*," the order of the terms is important.

$b - a$ means   "a subtracted from b"

$x - 7$ means   "7 subtracted from x"

The operation of subtraction is *not* commutative.

Quotients

When we divide a by b, the result is called the **quotient** of a and b. We call a the **dividend** and b the **divisor**. In algebra, we indicate division by using the symbol ÷ or a fraction bar.

The operation of division is *not* commutative. Just as with subtraction, the order in which division is performed makes a difference. For example, $12 \div 3$ is not the same as $3 \div 12$.

EXAMPLE 3

Write an algebraic expression for each word phrase.

    a. z subtracted from 13   **b.** 25 divided by R

Solution

    a. $13 - z$ **b.** $\dfrac{25}{R}$

Skills Review

Recall that the area of a rectangle is equal to the product of its length and its width. That is, $A = lw$.   The perimeter of a rectangle is equal to twice its length plus twice its width. That is, $P = 2l + 2w$.

  1. Delbert's living room is 20 feet long and 12 feet wide. How much oak baseboard does he need to border the floor? (Don't worry about doorways.)

  2. How much wood parquet tiling must he buy to cover the floor?

Decide which of the following phrases describe a perimeter and which describe an area.

  3. The distance you jog around the shore of a small lake

  4. The amount of Astroturf needed for the new football field

  5. The amount of grated cheese needed to cover a pizza

  6. The number of tulip bulbs needed to border a patio

Find the perimeter and area of each figure. All angles are right angles.

7.

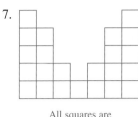

All squares are
1 centimeter on a side.

8.

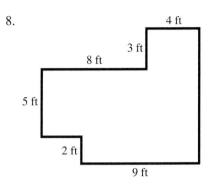

Answers: **1.** 64 feet **2.** 240 square feet **3.** perimeter **4.** area **5.** area
6. perimeter **7.** 32 centimeters; 23 square centimeters **8.** 44 feet;
90 square feet

Lesson

Writing Algebraic Expressions

Here are three steps that can help you translate a verbal description into an algebraic expression.

To Write an Algebraic Expression:

Step 1 Identify the unknown quantity, and write a short phrase to describe it.

Step 2 Choose a variable to stand for the unknown quantity.

Step 3 Use mathematical symbols to represent the relationship described.

The following example involves addition and multiplication.

EXAMPLE 4

Write an algebraic expression for each verbal description:

    a. 3 feet more than the length of the rug

    b. Twice the age of the building

    c. 30% of the money invested in stocks

Solution

    a. Steps 1–2 The length of the rug is unknown.

                length of the rug: l

        Step 3 The words "more than" indicate addition.

                $l + 3$

EXERCISE 1

a. Ten more than the number of students

Steps 1–2

Step 3

b. Five times the height of the triangle

Steps 1–2

Step 3

c. 4% of the original price

Steps 1–2

Step 3

d. Two and a quarter inches taller than last year's height

Steps 1–2

Step 3

---

b. Steps 1–2 The age of the building is unknown.

age of the building: a

Step 3 "Twice" means two times.

$2a$

c. Steps 1–2 The amount invested is unknown.

amount invested in stocks: s

Step 3 "30% of" means multiplying by 0.30.

$0.30s$

In Exercise 1, try your skill at writing algebraic expressions. (Answers to Exercises appear before the homework in each section.)

The next example involves subtraction and division.

EXAMPLE 5

Write an algebraic expression for each verbal description:

a. 5 square feet less than the area of the circle

b. The cost of the dinner split three ways

c. The ratio of your quiz score to 20

Solution

a. Steps 1–2 The area of the circle is unknown.

area of circle: A

Step 3 "Less than" indicates subtraction.

$A - 5$

b. Steps 1–2 The cost of the dinner is unknown.

cost of dinner: C

Step 3 "Split" means divided.

$\dfrac{C}{3}$

c. Steps 1–2 Your quiz score is unknown.

quiz score: s

Step 3 A "ratio" is a fraction.

$\dfrac{s}{20}$

Try writing algebraic expressions in Exercise 2 on page 41.

Evaluating Algebraic Expressions

An algebraic expression provides a pattern or rule for solving many versions of the problem it describes. In Example 1 of Section 1.3, we wrote an algebraic expression for Fernando's share of the rent: $R \div 4$. If the rent is $540, we replace R by 540:

$$R \div 4 = 540 \div 4 = 135$$

We find that Fernando's share is $135. The process of substituting a specific value for the variable into an algebraic expression is called **evaluating** the expression.

EXAMPLE 6

The Appliance Mart is having a storewide 15%-off sale. If the regular price of an appliance is p dollars, then the sale price s is given by

$$s = 0.85p$$

How much is a refrigerator that regularly sells for $600?

Solution

We substitute 600 for the regular price, p, in the expression.

$$s = 0.85p$$
$$= 0.85(600) = 510$$

The sale price is $510.

Now try Exercise 3.

Some Useful Algebraic Formulas

Algebraic expressions involving more than one variable appear in formulas used in many different fields. Here are some common formulas and an example for each.

> To find the distance traveled by an object moving at a constant speed or rate for a specified time, multiply the rate by the time:
>
> $$distance = rate \times time \qquad \boldsymbol{d = rt}$$

EXAMPLE 7

How far will a train moving at 80 miles per hour travel in 3 hours?

Solution

We evaluate the formula with $r = 80$ and $t = 3$:

$$d = rt$$
$$= (80)(3) = 240$$

The train will travel 240 miles.

EXERCISE 2

a. Sixty dollars less than the first-class airfare

Steps 1–2

Step 3

b. The quotient of the volume of the sphere and 6

Steps 1–2

Step 3

c. The ratio of the number of gallons of alcohol to 20

Steps 1–2

Step 3

d. The current population diminished by 50

Steps 1–2

Step 3

EXERCISE 3

Evaluate the algebraic expression in Example 6 to fill in the sale prices for various appliances in the following table:

p	s
120	
200	
380	
480	
520	

To find the profit earned by a business, subtract its costs from its revenue:
$$profit = revenue - cost \qquad \boldsymbol{P = R - C}$$

EXAMPLE 8

The owner of a sandwich shop spent $800 last week for labor and supplies. She received $1150 in revenue. What was her profit?

Solution

We evaluate the formula with $R = 1150$ and $C = 800$:
$$P = R - C$$
$$= 1150 - 800 = 350$$

The owner's profit was $350.

To find the interest earned on an investment, multiply the amount invested (the principal) by the interest rate and the length of the investment:
$$interest = principal \times interest\ rate \times time \qquad \boldsymbol{I = Prt}$$

EXAMPLE 9

How much interest will be earned on $800 invested for 3 years in a savings account that pays $5\frac{1}{2}\%$ interest?

Solution

First we write the interest rate, $5\frac{1}{2}\%$ in decimal form:
$$5\frac{1}{2}\% = 0.055$$

(See Appendix A.9 to review percents.) Then we evaluate the formula with $P = 800$, $r = 0.055$, and $t = 3$:
$$I = Prt$$
$$= 800 \times 0.055 \times 3 = 132$$

The account will earn $132 in interest.

To find a percent or part of a given amount, multiply the whole amount by the percent rate, expressed as a decimal:
$$part = percent\ rate \times whole \qquad \boldsymbol{P = rW}$$

EXAMPLE 10

How much ginger ale would you need to make 60 gallons of a fruit punch that is 20% ginger ale?

Solution

We evaluate the formula with $r = 0.20$ and $W = 60$:
$$P = rW$$
$$= 0.20 \times 60 = 12$$

You would need 12 gallons of ginger ale.

To find the average of a collection of values, divide the sum of the values by the number of values:

$$average\ value = \frac{sum\ of\ values}{number\ of\ values} \qquad A = \frac{S}{n}$$

EXAMPLE 11

Bert's quiz scores in chemistry are 15, 16, 18, 18, and 12. What is his average score?

Solution

We evaluate the formula with $S = 15 + 16 + 18 + 18 + 12$ and $n = 5$:

$$A = \frac{S}{n}$$
$$= \frac{15 + 16 + 18 + 18 + 12}{5} = 15.8$$

Bert's average score is 15.8.

ANSWERS TO 1.4 EXERCISES

1a. $10 + s$

1b. $5h$

1c. $0.04p$

1d. $h + 2.25$

2a. $f - 60$

2b. $\dfrac{V}{6}$

2c. $\dfrac{g}{20}$

2d. $p - 50$

3. 102, 170, 323, 408, 442

HOMEWORK 1.4

Each of the following words is related to one of the four arithmetic operations in Problems 1–4. Group the words under the correct operation.

times	sum of	take away	divided by
less than	twice	reduced by	(fraction) of
increased by	more than	product of	quotient of
exceeded by	ratio of	deducted from	total
difference of	minus	split	per

1. Addition   **2.** Subtraction   **3.** Multiplication   **4.** Division

Write the word phrases in Problems 5–14 as algebraic expressions.

5. Product of 4 and y **6.** Six more than x **7.** Twice b **8.** Two-thirds of s

9. 115% of g **10.** y divided by 7   **11.** t decreased by 5   **12.** g subtracted from h

13. The quotient of 7 and w **14.** The difference between v and 9

For Problems 15–22, use the three steps on page 39 to translate the phrase into an algebraic expression.

15. Three times the cost of a light bulb

16. The radius of the circle increased by 5 centimeters

17. Three-fifths of the savings account balance

18. 49% of the total

19. The price of the pizza divided by 6

20. The perimeter of the triangle diminished by 10 feet

21. The weight of the copper in ounces divided by 16

22. The ratio of 25 to the number of professors

Write algebraic expressions for the phrases in Problems 23–28. Each involves two variables.

23. The rebate deducted from the sale price

24. The distance traveled divided by the time elapsed

25. The product of the base and the height

26. The ratio of field goals scored to field goals attempted

27. The difference between the height of the roof and the height of the tree

28. The sum of the principal and the interest

In Problems 29 and 30, use evaluation to answer the questions.

29. The college bookstore charges a 25% markup over the wholesale price. If the wholesale price of a book is w dollars, then the bookstore's price p is given by

$$p = 1.25w$$

a. How much does the bookstore charge for a book whose wholesale price is $28?

b. Fill in the table with the bookstore's prices for various books.

w	12	16	20	30	36	40
p						

30. In 1 year, a new Futura automobile depreciates in value by 20% of its sale price. If the sale price of a new Futura is s dollars, then its value v after 1 year is given by

$$v = 0.80s$$

a. How much is a Futura that sells for $15,000 worth after 1 year?

b. Fill in the table with the values of various Futura models 1 year after purchase.

s	8000	10,000	12,000	16,000
v				

In Problems 31–36:
a. Use the formulas highlighted on pages 41, 42, and 43 to write an algebraic expression.
b. Evaluate the expression to answer the questions.

31. a. Write an expression for the distance traveled in t hours by a small plane flying at 180 miles per hour.

b. How far will the plane fly in 2 hours? In $3\frac{1}{2}$ hours? In half a day?

32. a. Suppose your father lends you $2000 to be repaid with 4% annual interest when you finish school. Write an expression for the amount of interest you will owe after t years.

b. How much interest will you owe if you finish school in 2 years? In 4 years? In 7 years?

33. a. The Earth Alliance made $6000 in revenue from selling tickets to Earth Day, an educational event for children. Write an expression for the Alliance's profit in terms of its costs.

  b. What was the profit if costs were $800? $1000? $2500?

34. a. A certain pesticide contains 0.02% by volume of a dangerous chemical. Write an expression for the amount of the chemical that enters the environment in terms of the number of gallons of pesticide used.

  b. How much of the chemical enters the environment if 400 gallons of the pesticide are used? 5000 gallons? 50,000 gallons?

35. a. Delbert keeps track of the total number of points he earns on homework assignments, each of which is worth 30 points. At the end of the semester, he has 540 points. Write an expression for Delbert's average homework score in terms of the number of assignments.

  b. What is Delbert's average score if there were 20 assignments? 25 assignments? 30 assignments?

36. a. Edgar's great-aunt plans to put $5000 in a trust fund for him until he turns 21, which will happen 3 years from now. She has a choice of several different accounts. Write an expression for the amount of interest the money will earn in 3 years if the account pays interest rate r.

  b. How much interest will $5000 earn in 3 years in an account that pays 8% interest? $10\frac{1}{4}$% interest? 12% interest?

Write an algebraic expression for the area or perimeter of each figure in Problems 37–42.

37. Area =

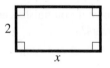

38. Area =

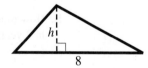

39. Perimeter =

40. Perimeter =

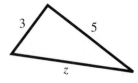

41. Area =

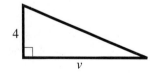

42. Perimeter =

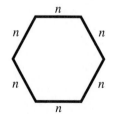

In Problems 43–50, fill in the blanks or answer the question.

43. A meaningful combination of numbers, variables, and operation symbols is called an _____.

44. When we add two numbers, the result is called the _____ and the two numbers are called the _____.

45. When we multiply two numbers, the result is called the _____ and the two numbers are called the _____.

46. Which operations obey the commutative laws? What do these laws say?

47. When we subtract two numbers, the result is called the _____.

48. When we divide two numbers, the result is called the _____.

49. What does it mean to *evaluate* an algebraic expression?

50. In each of the following formulas, explain what each of the variables represents.
a. $d = rt$

d. $P = rW$

b. $P = R - C$

e. $A = \dfrac{S}{n}$

c. $I = Prt$

f. $A = lw$

For Problems 51–54:
a. **Fill in the table. Plot the points and connect them with a smooth curve.**
b. **Write an algebraic expression for the second variable in terms of the first.**
c. **Evaluate your expression for the given values of x, and plot the new points. Do the new points lie on a continuation of your graph from part (a)?**

51.

x	0	5	15	20	25	30
y	15	20	30			

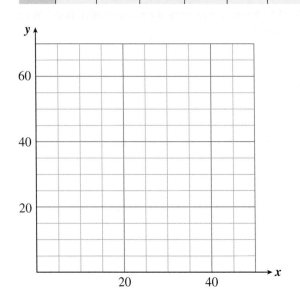

$x = 40; \quad x = 50$

52.

x	0	0.2	0.3	0.5	0.6	0.7
y	2	1.8	1.7			

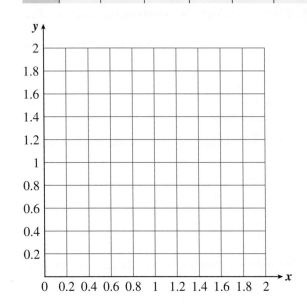

$x = 1; \quad x = 2$

53.

x	0	500	1000	2000	2500	3000
y	0	15	30	60		

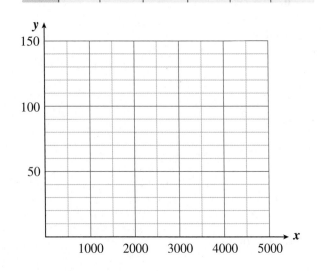

$x = 4500; \quad x = 5000$

54.

x	1	2	3	4	6	8
y	120	60	40			

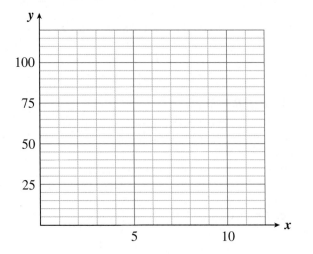

$x = 10; \quad x = 12$

1.5 Graphs of Equations

Reading

A graph is a way of visualizing an algebraic equation or a relationship between two variables. The graph has two **axes**, horizontal and vertical, and appropriate values for the two variables are displayed along the axes. Usually the first variable, or **independent variable**, is displayed on the horizontal axis and the second variable, or **dependent variable,** is displayed on the vertical axis. The graph itself shows how the two variables are related.

EXAMPLE 1

Francine is exactly 4 years older than Delbert. If we let F stand for Francine's age and D stand for Delbert's age, we can write the equation $F = D + 4$. The graph of this equation is shown in Figure 1.7. In this graph, D is the independent variable and F is the dependent variable.

Figure 1.7

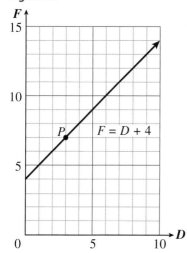

D	F
0	4
2	6
3	7
4	8
6	10
9	13

Each point on the graph has two **coordinates,** which designate the position of the point relative to the two axes. For example, the point labeled P in Figure 1.7 is located 3 units along the horizontal axis and 7 units up the vertical axis. This point represents the fact that when Delbert was 3 years old, Francine was 7 years old. The horizontal coordinate of P is 3, and its vertical coordinate is 7. We write the coordinates of point P inside parentheses, like this: $(3, 7)$. The notation $(3, 7)$ is called an **ordered pair** because the order in which the coordinates are listed makes a difference. *We always list the horizontal coordinate first, followed by the vertical coordinate.* (Try Exercise 1 now.)

How is all this related to the equation $F = D + 4$? The coordinates of each point on the graph are values for D and F that make the equation true. For example, the coordinates of the point P are $D = 3$ and $F = 7$. If we substitute these values into the equation, we get

$$F = D + 4$$
$$7 = 3 + 4$$

which is true. The ordered pair $(3, 7)$ is called a **solution** of the equation $F = D + 4$.

EXERCISE 2
a. Use the graph in Figure 1.7 to complete the table:

D	1	5	7	8
F				

b. By substituting its coordinates into the equation, verify that each point in your table represents a solution of the equation $F = D + 4$.

Each point on the graph represents a solution of the equation. By reading the coordinates of various points on the graph, we can construct a table of values for the equation. (Now Try Exercise 2 on the previous page.)

Skills Review

In Problems 1 and 2, what value corresponds to each labeled point?

1.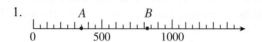

2.

3. a. Label the tick marks on this axis:

    b. On the axis in part a, plot 1400 and 8350.

4. a. Label an axis with 16 tick marks (not counting the one at 0), the highest being at 800.

    b. On the axis in part a, plot 132 and 614.

5. a. Label an axis in intervals of 40,000 from 0 to 600,000.

    b. On your axis, plot 250,000 and 472,600.

6. a. Label an axis in intervals of 0.5 from 0 to 5.

    b. On your axis, plot 1.3 and 3.77.

Answers: *1.* A: 350; B: 825 *2.* C: 700; D: 1400

3.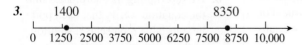

4.

5.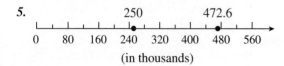

6.

Lesson

Graphing Equations

We can follow three steps to make a graph of an equation.

> **To Graph an Equation:**
>
> **Step 1**   Make a table of values.
> **Step 2**   Choose scales for the axes.
> **Step 3**   Plot the points, and connect them with a smooth curve.

EXAMPLE 2

Laura takes her daughter Stefanie berry-picking at a local strawberry farm. Laura can pick three baskets of strawberries in the time that it takes Stefanie to pick one basket.

   a. Choose variables and write an equation expressing this relationship.

   b. Graph the equation.

Solution

   a. Let L stand for the number of baskets that Laura has picked and S for the number of baskets that Stefanie has picked. These two variables are related by the equation

$$L = 3S$$

   b. We follow the steps suggested above.

   Step 1   We make a table of values. We choose some reasonable values for the number of baskets Stefanie might pick, as shown in the margin.

S	0	1	2	4	6
L					

   Then we find the corresponding values of L by using the equation $L = 3S$. For example, when $S = 1$,

$$L = 3(1) = 3$$

and when $S = 2$,

$$L = 3(2) = 6$$

The completed table appears in the margin.

S	0	1	2	4	6
L	0	3	6	12	18

   Step 2   Next, we choose scales for the axes of our graph. Values of S will appear along the horizontal axis and values of L along the vertical axis. Because the values of both variables are relatively small, we can scale both axes with intervals of length 1, as shown in Figure 1.8.

Figure 1.8

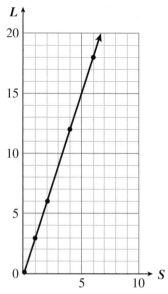

Step 3   We plot the points in the table as shown in Figure 1.8. Finally, we connect the points with a smooth curve. The points in this graph lie on a straight line.

Choosing Scales for the Axes

If the variables in an equation have very large (or very small) values, we must choose scales for the axes that fit these values. For most graphs, it is best to have between 10 and 20 tick marks on each axis. It is not necessary to use the same scale on both axes, as you can see in Example 3.

EXAMPLE 3

Corey's truck holds 20 gallons of gasoline and gets 18 miles to the gallon.

   a. Write an equation that relates the distance, d, that Corey can travel to the number of gallons of gas, g, in his truck.

   b. Graph the equation.

Solution

   a. Corey can travel 18 miles for every gallon of gas, so

$$d = 18g$$

   b. Step 1   First, we make a table of values for the equation. We choose values of g between 0 and 20, because Corey's truck holds at most 20 gallons of gas. We calculate the corresponding values of d using our equation.

g	2	5	8	10	15	20
d	36	90	144	180	270	360

Figure 1.9

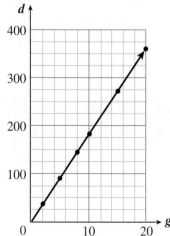

Step 2   Next, we choose scales for the axes. The largest value of g in the table is 20. If we would like 10 tick marks on the horizontal axis, the length of each interval should be $20 \div 10 = 2$ units, so we scale the horizontal axis in intervals of 2. To choose a scale for the vertical axis, we note that the largest value of d is 360. We round this up to 400 and use intervals of 25, which yields 16 tick marks on the vertical axis.

Step 3   Finally, we plot the points from our table of values. We must estimate the location of some of the points between tick marks. The graph is shown in Figure 1.9.

How do we know what scales to use on the axes? Depending on the size of the values to be graphed, intervals of convenient size, such as 5, 10, 25, 100, or 1000, are often used. It is a good idea to have no more than 15 or 20 tick marks on an axis; otherwise, the graph will be hard to read. One rule of thumb is to find the largest value for the variable in your table, round up to the nearest "convenient" number, then divide by the desired number of tick marks—say, ten.

EXERCISE 3

The Harris Aircraft company will give all its employees a 5% raise.

a. Write an equation that expresses each employee's raise, R, in terms of his or her present salary, S.

b. Graph your equation.

Step 1 Complete the table of values.

S	18,000	24,000	32,000	36,000
R				

Step 2 Choose scales for the axes on the grid, and label the axes.

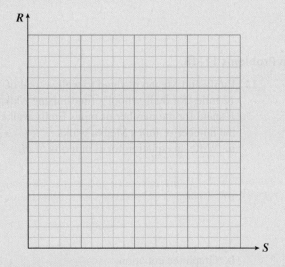

Step 3 Plot points and draw the graph.

ANSWERS TO 1.5 EXERCISES

1.

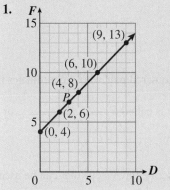

2.

D	1	5	7	8
F	5	9	11	12

$5 = 1 + 4$; $9 = 5 + 4$; $11 = 7 + 4$;
$12 = 8 + 4$

3a. $R = 0.05S$

3b.

S	18,000	24,000	32,000	36,000
R	900	1200	1600	1800

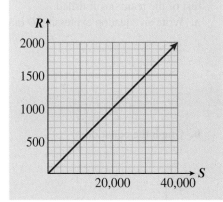

HOMEWORK 1.5

1. A _____ is a way of visualizing an algebraic equation.

2. The values of the variables are displayed on the _____ .

3. The independent variable is displayed on the _____ axis.

4. The position of a point on a graph is given by its _____ .

5. The notation (x, y) is called a(n) _____ .

6. In an ordered pair, the _____ coordinate is always listed first.

7. An ordered pair whose coordinates make an equation true is called a _____ of the equation.

8. The graph of an equation is a picture of its

_____ .

9. Describe three steps that can be used to graph an equation.

10. Explain how to choose scales for the axes of a graph.

Use the three steps on page 51 to graph the equations in Problems 11–18.

11. Delbert was late to swimming practice, and the rest of the team had already completed five laps when he started swimming. He stayed five laps behind during the rest of the workout. Let D stand for the number of laps Delbert has completed and T for the number of laps the rest of the team has finished.
a. Write an equation expressing D in terms of T.

12. Emily and Megan pledged to walk a total of 12 miles for their school's fund-raising walkathon. Let E stand for the number of miles Emily walks and M for the number of miles Megan walks.
a. Write an equation for E in terms of M.

b. Graph the equation.

b. Graph the equation.

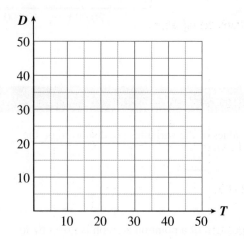

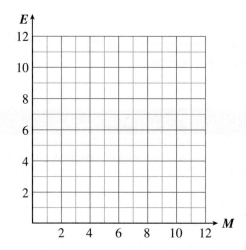

13. A bread recipe calls for exactly half as much rye flour as wheat flour. Let r stand for the number of cups of rye flour needed and w for the number of cups of wheat flour needed.
   a. Write an equation for r in terms of w.

   b. Graph the equation.

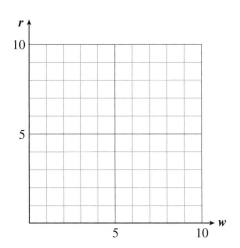

14. Wrapping paper costs $1.50 per roll at an after-holiday sale. Let r stand for the number of rolls Elma buys and p for the total price she pays (before tax).
   a. Write an equation for p in terms of r.

   b. Graph the equation.

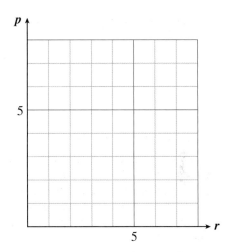

15. Uncle Ray's diet allows him to eat a total of 1000 calories for lunch and dinner. Let l stand for the number of calories in Uncle Ray's lunch and d for the number of calories in his dinner.
   a. Write an equation for d in terms of l.

   b. Graph the equation.

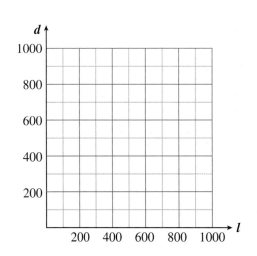

16. Ward and June together earn $40,000 a year. Let j stand for June's annual income and w for Ward's annual income.
   a. Write an equation for j in terms of w.

   b. Graph the equation.

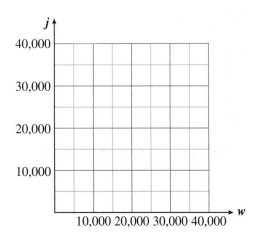

17. The Green Co-op plans to divide among its members its profit of $3600 from the sale of environmentally safe products. Let *m* stand for the number of members and *S* for the share of the profit each member will receive.

   a. Write an equation for *S* in terms of *m*.

18. The owners of a new art gallery want to install a large rectangular picture window to light the gallery's main room. To provide adequate lighting, the window should have an area of 240 square feet. The gallery owners must decide on the best dimensions for the window. Let *h* stand for the height of the window and *w* for its width.

   a. Write an equation for *h* in terms of *w*.

b. Graph the equation.

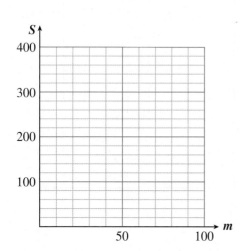

b. Graph the equation.

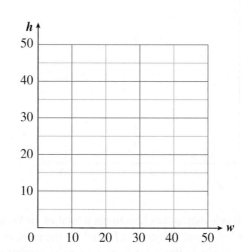

In Problems 19–22, decide whether the ordered pairs are solutions of the given equation.

19. $y = \dfrac{3}{4}x$

   a. (8, 6)   **b.** (12, 16)

   c. (2, 3)   **d.** $\left(6, \dfrac{9}{2}\right)$

20. $y = \dfrac{x}{2.5}$

   a. (1, 2.5)   **b.** (25, 10)

   c. (5, 2)   **d.** (8, 20)

21. $w = z - 1.8$

   a. $(10, 8.8)$ **b.** $(6, 7.8)$

   c. $\left(2, \dfrac{1}{5}\right)$ **d.** $(9.2, 7.4)$

22. $w = 120 - z$

   a. $(0, 120)$ **b.** $(65, 55)$

   c. $(150, 30)$ **d.** $(9.6, 2.4)$

In Problems 23–26, decide whether the ordered pairs are solutions of the equation whose graph is shown.

23. a. $(6.5, 3)$ **b.** $(0, 3.5)$

    c. $(8, 2)$ **d.** $(4.5, 1)$

24. a. $(8, 10)$ **b.** $(12, 1.5)$

    c. $(0.5, 4)$ **d.** $(9, 1.6)$

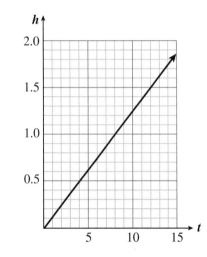

25. a. $(2, 6)$ **b.** $(4, 2)$

    c. $(10, 0)$ **d.** $(11, 7)$

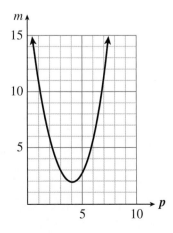

26. a. $(8, 3)$ **b.** $(3, 8)$

    c. $(4, 4)$ **d.** $(5, 10)$

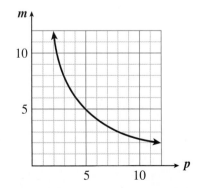

For each graph shown in Problems 27–30:

a. Decide which is the first variable and which is the second variable. Fill in the first column of the table correctly.

b. Fill in the missing values in the table. Choose your own points to complete the table.

c. Look for a pattern in the table, and write an equation for the second variable in terms of the first.

27.

	2	5		10	
	16		10		

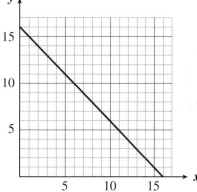

28.

	0	1			7	
			16	20		

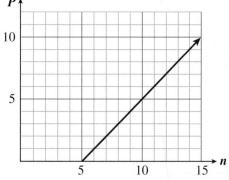

29.

	3		9		
	2	2.5			

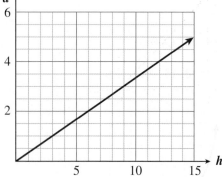

30.

	5	7			
			3	5	

For Problems 31–34, choose the equation that describes the given graph. Not all of the equations have a graph shown here, and a graph might have more than one equation. (*Hint:* Make a table of values from the graph. Which equation has the solutions in the table?)

a. $y = 150 - x$ b. $y = 15 - x$ c. $y = 0.2$ d. $y = \dfrac{x}{60}$

e. $y = \dfrac{x}{5}$ f. $y = 15 + x$ g. $y = 5x$ h. $y = \dfrac{60}{x}$

31.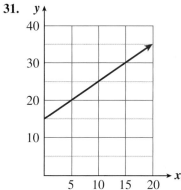

32.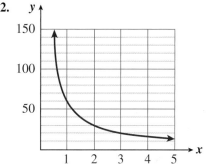

33.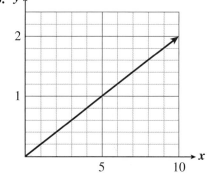

34.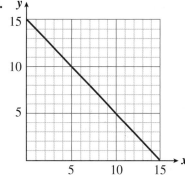

In Problems 35 and 36, fill in each table using the given equation. What do you notice? Explain.

35. a. $y = \dfrac{x}{8}$ **b.** $y = \dfrac{1}{8}x$ **c.** $y = 0.125x$ **36. a.** $y = \dfrac{3x}{5}$ **b.** $y = \dfrac{3}{5}x$ **c.** $y = 0.6x$

x	y
4	
8	
10	
16	

x	y
4	
8	
10	
16	

x	y
4	
8	
10	
16	

x	y
5	
10	
12	
15	

x	y
5	
10	
12	
15	

x	y
5	
10	
12	
15	

MIDCHAPTER REVIEW

In Problems 1–4, use complete sentences to answer the questions.

1. What do the heights of the bars in a histogram represent?

2. What is a *variable*? Give an example.

3. What is the difference between an *algebraic expression* and an *equation*?

4. What is the difference between *factors* and *terms*?

5. a. Write "*p* subtracted from *w*" as an algebraic expression.

   b. Evaluate your expression for $p = 5.6$ and $w = 56$.

6. a. Write "*m* divided by *g*" as an algebraic expression.

   b. Evaluate your expression for $m = 12$ and $g = 192$.

7. State a formula from this chapter for each of the variables below.
   a. Distance: $d =$

   b. Profit: $P =$

   c. Interest: $I =$

   d. Percent or part: $P =$

   e. Average value: $A =$

8. State formulas for the perimeter and area of a rectangle and for the area and circumference of a circle.

9. When constructing the graph of an equation, which variable do you display on which axis?

10. Write a paragraph explaining how the following terms are related to an equation in two variables: *solution, coordinates, ordered pair*.

11. The table shows the number of seats in the legislature for various countries and the number of those seats that were held by women in 1997. (For countries whose legislatures have more than one house, the figures are for the lower house.)

Country	Total number of seats	Number of seats held by women	Percent of seats held by women
Canada	295	63	
France	577	63	
Great Britain	659	120	
Israel	120	9	
Japan	500	23	
Mexico	500	71	
South Africa	400	100	
Sweden	349	141	
United States	435	51	

Source: *Time*, June 16, 1997.

a. Which country has the most seats in its legislature? Which has the fewest seats?

b. Which country has the most women in its legislature? Which has the fewest women?

c. Compute the percent of seats held by women in each country.

d. Which country has the highest percent of women in its legislature? Which has the lowest percent of women?

12. The bar graphs show the female literacy rate for ten countries and the population growth rate for the same ten countries.

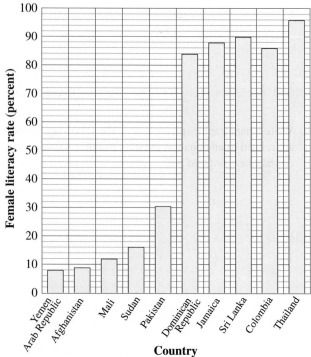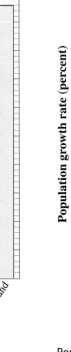

Female literacy rate, 1985

Population growth rate, 1988–2000

a. What are the female literacy rate and the population growth rate for Jamaica?

b. Which country has a 31% female literacy rate? A 2.9% population growth rate?

c. In what countries is the population growth rate 1.4%?

d. What country has the lowest female literacy rate? The lowest population growth rate?

e. Do these data suggest a connection between birth rate and female literacy rate?

13. The histogram shows the percent of the residents who have bachelor's degrees in the 50 states and the District of Columbia.

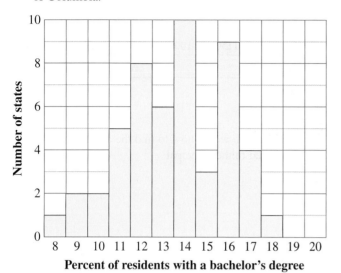

Percent of residents with a bachelor's degree

a. In how many states do 16% of the residents have bachelor's degrees?

b. What percent is common to six states?

c. What is the mode of the data? How many states have the modal percent?

d. In how many states do 9% or less of the residents have bachelor's degrees?

e. In how many states do 14% or more of the residents have bachelor's degrees?

f. Calculate the median and the mean percents.

14. Match each story with the graph that represents it. Create a story for the remaining graph.
  i. During a baseball game, the pitcher throws the ball, the batter hits it, it bounces, and a fielder catches it and holds it for a while.

  ii. In a tennis match, one player serves the ball, the other player returns it, and then the players volley back and forth.

  iii. A free-throw shooter bounces the basketball before taking her shot; then the ball bounces off the rim, and another player grabs it.

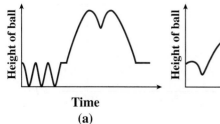

(a)

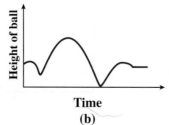

(b)

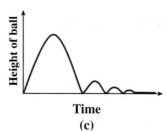

(c)

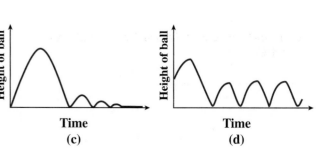

(d)

15. All living organisms contain a small amount of carbon-14, an isotope of carbon that gradually breaks down after the organism dies. By measuring how much of its carbon-14 has decayed, scientists can determine how old an object is. The graph shows how long it takes for a given fraction of the carbon-14 in an object to decay.

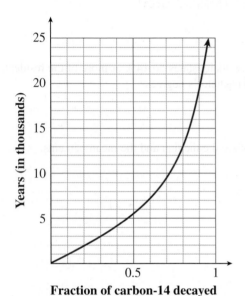

Fraction of carbon-14 decayed

  a. How old is a piece of reindeer bone if 80% of its carbon-14 has decayed?

  b. How much of the carbon-14 in an object will decay in 3000 years?

  c. The *half-life* of an isotope is the time it takes for half of the initial amount of the isotope to decay. What is the half-life of carbon-14?

16. Janel's mortgage payment is $1200 per month. Part of each payment is applied to the principal, and the rest pays the monthly interest on the loan. Fill in the table.

Interest	1150	1000	750	620	480
Calculation					
Principal					

  a. Explain in words how to find the amount that is applied to the principal.

  b. Write your explanation as a mathematical sentence:

    Principal =

  c. Let *P* stand for the amount that goes toward the principal and *I* for the monthly interest. Write an equation for *P* in terms of *I*.

  d. Plot the values from the table, and connect the data points with a smooth curve.

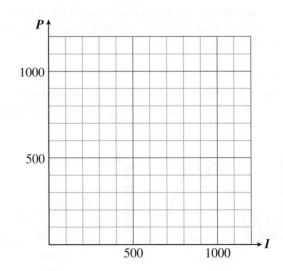

17. The graphs show the average annual earnings of male college and high school graduates from 1970 to 1990.

    a. What could a male college graduate expect to earn in 1972? What could a male high school graduate expect to earn in 1972?

    b. In what year could a high school graduate expect to earn $25,000? In what year could a college graduate expect to earn $25,000?

    c. What was the difference in earnings between a college graduate and a high school graduate in 1970? In 1985? In 1990?

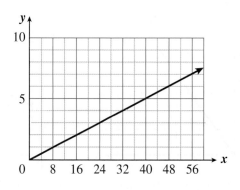

Average annual earnings of male college and high school graduates, 1970–1990

    d. Calculate the earnings of high school graduates as a percent of the earnings of college graduates at 5-year intervals.

Year	Average earnings		Percent
	High school graduates	College graduates	
1970	9567	13,264	
1975	13,542	17,477	
1980	19,469	24,311	
1985	23,853	32,822	
1990	26,653	39,238	

Source: U.S. Bureau of the Census, Digest of Educational Statistics, 1994.

18. a. Fill in the table using the graph.

    b. State a formula expressing *y* in terms of *x*.

x	16	32	40	56
y				

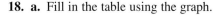

In Problems 19–22, choose variables for the unknown quantities and write an algebraic expression for each phrase.

19. Six degrees hotter than yesterday's temperature

20. The cost of the gasoline divided three ways

21. 8% of the total bill

22. Five inches less than the height of the triangle

23. a. Write an expression for the distance you can travel in 3 hours at a speed of r miles per hour.

   b. How far can you travel in 3 hours on a bicycle at a speed of 6 miles per hour? In a motorboat at a speed of 20 miles per hour?

24. a. Write an expression for the amount of interest earned in 2 years on $1500 deposited in an account that pays interest rate r.

   b. How much interest will $1500 earn in 2 years at $6\frac{1}{2}\%$ interest? At 8.3% interest?

25. On a typical evening, there are twice as many nonsmokers as smokers in a popular restaurant.

   a. Write an equation for the number of nonsmokers, n, in terms of the number of smokers, s.

   b. Make a table of values, and graph your equation.

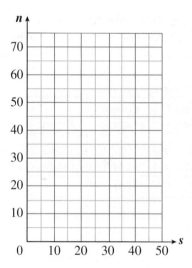

26. Match each equation with its graph.

   a. $y = 8 - x$

   b. $y = x + 8$

   c. $y = 8x$

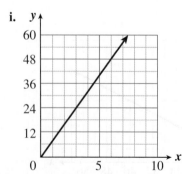

i.

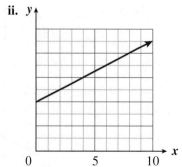

ii.

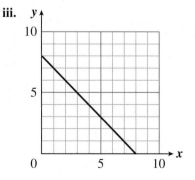

iii.

1.6 Solving Equations

Reading

An **equation** is just a statement that two quantities are equal. In Section 1.3, we studied equations in which one variable was defined in terms of another. For example, suppose Liz makes $6 an hour as a tutor in the Math Lab. Her weekly earnings, w, depend on the number of hours she works, h. The equation relating these two variables is

$$w = 6h$$

If we know the value of h for a given week, we can evaluate the expression on the right side of the equals sign to determine Liz's earnings that week. For example, if $h = 7$, then

$$w = 6(7) = 42$$

Liz makes $42 for 7 hours of work. When you fill in a table like the one in Exercise 1, you are evaluating the algebraic expression $6h$.

Now, suppose we know Liz's earnings for a certain week—that is, we know the value of w. Can we work backward to find the value of h, the number of hours Liz worked? For example, suppose Liz earned $54 last week. How many hours did she work? To answer this question, we substitute 54 for w in the equation $w = 6h$ and obtain

$$54 = 6h$$

We would like to find the value of h that makes this equation true. A value of a variable that makes an equation true is called a **solution** of the equation, and the process of finding this value is called **solving** the equation. In this section, we'll investigate some methods for solving equations.

First, try Exercise 2.

EXERCISE 1
Complete the table showing Liz's wages in terms of the number of hours she works.

h	3	5	8	10
w				

EXERCISE 2

a. Choose any number
for x: $x =$ _____ .

Multiply your
number by 5: $5x =$ _____ .

Divide the result

by 5: $\dfrac{5x}{5} =$ _____ .

Did you end up with your original number? Try multiplying and dividing your number by another number besides 5. Do you still end up with your original number? Multiplication and division are opposite, or *inverse*, operations, because one undoes the effects of the other.

b. Choose any number
for x: $x =$ _____ .

Add 4 to your
number: $x + 4 =$ _____ .

Subtract 4
from the
result: $x + 4 - 4 =$ _____ .

Did you end up with your original number? Try adding and subtracting another number besides 4. Do you still end up with your original number? Addition and subtraction are opposite, or inverse, operations, because one undoes the effects of the other.

Skills Review

Convert each percent to a decimal. (See Appendix A.9.)

1. 25%	2. 30%	3. 2%	4. 5%
5. 137%	6. 6.2%	7. 0.8%	8. 103.5%

Answers:
1. 0.25   *2.* 0.3   *3.* 0.02   *4.* 0.05   *5.* 1.37   *6.* 0.062   *7.* 0.008   *8.* 1.035

Lesson

Solving Equations by Trial and Error

Consider the equation from the reading assignment:

$$54 = 6h$$

If you experiment with different values of h, you will quickly see that the solution of the equation $54 = 6h$ is 9. If you look back at the table in Exercise 1, you'll see that $h = 8$ gives $w = 48$, which is too small, and $h = 10$ gives $w = 60$, which is too big. We can often use a table in this way to solve an equation by trial and error. Try this technique in Exercise 3.

Solving Equations by Algebraic Methods

Trial and error is a useful technique, especially for complicated equations, but it can be time-consuming. To solve simple equations, we can take advantage of inverse, or opposite, operations. Perhaps you already see a faster way to solve the equation in Exercise 3. An easy example will help us analyze the process.

EXAMPLE 1

Francine is exactly 4 years older than Delbert, so

$$F = D + 4$$

where D stands for Delbert's age and F stands for Francine's age. (We graphed this equation in Section 1.5.) Here is a table of values for the equation. You should find the missing values, and think about how you found each one.

D	5	7		10		18
F			12		19	

How do we find each value? When we are given a value of D, say $D = 5$, we *add* 4 to find the value of F. That is what the expression $D + 4$ says: Francine is 4 years older than Delbert. When we are given a value of F, say $F = 12$, we *subtract* 4 to find Delbert's age, because he is 4 years younger than Francine.

Try the similar problem in Exercise 4.

EXERCISE 3
One of Aunt Esther's Chocolate Dream cookies contains 42 calories, so d cookies contain c calories, where

$$c = 42d.$$

If Albert consumed 546 calories, how many cookies did he eat? Use trial and error to help you solve the equation

$$546 = 42d$$

d	c
5	
6	
7	
8	
9	
10	
11	
12	
13	
14	
15	

Solution:

EXERCISE 4

Fernando plans to share an apartment with three other students and split the rent equally.

a. Let r stand for the rent on the apartment and s for Fernando's share. Write an equation for s in terms of r.

b. Fill in the table.

r	260	300		360		480
s			80		105	

c. Explain how you found the unknown values of s.

Explain how you found the unknown values of r.

Now let's take another look at the equation from Example 1 relating Delbert's and Francine's ages:

$$F = D + 4$$

If Francine is 19 years old, we can write the equation as follows:

$$19 = D + 4$$

To find Delbert's age when Francine is 19, we must solve this equation for D. You can probably solve this particular equation in your head, but eventually you will encounter more complicated equations, and you'll need a strategy for finding their solutions.

To solve the equation algebraically, we will transform it into a new equation of the form $D = k$ (or $k = D$), where the number k is the solution. In other words, we must isolate D on one side of the equals sign. At present, we have $D + 4$ on one side of the equation, not D. We can isolate D by *subtracting* 4 from both sides, like this:

$$\begin{array}{r} 19 = D + 4 \\ -4 \quad\quad -4 \\ \hline 15 = D \end{array}$$

Subtract 4 from both sides of the equation.

On the right side of the equals sign, note that $D + 4 - 4$ is the same as D, because subtraction is the inverse, or opposite, operation for addition. The solution to the equation is 15. You can check that substituting 15 for D does make the equation true.

Check: $19 = D + 4$ Substitute 15 for D.
   $19 = 15 + 4$ True; the solution checks.

We have just used algebra to solve the equation $19 = D + 4$. Here is a general strategy for solving such equations:

To Solve an Equation Algebraically:

Step 1 Ask yourself what operation has been performed on the variable.

Step 2 Perform the opposite operation on *both* sides of the equation. Doing so should isolate the variable.

The following examples show how to solve equations involving each of the four arithmetic operations: addition, subtraction, multiplication, and division. Study each solution, and then complete the accompanying exercise. Don't do the exercises in your head! It is important to write down the steps of your solution.

EXERCISE 5

Solve $5 + y = 9$

EXAMPLE 2

Solve $x + 6 = 11$

$$x + 6 = 11$$ *Step 1* 6 is *added* to the variable.
$$\underline{-6 \quad -6}$$ *Step 2* We *subtract* 6 from both sides.
$$x \quad = \quad 5$$ The solution is 5.

Check: **5 + 6 = 11**

EXERCISE 6

Solve $x - 4 = 12$

EXAMPLE 3

Solve $n - 17 = 32$

$$n - 17 = 32$$ *Step 1* 17 is *subtracted* from the variable.
$$\underline{+17 \quad +17}$$ *Step 2* We *add* 17 to both sides.
$$n \quad = \quad 49$$ The solution is 49.

Check: **49 − 17 = 32**

EXERCISE 7

Solve $6z = 24$

EXAMPLE 4

Solve $12x = 60$

$$\frac{12x}{12} = \frac{60}{12}$$ *Step 1* The variable is *multiplied* by 12.
   Step 2 We *divide* both sides by 12.
$$x = 5$$ The solution is 5.

Check: **12(5) = 60**

EXERCISE 8

Solve $\dfrac{w}{3} = 6$

EXAMPLE 5

Solve $\dfrac{w}{7} = 21$

$$7\left(\frac{w}{7}\right) = 7(21)$$ *Step 1* The variable is *divided* by 7.
   Step 2 We *multiply* both sides by 7.
$$w = 147$$ The solution is 147.

Check: $\dfrac{147}{7} = 21$

Using Formulas

Using a formula often involves solving an algebraic equation. Here are the five formulas we studied in Section 1.4. Check that you remember what all of the variables in these formulas represent. If you do not, you should review the examples on pages 44, 45, and 46.

distance: $d = rt$

profit: $P = R - C$

interest: $I = Prt$

percent: $P = rW$

average: $A = \dfrac{S}{n}$

EXAMPLE 6

A savings account at Al's Bank earns 5% interest annually. If Jean's savings there earned $42.50 interest last year, what was her balance at the beginning of the year?

Solution

We will use the formula $I = Prt$. We are asked to find the principal, P. The values of I, r, and t are given, so we substitute those values into the formula:

$$I = P \cdot r \cdot t$$
$$42.50 = P(0.05)(1)$$

To find the value of P, we must solve the equation $42.50 = P(0.05)$. The variable, P, is multiplied by 0.05, so we will divide both sides of the equation by 0.05:

$$\frac{42.50}{0.05} = \frac{P(0.05)}{0.05}$$
$$850 = P$$

Jean had $850 in her account when the year began.

Now try Exercise 9, following the hints.

EXERCISE 9
The distance from Los Angeles to San Francisco is approximately 420 miles. How long will it take a car traveling at 60 miles per hour to go from Los Angeles to San Francisco?

First, decide which formula is appropriate.

Which of the variables in the formula is unknown?

What values are given for the other variables in the formula?

Substitute the known values into the formula. Solve the equation for the unknown variable.

Answer:

ANSWERS TO 1.6 EXERCISES

1. $18, $30, $48, $60

3. $d = 13$

4a. $s = \dfrac{r}{4}$

4b. $65, $75, $320, $90, $420, $120

4c. To find the values of s, divide r by 4. To find the values of r, multiply s by 4.

5. $y = 4$ **6.** $x = 16$

7. $z = 4$ **8.** $w = 18$

9. 7 hours

HOMEWORK 1.6

For Problems 1–4, fill in the table using the given equation, and explain how you found the unknown values for each table.

1. $q = 9 + t$

t	q
2	
4	
	15
	18
21	
	39

2. $m = h - 12$

h	m
15	
16	
	6
	8
23	
	15

3. $p = 5n$

n	p
0	
2	
	20
5	
	35
	55

4. $z = \dfrac{1}{2}w$

w	z
1	
3	
	3
	6
15	
	8.5

In Problems 5–10, decide whether the given value for the variable is a solution of the equation.

5. $x - 4 = 6$; $x = 10$

6. $\dfrac{a}{4} = 12$; $a = 3$

7. $4y = 28$; $y = 24$

8. $3.2t = 22.4$; $t = 7$

9. $\dfrac{0}{z} = 0$; $z = 19$

10. $\dfrac{24}{v} = 0$; $v = 0$

In Problems 11–28, solve the equation, and check your solution.

11. $x - 3 = 11$

12. $x + 4 = 18$

13. $10.6 = 7.8 + y$

14. $13.2 = a - 6.6$

15. $3y = 108$

16. $\dfrac{t}{4} = 27$

17. $42 = 3.5b$

18. $12y = 4$

19. $2.6 = \dfrac{a}{1.5}$

20. $\dfrac{v}{5} = 0$

21. $x - 4 = 0$

22. $\dfrac{v}{12} = 36$

23. $34x = 212$

24. $6.5b = 20.15$

25. $6z = 20$

26. $\dfrac{m}{1.8} = 1$

27. $9 = k + 9$

28. $8 = \dfrac{n}{8}$

For Problems 29–36:
a. Use a formula to write an equation for each problem.
b. Solve the equation, and answer the question.

29. Clive loaned his brother some money to buy a new truck, and his brother agreed to repay the loan in 1 year with 3% interest. Clive earned $75 in interest on the loan. How much did Clive loan his brother?

30. Sunshine Industries reported a first-quarter profit of $26,500 on its new line of solar-powered lawn sprinklers. The company's total cost for producing and marketing the sprinklers was $138,200. What was the revenue from the sprinklers?

31. Andy's average score on eight 50-point homework assignments was 38.25. How many homework points did Andy earn altogether?

32. Agnes drove 364 miles in 7 hours. How fast did she drive?

33. How long will it take a cyclist traveling at 13 miles an hour to cover 234 miles?

34. Forty-five percent of the voters voted for Senator Fogbank. If the senator received 540 votes, how many people voted?

35. A roll of carpet contains 400 square feet of carpet. If the roll is 16 feet wide, how long is the piece of carpet?

36. Brenda's recipe for fruit punch makes a beverage that is 36% ginger ale. If you have 9 quarts of ginger ale, how much of Brenda's fruit punch can you make?

Summarize what you learned in this section by answering the questions in Problems 37–42.

37. What is an equation?

38. What is a solution of an equation?

39. How can you check to see whether a given number is a solution to an equation?

40. a. Is $3 + 4 = 12$ an equation? Why or why not?

   b. Is $x + 4 = 12$ an equation? Why or why not?

41. Describe a trial-and-error method for solving an equation.

42. Describe a two-step strategy for solving an equation algebraically.

Use the following facts about angles to write and solve an equation to find the unknown angle in Problems 43–50.

- The sum of the angles in a triangle is 180°.
- A straight angle has a measure of 180°.
- Two angles whose sum is 180° are called **supplementary angles.**
- Two angles whose sum is 90° are called **complementary angles.**

43.

44.

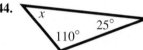

45.

46.

47.

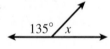

48.

49.

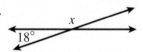

50.

1.7 Problem Solving with Algebra

Reading

The following example compares the algebraic methods for *evaluating an expression* and for *solving an equation*.

EXAMPLE 1

Several years ago, the Reedville City Council passed a resolution that 35% of the town's budget should be allotted to education.

    a. Choose variables, and write an equation for the amount allotted to education in terms of the total budget.

    b. If Reedville's total budget for next year is $1,800,000, how much will be allotted to education?

    c. If Reedville spent $875,000 on education last year, what was the town's total budget for that year?

Solution

    a. Let b stand for Reedville's total budget and s for the amount allotted to education. Then

$$(\text{amount for education}) = 35\% \text{ of total budget}$$
$$s = 0.35b$$

    b. To find the amount allotted for education, we *evaluate the expression* $0.35b$ for $b = 1{,}800{,}000$:

$$s = 0.35(1{,}800{,}000)$$
$$= 630{,}000$$

    The amount allotted to education is $630,000.

    c. To find the total budget, we substitute 875,000 for s to obtain the equation

$$875{,}000 = 0.35b$$

    Next we *solve the equation*. Because b is multiplied by 0.35, we must divide both sides of the equation by 0.35:

$$\frac{875{,}000}{0.35} = \frac{0.35b}{0.35}$$
$$2{,}500{,}000 = b$$

    The total budget last year was $2,500,000.

    We used two different algebraic techniques in Example 1: In part (b), the unknown quantity, s, was already isolated on one side of the equation. All we had to do was use arithmetic to simplify the other side of the equation. This process is called *evaluating an expression*. In part (c) of Example 1, we had to isolate the unknown quantity, b, by solving the equation.

Skills Review

Multiply or divide the fractions as indicated. (See Appendix A.2 to review products and quotients of fractions.)

1. $\dfrac{1}{3} \cdot \dfrac{2}{5}$ 2. $\dfrac{2}{7} \cdot \dfrac{2}{3}$ 3. $\dfrac{4}{3} \cdot \dfrac{1}{8}$

4. $\dfrac{3}{8} \cdot \dfrac{5}{9}$ 5. $\dfrac{1}{36} \div \dfrac{1}{9}$ 6. $\dfrac{4}{7} \div \dfrac{14}{5}$

7. $\dfrac{21}{5} \div \dfrac{7}{10}$ 8. $\dfrac{3}{20} \div \dfrac{9}{4}$

Answers: **1.** $\dfrac{2}{15}$ **2.** $\dfrac{4}{21}$ **3.** $\dfrac{1}{6}$ **4.** $\dfrac{5}{24}$ **5.** $\dfrac{1}{4}$ **6.** $\dfrac{10}{49}$ **7.** 6 **8.** $\dfrac{1}{15}$

Lesson

Solving Problems by Writing Equations

Many practical problems can be solved by first writing an equation that describes, or *models*, the problem and then solving the equation. We will use three steps to solve an applied problem.

Steps for Solving an Applied Problem:

Step 1 Identify the unknown quantity, and choose a variable to represent it.

Step 2 Find some quantity that can be described in two different ways, and write an equation using the variable to model the problem situation.

Step 3 Solve the equation, and answer the question in the problem.

EXAMPLE 2

Jerry needs an additional $35 for airfare to New York. The plane ticket costs $293. How much money does Jerry have?

Solution

We follow the three steps.

Step 1 First, we describe the unknown quantity in words and assign a variable to represent it.

   Amount of money Jerry has: m

Step 2 Next, we look for some quantity that can be described in two different ways; this will give us an equation. For this problem, the quantity is the airfare to New York.

$$m + 35 = 293$$

amount   amount   cost of
Jerry has   he needs   airfare

Step 3 Finally, we solve the equation and answer the question.

$$m + 35 = 293$$ Subtract 35 from both sides.
$$\underline{ -35 \quad -35}$$
$$m = 258$$

Jerry has $258.

_____

CAUTION!

The equation in Example 2 is about the airfare to New York. It expresses this quantity in two different ways. The equation is *not* about the amount of money Jerry has, which is what we want to find. This is typical of applied problems: The expressions on both sides of the equation that you write usually represent not the unknown quantity but some related quantity.

_____

Although you may be able to solve the problems in this section with arithmetic, the important thing is to learn how to write an algebraic equation, or model, for a problem. This skill will help you solve problems that are too difficult for arithmetic alone.

In Exercise 1, concentrate on writing an equation for the problem. Use the hints to help you solve the problem.

EXERCISE 1
A two-bedroom house costs $20,000 more than a one-bedroom house in the same neighborhood. The two-bedroom house costs $105,000. How much does the one-bedroom house cost?

Step 1 Choose a variable for the unknown quantity.

Cost of the one-bedroom house:

_____.

Step 2 Write an equation in terms of your variable.

_____ + _____ = _____

cost of one-   cost of two-
bedroom house   bedroom house

Step 3 Solve your equation.

The one-bedroom house costs

_____.

Now try Exercises 2 and 3 in the margin.

EXERCISE 2
A restaurant bill is divided equally among seven people. If each person pays $8.50, how much was the bill?

Step 1 Choose a variable for the unknown quantity. (What are you asked to find?)

Step 2 Write an equation. Express each person's share in two different ways.

Step 3 Solve your equation, and answer the question.

EXERCISE 3
Iris got a 6% raise. Her new salary is $21 a week more than her old salary. What was her old salary?

Step 1 Choose a variable for the unknown quantity.

Step 2 Write an equation. Express Iris's raise in two different ways.

Step 3 Solve your equation, and answer the question.

Figure 1.10

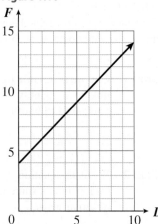

Solving Equations with Graphs

Figure 1.10 is the graph of the equation from Section 1.5, which gives Francine's age, F, in terms of Delbert's age, D:

$$F = D + 4$$

Recall that each point on the graph has two coordinates (D, F), which give a solution of the equation—that is, a pair of values for the variables that make the equation true. For example, the point $(2, 6)$ means that when Delbert was 2 years old, Francine was 6.

We can use the graph to answer two types of questions about the equation like $F = D + 4$:

1. Given a value of D, what is the corresponding value of F?
2. Given a value of F, what is the corresponding value of D?

The first of these tasks is another way of evaluating the algebraic expression $D + 4$, and the second is another way of solving an equation.

EXAMPLE 3

a. Use the graph of the equation $F = D + 4$ to evaluate the expression $D + 4$ for $D = 7$. Verify your answer algebraically.

b. Use the graph of the equation $F = D + 4$ to solve the equation $13 = D + 4$. Check your work by solving algebraically.

Solution

a. We locate the value $D = 7$ on the horizontal axis, as shown in Figure 1.11(a). We move vertically upward from that location to find the point on the graph whose D-coordinate is 7. This point is labeled A in the figure. Then we move horizontally from point A to the vertical axis to find the F-coordinate of point A. We see that the coordinates of point A are $(7, 11)$, which tells us that when $D = 7$, $F = 11$. To verify the answer algebraically, we substitute 7 for D in the equation:

$$F = D + 4 = 7 + 4 = 11$$

Figure 1.11

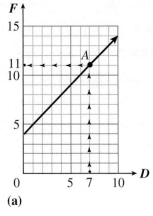 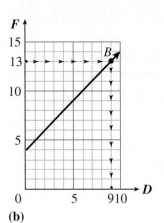

(a)   **(b)**

b. We locate the value $F = 13$ on the vertical axis, as shown in Figure 1.11(b). We move horizontally from that location to the point on the graph whose F-coordinate is 13. This point is labeled B in the figure. From point B, we move vertically downward to the horizontal axis to find the D-coordinate of the point. We see that the coordinates of point B are (9, 13), which tells us that when $F = 13$, $D = 9$. To verify the answer algebraically, we subtract 4 from both sides of the equation:

$$
\begin{aligned}
13 &= D + 4 \\
-4 \quad &\quad -4 \\
\hline
9 &= D
\end{aligned}
$$

Now try Exercise 4.

EXERCISE 4

Here is a graph of the equation from Example 1,

$$s = 0.35b$$

Note that both axes of the graph are scaled in thousands of dollars. Use the graph to estimate the answers to the questions in Example 1. Show directly on the graph how you obtain your estimates.

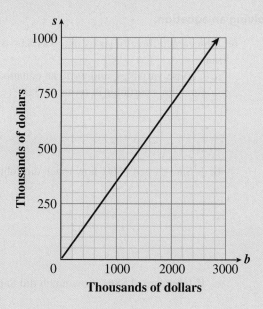

a. Evaluate the equation for $b = \$1,800,000$.

b. Solve the equation for $s = \$875,000$.

ANSWERS TO 1.7 EXERCISES

1. $85,000

2. $59.50

3. $350 a week

4a. $s \approx \$630,000$

4b. $b \approx \$2,500,000$

HOMEWORK 1.7

1. What is the difference between evaluating an expression and solving an equation? Use the formula $B = 1.08P$ as an example in your explanation.

2. What is the first step in solving an applied problem with algebra?

3. What are the next two steps in solving an applied problem with algebra?

4. Explain how you can use the graph of $y = x - 18.2$ to solve the following equation: $26.9 = x - 18.2$.

Problems 5–10 compare evaluating an expression and solving an equation.

5. A used car costs $3400 less than a new car of the same model.
   a. Choose variables, and write an equation for the cost of the used car in terms of the cost of the new car.

   b. If the new car costs $14,500, how much does the used car cost?

   c. If the used car costs $9200, how much does the new car cost?

6. John earned three times as much as Sam over the summer.
   a. Choose variables, and write an equation for John's earnings in terms of Sam's earnings.

   b. If Sam earned $560, how much did John earn?

   c. If John earned $861, how much did Sam earn?

7. The Dodgers won 60% of their games last season.
   a. Choose variables, and write an equation for the number of games the Dodgers won in terms of the number of games they played.

   b. If the Dodgers played 120 games, how many did they win?

   c. If the Dodgers won 96 games, how many did they play?

8. At a certain small college 65% of the students receive financial aid.
   a. Choose variables, and write an equation for the number of students who receive financial aid in terms of the total number of students at the college.

   b. If there are 1800 students enrolled at the college, how many receive financial aid?

   c. If 377 students receive financial aid, how many are enrolled at the college?

9. Sunshine Industries manufactures beach umbrellas. The company's profit on each umbrella is 18% of the selling price.
   a. Choose variables, and write an equation for the profit on each umbrella in terms of its selling price.

   b. If a beach umbrella sells for $60, what is the profit?

   c. If the profit on one umbrella is $7.20, what is the selling price?

10. Oranj-Aid is 8% orange juice.
   a. Choose variables, and write an equation for the amount of orange juice in a quantity of Oranj-Aid.

   b. How much orange juice is in 64 ounces of Oranj-Aid?

   c. How much Oranj-Aid would you have to drink to get 12 ounces of orange juice?

For Problems 11–14, write an algebraic equation and solve it. Follow the three steps on pages 79 and 80.

11. Martha paid $26 less for a suit at a discount store than her mother paid for the same suit at a boutique. If Martha paid $89 for the suit, how much did her mother pay?

12. Ruth needs $21 more to pay next semester's tuition. If tuition is $150, how much money does Ruth have?

13. Emily spends 40% of her monthly income on rent. If Emily's rent is $360 a month, how much is her monthly income?

14. Jo-Allyn bought a winter coat on sale for 70% of its original price. If she paid $126 for the coat, what was its original price?

Problems 15 and 16 use the algebraic formulas on page 73. Write an equation and solve it to answer the questions. Show your work!

15. a. Michael Johnson ran 200 meters in 19.32 seconds at the 1996 Olympics. What was his average speed? (Be sure to include units in your answer.)

16. a. The U.S. women's 4 × 100-meter relay team captured the gold medal at the 1996 Olympics with an average time per runner of 10.4875 seconds. What was the winning time for the team?

   b. Donovan Bailey of Canada set a world record when he ran the 100-meter dash in 9.84 seconds. What was his average speed?

   b. What was the average speed of the relay team?

   c. Gail Devers won the women's 100-meter dash in 10.94 seconds. What was her average speed? Who had the faster average speed, Gail Devers or the women's relay team?

   c. Who had the faster average speed in his race, Johnson or Bailey? (Can you explain this?)

Choose the appropriate equation to model each situation in Problems 17–22.

$$x + 7 = 26 \qquad 7x = 26 \qquad \frac{x}{7} = 26$$

$$x - 7 = 26 \qquad \frac{x}{26} = 7 \qquad \frac{26}{x} = 7$$

17. Sarah drove 7 miles farther to her high school reunion than Jenni drove. If Sarah drove 26 miles, how far did Jenni drive?

18. Lurline and Rozik live 26 miles apart. They meet at a theme park between their homes. If Lurline drove 7 miles to the park, how far did Rozik drive?

19. Doris is training for a triathlon. This week she averaged 26 miles per day on her bicycle. If she rode every day, what was her total mileage for the week?

20. Glynnis jogged the same route every day this week, for a total of 26 miles. How long is her route?

21. Astrid divided her supply of colored pencils among the 26 children in her class, and each child got 7 pencils. How many pencils did Astrid have?

22. Ariel lost 7 of the beads on her necklace, leaving 26. How many were there originally?

Use the graphs to solve each equation in Problems 23–26. (You may have to estimate some of your answers.) Check your work by solving algebraically.

23. Use the graph of $k = p - 20$ to solve these equations:

    a. $90 = p - 20$ **b.** $p - 20 = 40$

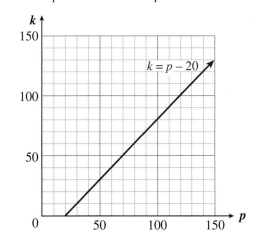

24. Use the graph of $y = x + 24$ to solve these equations:

    a. $x + 24 = 72$ **b.** $56 = x + 24$

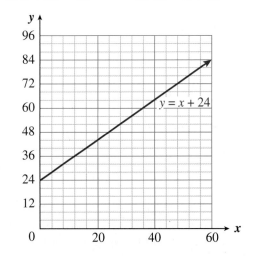

25. Use the graph of $d = 18g$ to solve these equations:

    a. $250 = 18g$ **b.** $18g = 210$

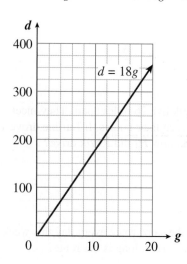

26. Use the graph of $y = \dfrac{x}{12}$ to solve these equations:

    a. $7.5 = \dfrac{x}{12}$ **b.** $\dfrac{x}{12} = 4.2$

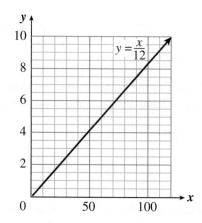

For Problems 27–30:

a. Give the coordinates of the indicated point.

b. Explain what those coordinates tell you in terms of the problem situation.

27. The graph gives the sales tax, T, in terms of the price, p, of an item, in dollars.

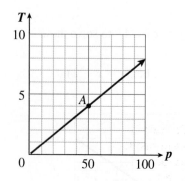

28. The graph gives the pressure, P, in atmospheres, of a gas that is compressed to a volume of V liters.

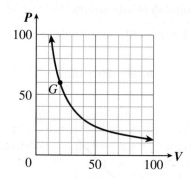

29. The graph gives the weight, *W*, in ounces, of a chocolate bunny that is *h* inches tall.

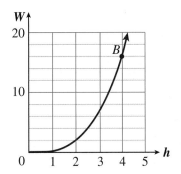

30. The graph gives the altitude, *a*, in meters, of a hot-air balloon at *t* minutes after takeoff.

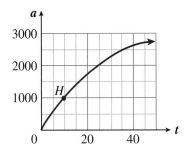

The first step in problem solving is to read the problem carefully so that you know exactly what you are trying to find. Problems 31–36 are "trick" problems.

31. Eve has seven apples. She eats all but three. How many apples are left?

32. Is it legal in Kentucky for a man to marry his widow's sister?

33. The perimeter of a rectangle is 38 centimeters. The rectangle's length is 3 centimeters more than twice its width. What is its perimeter?

34. How much dirt is in a hole 3 feet wide, 6 feet long, and 2 feet deep?

35. A bus picked up 12 passengers at the first stop on its route. At the next stop, it picked up 10 more passengers and let off 3. At the next stop, the bus let off 2 passengers and picked up 8. At the next stop, it picked up 11 people and let off 8. At the next stop, 5 people got on the bus, and at the last stop, 4 people got off. How many stops did the bus make?

36. As I was going to St. Ive's,
I met a man with seven wives.
Each wife had seven sacks;
Each sack had seven cats;
Each cat had seven kits.
Kits, cats, sacks, and wives:
How many were going to St. Ive's?

1.8 Order of Operations

Reading

The algebraic expressions we've studied so far use only one operation: addition, subtraction, multiplication, or division. However, it is not unusual for an algebraic expression to use two or more different operations. You must learn how to interpret such expressions. Specifically, you need to know which of the operations should be carried out first.

Addition and Multiplication

Suppose you would like to simplify a sum of three or more whole numbers, such as

$$2 + 5 + 8$$

It doesn't matter which addition you do first; you will get the same answer either way. Check this for yourself. (The parentheses show which part of the expression to simplify first.)

$$(2 + 5) + 8 = 7 + 8 \qquad \text{and} \qquad 2 + (5 + 8) = 2 + 13$$
$$= 15 \qquad\qquad\qquad\qquad\qquad = 15$$

Similarly, a product of three factors can be multiplied in any order. For example,

$$(3 \cdot 2) \cdot 4 = 6 \cdot 4 \qquad \text{and} \qquad 3 \cdot (2 \cdot 4) = 3 \cdot 8$$
$$= 24 \qquad\qquad\qquad\qquad\qquad = 24$$

These two facts illustrate the **associative laws** for addition and multiplication.

Associative Law for Addition

If a, b, and c are any numbers, then
$$(a + b) + c = a + (b + c)$$

Associative Law for Multiplication

If a, b, and c are any numbers, then
$$(a \cdot b) \cdot c = a \cdot (b \cdot c)$$

Subtraction and Division

The associative laws do *not* hold for subtraction or division. In Example 1, note that we get different answers depending on which operation we perform first.

EXAMPLE 1

a. Subtraction:
$$(20 - 8) - 5 = 12 - 5 = 7$$

but

$$20 - (8 - 5) = 20 - 3 = 17$$

b. Division:
$$(36 \div 6) \div 2 = 6 \div 2 = 3$$

but

$$36 \div (6 \div 2) = 36 \div 3 = 12$$

If there are no parentheses in an expression, how do we know which operations to perform first? In a string of additions and subtractions, we perform the operations *in order from left to right*. For example,

$$20 - 8 - 5 = 12 - 5 = 7$$

Similarly, we perform multiplications and divisions *in order from left to right*. For example,

$$36 \div 6 \div 2 = 6 \div 2 = 3$$

Try Exercise 1 now.

In this section, you'll learn how to simplify other types of expressions, including those that involve both additions and multiplications.

Skills Review

Add or subtract the fractions as indicated. (See Appendix A.3 to review addition of fractions.)

1. $\dfrac{3}{11} + \dfrac{5}{11}$ 2. $\dfrac{5}{8} - \dfrac{3}{8}$ 3. $\dfrac{1}{2} + \dfrac{1}{6}$

4. $\dfrac{7}{8} - \dfrac{3}{4}$ 5. $\dfrac{2}{3} - \dfrac{1}{4}$ 6. $\dfrac{1}{3} + \dfrac{2}{5}$

7. $\dfrac{1}{6} + \dfrac{1}{9} + \dfrac{7}{18}$ 8. $\dfrac{3}{4} + \dfrac{1}{14} - \dfrac{9}{28}$

Answers: *1.* $\dfrac{8}{11}$ *2.* $\dfrac{1}{4}$ *3.* $\dfrac{2}{3}$ *4.* $\dfrac{1}{8}$ *5.* $\dfrac{5}{12}$ *6.* $\dfrac{11}{15}$ *7.* $\dfrac{2}{3}$ *8.* $\dfrac{1}{2}$

Lesson

Which Operation Comes First?

How should we simplify the following expression?

$$4 + 6 \cdot 2$$

If we do the addition first, we get

$$(4 + 6) \cdot 2 = 10 \cdot 2 = 20$$

On the other hand, if we do the multiplication first, we get

$$4 + (6 \cdot 2) = 4 + 12 = 16$$

In order to avoid confusion, we always *perform multiplications and divisions before additions and subtractions*. Thus the correct way to simplify the expression $4 + 6 \cdot 2$ is

$$4 + 6 \cdot 2 = 4 + 12 \qquad \text{Multiply first; then add.}$$
$$= 16$$

Now try Exercise 2.

EXERCISE 1
Simplify each expression.
a. $30 - 17 - 5 + 4$

b. $72 \div 4 \cdot 3 \div 6$

EXERCISE 2
Simplify each expression. Perform multiplications before additions or subtractions.
a. $12 - 6\left(\dfrac{1}{2}\right)$

b. $2(3.5) + 10(1.4)$

What about more complicated expressions? So far, we have the following guidelines:

1. Perform all multiplications and divisions before any additions and subtractions.

2. Perform multiplications and divisions in order from left to right.

3. Perform additions and subtractions in order from left to right.

It can be helpful to group a longer expression into terms before beginning. Terms are separated by addition or subtraction symbols. Any multiplications and divisions that appear within a term should be performed first.

EXAMPLE 2

Simplify $6 + 2 \cdot 5 - 12 \div 3 \cdot 2$

Solution

We start by underlining each term of the expression. Then we simplify each term by doing all multiplications and divisions from left to right.

$$6 + \underline{2 \cdot 5} - \underline{12 \div 3 \cdot 2}$$
$$= 6 + 10 - 8$$
$$= 16 - 8 = 8$$

Note that we do $12 \div 3$ first, *not* $3 \cdot 2$:
$$12 \div 3 \cdot 2 = 4 \cdot 2 = 8.$$

Finally, we do additions and subtractions from left to right.

Now try Exercise 3.

Parentheses

Consider again the expression $4 + 6 \cdot 2$.   We know that we should perform the multiplication, $6 \cdot 2$,   first. What if we want to describe a situation in which the addition should be done first? In that case, we use parentheses to enclose the sum, $4 + 6$,   like this:

$$(4 + 6) \cdot 2$$

As part of the order of operations, we *perform any operations inside parentheses first.*

EXAMPLE 3

Simplify.

    a. $2 + 5(7 - 3)$ Subtract inside parentheses.
         $= 2 + 5(4)$ Multiply.
         $= 2 + 20 = 22$ Add.

    b. $6(10 - 2 \cdot 4) \div 4$ Multiply inside parentheses.
         $= 6(10 - 8) \div 4$ Subtract inside parentheses.
         $= 6(2) \div 4$ Multiply and divide in order from left to right.
         $= 12 \div 4 = 3$

Now try Exercise 4.

Fraction Bars

A fraction bar can have the same purpose as parentheses; it acts as a "grouping" device. In the order of operations, expressions that appear above or below a fraction bar are simplified along with expressions inside parentheses.

EXERCISE 3
Simplify

$12 + 24 \div 4 \cdot 3 + 16 - 10 - 4$

Begin by underlining each term.

EXERCISE 4
Simplify each expression. Begin by working inside parentheses.
a. $28 - 3(12 - 2 \cdot 4)$

b. $12 + 36 \div 4(9 - 2 \cdot 3)$

EXAMPLE 4

Simplify $\dfrac{24 + 6}{12 + 6}$

Solution

We must *not* try to "cancel" the sixes! We must begin by doing the additions above and below the fraction bar:

$$\frac{24 + 6}{12 + 6} = \frac{30}{18} = \frac{5}{3}$$

Now try Exercise 5.

Summary

Putting together all of the guidelines discussed above, we can summarize the order of operations as three steps.

Order of Operations

1. First, perform any operations that appear inside parentheses or above or below a fraction bar.
2. Next, perform all multiplications and divisions in order from left to right.
3. Finally, perform all additions and subtractions in order from left to right.

If a problem involves more than one type of grouping symbol (for example, both parentheses and brackets), we start by simplifying the expression inside the innermost grouping symbols and work outward.

EXAMPLE 5

Simplify $\;6 + 2[3(12 - 5) - 4(7 - 3)]$

Solution

In this expression, there are parentheses within square brackets. We begin by performing the operations within the parentheses:

$6 + 2[3(12 - 5) - 4(7 - 3)]$	Subtract inside parentheses.
$= 6 + 2[3(7) - 4(4)]$	Multiply inside brackets.
$= 6 + 2[21 - 16]$	Subtract inside brackets.
$= 6 + 2[5]$	Multiply, then add.
$= 6 + 10 = 16$	

Now try Exercise 6.

Using Your Calculator

Scientific calculators are programmed to perform multiplications and divisions before additions and subtractions. Try entering this expression into your calculator:

$$9 + 2 \cdot 5 - 3 \cdot 4$$

If you have a scientific calculator, it will automatically calculate the multiplications $2 \cdot 5$ and $3 \cdot 4$ first and then complete the calculation as

$$9 + 10 - 12 = 7$$

just as you would do by hand.

If your calculator has keys for parentheses, you can also enter expressions that contain parentheses or brackets. Try entering this expression from Example 3b:

$$6(10 - 2 \cdot 4) \div 4$$

If you do not have parentheses keys on your calculator, you can still use it to perform the operations step by step.

EXAMPLE 6

Use a calculator to simplify $2.4[25 - 3(6.7)] + 5.5$

Solution

We begin by entering the expression inside square brackets,

$$25 - 3 \times 6.7$$

to get 4.9. Then we enter

$$2.4 \times 4.9 + 5.5$$

to obtain the answer, 17.26.

CAUTION!

Calculators are *not* able to use a fraction bar as a grouping symbol. Consider the expression

$$\frac{24}{6 - 4}$$

which simplifies to $\frac{24}{2}$, or 12. If we enter this expression into a calculator as

$$24 \div 6 - 4$$

we get 0, which is not correct. This is because the calculator follows the order of operations and calculates $24 \div 6$ first. In order to tell the calculator that $6 - 4$ should be performed first, we must use parentheses and enter the expression as

$$24 \div (6 - 4)$$

We call this way of writing the expression the *on-line form*. This form uses parentheses instead of a fraction bar to indicate grouping. (If your calculator does not have parenthesis keys, you will have to carry out the simplification in steps, following the order of operations.)

In general, when using a calculator, you must *enclose in parentheses any expression that appears above or below a fraction bar*. Consequently, it is better to use the $\times$ symbol instead of parentheses for multiplication, as shown in Exercise 7.

EXERCISE 7

a. Write the expression
$$\frac{16.2}{(2.4)(1.5)}$$
in on-line form.

b. Use a scientific calculator to simplify the expression.
Hint: Enter 2.4×1.5 for the product in the denominator.

HOMEWORK 1.8

1. State the two associative laws.

2. Give examples to show that the associative laws do not hold for subtraction or division.

3. What is the purpose of parentheses and brackets in an algebraic expression?

4. What other symbol has a similar purpose?

5. True or false: In a string of multiplications and divisions, you should perform the multiplications before the divisions.

6. How do you convert an expression to on-line form?

7. How would you change the algebraic expression $20 - 2 \cdot 8 + 1$ if you wanted to indicate that:
a. The subtraction should be performed first?

b. The addition should be performed first?

c. The multiplication should be performed first?

8. Write two equivalent algebraic expressions for "the sum of 15 and 9, divided by the difference of 18 and 10":
a. Using a fraction bar

b. Using a division symbol, $\div$

9. Write an algebraic expression for the following instructions: "Multiply the sum of 5 and 7 by 4, and then divide by the difference of 10 and 8."

10. Underline the terms in this expression:

$$32 - 5 \cdot 4 + 36 \div 3 \cdot 4 - 8 \cdot 9 \div 6 + 6$$

Without performing the calculations, write down the steps you would use to simplify each expression in Problems 11 and 12.

11. $825 - 32(12) \div 4 + 2$

12. $2 \cdot 6 - 4 \cdot 18 + 2$

Find and correct the calculation error in Problems 13 and 14.

13. $(5 + 4) - 3(8 - 3 \cdot 2)$
$= 9 - 3(8 - 6)$
$= 6(2) = 12$

14. $4 + 4 \cdot 12 \div 2 \cdot 3$
$= 4 + 48 \div 6$
$= 4 + 8 = 12$

In Problems 15–36, simplify each expression by following the order of operations.

15. $2 + 4(3)$

16. $4(3 + 5)$

17. $15 - \dfrac{3}{4}(16)$

18. $\dfrac{1}{3} + \dfrac{2}{3} \cdot 2$

19. $6 \div \dfrac{1}{4} \cdot 3$

20. $6 \div \left(\dfrac{1}{4} \cdot 3\right)$

21. $3 + 3(2 + 3)$

22. $20 - 3(6 - 4)$

23. $\dfrac{1}{3} \cdot 12 - 3\left(\dfrac{5}{6}\right)$

24. $8 \cdot \dfrac{1}{5} + \dfrac{2}{5} \cdot 10$

25. $2 + 3 \cdot 8 - 6 + 3$

26. $24 \div 6 + 2 \cdot 8 \div 4$

27. $\dfrac{3(8)}{12} - \dfrac{6 + 4}{5}$

28. $\dfrac{7 \cdot 5 - 2 \cdot 4}{5 - 2}$

29. $28 + 6 \div 2 - 2(5 + 3 \cdot 2)$

30. $2 + 3(18 - 6) \div (4 + 10 \div 2)$

31. $3[3(3 + 2) - 8] - 17$

32. $36 - 8\left[5(15 - 11) - \dfrac{36}{2}\right]$

33. $\dfrac{3(3) + 5}{6 - 2(2)} + \dfrac{2(5) - 4}{9 - 4 - 2}$

34. $\dfrac{9(2) - 4}{9 - 2} + \dfrac{2 + 3(2)}{2 \cdot 3 - 2}$

35. $7[15 - 24 \div 2] - 9[5(4 + 2) - 4(6 + 1)]$

36. $6(13 - 3 \cdot 4) \div [18 - (1 + 2 \cdot 7)]$

For Problems 37–42, use your calculator to simplify each expression. Round your answers to three decimal places if necessary.

37. $\dfrac{6.4 + 3.5}{3.6(3.2)}$

38. $8.46 - 12.8 \div (3.98 + 1.4)$

39. $\dfrac{26.2 - 9.1}{8.4 \div 7.7} + 5.1(6.9 - 1.6)$

40. $\dfrac{2.3(9.7 + 3.2)}{1.2(5.9 - 1.3)}$

41. $\dfrac{1728(847 - 603)}{216(98 - 38)} + 6(876 - 514)$

42. $\dfrac{3(19 + 18) + 21(93 - 47)}{140 + 5(98 - 91)}$

For Problems 43 and 44, use trial and error to solve each equation. Use the suggested values in the tables to get started.

43. $y = 1.4 + 2.5(x - 1.8) = 3.9$

x	2	2.5	3	3.5	4	4.5	5
y							

x							
y							

44. $y = \dfrac{4x + 18}{8} - 4.2 = 6.55$

x	5	10	15	20	25	30
y						

x						
y						

In Problems 45–54, evaluate each pair of expressions mentally.

45. a. $8 + 2 \cdot 5$ **b.** $(8 + 2) \cdot 5$

46. a. $10 - 6 \div 2$ **b.** $(10 - 6) \div 2$

47. a. $\dfrac{24}{2 + 6}$ **b.** $\dfrac{24}{2} + 6$

48. a. $\dfrac{8 + 4}{2}$ **b.** $8 + \dfrac{4}{2}$

49. a. $(9 - 4) - 3$ **b.** $9 - (4 - 3)$ **50. a.** $27 \div (9 \div 3)$ **b.** $(27 \div 9) \div 3$

51. a. $6 \cdot 8 - 6$ **b.** $6(8 - 6)$ **52. a.** $24 \div 3 \cdot 2$ **b.** $24 \div (3 \cdot 2)$

53. a. $\dfrac{36}{6(3)}$ **b.** $\dfrac{36}{6}(3)$ **54. a.** $20 - 8 + 2$ **b.** $20 - (8 + 2)$

55. Think of the region shown as a rectangle with a smaller rectangle removed, and write an expression for its area. Then simplify your expression to find the area. (The measurements given are in centimeters.)

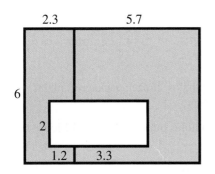

56. Think of the region shown as two triangles, and write an expression for its area. Then simplify your expression to find the area. (The measurements given are in inches.)

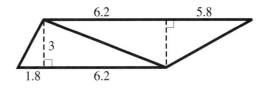

1.9 More Algebraic Expressions

Reading

In this section, we'll write algebraic expressions that involve two or more operations. First, let's review expressions having one operation.

Skills Review

Choose the correct algebraic expression for each of the following situations.

$$n + 12 \qquad 12n \qquad \frac{n}{12}$$

$$\frac{12}{n} \qquad 12 - n \qquad n - 12$$

1. Helen bought n packages of tulip bulbs. If each package contains 12 bulbs, how many bulbs did she buy?

2. Henry bought a package of n gladiolus bulbs, as well as 12 loose bulbs. How many bulbs did he buy?

3. Together, Karen and Dave sold 12 tickets to the spring concert. If Karen sold n tickets, how many did Dave sell?

4. Together, Karl and Diana collected n used books for the book sale. If Karl collected 12 books, how many did Diana collect?

5. Greta made n dollars last week. If she worked for 12 hours, how much did she earn per hour?

6. Gert jogged for 12 minutes. If she jogged n miles, how many minutes did it take her to jog 1 mile?

Answers: ***1.*** $12n$ ***2.*** $n + 12$ ***3.*** $12 - n$ ***4.*** $n - 12$ ***5.*** $\dfrac{n}{12}$ ***6.*** $\dfrac{12}{n}$

The order of operations you learned in Section 1.8 applies to variables as well as to constants. For example, the expression

$$5 + 2x$$

tells us to multiply x by 2 and then add 5 to the result. Thus, to evaluate the expression for $x = 8$, for example, we write

$$5 + 2(8) = 5 + 16$$
$$= 21$$

where we perform the multiplication first.

In the next example, note how parentheses change the meaning of an expression.

EXAMPLE 1

Write algebraic expressions for each of the following phrases. Then evaluate each phrase for $x = 2$ and $y = 6$.

    a. Three times the sum of x and y **b.** The sum of $3x$ and y

Solution

    a. $3(x + y)$ **b.** $3x + y$
        $3(2 + 6) = 3(8) = 24$ $3(2) + 6 = 6 + 6 = 12$

Now try Exercise 1.

Lesson

Use the order of operations to help you do Exercises 2 through 6. In each case, explain why your choice is correct.

EXERCISES

2. Which expression represents "6 times the sum of x and 5"?

    a. $6x + 5$ **b.** $6(x + 5)$

3. Which expression represents "$\frac{1}{2}$ the difference of p and q"?

    a. $\frac{1}{2}(p - q)$ **b.** $\frac{1}{2}p - q$

4. Which expression represents "4 less than the product of 6 and w"?

    a. $6w - 4$ **b.** $4 - 6w$

5. Which expression represents "2 less than the quotient of 10 and z"?

    a. $\dfrac{10}{z} - 2$ **b.** $2 - \dfrac{10}{z}$

6. Explain why $12x \div 3y$ is not the same as $\dfrac{12x}{3y}$.

EXERCISE 1
Write algebraic expressions for the following phrases.

a. The sum of 3 times t and 1, divided by 2

b. The ratio of 25 to the sum of p and 6

Writing Algebraic Expressions

In Exercise 7, you will write an algebraic expression to describe a situation. Use numerical calculations to help you.

EXERCISE 7

Neda decides to order some photo albums as gifts. Each album costs $12, and the shipping cost is $4.

a. What is Neda's bill if she orders 3 albums? 5 albums?

b. Describe in words how you calculated your answers to part (a).

c. Fill in the table.

Number of albums	2	3	4	6	8	10	15
Calculation	12(2) + 4						
Neda's bill	28						

d. Write an algebraic expression for Neda's bill if she orders a albums.

e. Write an equation that gives Neda's bill, B, in terms of a.

EXERCISE 8

a. Emily bought five rosebushes for her garden. Each rosebush cost $9 plus tax. If the tax on one rosebush is t, write an expression for the total amount Emily paid.

b. Megan would like to buy a kayak on sale. She calculates that the kayak she wants costs $40 less than 3 weeks' salary. Megan makes w dollars per week. Write an expression for the price of the kayak.

If you have trouble writing an algebraic expression, choosing a specific numerical value for the variable can clarify things.

EXAMPLE 2

Alida keeps $100 in cash from her weekly paycheck and deposits 40% of the remainder in her savings account. If the amount of Alida's paycheck is p, write an expression for the amount she deposits in savings.

Solution

Suppose Alida's paycheck is $500. First, she subtracts $100 from that amount to get $500 - 100$, and then she takes 40% of the remainder for her savings:

$$0.40(500 - 100)$$

If her paycheck is p dollars, we perform the same operations on p instead of on 500. The expression is thus

$$0.40(p - 100)$$

Try Exercise 8. Start by using a specific value for the variable if that helps.

Evaluating Algebraic Expressions

In Exercises 9 and 10, fill in the tables to evaluate each expression in two steps. Note how the order of operations differs in parts (a) and (b).

Some algebraic expressions involve more than one variable. To evaluate these expressions, we substitute the given values for the variable and then use the order of operations to simplify.

EXAMPLE 3

When a company purchases a piece of equipment such as a computer or a photocopier, the value of the equipment depreciates over time. One way to calculate the value of the equipment is to use the formula

$$V = C\left(1 - \frac{t}{n}\right)$$

where C is the original cost of the equipment, t is the number of years since it was purchased, and n stands for the useful lifetime of the equipment in years. Find the value of a 4-year-old photocopier if it has a useful lifetime of 6 years and cost $3000 when new.

Solution

We evaluate the expression $C\left(1 - \frac{t}{n}\right)$ for $C = 3000$, $t = 4$, and $n = 6$.

$$C\left(1 - \frac{t}{n}\right) = 3000\left(1 - \frac{4}{6}\right)$$

$$= 3000\left(\frac{1}{3}\right) = 1000$$

The 4-year-old photocopier is worth $1000.

In Example 3, can you explain why $1 - \frac{4}{6} = \frac{1}{3}$? If not, please review Appendix A.3 on adding and subtracting fractions.

Now try Exercise 11.

Graphing Equations

Example 4 demonstrates how to graph an equation that includes two operations.

EXAMPLE 4

At 6 A.M. the temperature was 50°F, and it has been rising by 4°F every hour.

    a. Write an equation for the temperature, T, after h hours.

    b. Graph the equation.

Solution

    a. We set $h = 0$ at 6 A.M. At that time, $T = 50$. Because the temperature is rising 4°F every hour, we multiply the number of hours by 4 and add that to 50 to get the new temperature. Thus our equation is
       $T = 50 + 4h$.

EXERCISE 9

a. $8 + 3t$

t	$3t$	$8 + 3t$
0		
2		
7		

b. $3(t + 8)$

t	$t + 8$	$3(t + 8)$
0		
2		
7		

EXERCISE 10

a. $6 + \dfrac{x}{2}$

x	$\dfrac{x}{2}$	$6 + \dfrac{x}{2}$
4		
8		
9		

b. $\dfrac{6 + x}{2}$

x	$6 + x$	$\dfrac{6 + x}{2}$
4		
8		
9		

EXERCISE 11

Evaluate $8(x + xy)$ for $x = \dfrac{1}{2}$ and $y = 6$.

b. Step 1   We first make a table of values for the equation. We choose some values for h, as shown in the first column of the table. For each value of h, we evaluate the formula for T and record these values in the second column.

Ordered pairs

h	T
0	50
1	54
3	62
5	70

$T = 50 + 4(0) = 50 + \ 0 = 50 \qquad (0, 50)$

$T = 50 + 4(1) = 50 + \ 4 = 54 \qquad (1, 54)$

$T = 50 + 4(3) = 50 + 12 = 62 \qquad (3, 62)$

$T = 50 + 4(5) = 50 + 20 = 70 \qquad (5, 70)$

Step 2   Each ordered pair from the table corresponds to a point on the graph. We use the values in the table to help us choose appropriate scales for the axes so that all the points will fit on the graph. For this graph, we use increments of 1 on the horizontal or h-axis and increments of 10 on the vertical or T-axis.

Step 3   Finally, we plot the points on the grid and connect them, as shown in Figure 1.12. (All the points should lie on a straight line.)

Figure 1.12

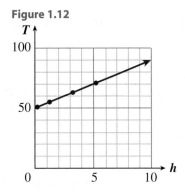

HOMEWORK 1.9 <www

1. Salewa saved $5000 to live on while going to school full-time. She needs $200 per week for living expenses.

    a. How much of Salewa's savings will be left after 3 weeks? After 12 weeks?

b. Describe in words how you calculated your answers to part (a).

    c. Fill in the table.

Number of weeks	2	4	5	6	10	15	20
Calculation	5000 − 200(**2**)						
Savings left	4600						

    d. Write an algebraic expression for Salewa's savings after w weeks.

    e. Write an equation that gives Salewa's savings, S, in terms of w.

2. As a student at City College, Delbert pays tuition that consists of a $50 registration fee plus $15 for each unit he takes.

    a. How much will Delbert owe the college if he takes 6 units? If he takes 10 units?

b. Describe in words how you calculated your answers to part (a).

    c. Fill in the table.

Number of units	3	5	8	9	12	15	16
Calculation	50 + 15(**3**)						
Tuition	95						

    d. Write an algebraic expression that gives Delbert's tuition if he takes u units.

    e. Write an equation that gives Delbert's tuition, T, in terms of u.

3. To calculate how much state income tax she owes, Francine subtracts $2000 from her income and then takes 12% of the result.
   a. What is Francine's state income tax if her income is $8000? If her income is $10,000?

   b. Describe in words how you calculated your answers to part (a).

   c. Fill in the table.

Income	Calculation	State tax
5000	0.12 (**5000** − 2000)	**360**
7000		
12,000		
15,000		
20,000		
24,000		
30,000		

   d. Write an algebraic expression for Francine's state income tax if her income is I.

   e. Write an equation that gives Francine's tax, T, in terms of I.

4. The Youth Alliance raises funds for four different charities and divides the proceeds equally among them. The group has already raised $900 and has one more fundraising event, a car wash, to complete.
   a. How much will each charity receive if the Youth Alliance makes $300 on the car wash? If it makes $400?

   b. Describe in words how you calculated your answers to part (a).

   c. Fill in the table.

Car wash proceeds	Calculation	Each charity's share
100	(900 + **100**) ÷ 4	**250**
200		
500		
600		
800		
1000		
1100		

   d. Write an algebraic expression for each charity's share if the Youth Alliance raises d dollars at the car wash.

   e. Write an equation that gives each charity's share, S, in terms of d.

For Problems 5–8, fill in the table to evaluate the expression in two steps.

5. $5z - 3$

z	$5z$	$5z - 3$
2		
4		
5		

6. $\dfrac{m - 3}{4}$

m	$m - 3$	$\dfrac{m - 3}{4}$
6		
7		
9		

7. $2(12 + Q)$

Q	$12 + Q$	$2(12 + Q)$
0		
4		
8		

8. $1 - \dfrac{1}{w}$

w	$\dfrac{1}{w}$	$1 - \dfrac{1}{w}$
1		
2		
3		

In Problems 9–18:
a. Choose variables for the unknown quantities, and write an algebraic expression.
b. Evaluate the expression for the given value(s).

9. a. Three inches less than twice the width

   b. The width is 13 inches.

10. a. One hour more than four times the plane's travel time

   b. The plane traveled for 2 hours.

11. a. Twenty dollars more than 40% of the principal

   b. The principal is $500.

12. a. Fifteen miles less than 1.6 times the distance to the city

   b. The distance to the city is 30 miles.

13. a. One-third of the total number of cars and trucks

    b. There are 7 cars and 5 trucks.

14. a. Twelve times the sum of rent and utilities

    b. Rent is $450, and utilities are $36.

15. a. Half the women and two-thirds of the men

    b. There are 18 women and 12 men.

16. a. The ratio of hits plus walks to times at bat

    b. Darryl had 12 hits and 3 walks in 40 times at bat.

17. a. Eighty percent of the difference between Norm's verbal and math SAT scores

    b. Norm's verbal score was 680, and his math score was 655.

18. a. Six percent of the sum of your regular salary and your overtime pay

    b. Your regular salary was $360, and your overtime pay was $80.

Evaluate the expressions in Problems 19–28 for the given values.

19. $2y + x$ for $x = 8$ and $y = 9$

20. $8(s - t)$ for $s = 27$ and $t = 15$

21. $4a + 3b$ for $a = 8$ and $b = 7$

22. $\dfrac{w + 2z}{w}$ for $w = 9$ and $z = 3$

23. $\dfrac{a}{b} - \dfrac{b}{a}$ for $a = 8$ and $b = 6$

24. $v + gt$ for $v = \dfrac{9}{4}$, $g = 16$,   and $t = \dfrac{3}{2}$

25. $\dfrac{24 - 2x}{2 + y} - \dfrac{4x + 1}{3y}$ for $x = 4$ and $y = 6$

26. $\dfrac{h(b + c)}{2}$ for $h = 5.7$, $b = 8.1$,   and $c = 2.9$

27. $\dfrac{a}{1 - r}$ for $a = 6$ and $r = 0.2$

28. $\dfrac{1 + e}{1 - e}$ for $e = 0.4$

29. The perimeter of a rectangle of length l and width w is given by $P = 2l + 2w$. Find the perimeter of a rectangular meeting hall with dimensions 8.5 meters by 6.4 meters.

30. The annual inventory cost of a small appliance store is given by $I = 6s + 18r$, where s is the average number of items in stock and r is the number of times per year that merchandise is reordered. Find the annual inventory cost if the store keeps 150 appliances in stock and reorders four times annually.

31. The area of a trapezoid with bases B and b and height h is given by $A = \frac{1}{2}(B + b)h$. Find the area of a trapezoid whose bases are 9 centimeters and 7 centimeters and whose height is 3 centimeters.

32. The volume of a pyramid with a rectangular base of length l and width w and with height h is given by $V = \frac{1}{3}(lw)h$. Find the volume of a pyramid whose base is 4 feet long and 5 feet wide and whose height is 2 feet.

33. If a manufacturer produces n items at a cost of c dollars each and sells them at a price of p dollars each, the company's profit is given by $P = n(p - c)$. Find the profit earned by a manufacturer of bicycle equipment by selling 300 bicycle helmets that cost \$32 each to produce and sell for \$50 apiece.

34. The surface area of an open box with a square base is given by $A = s(4h + s)$, where s stands for the length of one side of the base and h stands for the height of the box. Find the surface area of such a box whose base length is 2.4 feet and whose height is 1.6 feet.

35. The temperature in degrees Celsius (°C) is given by $C = \frac{5}{9}(F - 32)$, where F stands for the temperature in degrees Fahrenheit (°F). Find normal body temperature in degrees Celsius if normal temperature is 98.6° Fahrenheit.

36. The temperature in degrees Fahrenheit is given by $F = \frac{9}{5}C + 32$, where C stands for the temperature in degrees Celsius. Find the temperature outside in degrees Fahrenheit if it is 17.5° Celsius.

For Problems 37–42:
a. Write an equation relating the variables.
b. Complete the table of values.

37. Greta's math notebook has 100 pages, and she uses 6 pages per day on average for her notes and homework. How many pages, P, will she have left after d days?

d	0	2	5	10	15
P					

38. Asa has typed 220 words of his term paper and is still typing at a rate of 20 words per minute. How many words, W, will Asa have typed after m more minutes?

m	0	5	10	15	30
W					

39. Delbert shares a house with four roommates. He pays $200 rent per month, plus his share of the utilities. If the utilities cost U dollars, how much money, M, will Delbert owe?

U	20	40	80	100	200
M					

40. For her roommate's shower, Francine spent $50 on a gift plus her share of the cost of the party. If the party cost P dollars and 12 people contributed (including Francine), what was the total amount, T, that Francine spent?

P	60	84	90	120	150
T					

41. Estelle is office manager at her company. She reserved $40 from the sunshine fund for a party and used the rest to buy new desk sets for six people. If the sunshine fund originally contained S dollars, how much, D, can Estelle spend on each desk set?

S	70	100	112	130	160
D					

42. Roberto's school provided him with s square feet of poster paper to distribute to the students in his class. Roberto contributed 8 square feet of his own paper. If there are 12 students in Roberto's class, how much poster paper, p, does each student get?

s	16	22	28	40	46
p					

For Problems 43–46:
a. Write an algebraic expression.
b. Make a table showing at least six pairs of values for the expression.

43. The ratio of a number to 4 more than the number

44. The quotient of a number and 3 less than twice the number

45. The sum of a and 5, times the difference of a and 5

46. The product of 4 more than x and 4 less than x

Write an algebraic formula for calculating each quantity described in Problems 47–50.

47. To calculate the recommended weight, W, for an adult over age 25: Record the adult's current age, A, and his or her weight at age 25. (This is the "ideal" weight, I.) Calculate the difference between the adult's current age and 25. Add $\frac{1}{2}$ pound for every year over age 25 to the ideal weight.

48. To calculate the recommended weight, R, for a woman over 64 inches tall: Record her height, H, and the recommended standard weight, W, for a woman her age. Calculate the difference between her height and 64 inches. Add 5 pounds for every inch taller than 64 inches to the recommended weight.

49. To calculate income tax owed, T: Subtract deductions, D, from income, I, to get the adjusted income. Take 8% of the adjusted income to get the base tax. Add $300 to the base tax.

50. To calculate the amount of a phone bill, T: Subtract long-distance charges, L, from this month's amount, A. Take 2% of that difference to get the surcharge. Add the surcharge to this month's amount.

Write an algebraic expression for the area or perimeter of each figure in Problems 51–54.

51. Perimeter =

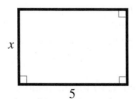

52. Area =

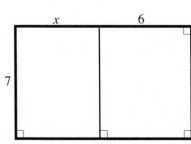

53. Area =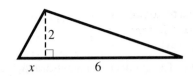

54. Perimeter =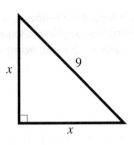

Equations with Two or More Operations

Reading

In Section 1.6, we used two steps for solving an equation: First, think about what operation is performed on the variable. Then perform the *opposite* operation on *both* sides of the equation. This has the effect of isolating the variable on one side of the equation. Before continuing, you should try the following review problems.

Skills Review

Solve each equation. If you can, do so mentally (without using pencil and paper).

1. $\dfrac{u}{3} = 6$ 2. $\dfrac{v}{5} = 10$ 3. $7 = 3 + s$ 4. $6 = t + 4$

5. $a - \dfrac{1}{3} = \dfrac{2}{3}$ 6. $\dfrac{1}{2} = b - \dfrac{1}{2}$ 7. $20 = 5m$ 8. $4n = 36$

9. $\dfrac{1}{4}p = 8$ 10. $\dfrac{1}{2}q = 4$ 11. $7t = 5$ 12. $3 = 8w$

Answers: **1.** 18 **2.** 50 **3.** 4 **4.** 2 **5.** 1 **6.** 1 **7.** 4 **8.** 9 **9.** 32

10. 8 **11.** $\dfrac{5}{7}$ **12.** $\dfrac{3}{8}$

In Section 1.9, we wrote and evaluated algebraic expressions having two or more operations. In this section, we'll see how to solve equations that arise from such expressions. For example, in Problem 2 of Homework 1.9, you wrote the expression

$$T = 50 + 15u$$

for Delbert's college tuition consisting of a $50 registration fee plus $15 per unit. Now suppose Delbert would like to know how many units he can take if he has $290 saved for tuition. We can answer this question by solving the equation

$$290 = 50 + 15u$$

where we have substituted $290 for the tuition, T.

Think about the expression $50 + 15u$. How would you evaluate this expression if you were given a value for u? Following the order of operations, you would

1. first multiply by 15
2. then add 50.

In order to solve the equation $290 = 50 + 15u$,   we must reverse these two steps to "undo" the operations and isolate the variable. We first subtract 50 from both sides of the equation:

$$
\begin{array}{r}
290 = 50 + 15u \\
-\,50 \quad -\,50 \\
\hline
240 = 15u
\end{array}
$$

This isolates the term that contains the variable, $15u$. Then we divide both sides of the equation by 15:

$$\frac{240}{15} = \frac{15u}{15}$$

$$16 = u$$

With \$290, Delbert can take 16 units. We can check the solution by substituting 16 for u in the original equation.

$$50 + 15u = 290$$

$$50 + 15(16) = 50 + 240 = 290$$

Because a true statement results, the solution checks.

Note how we reversed the operations performed on u in the equation:

Operations performed on u	*Steps for solution*
1. Multiply by 15	1. Subtract 50
2. Add 50	2. Divide by 15

In general, to solve an equation that involves two or more operations, we "undo" those operations *in reverse order*. Try Exercise 1 now.

Lesson

In the reading assignment we saw that to solve an equation, you must undo the operations performed on the variable in the reverse order. Try Exercises 2–4 to practice undoing operations.

EXERCISE 3

Consider the equation $3n - 5 = p$

Refer to Exercise 2a to help you answer these questions:

a. Let $n = 2$.

Explain how to find p in two steps.

b. Let $p = 7$.

Explain how to find n in two steps.

EXERCISE 1

a. If you put on socks and then put on shoes, what must you do to reverse the process?

b. You leave home and bicycle north for 3 miles and then east for 2 miles. Suddenly you notice that you have dropped your wallet. How should you retrace your route?

EXERCISE 2

Fill in the table with the correct values.

a.

n	$3n$	$3n - 5$
2		
5		
		7
		22

b.

m	$\dfrac{m}{4}$	$\dfrac{m}{4} + 1$
8		
12		
		6
		2

EXERCISE 4

Consider the equation $\dfrac{m}{4} + 1 = h$

Refer to Exercise 2b to help you answer these questions:

a. Let $m = 8$.

Explain how to find h in two steps.

b. Let $h = 6$.

Explain how to find m in two steps.

Solving Equations

Here are some examples of solving equations. First, we use the order of operations to analyze the expression containing the variable and to plan our approach. Then we carry out the steps of the solution. Study the examples, and then try the corresponding exercises.

EXAMPLE 1

Solve $4x - 5 = 7$

Solution

Operations performed on x	Steps for solution
1. Multiply by 4	1. Add 5
2. Subtract 5	2. Divide by 4

$$4x - 5 = 7$$
$$4x - 5 + 5 = 7 + 5 \qquad \text{Add 5 to both sides.}$$
$$4x = 12$$
$$\frac{4x}{4} = \frac{12}{4} \qquad \text{Divide both sides by 4.}$$
$$x = 3$$

The solution is 3.

Check: $4(\mathbf{3}) - 5 = 7$ (True)

EXERCISE 5

Solve $3x + 2 = 17$

EXAMPLE 2

Solve $\dfrac{3t}{4} = 6$

Solution

Operations performed on t	Steps for solution
1. Multiply by 3	1. Multiply by 4
2. Divide by 4	2. Divide by 3

$$\frac{3t}{4} = 6$$

$$4\left(\frac{3t}{4}\right) = (6)4 \quad \text{Multiply both sides by 4.}$$

$$3t = 24$$

$$\frac{3t}{3} = \frac{24}{3} \quad \text{Divide both sides by 3.}$$

$$t = 8$$

The solution is 8.

Check: $\dfrac{3(8)}{4} = 6$ (True)

EXAMPLE 3

Solve $5(z - 6) = 65$

Solution

Operations performed on z	Steps for solution
1. Subtract 6	1. Divide by 5
2. Multiply by 5	2. Add 6

$$5(z - 6) = 65$$

$$\frac{5(z - 6)}{5} = \frac{65}{5} \quad \text{Divide both sides by 5.}$$

$$z - 6 = 13$$

$$z - 6 + 6 = 13 + 6 \quad \text{Add 6 to both sides.}$$

$$z = 19$$

The solution is 19.

Check: $5(19 - 6) = 65$ (True)

EXERCISE 6

Solve $\dfrac{4a}{5} = 12$

EXERCISE 7

Solve $\dfrac{b + 3}{4} = 6$

Problem Solving

As you saw in Section 1.7, algebraic equations are often used to solve applied problems.

EXAMPLE 4

Mitch bought a compact disc player for $269 and a number of compact discs at $14 each.

a. Write an equation for Mitch's total bill, B, in terms of the number of discs, d, that he bought.
b. If his total bill was $367, how many discs did Mitch buy?

Solution

a. We must add the cost of the compact discs to the cost of the player:

$$B = 269 + 14d$$

b. We substitute 367 for B and solve for d:

$$269 + 14d = 367$$

$$-269 \qquad -269 \qquad \text{Subtract 269 from both sides.}$$

$$14d = 98$$

$$\frac{14d}{14} = \frac{98}{14} \qquad \text{Divide both sides by 14.}$$

$$d = 7$$

Mitch bought 7 compact discs.

HOMEWORK 1.10

In Problems 1–4, fill in each table. Use two steps for each row; fill in the middle column first.

1.

x	$2x$	$2x + 4$
3		
6		
		14
		20

2.

a	$3a$	$\frac{3a}{2}$
4		
8		
		9
		18

3.

q	$q - 3$	$5(q - 3)$
3		
		10
4		
		20

4.

w	$w + 3$	$\frac{w + 3}{2}$
7		
		4
1		
		3

5. Consider the equation $y = 2x + 4$. Refer to the table in Problem 1 to help you answer these questions:
   a. Let $x = 3$. Explain how to find y in two steps.

   b. Let $y = 14$. Explain how to find x in two steps.

6. Consider the equation $H = \dfrac{3a}{2}$. Refer to the table in Problem 2 to help you answer these questions:
   a. Let $a = 4$. Explain how to find H in two steps.

   b. Let $H = 9$. Explain how to find a in two steps.

7. Consider the equation $5(q - 3) = R$. Refer to the table in Problem 3 to help you answer these questions:
   a. Let $q = 3$. Explain how to find R in two steps.

   b. Let $R = 10$. Explain how to find q in two steps.

8. Consider the equation $B = \dfrac{w + 3}{2}$. Refer to the table in Problem 4 to help you answer these questions:
   a. Let $w = 7$. Explain how to find B in two steps.

   b. Let $B = 4$. Explain how to find w in two steps.

9. Delbert is thinking of a number. If he multiplies the number by 5 and then subtracts 6, the result is 29. What is the number?

10. Francine is thinking of another number. If she adds 12 to the number and then divides by 4, the result is 15. What is the number?

Solve the equations in Problems 11–30.

11. $6x - 13 = 5$

12. $7p + 6 = 13$

13. $\dfrac{2a}{5} = 8$

14. $15 = \dfrac{3y}{5}$

15. $\dfrac{x}{4} + 2 = 3$

16. $7 = \dfrac{z}{6} - 4$

17. $6x + 5 = 5$

18. $2z - 8 = 0$

19. $24 = 4(p - 7)$

20. $3(h + 8) = 27$

21. $0 = \dfrac{5z}{7}$

22. $\dfrac{36t}{81} = 8$

23. $\dfrac{k + 4}{5} = 9$

24. $\dfrac{q - 4}{6} = 7$

25. $\dfrac{2x}{3} - 5 = 7$

26. $1 = \dfrac{4a}{5} - 7$

27. $7 = \dfrac{4b - 3}{3}$

28. $12 = \dfrac{7u + 4}{5}$

29. $11.8w - 37.8 = 120.32$

30. $9.7(2.6 + v) = 58.2$

In Problems 31–34, use the graphs to answer the questions.

31. Gregory purchases stereo equipment on a monthly installment plan. After m months, Gregory will still owe a balance of B dollars.

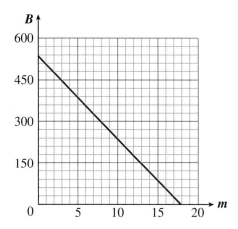

a. What is the price of the stereo equipment?

b. How much does Gregory pay each month?

c. How much will Gregory owe after 6 months?

d. How many monthly payments will Gregory make?

e. Write an algebraic equation for B in terms of m.

f. Write an equation and verify your answer to part (d).

32. Esther hires a plumber to work at her house. After the plumber has worked h hours, Esther owes him F dollars.

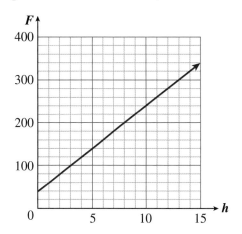

a. What is the plumber's initial fee, just for coming to the house?

b. How much does the plumber charge per hour?

c. How much will Esther owe the plumber after 3 hours of work?

d. If the plumber's fee is $160, how many hours did he work?

e. Write an algebraic equation for F in terms of h.

f. Write and solve an equation to verify your answer to part (d).

33. Francine's new puppy weighed 3.8 pounds when she brought it home, and it should gain approximately 0.6 pound per week. The graph shows the puppy's weight, W, after t weeks.

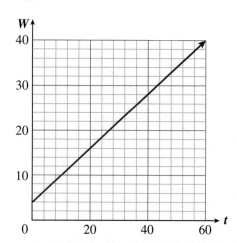

a. Estimate the puppy's weight after 8 weeks.

b. Estimate how long it will take the puppy's weight to reach 26 pounds.

c. Write an algebraic equation for W in terms of t.

d. Write and solve an equation to verify your answer to part (b).

34. Marvin's telephone company charges $13 a month plus $0.15 per minute. The graph shows Marvin's phone bill, P, if he talks for m minutes.

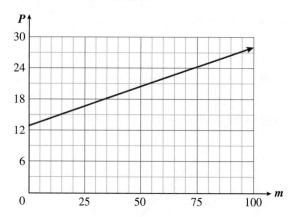

a. Estimate Marvin's phone bill if he talks for 45 minutes.

b. Estimate the number of minutes Marvin has talked if his phone bill is $25.

c. Write an algebraic equation for P in terms of m.

d. Write and solve an equation to verify your answer to part (b).

In Problems 35–38:
a. **Write an equation relating the variables.**
b. **Graph your equation.**
c. **Write and solve an equation to answer the question.**
d. **Verify your solution from part (c), using your graph.**

35. Lori's history test had one essay question and some short-answer questions. Lori scored 20 points on the essay, and she gets 4 points for each correct short answer. Write and graph an equation for Lori's score, s, if she gives x correct short answers. If Lori's score was 76, how many correct answers did she give?

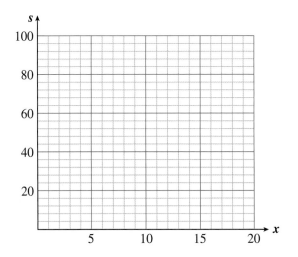

36. Yakov has saved $300 and plans to add $20 per week to his savings. Write and graph an equation for Yakov's savings, S, after w weeks. How long will it take Yakov's savings to grow to $560?

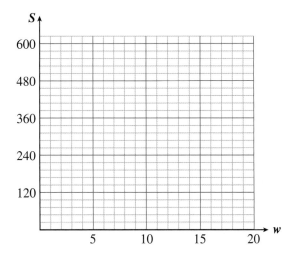

37. Gary employs three part-time waiters in his restaurant, and each waiter earns $200 per week. He plans to give them all a raise. Write and graph an equation for Gary's new weekly payroll, P, if he raises each salary by r dollars. If Gary can afford $654 a week for salaries, what raise can he give to each waiter?

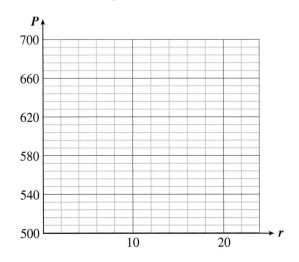

38. Heidi planted a live 6-foot Christmas tree, and it is growing 4.5 inches per year. Write and graph an equation for the height, h, of the tree after y years. (*Hint:* How many inches are in 6 feet?) How long will it take the tree to grow to a height of 9 feet?

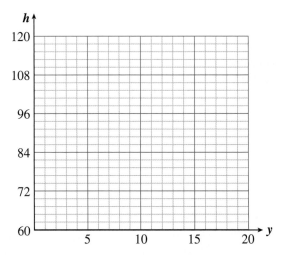

Solve Problems 39–44 in three steps:
a. **Choose a variable for the unknown quantity.**
b. **Write an equation that involves the variable.**
c. **Solve your equation, and answer the question.**

39. Simona bought a car for $10,200. She paid $1200 down and will pay the rest in 36 monthly installments. How much is each installment? (*Hint:* Write an equation about the total amount Simona pays.)

40. For English class, Earlene has 10 days to read *A Tale of Two Cities*, which is 417 pages long. If she reads 75 pages the first day, how many pages must she read on each of the remaining 9 days? (*Hint:* Write an equation about the total number of pages Earlene must read.)

41. Every school day, each member of the cross-country team runs a certain distance d, which is assigned on the basis of ability, plus a half mile around the track. Greta runs 22.5 miles per week. What is her assigned daily distance? (*Hint:* Write an equation about the total distance Greta runs in one school week.)

42. Corey is planning a rafting trip for a group of tourists. He must plan on bringing 2 gallons of drinking water for each tourist as well as for each of the three guides. If the supply raft can carry 24 gallons of water, how many tourists can go on the trip? (*Hint:* Write an equation about the amount of water that must be taken.)

43. The Tree People planted new trees in an area that was burned by brush fires. Sixty percent of the seeds sprouted, but gophers ate 38 of the new sprouts. That left 112 new saplings. How many seeds were planted? (*Hint:* Write an equation about the number of seeds that survived.)

44. The treasurer for the Botanical Society would like to raise the annual dues by $15. That would allow the society to collect $1760 in dues next year. If the society has 32 members, what were the dues this year? (*Hint:* Write an equation about next year's dues.)

Use one of the formulas below to solve Problems 45–48.

> *Perimeter of a rectangle:*   **P = 2*l* + 2*w***
>
> *Area of a triangle:* $A = \dfrac{bh}{2}$

45. A farmer has 500 yards of fencing material to enclose a rectangular pasture. He wants the pasture to be 75 yards wide. How long will it be?

46. Inez has 124 centimeters of wood molding to make a rectangular picture frame. If the frame must be 8 centimeters wide, how long will it be?

47. A triangular sail requires 12 square meters of fabric. If the base of the sail measures 4 meters, how tall is the sail?

48. A city park has the shape of a triangle and covers 30,000 square feet. If the base of the triangle measures 150 feet, what is the height of the triangle?

In Problems 49–52, find the value of *x* in each figure.

49. Area = 36

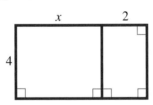

50. Area = 56

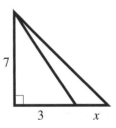

51. Area = 20

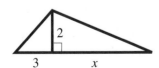

52. Area = 70

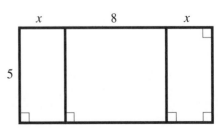

CHAPTER SUMMARY AND REVIEW

Glossary: Write a sentence explaining the meaning of each term.

bar graph	equation	line graph	solution
histogram	solve	axis	evaluate
variable	independent variable	algebraic expression	dependent variable
term	coordinates	factor	ordered pair

Properties and formulas

1. State the commutative laws, and explain their use.

2. State the associative laws, and explain their use.

3. State the six useful algebraic formulas introduced in this chapter, and explain what the variables stand for.

4. State and explain the order of operations for simplifying algebraic expressions.

Techniques and procedures

5. Give three steps for writing an algebraic expression.

6. Give three steps for graphing an equation.

7. Explain how to solve an equation algebraically.

8. Give three steps for solving an applied problem.

Review problems

9. The bar graph summarizes the participation of 18- to 21-year-olds in recent U.S. presidential elections, showing both percent registering to vote and percent actually voting.

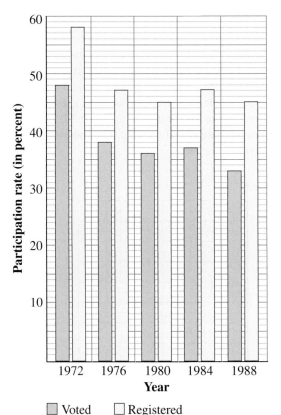

Participation of 18- to 21-year-olds in presidential elections

a. What variable is displayed on the horizontal axis?

b. What two variables are displayed by the heights of the bars?

c. What percent of 18- to 21-year-olds were registered in 1976? What percent of 18- to 21-year-olds voted in that election?

d. In what year did 36% of 18- to 21-year-olds vote?

e. Do you detect an overall trend in the percent of young people who participate in presidential elections?

10. The bar graph shows fertility trends in the developing world, by region.

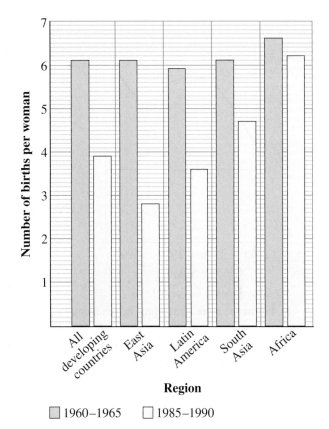

Fertility trends in the developing world, by region

a. What variable is displayed on the horizontal axis?

b. What two variables are displayed by the heights of the bars?

c. What was the average birth rate for women in South Asia from 1960 to 1965? From 1985 to 1990?

d. In what region of the developing world was the birth rate the lowest from 1985 to 1990?

e. Which region experienced the greatest decline in birth rate from 1960 to 1990? Which region experienced the smallest decline in birth rate during that period?

11. The line graph shows the number of bachelor's and master's degrees awarded per 100 students at California State University, Northridge (CSUN).

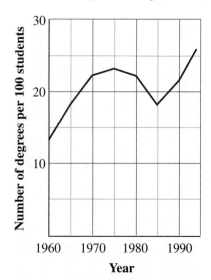

Year

Source: Los Angeles Times, January 7, 1996.

a. Fill in the table by reading values from the graph.

Year	Number of degrees per 100 students
1960	
1965	
1970	
1975	
1980	
1985	
1990	
1994	

b. In what years did the number of degrees awarded per 100 students decline?

c. If CSUN's enrollment in 1994 was approximately 25,000 students, about how many degrees did the university award in that year?

12. a. The table on page 8 shows how long a nest egg of $100,000 will last under different conditions. Locate the row that indicates annual withdrawals starting at $10,000. How long will the money last at 8% interest?

b. Make a line graph showing how long the money lasts (with $10,000 initial withdrawal) at different interest rates. Label the horizontal axis to show the variable "Interest rate" and the vertical axis to show the variable "Time." Mark off both axes in intervals appropriate for the data in the table. Make sure your intervals are equally spaced. Plot the data points given in the table.

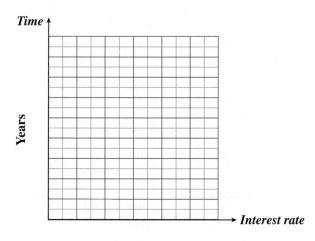

Percent

c. Do you think the lifetime of your nest egg increases smoothly as the interest rate increases? If so, connect the data points on your graph with a smooth curve. Otherwise, use line segments.

13. Martha's car can travel 22 miles on a gallon of gasoline. Fill in the table with the distances Martha can drive on various amounts of gasoline.

Gallons of gas	3	5	8	10	11	12
Calculation						
Miles						

a. Explain in words how to find the distance Martha can drive.

b. Write your explanation as a mathematical sentence:

Miles Martha can drive =

c. Let g stand for the number of gallons of gas in Martha's car and m for the number of miles she can drive. Write an equation for m in terms of g.

d. Plot the values from your table, and connect the data points with a smooth curve.

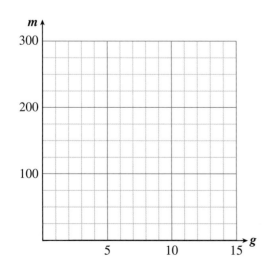

14. Mariel has a 200-page history assignment to read. Fill in the table with the number of pages Mariel has left to read after she has completed each given number of pages.

Pages read	20	50	85	110	135	180
Calculation						
Pages left						

a. Explain in words how to find the number of pages left to read.

b. Write your explanation as a mathematical sentence:

Pages left to read =

c. Let r stand for the number of pages Mariel has already read and l for the number of pages she has left to read. Write an equation for l in terms of r.

d. Plot the values from your table, and connect the data points with a smooth curve.

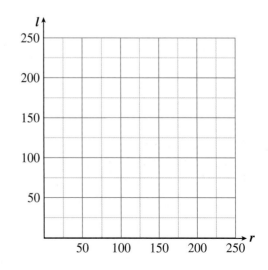

15. Find the pattern and fill in the table. Then write an equation that expresses the second variable in terms of the first.

x	0.5	1.0	1.5	2.0	4.0	6.0	7.5
y	0.125	0.25	0.375	0.5			

16. Use the graph to enter values in the table. Then write an equation that expresses the second variable in terms of the first.

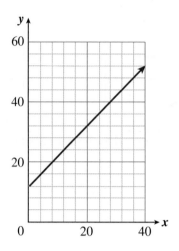

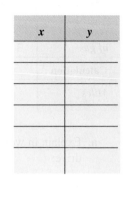

x	y

In Problems 17–20, write an algebraic expression for each phrase.

17. Five greater than *z*

18. Twenty-eight percent of *t*

19. Sixty dollars less than the first-class airfare

20. The quotient of the volume of the sphere and 6

21. a. Write an expression for the distance traveled in *t* seconds by a car moving at 88 feet per second.

b. How far will the car travel in 5 seconds? In half a second? In half a minute?

22. a. Write an expression for the amount of interest earned in *t* years on $500 deposited in an account that pays 7% annual interest.

b. How much interest will the account earn in 1 year? In 2 years? In 5 years?

23. Rachel has saved $60 to help pay for her books this semester. Let b stand for the price of her books and n for the amount she still needs.

   a. Write an equation for n in terms of b.

   b. Fill in the table.

b	100	120	150	180	200
n					

   c. Choose appropriate scales for the axes, and graph your equation.

24. Matt wants to travel 360 miles to visit a friend over spring break. He is deciding whether to ride his bike or drive. If he travels at an average speed of r miles per hour, then the trip will take t hours, where

$$t = \frac{360}{r}$$

   a. Fill in the table.

r	10	20	30	40	60	80	90
t							

   b. Choose appropriate scales for the axes, and graph the equation.

25. Decide which ordered pairs are solutions of the equation $y = \dfrac{5}{2}x$.

   a. (4, 10)   **b.** (5, 2)

   c. (1.8, 4.5)   **d.** (3, 7.5)

26. Decide which ordered pairs are solutions of the equation whose graph is shown.

   a. (5, 0)   **b.** (2, 8)

   c. (8, 3)   **d.** (6, 8)

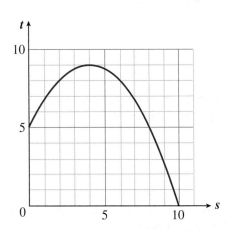

In Problems 27 and 28, decide whether the given values are solutions of the equation.

27. $\dfrac{t}{6} = 3$

   a. $t = 18$ **b.** $t = 2$

28. $8z = 36.8$

   a. $z = 28.8$ **b.** $z = 4.6$

In Problems 29–34, solve each equation.

29. $12 = 3 + y$

30. $2.7 = t - 1.8$

31. $12.2a = 4.88$

32. $8 = \dfrac{y}{4}$

33. $\dfrac{w}{2} = 0$

34. $9.5 = \dfrac{b}{0.6}$

For Problems 35–40, write and solve an equation to answer the question.

35. Staci invested some money in a T-bill account that pays $9\frac{1}{2}\%$ interest, and 1 year later the account had earned $171 in interest. How much did Staci deposit in the account?

36. A jet flew 2800 miles at a speed of 560 miles per hour. How long did the trip take?

37. Fifty-three percent of the ballots cast in a recent election were for candidate Phil I. Buster. If candidate Buster received 106,000 votes, how many people voted?

38. Salim's test average was 3.2 points below the minimum required for an A. If Salim's test average was 89.3 points, what was the minimum required for an A?

39. Ivan has a new puppy whose adult weight will be about 85 pounds. When the ratio of the puppy's weight to its adult weight is 0.7, Ivan should take the puppy to the vet for shots. How much will the puppy weigh when it is ready for shots?

40. If a rectangular garden plot is 12 feet wide, how long must it be to provide 180 square feet of gardening space?

For Problems 41 and 42, solve each equation graphically.

41. Use the graph of $y = 1.25x$ to solve each equation.

    a. $1.25x = 40$ **b.** $30 = 1.25x$

42. Use the graph of $y = \dfrac{24}{x}$ to solve each equation.

    a. $\dfrac{24}{x} = 6$ **b.** $1.5 = \dfrac{24}{x}$

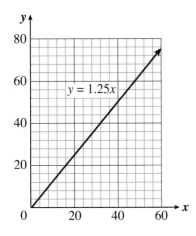

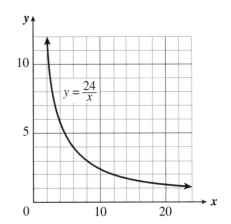

In Problems 43–46, simplify the expressions by following the order of operations.

43. $6 + 18 \div 6 - 3 \cdot 2$ **44.** $\dfrac{2}{3} + \dfrac{1}{3} \cdot 4$ **45.** $36 \div (9 - 3 \cdot 2) \cdot 2 - 24 \div 4 \cdot 3$ **46.** $\dfrac{1.2(7.7)}{4.3 - 2.3}$

Write an algebraic expression for each phrase in Problems 47–50.

47. Four pages longer than half the length of the first draft

48. \$100 short of twice last year's price

49. Three times the sum of the length and 5.6

50. Seventy-three percent of the difference of Robin's weight and 100 pounds

Evaluate the expressions in Problems 51–54 for the given values.

51. $mx + b$ for $m = \dfrac{1}{2}$, $x = 3$, $b = \dfrac{5}{2}$

52. $\dfrac{1}{m} - \dfrac{1}{n}$ for $m = 4$, $n = 6$

53. $\dfrac{3w + z}{z}$ for $w = 8$, $z = 6$

54. $2(l + w)$ for $l = \dfrac{1}{3}$, $w = \dfrac{1}{6}$

55. The total amount of money in an account is given by $P(1 + rt)$,   where P is the initial investment, r is the annual interest rate, and t is the number of years the money has been in the account. How much money is in an account that earns 10% interest annually if $1000 was invested 3 years ago?

56. If an object is thrown downward with initial speed v, in feet per second, then in t seconds it will fall a distance $t(16t + v)$,   in feet. How far will a penny fall in 2 seconds if it is thrown at a speed of 10 feet per second?

For Problems 57 and 58, write an equation relating the variables, make a table of values, and graph your equation.

57. A new computer workstation for a graphics design firm costs $2000 and depreciates in value by $200 every year. Write and graph an equation that gives the value, V, of the workstation after t years.

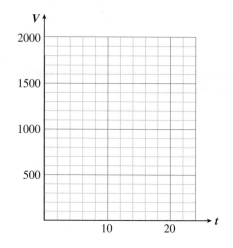

58. Ranwa teaches aerobic dance. She is paid $5 for an hour-long class, plus $2 for each student in the class. Write and graph an equation that gives Ranwa's pay, P, if there are s students in the class.

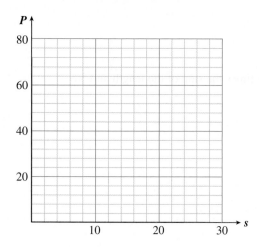

Solve the equations in Problems 59–62.

59. $3x - 4 = 1$ 　　　　**60.** $1.2 + 0.4z = 3.2$ 　　　　**61.** $\dfrac{7v}{8} - 3 = 4$ 　　　　**62.** $13 = \dfrac{2}{7}x + 13$

For Problems 63 and 64, write and solve an equation to answer the question.

63. Customers can pick their own fruit at Sunny Orchard. There is a \$5.75 entrance fee, and the fruit costs \$4.50 a bushel. Bahn spent \$19.25 at the orchard. How many bushels of fruit did he pick?

64. The perimeter of a rectangle is the sum of twice its width and twice its length. If 150 meters of fence enclose a rectangular yard that has a length of 45 meters, what is the width of the yard?

Linear Equations

2.1 Adding Signed Numbers

We begin our study of linear equations by reviewing the rules for operations on positive and negative numbers, sometimes called *signed numbers*. The reading assignment discusses four topics we'll need for this review: (1) number lines, (2) absolute value, (3) order symbols, and (4) inequalities.

Reading

Number Lines

There are many situations in which it makes sense to consider numbers less than zero, or **negative numbers**. For example, a temperature of $-10°$F (minus 10 degrees Fahrenheit) means 10 degrees below zero. Death Valley in California lies 280 feet below sea level, and we say its elevation is -280 feet. A checking account that is overdrawn by $20 has a balance of $-\$20$.

Figure 2.1

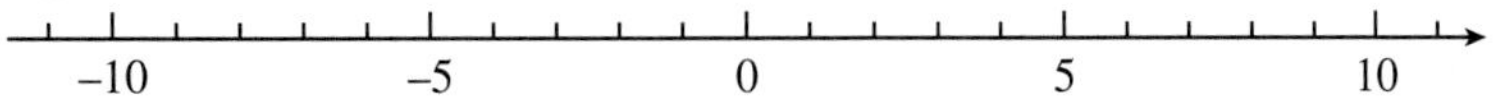

Numbers greater than zero are called **positive numbers**. We use a **number line**, as shown in Figure 2.1, to illustrate relationships among positive numbers, negative numbers, and zero. To the right of zero, we mark the numbers 1, 2, 3, 4, . . . at evenly spaced intervals. These numbers are called the **natural numbers**, or the **counting numbers**. The **whole numbers** include the natural numbers and zero: 0, 1, 2, 3, ... The natural numbers, zero, and the negatives of the natural numbers are called the **integers**:

$$\ldots, -3, -2, -1, 0, 1, 2, 3, \ldots$$

Fractions such as $\frac{2}{3}$, $-\frac{5}{4}$, and 3.6 lie between integers on the number line.

If we want to indicate a particular number, we place a dot at the appropriate position on the number line. The dot is called the **graph** of the number.

EXAMPLE 1

The graphs of 6, -6, $-2\frac{1}{3}$, 3, and $\frac{19}{4}$ are shown in Figure 2.2.

Figure 2.2

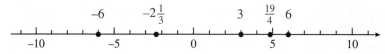

Absolute Value

In Example 1, the numbers 6 and -6 are called **opposites**. Observe in Figure 2.2 that even though the graphs of 6 and -6 lie on opposite sides of 0 on the number line, each is the same *distance* (6 units) from 0. The distance between a number and 0 on the number line is called the **absolute value** of the number.

We use vertical bars to denote the absolute value of a number. Thus

$$|6| = 6 \quad \text{and} \quad |-6| = 6$$

Because distances are always positive, *the absolute value of a number is never negative.*

EXAMPLE 2

Simplify each expression.

    a. $|-3|$ **b.** $-|3|$ **c.** $-(-3)$ **d.** $-|-3|$

Solution

    a. -3 is 3 units from 0, so $|-3| = 3$.

    b. $-|3|$ is the opposite of $|3|$, so $-|3| = -3$.

    c. The opposite of -3 is $+3$, so $-(-3) = 3$.

    d. $-|-3|$ is the opposite of $|-3|$, so $-|-3| = -3$.

A variable such as x can represent a positive number, a negative number, or zero.

EXAMPLE 3

Suppose x represents -8. Evaluate each expression.

a. $-x$ **b.** $|x|$ **c.** $|-x|$

Solution

a. $-x = -(-8) = 8$ **b.** $|x| = |-8| = 8$ **c.** $|-x| = |-(-8)| = 8$

Absolute-value bars act as grouping symbols in the order of operations. Operations within absolute-value bars should be performed first.

Now try Exercise 1.

Order Symbols

Which is greater, -5 or -10? A temperature of $-10°$ is colder than a temperature of $-5°$, and a checking account balance of $-\$10$ is less than a balance of $-\$5$. It seems that -5 is greater than -10. As we move from left to right on a number line, the numbers increase. For example, the graph of -6 lies to the left of the graph of -2 in Figure 2.3. Therefore, -6 *is less than* -2, or -2 *is greater than* -6.

Figure 2.3

We use special symbols to indicate order:

$<$ means   "is less than"
$>$ means   "is greater than"

For example,

$-6 < -2$ is read   "-6 is less than -2"
$-2 > -6$ is read   "-2 is greater than -6"

The small ends of the symbols $<$ and $>$ always point to the smaller number.

$3 < 5$ $5 > 3$
points to the _____| |_____ points to the
smaller number   smaller number

Now try Exercise 2.

EXAMPLE 4

a. Which is the lower altitude, -81 feet or -94 feet?

b. Express the relationship between the altitudes using one of the order symbols.

Solution

a. Negative altitudes correspond to feet below sea level, and 94 feet is farther below sea level than 81 feet is. Therefore, -94 feet is the lower altitude.

b. $-94 < -81$, or $-81 > -94$

EXERCISE 1
Simplify each expression.
a. $|6| - |-3|$

b. $|8 - 3| - 2$

EXERCISE 2
Replace the comma in each pair by the proper symbol: $<$ or $>$.
a. $1, -3$

b. $-9, -6$

Inequalities

A statement that uses one of the symbols $<$ or $>$ is called an **inequality**. Examples of inequalities are

$$-1 > -3 \quad \text{and} \quad x < 2$$

Unlike the equations we have studied, which have at most one solution, an inequality can have infinitely many solutions. For example, the solutions of the inequality $x < 2$ include 1, 0, -1, -2, and all the other negative integers, as well as fractions less than 2, such as $1\frac{3}{5}$, $\frac{2}{3}$, and $-\frac{17}{8}$. In fact, *all* the numbers to the left of 2 on the number line are solutions of $x < 2$. Because we cannot list all these solutions, we often represent them on a number line, as shown in Figure 2.4.

Figure 2.4

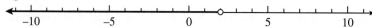

The solid line extending to the left of 2 in Figure 2.4 indicates that all the numbers less than 2 are solutions of the inequality. The open circle at 2 shows that 2 is not a solution (because 2 is not less than 2).

Now try Exercise 3.

EXERCISE 3

a. List three solutions of the inequality $x > -4$.

b. Graph all the solutions on the number line.

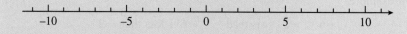

Skills Review

Determine which of the two decimal numbers in each pair is greater. If the two numbers are equal, say so. (To review decimal numbers, see Appendix A.5.)

1. 0.3, 0.03   2. 0.9, 0.10   3. 0.444, 0.45   4. 0.2222, 0.2

5. 3.20, 3.2   6. 7.01, 7.010   7. 2.09, 2.90   8. 9.999, 10.998

Answers: *1.* 0.3 *2.* 0.9 *3.* 0.45 *4.* 0.2222 *5.* equal *6.* equal
7. 2.90 *8.* 10.998

Lesson

We will use number lines to investigate addition of signed numbers. Then we'll generalize the results to formulate a rule for addition.

Adding Two Numbers with the Same Sign

Case 1: The sum of two positive numbers

To illustrate the sum $5 + 3$ on a number line, we begin by graphing the first number, $+5$. (See Figure 2.5.) From that point, we move 3 units in the positive direction, or to the right. This brings us to the sum, which is $+8$. Thus $5 + 3 = 8$.

Figure 2.5

Case 2: The sum of two negative numbers

Think of negative numbers as representing debts. The sum of two debts is an even bigger debt, so we expect the sum of two negative numbers to be a negative number. Thus we guess that the sum $(-5) + (-3)$ equals -8. To illustrate this sum on a number line, we first graph -5, as shown in Figure 2.6. Then we move 3 units in the negative direction, or to the left. This brings us to -8, as expected: $(-5) + (-3) = -8$.

Figure 2.6

Now try Exercise 4.

EXERCISE 4

For each addition illustrate the sum on the number line, and give the answer.

a. $2 + 4$

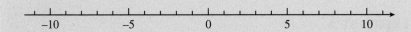

b. $(-4) + (-7)$

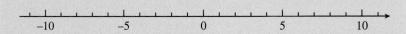

c. $(-6) + (-3)$

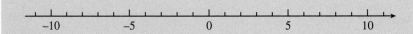

From Exercise 4, we see that

> The sum of two positive numbers is positive.

and

> The sum of two negative numbers is negative.

Thus the *sign* of the sum is the same as the sign of the two terms. To find the *value* of a sum, we can add the absolute values, or unsigned parts, of the two terms. For example, to add $(-4) + (-7)$, we add the numbers 4 and 7 without their signs, to get 11, then make the sum negative:

$$(-4) + (-7) = -11$$

Now try Exercise 5.

Adding Two Numbers with Opposite Signs

Next, we'll consider adding two numbers with opposite signs. First, think in terms of debits and credits. Suppose you have debts of $9 and assets of $6. What is your net worth? We can model this situation by the sum

$$-9 + (+6)$$

where debts are represented by negative numbers and assets by positive numbers. If you use your $6 to pay off part of your debts, you will still owe $3, so your net worth is $-\$3$. It makes sense that if your debts are greater than your assets, your net worth is negative.

Look at the same problem on the number line in Figure 2.7. We start by graphing -9 and then move 6 units in the positive direction, which brings us to the sum, -3.

Figure 2.7

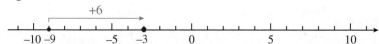

Try the sums in Exercise 6.

EXERCISE 5

Find the following sums without using a number line.

a. $(-9) + (-9)$

b. $(-14) + (-11)$

EXERCISE 6

Use the number lines to find each sum.

a. $(+5) + (-3)$

b. $(-7) + (+2)$

c. $(-5) + (+9)$

In Exercise 6, note that the *sign* of the sum is the same as the sign of the number with the *larger absolute value*. (Can you explain this in terms of debits and credits?) Also note that we can find the *value* of the sum by *subtracting* the absolute values of the two numbers. For example, to add $(-5) + (+9)$, we subtract 5 from 9 to get 4 and then make the answer positive, because $+9$ has a larger absolute value than -5. Thus

$$(-5) + (+9) = 4$$

Now try Exercises 7 and 8.

We can now formulate two rules for adding signed numbers.

Rules for Adding Integers

1. To add two numbers with the same sign, add their absolute values. The sum has the same sign as the numbers.
2. To add two numbers with opposite signs, subtract their absolute values. The sum has the same sign as the number with the larger absolute value.

Now try Exercise 9.

ANSWERS TO 2.1 EXERCISES

1a. 3 **1b.** 3 **2a.** $1 > -3$

2b. $-9 < -6$ **3a.** $-2, 0, 3$ **3b.**

4a. 6 **4b.** -11 **4c.** -9

5a. -18 **5b.** -25 **6a.** 2 **6b.** -5 **6c.** 4

7. Subtract 2 from 7 to get 5; then make the answer negative.

8a. -8 **8b.** 6 **9a.** -9 **9b.** -5 **9c.** 12 **9d.** 8

9e. 0 **9f.** -10

EXERCISE 7

Give a similar explanation for the answer to Exercise 6b.

EXERCISE 8

Compute the following sums without using number lines.

a. $4 + (-12)$

b. $15 + (-9)$

EXERCISE 9

Use the rules for adding integers to find the following sums.

a. $-6 + (-3)$

b. $-8 + (+3)$

c. $-7 + 19$

d. $18 + (-10)$

e. $5 + (-5)$

f. $-5 + (-5)$

Graph the numbers in Problems 1 and 2 on the number lines.

1. $0, -3, 3, \dfrac{1}{2}, -\dfrac{5}{3}, 5\dfrac{1}{4}$

2. $-4.5, -2, -1.4, 0.6, 2$

Simplify each expression in Problems 3–12.

3. $|6|$

4. $|-10|$

5. $-|9|$

6. $-|-6|$

7. $|-(-8.5)|$

8. $|-8| - |-4|$

9. $3|-6|$

10. $2 + 5|-3|$

11. $|7 - 3| - 2|-2|$

12. $|-7||12 - 5|$

In Problems 13–22, replace the comma in each pair of numbers by the proper symbol: <, >, or =.

13. $0, -4$

14. $3, -7$

15. $-5, -9$

16. $-3, -3\dfrac{3}{4}$

17. $13.6, 13.66$

18. $-18.4, -19.6$

19. $-2, |-2|$

20. $|-3.8|, 2.4$

21. $-|-4.1|, -5.8$

22. $-(-6), -|-6|$

23. For each value of x given below, find the value of $-x$.
   a. 14

   b. -8

   c. -21

   d. 17

24. Complete each statement:
   a. If x is a positive number, then $-x$ is a

   _____ number.

   b. If x is a negative number, then $-x$ is a

   _____ number.

25. For each value of x given below, find the value of $-(-x)$.

   a. 9

   b. -4

   c. -13

   d. 15

26. Complete this statement: For any value of x, $-(-x) =$

   _____.

27. Write each statement using $>$ or $<$.

   a. p is positive

   b. n is negative

28. Are there any values of x for which $x < x$?

29. Explain the difference between the opposite of a number and the absolute value of a number. Use examples in your explanation.

30. Which two of the following statements say the same thing?

   a. $b > 3$ **b.** $3 > b$ **c.** $3 < b$

In Problems 31–36:
a. Give three solutions to the inequality.
b. Graph all the solutions on a number line.

31. $x > -3$

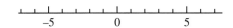

32. $x < -1$

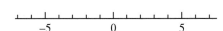

33. $x < 4$

34. $x > 1$

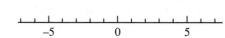

35. $-2 > x$

36. $-5 < x$

In Problems 37–48, add the given numbers.

37. $5 + (-3)$ **38.** $-5 + 3$ **39.** $-5 + (-3)$ **40.** $-12 + (-18)$

41. $-12 + 18$ **42.** $12 + (-18)$ **43.** $-47 + 22$ **44.** $-37 + (-52)$

45. $6.8 + (-2.7)$ **46.** $-2.5 + (-3.1)$ **47.** $-\dfrac{5}{6} + \dfrac{2}{3}$ **48.** $\dfrac{2}{9} + \left(-\dfrac{5}{6}\right)$

In Problems 49–56, add the numbers from left to right.

49. $-4 + (-5) + 6$ **50.** $-5 + 3 + (-4)$ **51.** $-8 + (-8) + (-6)$

52. $-4 + (-7) + (-7)$ **53.** $9 + (-12) + 7 + (-15)$ **54.** $6 + (-14) + 12 + (-17)$

55. $-26 + 13 + (-11) + (-32) + 16 + 20$

56. $-35 + (-5) + 28 + (-21) + 13 + (-14)$

57. Thelma's failing company is worth $-\$1000$. She decides to merge with Louise's company, which is worth $\$1500$. What is the combined worth of the two companies?

58. Rick's bank account is overdrawn by $\$219$. What will the new balance be if he deposits $\$196$?

59. Emily is at an elevation of -87 feet when she begins climbing a mountain. What is her elevation after she ascends 127 feet?

60. A helicopter takes off from New Orleans (elevation -5 feet) and climbs 855 feet. What is the helicopter's new elevation?

In Problems 61–64, calculate the net gain or loss on each stock over two days of trading.

		Tuesday	*Wednesday*			*Tuesday*	*Wednesday*
61.	Megacorp	down 2	down $1\frac{1}{2}$	**62.**	Tronics, Inc.	down 3	up $\frac{5}{8}$
63.	Technico	up $2\frac{1}{2}$	down $\frac{3}{4}$	**64.**	Envirogreen	down 1	up $2\frac{1}{8}$

65. The bar graph shows the percent change in the price of nine cereals between June 1996 and June 1997.

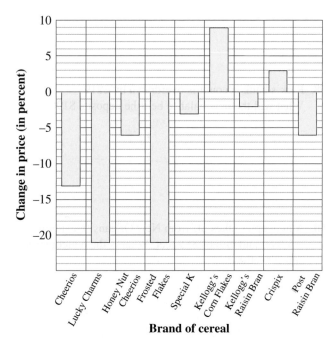

Change in price of various cereals, June 1996 to June 1997
Source: Consumer Reports, October 1997.

a. By what percent did the price of Crispix change during this time period? The price of Special K?

b. Which cereals showed an increase in price?

c. Which two cereals showed the greatest percent decrease in price? By how much did their prices decrease?

d. Although Lucky Charms and Frosted Flakes showed the same percent change in price, their prices changed by different amounts. How is this possible?

66. The bar graph shows the percent change in the crime rate in ten cities between 1995 and 1996.

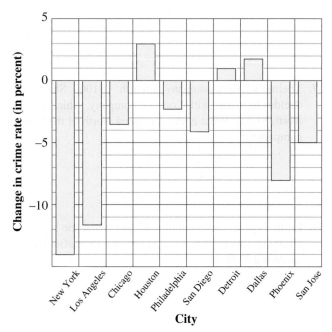

Change in crime rate, 1995 to 1996
Source: Los Angeles Times, June 7, 1997.

a. What was the percent change in crime rate in San Diego during this time period? In Dallas?

b. Which cities showed an increase in crime?

c. Which city showed the greatest decrease in crime? How large was the drop?

d. Although San Jose showed a greater percent change in its crime rate than Chicago, Chicago actually had a greater drop in number of crimes. How is this possible?

2.2 Subtracting Signed Numbers

In Section 2.1, we reviewed the rules for adding signed numbers. In this section, we see how to subtract signed numbers. It turns out that you won't have to learn any new rules for subtracting, because every subtraction problem can be converted into an addition problem with the same answer.

Reading

Subtracting a Positive Number

In this reading assignment, you should convince yourself that subtracting a positive number gives the same result as adding the negative number with the same absolute value. Begin by comparing these two problems:

$$9 + (-5)$$
$$9 - (+5)$$

We can illustrate both of these problems on a number line. The first problem is an addition problem, and we graph it as we did in Section 2.1. We first plot 9 on the number line and then move 5 units in the negative direction, or to the left, as shown in Figure 2.8. The sum is 4.

Figure 2.8

Now consider the second problem, $9 - (+5)$. This is really just a familiar subtraction problem, $9 - 5$. To illustrate this subtraction, we again start with 9 and move 5 units to the left, to end up at 4. The graph is identical to the one in Figure 2.8. Thus

$$9 + (-5) = 4 \qquad \text{and} \qquad 9 - (+5) = 4$$

From this example, you can see that *subtracting a positive number gives the same result as adding the negative number with the same absolute value.* Thus, to subtract a positive number, we add its opposite.

Try the following exercises using number lines. You should get the same answer for each pair of problems.

EXERCISE 1
a. $3 + (-10)$

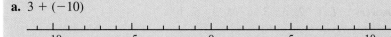

b. $3 - (+10)$

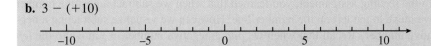

EXERCISE 2

a. $-6 + (-3)$

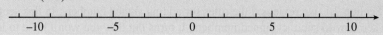

b. $-6 - (+3)$

Skills Review

Choose the correct equation for each situation.

a. $5x - 8 = 30$	**b.** $\dfrac{x}{5} - 30 = 8$	**c.** $\dfrac{x - 30}{5} = 8$
d. $5x + 8 = 30$	**e.** $\dfrac{x}{5} + 30 = 8$	**f.** $\dfrac{x + 30}{5} = 8$

1. Ilciar has earned a total of 30 points on the first four quizzes in his biology class. What must he earn on the fifth quiz to end up with an average of 8?

2. Jocelyn ordered five exotic plants from a nursery. She paid a total of $30, including an $8 shipping fee. How much did she pay for each plant?

3. The five members of the chess team pitched in to buy new chess boards. They used $30 from their treasury, and each member donated $8. How much did the new boards cost?

4. Hemman bought five tapes on sale, and he cashed in a gift certificate worth $8. He then owed the clerk $30. How much did each tape cost?

5. Nirusha and four other people won the office baseball pool. After spending $30 of her share, Nirusha had $8 left. What was the total amount in the pool?

Answers: *1.* f *2.* d *3.* c *4.* a *5.* b

Lesson

Subtracting a Negative Number

In the reading assignment, you learned that when we subtract a positive number we move to the left on the number line. How can we visualize subtracting a negative number? Recall that when we *add* a negative number, we move to the *left* on the number line. There are only two directions to choose from, so when we *subtract* a negative number, we must move to the *right*.

Compare the two problems illustrated in Figure 2.9. The first is an example of adding -3, and the second is an example of subtracting -3.

Figure 2.9

(a) Addition: $5 + (-3)$ Move 3 units to the left.

(b) Subtraction: $5 - (-3)$ Move 3 units to the right.

The difference $5 - (-3)$ in Figure 2.9b has the same value as the sum $5 + (+3)$,   namely 8. In general, *subtracting a negative number gives the same result as adding the positive number with the same absolute value*. Thus, to subtract a negative number, we add its opposite.

At first, it may be difficult to understand why subtracting a negative number is the same as adding a positive number. Try thinking in terms of debts and assets. For example, suppose you have $25 in your wallet, and you have $5 in your hand to pay off a debt you owe a friend. If, instead of taking your $5, your friend cancels (or "subtracts") the debt, you are actually $5 richer:

$$25 - (-5) = 30$$

For Exercise 3, remember that to subtract a negative number, we move to the *right* on the number line.

EXERCISE 3

Use the number lines to illustrate each subtraction problem.

a. $2 - (-6)$

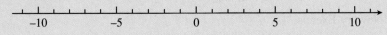

b. $-7 - (-4)$

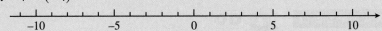

The Subtraction Rule

From Exercises 1–3, you can see that to subtract any number, positive or negative, we add its opposite. We can do this in steps, as follows.

> **To Subtract *b* from *a*:**
> **1.** Change the sign of *b*.
> **2.** Change the subtraction to addition.
> **3.** Proceed as in addition.

It is not necessary to use the subtraction rule for all subtraction problems. For example, we can compute the difference $11 - 5$ just as we did in arithmetic:

$$11 - 5 = 6$$

However, for differences that involve negative numbers, it is usually easier to use the subtraction rule.

EXAMPLE 1

Use the subtraction rule to rewrite each of the subtraction problems in Exercises 1–3 as an addition problem.

a. $3 - (+10) = 3 + (-10) = -7$ →change to addition / →change sign

b. $-6 - (+3) = -6 + (-3) = 9$ →change to addition / →change sign

c. $2 - (-6) = 2 + (+6) = 8$ →change to addition / →change sign

d. $-7 - (-4) = -7 + (+4) = 3$ →change to addition / →change sign

Now try Exercise 4.

Order of Operations: Addition and Subtraction

Remember that additions and subtractions should be performed in order from left to right.

EXAMPLE 2

Simplify $7 - 9 + 12 - 6$.

Solution

Rewrite the subtraction $7 - 9$ as an addition; then add from left to right.

$$7 - 9 + 12 - 6 = 7 + (-9) + 12 - 6 \quad \text{The sum of 7 and } -9 \text{ is } -2.$$
$$= -2 + 12 - 6 \quad \text{The sum of } -2 \text{ and 12 is 10.}$$
$$= 10 - 6 = 4$$

In Example 2, note that the expression $7 - 9$ can be interpreted as either

$$7 - (+9) \qquad \text{or} \qquad 7 + (-9)$$

Both give the same result, -2. Because addition is usually easier to perform than subtraction, it is usually more efficient to regard an expression such as

EXERCISE 4
Rewrite each subtraction problem as an addition; then compute the answer.

a. $3 - (-9)$

b. $-4 - (-7)$

$7 - 9 + 12 - 6$ as a *sum* of positive and negative numbers. Try this yourself: Add 7 and −9 and +12 and −6. You should get 4, the same answer as before.

In Exercise 5, rewrite all the operations as equivalent additions; then compute the sum.

The rules for the order of operations also hold for negative numbers. In problems that contain brackets, we perform the operations inside brackets first.

EXAMPLE 3

Simplify $4 - 3 - [-6 + (-5) - (-2)]$.

Solution

We perform the operations inside brackets first. Then we simplify by rewriting the problem as an addition.

$$4 - 3 - [-6 + (-5) - (-2)]$$
$$= 4 - 3 - [-6 - 5 + 2] \quad -6 - 5 = -11; \quad -11 + 2 = -9$$
$$= 4 - 3 - [-9] \quad \text{Rewrite as an addition.}$$
$$= 4 - 3 + 9 = 10 \quad \text{Add from left to right.}$$

EXERCISE 5
Simplify $5 - (+7) - 3 - (-2)$

ANSWERS TO 2.2 EXERCISES
1. −7 **2.** −9 **3a.** 8
3b. −3 **4a.** 12 **4b.** 3
5. −3

HOMEWORK 2.2

In Problems 1–8, rewrite each subtraction as an addition, and give the answer.

1. $4 - 8$

2. $5 - 7$

3. $3 - (-9)$

4. $6 - (-2)$

5. $-8 - (-6)$

6. $-5 - (-3)$

7. $-6 - 5$

8. $-2 - 7$

In Problems 9–24, add or subtract as indicated.

9. $12 + (-6)$

10. $24 + (-17)$

11. $6 - (-4)$

12. $8 - (-7)$

13. $-2 - 8$

14. $-11 - 5$

15. $-7 + 9$

16. $-12 + 5$

17. $-14 - (-3)$

18. $-18 - (-10)$

19. $-5 + (-4)$

20. $-3 + (-8)$

21. $-6 - (-6)$ **22.** $-9 + 9$ **23.** $-4 - 4$ **24.** $5 - (-5)$

25. The temperature at noon was 2°C, and it dropped 8°C over the next 12 hours. What was the temperature at midnight?

26. The Falcons gained 4 yards on their first play but lost 12 yards on their second play. What was their net change in position after two plays?

27. A tourist in California can travel from Death Valley, at an elevation of -282 feet, to Mt. Whitney, at 14,494 feet. What is the tourist's net change in elevation?

28. How much higher is the Valdes Peninsula (elevation -131 feet) than the Dead Sea (elevation -1312 feet)?

29. Although his account is overdrawn by $24.20, Nelson writes a check for $11.20. What is his new balance?

30. Socrates died in 399 B.C. at the age of 70. In what year was he born?

In Problems 31–34, compute each sum or difference in parts (a) and (b), and decide which is easier. Then decide how to simplify the expression in part (c).

31. a. $15 - (+5)$ **b.** $15 + (-5)$ **c.** $15 - 5$

32. a. $8 - (+12)$ **b.** $8 + (-12)$ **c.** $8 - 12$

33. a. $-6 - (+2)$ **b.** $-6 + (-2)$ **c.** $-6 - 2$

34. a. $-3 - (+7)$ **b.** $-3 + (-7)$ **c.** $-3 - 7$

Simplify each expression in Problems 35–40 by adding the signed numbers.

35. $6 - 3 + 4 - 5$

36. $2 + 5 - 8 - 1$

37. $13 - 6 - 12 + 17$

38. $-23 + 28 - 14 + 21$

39. $120 - 80 + 20 - 40$

40. $-34 - 52 + 68 - 21$

Simplify each expression in Problems 41–46 by following the rules for addition and subtraction.

41. $3 + 2 - (-4) + (-7)$

42. $-6 + 5 + (-3) - (-8)$

43. $12 - (-7) - (-2) - 4$

44. $-11 - 2 - (-4) - (-3)$

45. $21 + (-15) - (-2) - 7$

46. $-14 - (-16) - 4 + (-7)$

Simplify each expression in Problems 47–54 by following the order of operations.

47. $-18 - [8 - 12 - (-4)]$

48. $-6 + [-7 + (-6) - 2]$

49. $3 - (-6 + 2) + (-1 - 4)$

50. $-3 + (4 - 6) - (-8 + 4)$

51. $-7 + [8 - (-2)] - [6 + (-4)]$

52. $2 - [-4 + (-3)] - [7 + (-3)]$

53. $0 - [5 - (-1)] + [-6 - 3]$ **54.** $0 + [3 - (-4)] - [-6 + 4]$

For Problems 55–64, compute the answer mentally.

55. $2 - 7$ **56.** $9 - 5$ **57.** $-2 - 7$ **58.** $-9 - 5$

59. $-2 + 7$ **60.** $-9 + 5$ **61.** $2 - (-7)$ **62.** $9 - (-5)$

63. $-2 - (-7)$ **64.** $-9 - (-5)$

Evaluate the expressions in Problems 65–68.

65. $15 - x - y$ for $x = -6$, $y = 8$ **66.** $-a + 3 - b$ for $a = -2$, $b = 5$

67. $p - (4 - m)$ for $p = -2$, $m = -6$ **68.** $7 - (c + k)$ for $c = -3$, $k = -9$

In Problems 69–74, find three values of x that satisfy the inequality.

69. $x + 2 > 5$ **70.** $x - 3 < 6$ **71.** $x - 4 < -2$

72. $x + 8 > -1$ **73.** $-12 > x - 9$ **74.** $2 < x - 6$

75. The ten students in a small class receive the following scores on a 10-point quiz:

$$4, 5, 6, 7, 7, 8, 8, 9, 10, 10$$

a. Compute the mean quiz score, denoted by $\bar{x}$. (See Appendix A.11 to review the definition of *mean*.)

b. For each quiz score, compute the *deviation from the mean*, $x - \bar{x}$, and fill in the table.

Score, x	Deviation from the mean, $x - \bar{x}$
4	$4 - \bar{x} =$
5	
6	
7	
7	
8	
8	
9	
10	
10	

c. Compute the average of the deviations from the mean. Can you explain why your answer makes sense?

76. Repeat Problem 75 for the following scores received by 12 students on a 20-point quiz:

$$20, 19, 17, 17, 16, 15, 15, 14, 12, 12, 10, 7$$

Score, x	Deviation from the mean, $x - \bar{x}$
20	$20 - \bar{x} =$
19	
17	
17	
16	
15	
15	
14	
12	
12	
10	
7	

77. The graph shows Mexico's economic growth rate from the first quarter of 1995 through the third quarter of 1997.

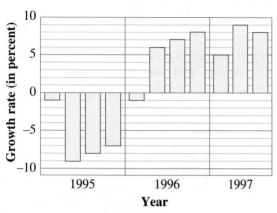

Mexico's economic growth, 1995–1997

Source: Los Angeles Times, November 19, 1997.

    a. What was Mexico's economic growth rate in the second quarter of 1995?

    b. When was the growth rate approximately 5%?

    c. What was the change in growth rate from the last quarter of 1995 to the first quarter of 1996?

    d. What was the change in growth rate from the first quarter of 1995 to the second quarter of 1995?

    e. What was the change in growth rate from the first quarter of 1996 to the second quarter of 1996?

78. The graph shows the growth rate of the U.S. gross national product (GNP) for the years 1978 to 1988.

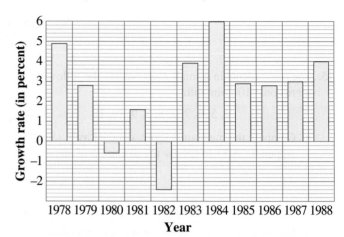

Growth in U.S. GNP, 1978–1988

    a. What was the growth rate in 1985?

    b. In what year was the growth rate approximately 3.9%?

    c. In what two years was the growth rate negative? What was the growth rate in those years?

    d. What was the change in the growth rate from 1979 to 1980? From 1980 to 1981?

    e. In what year did the GNP experience the greatest increase in growth rate over the previous year?

Multiplying and Dividing Signed Numbers

In this section, we consider products and quotients of signed numbers. In the reading assignment, we review the connection between multiplication and division and consider quotients that involve zero.

Reading

Recall that the operation of division is the opposite of multiplication. We calculate a quotient by asking what number, when multiplied by the denominator, will yield the numerator. For example, we know that

$$\frac{12}{3} = 4 \qquad \text{because} \qquad 3 \cdot 4 = 12$$

In general, if $\dfrac{a}{b} = c$, then $bc = a$ (as long as $b \neq 0$).

Try Exercise 1 now.

Thus every division problem can be converted into an equivalent multiplication problem. For example,

$$\frac{400}{16} = 25 \qquad \text{is equivalent to} \qquad 16 \cdot 25 = 400$$

The numerator in the division problem becomes the product in the multiplication problem.

Now try Exercise 2.

Quotients Involving Zero

In algebra, you will encounter quotients whose numerator is equal to zero. For example, what is the meaning of an expression such as $\dfrac{0}{2}$? We can rewrite this quotient as an equivalent multiplication problem:

$$\frac{0}{2} = ? \qquad \text{is equivalent to} \qquad 2 \cdot ? = 0$$

What number can we substitute for the question mark to make $2 \cdot ? = 0$ a true statement? The only solution for this equation is 0. Thus, replacing the question marks by 0, we have

$$\frac{0}{2} = 0 \qquad \text{because} \qquad 2 \cdot 0 = 0$$

In fact, 0 divided by any (nonzero) number is 0.

Now, consider a quotient with zero in the denominator, such as $\dfrac{2}{0}$.

$$\frac{2}{0} = ? \qquad \text{is equivalent to} \qquad 0 \cdot ? = 2$$

What number can we substitute for the question mark to make $0 \cdot ? = 2$ a true statement? This equation has no solution, because 0 times any number is 0, not 2. Thus there is no answer for the division problem $\dfrac{2}{0}$. We say that the quotient $\dfrac{2}{0}$ is *undefined*. In general, *we cannot divide any number by zero.*

EXERCISE 1

Use the relationship between products and quotients to complete each statement. No calculation is necessary.

a. $\dfrac{8190}{26} = \underline{\hspace{2cm}}$

   because $26 \cdot 315 = 8190$.

b. $62 \cdot \underline{\hspace{2cm}} = 83.7$

   because $\dfrac{83.7}{62} = 1.35$.

EXERCISE 2

Rewrite each division problem as a multiplication problem.

a. $\dfrac{144}{64} = 2.25$

b. $36 \div \dfrac{3}{8} = 96$

To summarize, we state the following result.

> If a is any nonzero number, then
>
> $$\frac{0}{a} = 0 \quad \text{and} \quad \frac{a}{0} \text{ is undefined.}$$

Try Exercise 3.

Skills Review

Calculate each sum or difference from left to right.

1. $(-2) + (-2) + (-2)$
2. $(-3) + (-3) + (-3)$
3. $(-1) + (-1) + (-1) + (-1) + (-1)$
4. $(-6) + (-6) + (-6) + (-6)$
5. $28 - 7 - 7 - 7 - 7$
6. $18 - 6 - 6 - 6$
7. $-6 - (-2) - (-2) - (-2)$
8. $-9 - (-3) - (-3) - (-3)$

Answers: 1. -6 *2.* -9 *3.* -5 *4.* -24 *5.* 0 *6.* 0 *7.* 0 *8.* 0

Lesson

Products

You may recall that multiplication is really repeated addition. For example,

$$3(2) \quad \text{means} \quad \underbrace{2 + 2 + 2}_{\text{the sum of three 2s}}$$

Similarly,

$$3(-2) \quad \text{means} \quad \underbrace{-2 + (-2) + (-2)}_{\text{the sum of three } -2\text{s}}$$

which is -6, so $3(-2) = -6$.   You can think of this in terms of money: If you owe three different people \$2 each, then you are \$6 in debt.

Now try Exercise 4.

Because multiplication is commutative, it is also true that $-2(3) = -6$ and that $-4(5) = -20$.   Use these examples to complete the following statement:

The product of a positive number and a negative number is a _____ *number.*

EXAMPLE 1

Use your calculator to verify each product.

a. $(-4)(7) = -28$ **b.** $5(-2.6) = -13$ **c.** $\dfrac{-3}{4} \cdot \dfrac{1}{2} = \dfrac{-3}{8}$

EXERCISE 3

Find each quotient, if it exists.

a. $\dfrac{0}{18}$

b. $\dfrac{13}{0}$

c. $6 \div 0$

d. $0 \div 8$

EXERCISE 4

Write the product $5(-4)$ as a repeated addition, and compute the product.

Now let's investigate the product of two negative numbers. First, study the sequence of products in the left column; then complete the similar list in the right column.

$$
\begin{array}{ll}
4(-2) = -8 & 4(-5) = \\
3(-2) = -6 & 3(-5) = \\
2(-2) = -4 & 2(-5) = \\
1(-2) = -2 & 1(-5) = \\
0(-2) = 0 & 0(-5) = \\
-1(-2) = 2 & -1(-5) = \\
-2(-2) = 4 & -2(-5) = \\
-3(-2) = 6 & -3(-5) =
\end{array}
$$

On the basis of your calculations above, complete the following conjecture, or educated guess, about the product of two negative numbers.

The product of two negative numbers is a _____ number.

EXAMPLE 2

Use your calculator to verify each product.

a. $(-6)(-3) = 18$ **b** $(-2.1)(-3.4) = 7.14$ **c.** $\dfrac{-5}{3} \cdot \dfrac{-3}{10} = \dfrac{1}{2}$

Because we know that the product of two positive numbers is a positive number, we can now summarize our results about the products of signed numbers.

Products of Signed Numbers

1. The product of two numbers with opposite signs is a negative number.

2. The product of two numbers with the same sign is a positive number.

Now try Exercise 5.

Quotients

Now that you know the rules for multiplying signed numbers, you can easily discover the rules for quotients of signed numbers. Try Exercise 6 now.

From Exercise 6, you can see that the same rules hold for division of signed numbers as for multiplication.

Quotients of Signed Numbers

1. The quotient of two numbers with opposite signs is a negative number.

2. The quotient of two numbers with the same sign is a positive number.

EXAMPLE 3

Use your calculator to verify each quotient.

a. $\dfrac{-12}{3} = -4$ **b.** $(-6.4) \div (-2) = 3.2$ **c.** $\dfrac{3}{4} \div \left(\dfrac{-2}{3}\right) = \dfrac{-9}{8}$

Now try Exercise 7.

EXERCISE 5
Compute the following products.

a. $4(-2)$

b. $(-3)(-3)$

c. $-5(3)$

EXERCISE 6
Find each quotient by rewriting the division as an equivalent multiplication problem.

a. $\dfrac{6}{3} =$ _____

because _____.

b. $\dfrac{6}{-3} =$ _____

because _____.

c. $\dfrac{-6}{3} =$ _____

because _____.

d. $\dfrac{-6}{-3} =$ _____

because _____.

EXERCISE 7
Compute each quotient.

a. $\dfrac{-25}{-5}$

b. $\dfrac{32}{-8}$

c. $\dfrac{-27}{9}$

Order of Operations

Compare the expressions in Example 4. Be sure to notice how parentheses and minus signs are used.

EXAMPLE 4

Simplify each expression.

    a. $3(-8)$ **b.** $3 - (-8)$ **c.** $3 - 8$ **d.** $-3 - 8$

Solution

    a. This expression is a product: $3(-8) = -24$.

    b. This is a subtraction: $3 - (-8) = 3 + 8 = 11$.

    c. This is an addition: $3 - 8 = 3 + (-8) = -5$.

    d. This is also an addition: $-3 - 8 = -3 + (-8) = -11$.

In Exercise 8, remember to perform multiplications and divisions before additions and subtractions. When should you perform operations inside parentheses?

When we evaluate an algebraic expression at a negative number, it is a good idea to enclose the negative number in parentheses. This will help prevent you from confusing multiplication with subtraction.

EXAMPLE 5

Evaluate $2x - 3xy$ for $x = -5$ and $y = -2$.

Solution

Substitute -5 for x and -2 for y; then follow the order of operations.

$$2x - 3xy = 2(-5) - 3(-5)(-2) \quad \text{Do multiplications first.}$$
$$= -10 - 3(10)$$
$$= -10 - 30 = -40$$

EXERCISE 8
Simplify each expression.

a. $-6(-2)(-5)$

b. $-6(-2) - 5$

c. $-6(-2 - 5)$

d. $-6 - (2 - 5)$

ANSWERS TO 2.3 EXERCISES

1a. 315 **1b.** 1.35

2a. $64 \cdot 2.25 = 144$ **2b.** $\dfrac{3}{8} \cdot 96 = 36$

3a. 0 **3b.** Undefined **3c.** Undefined **3d.** 0

4. $-4 + (-4) + (-4) + (-4) + (-4) = -20$

5a. -8 **5b.** 9 **5c.** -15

6a. 2 **6b.** -2 **6c.** -2 **6d.** 2

7a. 5 **7b.** -4 **7c.** -3

8a. -60 **8b.** 7 **8c.** 42 **8d.** -3

HOMEWORK 2.3

In Problems 1–12, multiply or divide as indicated.

1. $(-8)(-4)$

2. $7(-9)$

3. $\dfrac{12}{-4}$

4. $\dfrac{-36}{9}$

5. $-20 \div (-5)$

6. $-5 \div 20$

7. $-6(-1)(3)$

8. $-4(-7)(-1)$

9. $\dfrac{-8}{0}$

10. $\dfrac{0}{-4}$

11. $(-5)(0)(6)$

12. $(-3)(-2)(-1)(0)$

13. Whitney is climbing down a sheer cliff. She has spaced pitons vertically every 6 meters. What is her net change in elevation after she has descended to the eighth piton from the top?

14. The bank erroneously charged Utako with five withdrawals of $400 each. If her balance is actually $0, what does the bank believe her balance is?

15. The temperature on the ice planet Hoth dropped 115° in just 4 hours. What was the average change in temperature per hour?

16. Bertha lost 30 pounds in 8 weeks. What was Bertha's average change in weight per week?

17. The U.S. balance book shows an entry of $-\$100,000,000,000$ representing the failure of numerous savings and loan institutions. If this amount were to be divided equally among the 100 million adult U.S. citizens, how much of the balance would be assigned to each?

18. Petra implements a scheme that successfully cuts her firm's $250,000 debt in half. What is the resulting balance?

19. Find each product.

   a. $(-2)(3)(4)$ **b.** $(-2)(-3)(4)$

   c. $(-2)(-3)(-4)$ **d.** $(-2)(-3)(4)(2)$

   e. $(-2)(-3)(-4)(2)$ **f.** $(-2)(-3)(-4)(-2)$

20. Use your results from Problem 19 to complete these statements:

   a. The product of an odd number of negative numbers is _____.

   b. The product of an even number of negative numbers is _____.

In Problems 21 and 22, perform the indicated operations.

21. a. $-12 - 4$ **b.** $-12(-4)$

   c. $-12 - (-4)$ **d.** $-12 + (-4)$

22. a. $-5 - 9$ **b.** $-5(-9)$

   c. $-5 - (-9)$ **d.** $-5 + (-9)$

23. Answer the following questions about Problems 21 and 22:

   a. Which two parts have the same answer?

   b. Which part is a multiplication problem?

24. Complete these statements:

   a. The sum of two negative numbers is always _____.

   b. The product of two negative numbers is always _____.

In Problems 25 and 26, perform the indicated operations.

25. a. $-3 - 3$ **b.** $-3(-3)$

   c. $-3 - (-3)$ **d.** $-3 + (-3)$

   e. $-3 \div (-3)$

26. a. $8 - 8$ **b.** $8(-8)$

   c. $8 - (-8)$ **d.** $8 + (-8)$

   e. $8 \div (-8)$

27. a. Find two numbers whose sum equals zero.

    b. Find two numbers whose difference equals zero.

    c. Find two numbers whose product equals zero.

    d. Find two numbers whose quotient equals zero.

28. a. Find two numbers whose product is 1.

    b. Find two numbers whose product is -1.

    c. Find two numbers whose quotient is 1.

    d. Find two numbers whose quotient is -1.

Use the order of operations to simplify each expression in Problems 29–40.

29. $-2(-3) - 4$

30. $3(-2) + 1$

31. $5(-4) - 3(-6)$

32. $6 - 2(-5) + 7$

33. $(-4 - 3)(-4 + 3)$

34. $-6 - 3(-4 - 2)$

35. $-3(8) - 6(-2) - 5(2)$

36. $2(-8) - [4(-3) - 2]$

37. $\dfrac{15}{-3} - \dfrac{4 - 8}{8 - 12}$

38. $\dfrac{4 - 2(-5)}{-4 + 3(-1)}$

39. $\dfrac{2(-3) - 4(-8)}{-4 - (-2)(-3)}$

40. $\dfrac{-4(-3) - 4(-5)}{-4(-3 - 5)}$

41. Simplify each expression.
  a. $-3(-4)(-5)$ **b.** $-3(-4) - 5$

  c. $-3(-4 - 5)$ **d.** $-3 - (-4 - 5)$

  e. $-3 - (-4)(-5)$ **f.** $(-3 - 4)(-5)$

42. Simplify each expression.
  a. $24 \div 6 - 2$ **b.** $24 \div (-6 - 2)$

  c. $24 \div (-6) - 2$ **d.** $24 - 6 \div (-2)$

  e. $24 \div (-6) \div (-2)$ **f.** $24 \div (-6 \div 2)$

43. a. Use your calculator to verify that
  $$-\frac{2}{5} = \frac{-2}{5} = \frac{2}{-5}.$$

  b. Does $-\dfrac{2}{5} = \dfrac{-2}{-5}$?

44. a. Use your calculator to verify that
  $$-\frac{9}{4} = \frac{-9}{4} = \frac{9}{-4}.$$

  b. Does $-\dfrac{9}{4} = \dfrac{-9}{-4}$?

45. Evaluate the following expressions for $x = 5$. What do you notice?
  a. $\dfrac{-3}{4}x$ **b.** $\dfrac{-3x}{4}$ **c.** $-0.75x$

46. Evaluate each of the following expressions for $x = -6$. What do you notice?
  a. $\dfrac{-8}{5}x$ **b.** $\dfrac{-8x}{5}$ **c.** $-1.6x$

47. Does $\dfrac{-8}{5}x = \dfrac{-8}{5x}$? Support your answer with examples.

48. Does $\dfrac{-3}{4}x = \dfrac{-3}{4x}$? Support your answer with examples.

Evaluate the expressions in Problems 49–58.

49. $2x + 5$ for $x = -3$

50. $-4x - 3$ for $x = -2$

51. $12x - 3xy$ for $x = -3$, $y = 2$

52. $ab(6 - 4a)$ for $a = -6$, $b = -2$

53. $\dfrac{y - 3}{x - 4}$ for $x = -9$, $y = 2$

54. $\dfrac{56 - h}{t}$ for $h = 200$, $t = 3$

55. $\dfrac{5}{9}(F - 32)$ for $F = -13$

56. $\dfrac{9}{5}C + 32$ for $C = -20$

57. $\dfrac{1}{2}t(t - 1)$ for $t = \dfrac{2}{3}$

58. $\dfrac{1}{3}(h + 1)(h - 1)$ for $h = -\dfrac{2}{5}$

59. On a notoriously difficult math contest, there are ten questions. You get 2 points for every correct answer but lose 1 point for each wrong answer. A blank answer gets 0 points. Miranda's class earned the following scores: two people scored -5, four people scored -3, one person scored -2, one person scored 3, one person scored 6, and three people scored 12. What was the average score in Miranda's class?

60. Last month, Delbert's checking account posted the following daily balances: $26 for six days, $19 for eight days, $-\$40$ for five days, $-\$20$ for three days, $0 for six days, and $15 for three days. What was Delbert's average daily balance last month?

In Problems 61–66, find three solutions for each inequality.

61. $4x < 7$

62. $3x > 10$

63. $-2x > 8$

64. $-6x < 12$ **65.** $-3x < -6$ **66.** $-5x > -15$

2.4 Graphs of Linear Equations

In this section, we study equations whose graphs are straight lines. These equations are called **linear equations**, and they can be written in the form

$$Ax + By = C$$

where A, B, and C are constants, and A and B are not both zero.

Reading

The graphs we have considered so far have used only positive values of the variables. However, many graphs include negative values as well. To construct such a graph, we begin by drawing a pair of perpendicular number lines for the horizontal and vertical axes, as shown in Figure 2.10. We use x for the independent variable and y for the dependent variable. Consequently, we refer to the horizontal axis as the **x-axis** and to the vertical axis as the **y-axis**. The two axes intersect at the zero point of each number line. Thus the coordinates of this intersection point, which is called the **origin**, are (0, 0).

The two axes divide the plane into four **quadrants**, or regions. Points with positive x-coordinates lie to the right of the vertical axis, and points with negative x-coordinates lie to the left. Points with positive y-coordinates lie above the horizontal axis, and points with negative y-coordinates lie below. The grid constructed in this way is called a **Cartesian coordinate system**, after the French mathematician and philosopher René Descartes, who devised it.

Figure 2.10

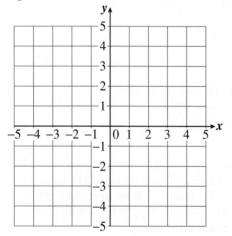

EXAMPLE 1

Plot the given points on a Cartesian coordinate system.

a. (2, 3) **b.** (−2, 1) **c.** (0, −2) **d.** (3, −1) **e.** (−4, −3)

Solution

For each point, the first coordinate gives the location in the horizontal, or x-direction, and the second coordinate gives the location in the vertical, or y-direction.

a. Starting from the origin, we count 2 units *right* and then 3 units *up* to locate the point.

b. Starting from the origin, we count 2 units *left* and then 1 unit *up*.

c. Starting from the origin, we count 0 units in the horizontal direction (in other words, we stay on the vertical axis) and then 2 units *down*.

d. Starting from the origin, we count 3 units *right* and then 1 unit *down*.

e. Starting from the origin, we count 4 units *left* and then 3 units *down*.

All the points are shown in Figure 2.11.

Figure 2.11

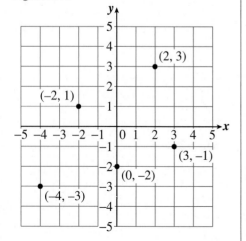

Now try Exercise 1.

Skills Review

Decide whether each ordered pair is a solution of the given equation. (See Section 1.5.)

1. $y = x - 3$
   a. $(5, 2)$ b. $(-2, 1)$
   c. $(-2, 6)$ d. $(1.4, -1.6)$

2. $y = -5x$
   a. $(2, -3)$ b. $(0, 0)$
   c. $(-5, 1)$ d. $\left(\dfrac{-1}{3}, \dfrac{5}{3}\right)$

3. $y = \dfrac{x}{-3}$
   a. $(2, -6)$ b. $\left(5, -\dfrac{5}{3}\right)$
   c. $(-9, -3)$ d. $(3, -1)$

4. $y = -5 + x$
   a. $(0, 5)$ b. $(-5, 25)$
   c. $(-2, -7)$ d. $(2.4, -2.6)$

Answers: 1. a. Yes *b.* No *c.* No *d.* Yes *2. a.* No *b.* Yes
c. No *d.* Yes *3. a.* No *b.* Yes *c.* No *d.* Yes *4. a.* No *b.* No
c. Yes *d.* Yes

Lesson

In this lesson, we use a Cartesian coordinate system to create graphs that show both positive and negative values.

Activity 1

The temperature in Nome was $-12°F$ at noon. It has been rising at a rate of $2°F$ per hour all day.

a. Fill in the table. T stands for the temperature, and h stands for the number of hours after noon. Negative values of h represent hours before noon.

b. Write an equation for the temperature, T, after h hours.

c. Graph the equation from part b on the grid in Figure 2.12, using the values in the table.

Figure 2.12

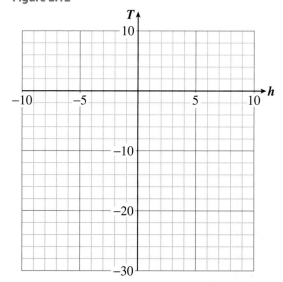

EXERCISE 1
Give the coordinates of each point.

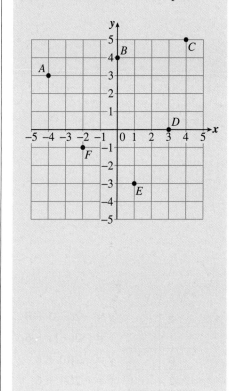

h	T
-3	
-2	
-1	
0	
1	
2	
3	

Use your graph to answer the following questions:

d. What was the temperature 8 hours before noon?

e. When will the temperature reach $-4°F$? When will the temperature reach $4°F$?

f. How much did the temperature change between 2 P.M. and 6 P.M.? _____

In Section 1.5, we graphed equations using three steps:

1. Make a table of values.
2. Choose scales for the axes.
3. Plot the points, and connect them with a smooth curve.

We can use these same steps to graph a linear equation.

Activity 2
Graph the equation $y = -2x + 6$.

Solution

a. Choose values for x and make a table of values. Be sure to choose both positive and negative x-values. (One possible selection of x-values appears in the table below.) To find the y-value for each point, substitute the x-value into the equation and evaluate.

b. Look at the values in the table to help you choose scales for the axes. You need to choose scales that include the largest and smallest values in your table. For this graph, we have chosen a scale from -10 to 20 on the y-axis.

c. Plot your points on the grid in Figure 2.13, and connect them with a smooth curve. All the points should lie on one straight line. _____

x	y
-3	$y = -2(-3) + 6$
-1	$y = -2(-1) + 6$
0	$y = -2(0) + 6$
2	$y = -2(2) + 6$
4	$y = -2(4) + 6$

Figure 2.13

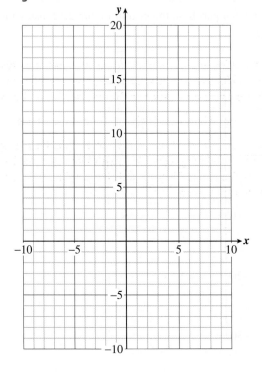

Recall that the graph of an equation is a picture of the solutions of the equation. That is, each point on the graph represents a solution. If we substitute the coordinates of the point into the equation, a true statement results. For example, the point $(7, -8)$ lies on the graph in Figure 2.13. If we substitute $x = 7$ and $y = -8$ into the equation $y = -2x + 6$, we should get a true statement:

$$y = -2x + 6$$
$$-8 = -2(7) + 6$$
$$-8 = -14 + 6 \quad \text{True.}$$

Now try Exercises 2 and 3.

EXERCISE 2
Use the graph in Figure 2.13 to answer these questions:

a. Find the value of $-2x + 6$ when $x = -5$.

b. Find the x-value for which $-2x + 6 = -10$.

EXERCISE 3
Which of the following points represent solutions to the equation whose graph is shown?
a. $(-3, -2)$

b. $(-6, -4)$

c. $(-4, 0)$

d. $(3, -6)$

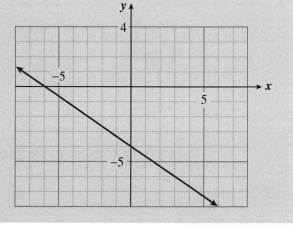

If the coefficient of x is a fraction, we can make our task easier by choosing multiples of the denominator for the x-values. That way we won't have to work with fractions to find the corresponding y-values. In the equation in Exercise 4, the coefficient of x is $-\dfrac{2}{3}$, so we choose multiples of 3 for the x-values.

Whenever you graph a linear equation, you should extend the graph far enough to cross both the x-axis and the y-axis. The points where the graph crosses the axes are important for applications. We will study these points and other properties of linear equations in later sections.

Now try Exercise 4.

EXERCISE 4

Graph $y = -\dfrac{2}{3}x - 4$.

a. Fill in the table of values.

x	y
−9	
−3	
0	
3	
6	

$y = -\dfrac{2}{3}(-9) - 4$

$y = -\dfrac{2}{3}(-3) - 4$

$y = -\dfrac{2}{3}(0) - 4$

$y = -\dfrac{2}{3}(3) - 4$

$y = -\dfrac{2}{3}(6) - 4$

b. Choose appropriate scales and label the axes on the grid.

c. Plot the points and connect them with a straight line.

ANSWERS TO 2.4 EXERCISES

1. $A(-4, 3)$, $B(0, 4)$, $C(4, 5)$, $D(3, 0)$, $E(1, -3)$, $F(-2, -1)$

2a. 16 **2b.** 8

3. $(-3, -2)$ and $(3, -6)$

4.

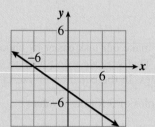

HOMEWORK 2.4

1. Delbert inherited $5000 and has been spending money at the rate of $100 per day. Right now, he has $2000 left.
   a. Fill in the table. *B* stands for Delbert's balance *d* days from now. Negative values of *d* represent days in the past.

d	−15	−5	0	5	15	20	25
B							

   b. Write an equation for Delbert's balance, *B*, after *d* days.

   c. Graph your equation using the values in the table.

   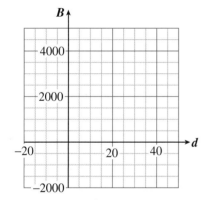

   Use your graph to answer the following questions:
   d. What was Delbert's balance 15 days ago?

   e. When will Delbert's balance reach $500?

   f. How much does Delbert spend from the beginning of day −5 to the beginning of day 40?

2. Francine borrowed money from her mother, and she currently owes her mother $750. She has been paying off the debt at a rate of $50 per month.
   a. Fill in the table. *F* stands for Francine's financial status, and *m* is the number of months from now. Negative values of *m* represent months in the past. (Francine's current financial status is −$750.)

m	−5	−2	0	2	6	10	12
F							

   b. Write an equation for Francine's financial status, *F*, in terms of *m*.

   c. Graph your equation using the values in the table.

   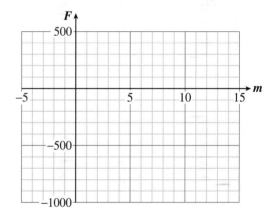

   Use your graph to answer the following questions:
   d. What will Francine's financial status be 7 months from now?

   e. When was Francine's financial status −$900?

   f. How much does Francine pay her mother from month 2 to month 9?

3. Jayme parks in the garage of her office building, 45 feet underground on floor -5. Each floor of the building is 9 feet tall.

  a. Fill in the table. *F* stands for the floor, and *E* is the elevation on that floor. Negative elevations are below ground level.

F	−5	−2	0	1	2	4	6
E							

  b. Write an equation for the elevation, *E*, on floor *F*.

  c. Graph your equation using the values in the table.

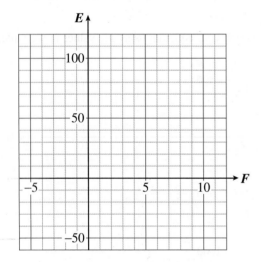

Use your graph to answer the following questions:

  d. What is the elevation on floor -3?

  e. On what floor is the elevation 72 feet?

  f. What is Jayme's change in elevation when she takes the elevator from her parking level to her job on the tenth floor?

4. Ryan is cooling a chemical compound in the laboratory at a rate of 5°F per hour. Right now the temperature of the compound is 35°F.

  a. Fill in the table. *T* stands for the temperature of the compound, and *h* is the number of hours from now.

h	−5	−3	−1	0	2	4	5
T							

  b. Write an equation for the temperature, *T*, of the compound after *h* hours.

  c. Graph your equation using the values in the table.

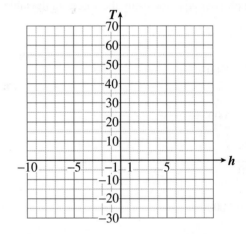

Use your graph to answer the following questions:

  d. When was the compound at room temperature, 70°F?

  e. What will the temperature of the compound be 10 hours from now?

  f. How much does the temperature drop between the third hour and the ninth hour?

In Problems 5–12, make a table of values and graph each equation. Extend the line far enough to cross both axes.

5. $y = x + 3$

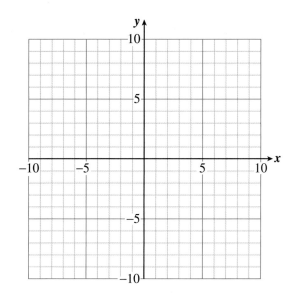

6. $y = -4 - x$

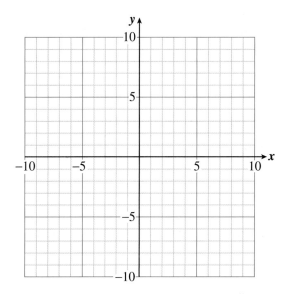

7. $y = 2x + 1$

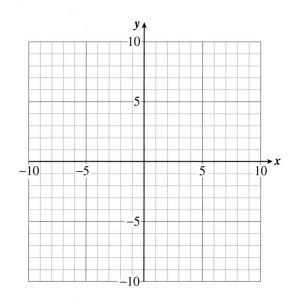

8. $y = 3x - 1$

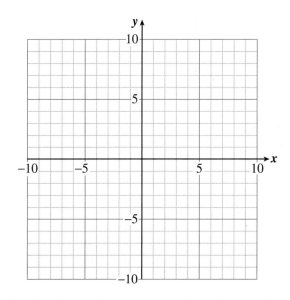

9. $y = -\dfrac{1}{2}x - 5$

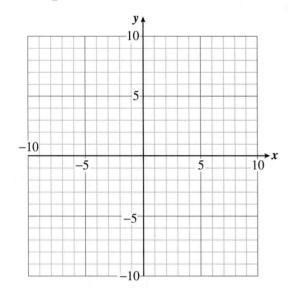

10. $y = \dfrac{3}{2}x + 2$

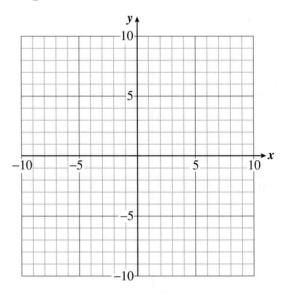

11. $y = \dfrac{5}{4}x - 4$

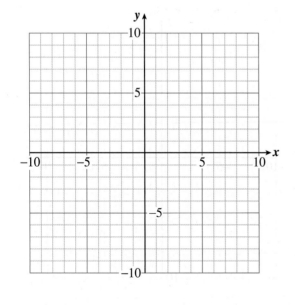

12. $y = \dfrac{-3}{4}x + 2$

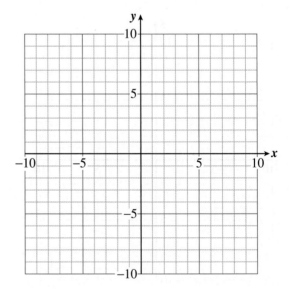

In Problems 13–16, each figure shown is the graph of an equation. Decide which of the given points are solutions of the equation.

13.

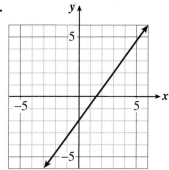

a. $(2, 0)$ b. $(3, 2)$
c. $(-1, -2)$ d. $(-3, -6)$

14.

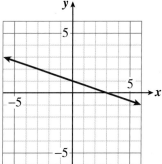

a. $(0, 3)$ b. $(3, 0)$
c. $(-2, -1)$ d. $(-3, 2)$

15.

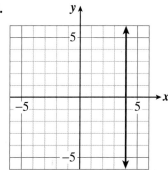

a. $(4, -2)$ b. $(4, 4)$
c. $(0, 4)$ d. $(-1, -4)$

16.

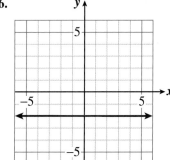

a. $(-2, 0)$ b. $(3, -2)$
c. $(2, -2)$ d. $(-1, 2)$

Match each equation in Problems 17–20 with the correct graph.

17. $y = 12 - 3x$ **18.** $y = 3x + 2$ **19.** $y = 2x - 6$ **20.** $y = -8 + \dfrac{1}{2}x$

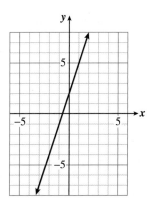

(a)

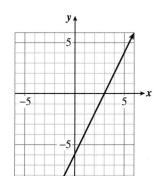

(b)

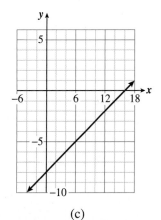

(c)

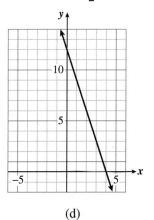

(d)

21. Here is the graph of $y = 2x + 6$.

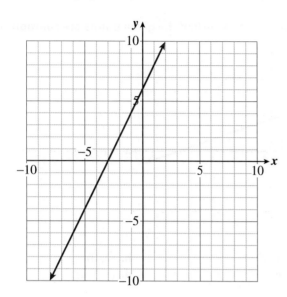

a. Use the graph to evaluate the expression
$2x + 6$ for $x = -5$.

b. Find a point on the graph where $y = -4$.   What is
the x-value of the point?

c. Verify that the coordinates of your point in part (b)
satisfy the equation of the graph.

d. Use the graph to find an x-value that produces a
y-value of 8.

e. Find two points on the graph for which $y > -4$.
What are their x-values?

22. Here is the graph of $y = -3x + 4$.

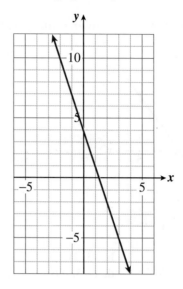

a. Use the graph to evaluate the expression $-3x + 4$
for $x = -2$.

b. Find a point on the graph with $y = 10$.   What is
the x-value of the point?

c. Verify that the coordinates of your point in part (b)
satisfy the equation of the graph.

d. Use the graph to find an x-value that produces a
y-value of -2.

e. Find two points on the graph for which $y < 10$.
What are their x-values?

Plot each pair of points in Problems 23 and 24. Then find the distance between the two points.

23.

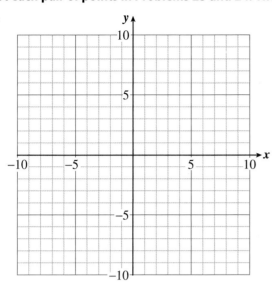

a. $A(-6, 8)$, $B(-6, 3)$

b. $C(1, 5)$, $D(1, -7)$

c. $E(7, -2)$, $F(7, -8)$

24.

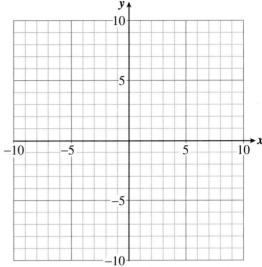

a. $P(-2, 8)$, $Q(5, 8)$

b. $R(-9, 2)$, $S(-3, 2)$

c. $T(1, -4)$, $U(6, -4)$

25. Subtract the y-coordinates of each pair of points in Problem 23, and take the absolute value of the result. Compare to your answers for Problem 23.

26. Subtract the x-coordinates of each pair of points in Problem 24, and take the absolute value of the result. Compare to your answers for Problem 24.

27. a. What can you say about the x-coordinates of two points that lie on the same vertical line?

28. a. What can you say about the y-coordinates of two points that lie on the same horizontal line?

b. Explain how to find the distance between two points that lie on the same vertical line.

b. Explain how to find the distance between two points that lie on the same horizontal line.

2.5 Solving Linear Equations

Reading

Recall that a solution to an equation is a value of the variable that makes the equation true.

Skills Review

Determine whether the given value is a solution of the equation. (See Section 1.6.)

1. $u + 7 = 5, \quad u = -2$

2. $-2 = v + 1, \quad v = -3$

3. $-7s = 21, \quad s = 3$

4. $-17 = -\dfrac{1}{2}t, \quad t = -34$

5. $a - 2 = -7, \quad a = -9$

6. $-8 = b - 6, \quad b = 2$

7. $-4 = \dfrac{p}{-5}, \quad p = -20$

8. $\dfrac{q}{10} = -20, \quad q = -2$

Answers: **1.** Yes **2.** Yes **3.** No **4.** No **5.** No **6.** No **7.** No **8.** No

In Chapter 1, we solved equations by writing simpler equations with the same solution as the original. To create the simpler equations, we isolated the variable using one or more of the following steps:

1. Add or subtract the same number on both sides of an equation.
2. Multiply or divide both sides of the equation by the same number, as long as that number is not zero.

This strategy works because of the four *properties of equality*, one for each of the four arithmetic operations.

Addition and Subtraction Properties of Equality

If the same quantity is added to or subtracted from both sides of an equation, the solution is unchanged. In symbols:

If $a = b$, then $a + c = b + c$ and $a - c = b - c$

Multiplication and Division Properties of Equality

If both sides of an equation are multiplied or divided by the same nonzero quantity, the solution is unchanged. In symbols:

If $a = b$, then $ac = bc$ and $\dfrac{a}{c} = \dfrac{b}{c}$

We can use the same techniques to solve an equation that includes signed numbers.

EXAMPLE 1

Solve $8 - 3x = -10$.

Solution

The left side of the equation has two terms: 8 and $-3x$. We want to isolate the term containing the variable, so we subtract 8 from both sides.

$$\begin{array}{r} 8 - 3x = -10 \\ \underline{-8 \qquad\ \ -8} \\ -3x = -18 \end{array}$$

Next, we divide both sides by -3:

$$\frac{-3x}{-3} = \frac{-18}{-3}$$
$$x = 6$$

Thus the solution is 6.

Check: $8 - 3(6) = 8 - 18 = -10$

A negative fraction such as $-\dfrac{3}{5}$ can be written in *standard form* with the minus sign in the numerator: $\dfrac{-3}{5}$. You will need to know this fact to complete Exercise 1. Follow the suggested steps to solve the equation.

Lesson

Solving Equations Graphically

In Example 1, we used algebra to solve the equation $8 - 3x = -10$.   We can also use a graph to solve an equation. Figure 2.14 shows a graph of the equation $y = 8 - 3x$.   Remember that each point on the graph represents a solution of the equation. To solve the equation for a particular y-value, we need only locate the corresponding point on the graph.

Figure 2.14

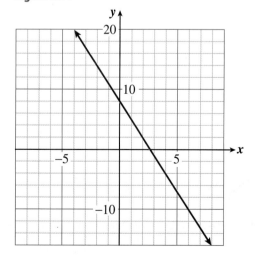

EXERCISE 1

Solve $-6 - \dfrac{2x}{3} = 8$.

Step 1   Add 6 to both sides.

Step 2   Rewrite the fraction $-\dfrac{2x}{3}$ in standard form.

Step 3   Multiply both sides by 3.

Step 4   Divide both sides by -2.

Step 5   Check your solution.

EXERCISE 2
The point (0, 8) also lies on the graph in Figure 2.14. This point gives the solution to a certain equation in x. Write the equation, and give its solution.

EXERCISE 3
The point $(-12, 2)$ lies on the graph of $y = -6 - \dfrac{2x}{3}$.

a. Using the information given, solve the equation $-6 - \dfrac{2x}{3} = 2$ mentally.

b. Verify your solution algebraically.

EXERCISE 4
Use the graph in Figure 2.14 to solve the equation $8 - 3x = 14$ as follows:

Step 1   Locate the point on the graph whose y-coordinate is 14.

Step 2   Find the x-coordinate of the point.

Step 3   Check that your x-value is a solution for $8 - 3x = 14$.

EXAMPLE 2

Use the graph of $y = 8 - 3x$ to solve the equation $-10 = 8 - 3x$.

Solution
We locate the point on the graph in Figure 2.14 whose y-coordinate is -10. You can see that the x-coordinate of this point is 6. Thus the point $(6, -10)$ lies on the graph. When we substitute $x = 6$ and $y = -10$ into the equation of the graph, we get

$$-10 = 8 - 3(6)$$

The fact that this is a true statement tells us that $x = 6$ is the solution of the equation $-10 = 8 - 3x$. We found the same solution, $x = 6$, algebraically in Example 1.

Now try Exercises 2 and 3.

When we use a graph to evaluate an algebraic expression, we start with a given x-value, locate the corresponding point on the graph, and find the y-coordinate of that point. When we use a graph to solve an equation, we do just the opposite: We start with the given y-value, locate the corresponding point on the graph, and find the x-coordinate of that point.

Now try Exercise 4.

Estimating Solutions from a Graph

Sometimes we can only *estimate* the solution of an equation from a graph. The scales on the axes may not be detailed enough for us to read the exact value of the x-coordinate, so we cannot be sure we have found the exact solution of the equation. Nonetheless, we can make an estimate that will be approximately correct.

EXAMPLE 3

The graph of $y = 9.6 - 2.4x$ is shown in Figure 2.15. Use the graph to solve the equation $9.6 - 2.4x = -3$.

Figure 2.15

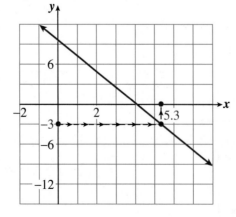

Solution

Locate the point on the graph whose y-coordinate is -3. Then find the x-coordinate of that point. The x-coordinate is somewhere between 5 and 6, but it is difficult to read its value exactly. We might estimate that the x-coordinate is 5.3. To check whether this is a good approximation for the solution, we substitute $x = 5.3$ into the equation:

$$y = 9.6 - 2.4(5.3) = -3.12$$

The y-value for $x = 5.3$ is not exactly -3, but it is close. Therefore, we can conclude that the solution is approximately 5.3.

The symbol $\approx$ indicates that two quantities or expressions are approximately equal. Thus we can write our solution to Example 3 as $x \approx 5.3$.

Now try Exercises 5 and 6.

Applied Problems

Problem solving often involves signed numbers, either in the equation that models the problem or in its solution, or both.

EXAMPLE 4

The trout population in Clear Lake has been decreasing by approximately 60 fish per year, and this year there are about 430 trout in the lake. If the population drops below 100, the Park Service will have to restock the lake.

a. Write an equation for the population, P, of trout x years from now, if the current trend continues.

b. When will the Park Service have to restock the lake?

c. Graph your equation for P, and illustrate your answer to part (b) on the graph.

Solution

a. The population starts at 430 this year and decreases by 60 fish in each following year. Thus $P = 430 - 60x$.

b. We would like to find the value of x when $P = 100$. We substitute 100 for P and solve the equation for x.

$$\begin{array}{rl} 100 = & 430 - 60x \qquad \text{Subtract 430 from both sides.} \\ \underline{-430 \quad -430} & \\ -330 = & -60x \end{array}$$

$$\begin{array}{l} \dfrac{-330}{-60} = \dfrac{-60x}{-60} \qquad \text{Divide by } -60. \\ 5.5 = x \end{array}$$

The Park Service will have to restock the lake with trout in 5.5 years if the trout population continues to decline at the current rate.

EXERCISE 5

Use algebra to find an exact solution for the equation in Example 3.

EXERCISE 6

a. Use the graph in Figure 2.15 to estimate the solution to the equation $9.6 - 2.4x = 6$.

b. Use the equation to check the accuracy of your approximation.

c. Figure 2.16 shows the graph of the equation $P = 430 - 60x$. To solve the equation $100 = 430 - 60x$, we locate the point on the graph where $P = 100$ and read its x-coordinate, which is about 5.5.

Figure 2.16

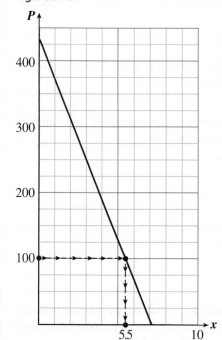

HOMEWORK 2.5

Solve the equations in Problems 1–18.

1. $x - 9 = -4$

2. $y + 6 = -2$

3. $-9z = 12$

4. $-3w = -15$

5. $\dfrac{-a}{4} = 8$

6. $36 = \dfrac{-3b}{5}$

7. $9 - x = 3$

8. $-3 - x = -5$

9. $3c - 7 = -13$

10. $-3 + 4x = -15$

11. $-5 = -2 - 3t$

12. $2 = 30 - 7w$

13. $1 - \dfrac{b}{3} = -5$

14. $-8 = -2 - \dfrac{z}{4}$

15. $\dfrac{3y}{5} + 2 = -4$

16. $4 = 1 - \dfrac{3x}{7}$

17. $\dfrac{5x}{2} + 10 = 0$

18. $-4 + \dfrac{3x}{2} = -4$

19. What is the solution of the equation $-x = -3$?

20. What is the solution of the equation $-x = 6$?

In Problems 21–24, find the error in each incorrect solution, and then write a correct solution.

21. $6 - 3x = -12$
$-3x = -6$
$x = 2$

22. $-5 - 2x = -17$
$-2x = -12$
$x = -6$

23. $-2 + \dfrac{2}{3}x = -4$
$-2 + 2x = -12$
$2x = -10$
$x = -5$

24. $-1 - \dfrac{3}{4}x = 8$
$\dfrac{3}{4}x = 9$
$x = 12$

In Problems 25–28, use the graph to solve each equation. (You may have to estimate some solutions.) Then solve the equations algebraically.

25.

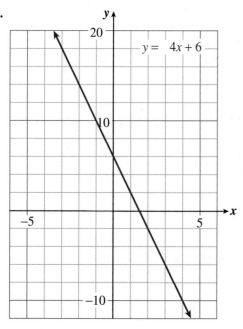

a. $-4x + 6 = 18$

b. $-4x + 6 = 8$

c. $-4x + 6 = -6$

26.

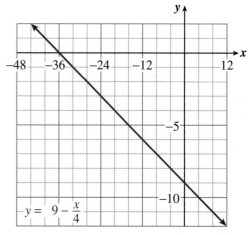

a. $-9 - \dfrac{x}{4} = -10$

b. $-9 - \dfrac{x}{4} = -5$

c. $-9 - \dfrac{x}{4} = 1$

27.

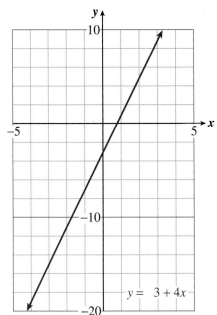

$y = 3 + 4x$

a. $-3 + 4x = 5$

b. $-3 + 4x = -6$

c. $-3 + 4x = -15$

28.

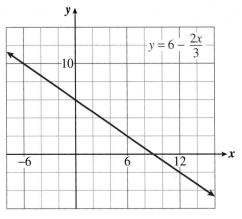

$y = 6 - \dfrac{2x}{3}$

a. $6 - \dfrac{2x}{3} = 10$

b. $6 - \dfrac{2x}{3} = 0$

c. $6 - \dfrac{2x}{3} = -3$

In Problems 29–32, use the graph to estimate the solution of the equation. Then use a calculator to help you solve the equation algebraically. How close to the actual answer was your estimate from the graph?

29. $37.21 - 8.4t = 24.61$

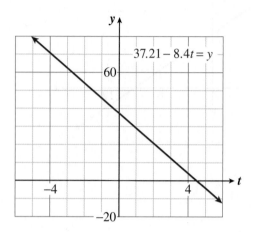

30. $-71.1 = 2.28p - 153.2$

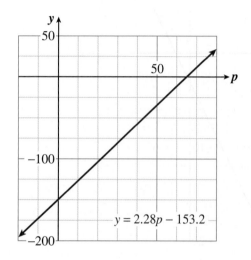

31. $-26.4 = -3.65 + 9.1x$

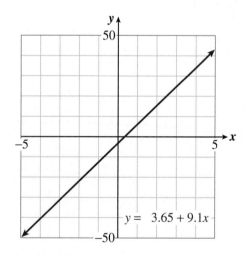

32. $320 - 1.8w = -544$

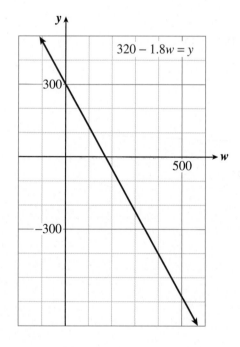

33. On a 100-point test, Lori loses 5 points for each wrong answer.
   a. Write an equation for Lori's score, s, if she gives x wrong answers.

   b. Complete the table of values, and graph your equation.

x	2	5	6	12
s				

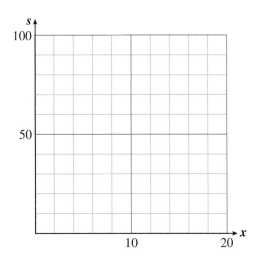

   c. Write an equation and solve it algebraically to answer the following question: If Lori's score is 65, how many wrong answers did she give?

   d. Use your graph to verify the solution to the equation you wrote in part (c).

34. The water in Silver Pond is 10 feet deep, but the water level is dropping at a rate of $\frac{1}{2}$ inch per week.
   a. Write an equation for the depth, d, of the pond after w weeks. (*Hint:* Convert all units to inches.)

   b. Complete the table of values, and graph your equation.

w	12	24	60	96
d				

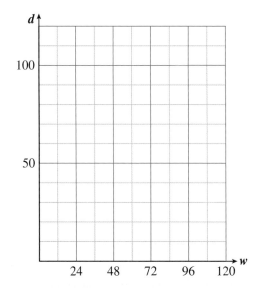

   c. Write an equation and solve it algebraically to answer the following question: How long will it take until the pond is 8 feet deep?

   d. Use your graph to verify the solution to the equation you wrote in part (c).

35. Larry bought a 10-pound box of laundry detergent, and every week he uses $\frac{1}{4}$ pound for the laundry.

a. Write an equation for the amount of detergent, D, left after w weeks.

b. Complete the table of values, and graph your equation.

w	2	8	10	28
D				

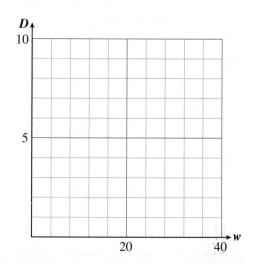

c. Write an equation and solve it algebraically to answer the following question: How long will it take until Larry has only $3\frac{1}{2}$ pounds of detergent left?

d. Use your graph to verify the solution to the equation you wrote in part (c).

36. A new computer workstation for a graphic design firm costs $2000 and depreciates in value $200 every year.

a. Write an equation for the value, V, of the workstation after t years.

b. Complete the table of values, and graph your equation.

t	2	4	5	8
V				

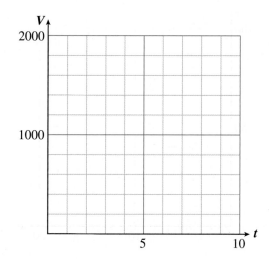

c. Write an equation and solve it algebraically to answer the following question: When will the workstation be worth only $600?

d. Use your graph to verify the solution to the equation you wrote in part (c).

Solve Problems 37–40 by writing and solving an equation.

37. During a 4-day warming trend, the temperature rose from −6°F to 26°F. What was the average change in temperature per day?

38. Today, the temperature is 56°F. If the temperature drops by 4°F per day, how long will it be before the first freeze (32°F)?

39. Eric is on a diet to reduce his current weight of 196 pounds to 162 pounds. If he loses 4 pounds per week, how long will it take him to reach his desired weight?

40. Linda wants to weigh 128 pounds for her graduation in 7 weeks. If she weighs 149 pounds now, how much must she lose each week to meet her goal?

For Problems 41–44, you will need to use the formula

$$\text{profit} = \text{revenue} - \text{cost}$$

41. A new company had $600,000 in revenue last year but ended up with a net loss of $45,000. What were the company's expenses?

42. A landscape design firm spent $240,000 last year and ended up with a net loss of $60,000. What was the firm's revenue?

43. It costs Mesa Airlines $1800 to provide a commuter flight from San Diego to Phoenix. There are 20 seats on the airplane. How much should Mesa charge for a ticket in order to make a profit of $900 on a full flight?

44. Mesa Airlines lost $520 on a flight from Modesto to Bakersfield. The flight cost Mesa $1200, and there were eight passengers on board. How much was the fare for the flight?

45. Here is is a mystery graph. Its equation is
$y = mx + k$,   where *m* and *k* are numbers, but you
don't know which numbers they are! Use the graph to
answer the questions that follow:

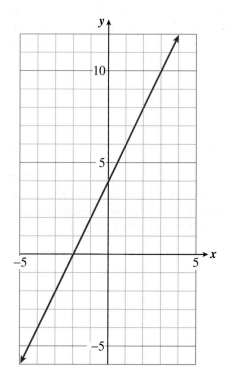

a. Find a value of *x* for which $mx + k = 6$.

b. Find at least three values of *x* for which
$mx + k < 4$.

c. Find at least three values of *x* for which
$mx + k > -2$.

46. Here is another mystery graph. Its equation is
$y = nx + q$,   where *n* and *q* are numbers. Use the
graph to decide whether each equation is true or false.

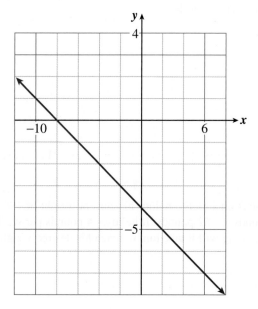

a. $4n + q = -6$

b. $-2n + q = 3$

c. $2n + q < -4$

d. $-10n + q < 0$

MIDCHAPTER REVIEW

Use complete sentences to answer the questions in Problems 1–4.

1. Explain the terms *natural numbers*, *whole numbers*, and *integers*.

2. A classmate claims that the *opposite* of a number and the *absolute value* of a number are the same and uses $x = -3$ as an example. Do you agree? Give examples of your own.

3. You have probably heard people say, "Two negatives make a positive."
   a. For which of the four arithmetic operations is this statement always true?

   b. For which operations is it always false?

   c. For which operations is it sometimes true and sometimes false?
   Make up an example for each case.

4. In a Cartesian coordinate system, the axes divide the plane into four ———————————. The point $(0, 0)$ is called the ———————————.

Simplify the expressions in Problems 5–14.

5. a. $6 - 2(-4)$ b. $6 - 2|-4|$

6. a. $-3|5 - 9|$ b. $-3|5| - |9|$

    c. $6 - (2 - 4)$ d. $6 - |2 - 4|$

    c. $|-3||-5| - 9$ d. $|-3 - 5||5 - 9|$

7. $-48 + 37 - 25 - 54$

8. $-7.9 + (-2) - (-5) - 2.7$

9. $-5[4 - 2(3)] + 6$

10. $3(-2) - 2[(6 - 8) + 5(-3)]$

11. $\dfrac{2 - (-9)(-4)}{1 - 2(-8)}$

12. $\dfrac{-8 - (-2)(-4)}{4 - 3(-2)}$

13. $\dfrac{-3}{5} + \dfrac{1}{2}\left[-\dfrac{1}{3} - \left(\dfrac{-1}{3}\right)\right]$

14. $\left(\dfrac{-5}{4} + \dfrac{1}{4}\right)\left[\dfrac{7}{3} - \left(\dfrac{-2}{3}\right)\right]$

Evaluate the expressions in Problems 15–18.

15. $(m + n)(m - n)$ for $m = -8,\ \ n = -2$

16. $-\dfrac{3k}{2 - l}$ for $k = -7,\ \ l = -4$

17. $1.8C + 32$ for $C = -23.6$

18. $\dfrac{5(F - 32)}{9}$ for $F = -13.27$

19. Francine's checking account was overdrawn by $13.26 when she wrote a check for $15.00. What is her new balance?

20. Stefanie weighed 3.45 kilograms at birth. Two days later, she weighed 3.30 kilograms. What was the net change in her weight?

Find three solutions for each inequality in Problems 21 and 22.

21. $x - 3 < -12$

22. $-3x < -12$

23. The water level in the city reservoir was 30 feet below normal four days ago, when the city began diverting water from a nearby river. The level is rising at a rate of 2 feet per day.

a. Fill in the table. W stands for the water level d days from today. Negative values of W represent water levels below normal.

d	-3	0	4	8	12
W					

b. Write an equation for W in terms of d.

c. Graph your equation on the grid to the right.

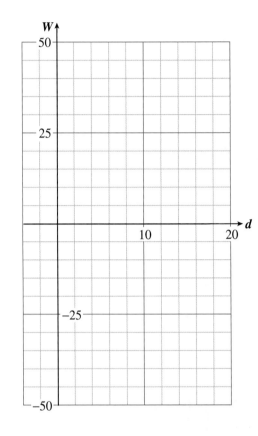

d. What will the water level be 1 week from now?

e. When will the water reach its normal level?

f. How much will the water level change from the 6th day to the 14th day?

24. Decide which of the given points are solutions of the equation whose graph is shown.

    a. $(-3, -1)$ **b.** $(-9, -7)$

    c. $(-4.5, 5)$ **d.** $(6, -3)$

    e. $(-6, 3)$ **f.** $(9, 3)$

25. What is the total distance you would travel if you started at the point $(-5, -6)$, moved to the point $(-5, 7)$, and then moved to the point $(8, 7)$?

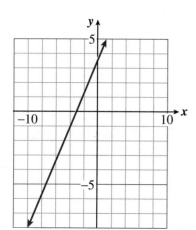

26. Here is a graph of $y = \dfrac{-2}{3}x + 9$.

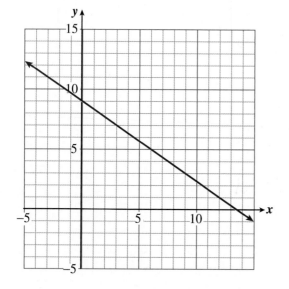

  a. Use the graph to evaluate the expression

$$\frac{-2}{3}x + 9 \quad \text{for} \quad x = 3.$$

  b. Find a point on the graph with $y = 11$. What is the x-value of the point? Now solve the equation

$$\frac{-2}{3}x + 9 = 11.$$

  c. Find a point on the graph with $y = 3$. What is the x-value of the point? Now solve the equation

$$\frac{-2}{3}x + 9 = 3.$$

For Problems 27 and 28, make a table of values and graph each equation.

27. $y = -2x + 8$

28. $y = \dfrac{-2}{3}x - 6$

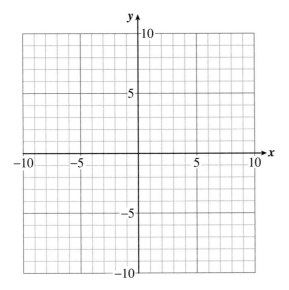

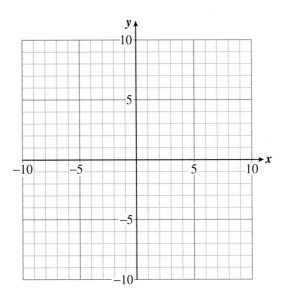

Solve each equation in Problems 29–32 algebraically.

29. $-4 = -3c + 2$

30. $30 - 7x = 2$

31. $7 - \dfrac{2y}{3} = -5$

32. $\dfrac{3x}{2} + 3 = 2$

33. a. Use the graph of $y = 8.8 - 2.4x$ to estimate the solution of $8.8 - 2.4x = 20$.

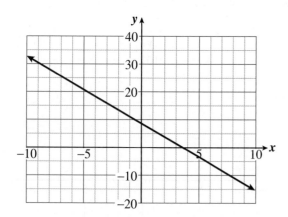

b. Solve the equation in part (a) algebraically. How close to the actual solution was your estimate from the graph?

34. Beryl is sailing in a hot-air balloon at an altitude of 500 feet. She begins a slow descent at the rate of 15 feet per minute.

a. Write an equation for Beryl's altitude, h, after m minutes.

b. Complete a table of values, and graph your equation.

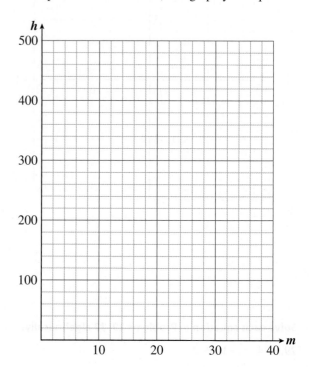

c. When will Beryl reach an altitude of 230 feet? Write and solve an equation to answer this question; then verify the solution on your graph.

In Problems 35 and 36, write and solve an equation.

35. Today, the high temperature in Cedar Rapids is 16°F, and one week ago, the high temperature was −5°F. What was the average change in temperature per day over the past week?

36. Last month, a dip in food prices caused a change of −1.5 in the consumer price index (CPI). This was 60% of the net change in the CPI for the month. What was the net change in the CPI last month?

2.6 Solving Linear Inequalities

Reading

Before beginning the reading assignment, try the following review problems about inequality symbols.

Skills Review

Replace the comma in each pair of numbers by the proper symbol: <, >, or =. (See Section 2.1 to review inequality symbols.)

1. −4, 3   2. 9, −10   3. −6, −5   4. −3, −10

5. $\dfrac{-3}{4}, \dfrac{3}{-4}$ 6. $\dfrac{-5}{-7}, \dfrac{5}{7}$ 7. −2(−3), −5   8. −12, −6(−2)

Answers: 1. < *2.* > *3.* < *4.* > *5.* = *6.* = *7.* > *8.* <

Properties of Inequalities

In Section 2.5, we reviewed some properties of equality and used them to solve linear equations. In this section, we establish three properties of inequalities. We'll begin with an inequality that we know is true:

$$3 < 5$$

If we add or subtract the same quantity on both sides of the inequality, it is still true. For example, let's add 4 to both sides of the inequality:

$$3 + 4 < 5 + 4$$
$$7 < 9$$

Adding 4 merely shifts the two values to the right on the number line, without changing their relative position. (See Figure 2.17.)

Figure 2.17

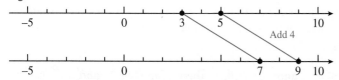

Now suppose we subtract 6 from both sides of our original inequality:

$$3 - 6 < 5 - 6$$
$$-3 < -1$$

The resulting inequality is still true. The number lines in Figure 2.18 illustrate this calculation.

Figure 2.18

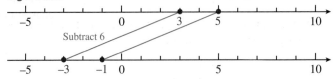

We can state these results as a property of inequalities:

> **1.** If we add or subtract the same quantity on both sides of an inequality, the truth of the statement is unchanged.
>
> In symbols:
>
> **If $a < b$ then $a + c < b + c$ and $a - c < b - c$**

What if we multiply both sides of an inequality by the same number? First let's consider the result of multiplying by a *positive* number. We multiply both sides of the inequality $3 < 5$ by 2:

$$2(3) < 2(5)$$
$$6 < 10$$

Again, the new inequality is still true. This calculation is illustrated on the number lines in Figure 2.19.

Figure 2.19

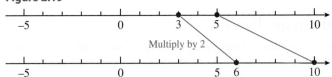

What happens with division by a positive number? If we divide both sides of the inequality $3 < 5$ by 4,   we find that

$$\frac{3}{4} < \frac{5}{4}$$

Once again, this is a true statement. We can now state a second property of inequalities.

> **2.** If we multiply or divide both sides of an inequality by the same *positive* quantity, the truth of the statement is unchanged.
>
> In symbols:
>
> **If $a < b$ and $c > 0$,   then $ac < bc$ and $\dfrac{a}{c} < \dfrac{b}{c}$**

Finally, consider what happens if we multiply both sides of the inequality $3 < 5$ by a *negative* number, say -2:

$$-2(3) < -2(5)$$
$$-6 < -10 \quad \text{False.}$$

This time the new inequality is *false*! To make the new inequality a true statement, we must *reverse the direction* of the inequality:

$$3 < 5$$
$$-2(3) > -2(5)$$
$$-6 > -10$$

The result is illustrated in Figure 2.20. Note that the order of the two numbers is reversed when we multipy by a negative number. The same is true if we divide by a negative number.

Figure 2.20

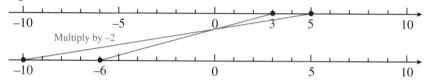

We state this result as a third property of inequalities.

---

3. If we multiply or divide both sides of an inequality by the same *negative* quantity, we must reverse the direction of the inequality.

In symbols:

$$\text{If} \quad a < b \quad \text{and} \quad c < 0, \quad \text{then} \quad ac > bc \quad \text{and} \quad \frac{a}{c} > \frac{b}{c}$$

In the lesson, we'll see how these three properties can help us solve inequalities that involve variables.

Now try Exercise 1.

Lesson

Solving Inequalities Graphically

In Section 2.1, you learned that an inequality may have infinitely many solutions. In this section we investigate some techniques for finding these solutions.

We can use graphs or tables of values to solve inequalities. Consider the inequality

$$2x + 1 < 7$$

To solve this inequality, we must find all values of x for which $2x + 1$ is less than 7. One way to do this is to evaluate $2x + 1$ for different values of x and see which values give the desired outputs. The outcome of such a search is recorded in the following table:

x	-3	-2	-1	0	1	2	3	4
$2x + 1$	-5	-3	-1	1	3	5	7	9

EXERCISE 1
Fill in the correct symbol, $>$ or $<$, in each statement.

a. If $x > 8$, then $x - 7$ ____ 1.

b. If $x < -4$, then $3x$ ____ -12.

c. If $x > -2$, then $-9x$ ____ 18.

Figure 2.21

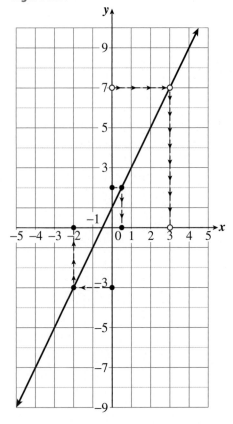

From the table, it appears that values of x less than 3 produce outputs less than 7. However, we have not tried *every* possible x-value; in particular, we cannot be sure from the table alone what happens for noninteger values of x. Another way to solve an inequality is to look at a graph.

Figure 2.21 shows the graph of $y = 2x + 1$. We are interested in outputs, or y-values, less than 7. These values are marked with a heavy line on the y-axis. Let's see which x-values produce these outputs. We choose two different outputs in the target region, say, and -3 and 2. We read the graph "backwards," as shown by the arrows, to find the corresponding x-values. We see that an x-value of -2 produces the output -3, and an x-value of $\frac{1}{2}$ produces the output 2. If we tried more outputs less than 7, we would find that all their corresponding x-values are less than 3. In other words, the set of all x-values that satisfy

$$2x + 1 < 7$$

is

$$x < 3$$

This is the solution we are looking for. It includes *all* numbers less than 3, not just the integers. We can indicate the solution set graphically by drawing a red line on the x-axis, as shown in Figure 2.21.

Two other inequality symbols occur in applications:

$\geq$	means	"greater than or equal to"
$\leq$	means	"less than or equal to"

For example, the graph of all solutions of the inequality

$$x \geq -2$$

is shown in Figure 2.22.

Figure 2.22

We use a solid dot at -2 to show that $x = -2$ is included among the solutions.

Figure 2.23

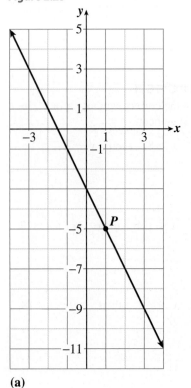

(a)

EXAMPLE 1

The graph of $y = -2x - 3$ is shown in Figure 2.23(a). Use the graph to solve the following:

a. $-2x - 3 = -5$ **b.** $-2x - 3 \geq -5$

Solution

a. Locate the point in Figure 2.23(a) whose y-coordinate is -5; this point is labeled P. The x-coordinate of point P is the x-value that satisfies the equation $-2 - 3 = -5$. From the graph, we see that the x-coordinate of point P is 1. You can verify algebraically that $x = 1$ is the solution of the equation $-2x - 3 = -5$.

b. The solutions of the inequality $-2x - 3 \geq -5$ are given by points on the graph whose y-coordinates are greater than or equal to -5. These points are shown by the red portion of the graph in Figure 2.23(b). We can find the x-coordinates of several of these points by following the arrows in the figure. In fact, all of the indicated points have x-coordinates that are less than or equal to 1.

In Example 1, we had to use inputs *less* than 1 to obtain outputs *greater* than -5. This is because the graph in Figure 2.23 is decreasing as we move from left to right. (Can you think of a way you can tell from the equation $y = -2x - 3$ that its graph will be decreasing?)

Figure 2.23

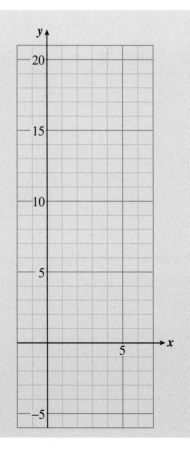

(b)

EXERCISE 2

Solve the following inequality using a graph:

$$15 - 3x < 9$$

Follow these steps:

Step 1 Begin by filling in the table of values for the expression $15 - 3x$.

Step 2 Next, graph the equation $y = 15 - 3x$. (Use the values from the table.)

Step 3 Mark on the y-axis all the points that have y-coordinates less than 9.

Step 4 Now mark the points *on the graph* that have y-coordinates less than 9. (These are also the points for which $15 - 3x < 9$, because $y = 15 - 3x$ for all points on the graph.)

Step 5 Finally, find the x-coordinates of all the points you marked on the graph. To do this, drop straight down from each point on the graph to the x-axis.

Mark the x-axis with these values. These are the solutions of the inequality.

x	$15 - 3x$
-2	
-1	
0	
1	
2	
3	
4	
5	
6	
7	

Solving Inequalities Algebraically

We can also solve inequalities algebraically. The rules for solving an inequality are very similar to the rules for solving an equation, with one important difference. In the reading assignment, we developed the following rules for solving inequalities.

Solving Inequalities

1. We can add or subtract the same quantity on both sides of an inequality.
2. We can multiply or divide both sides by the same positive number.
3. If we multiply or divide both sides by a *negative* number we must *reverse* the direction of the inequality.

EXERCISE 3

Use one of the three rules stated above to solve each inequality. The first one is completed for you.

a. $x - 8 < 3$

$$\underline{\quad +8 \quad +8\quad}$$ What should you do to isolate x? (Add 8 to both sides.)

$x \qquad < \quad 11$ Should you reverse the direction of the inequality? (No.)

Decide whether your answer is reasonable:

Is 10 a solution? $10 - 8 < 3$ is true, so 10 is a solution.
Is 12 a solution? $12 - 8 < 3$ is false, so 12 is not a solution.

b. $\dfrac{x}{4} \geq -2$ What should you do to isolate x?
Should you reverse the direction of the inequality?

Decide whether your answer is reasonable:

Is -12 a solution?

Is -5 a solution?

c. $-5x > 20$ What should you do to isolate x?
Should you reverse the direction of the inequality?

Decide whether your answer is reasonable:

Is -10 a solution?

Is 2 a solution?

Example 2 illustrates solving an inequality with several steps.

EXAMPLE 2

Solve $-3x + 1 > 7$,   and graph the solutions on a number line.

Solution

Just as we do when solving an equation, we must isolate x on one side of the inequality.

$$-3x + 1 - 1 > 7 - 1 \quad \text{\small Subtract 1 from both sides.}$$
$$-3x > 6$$

$$\frac{-3x}{-3} < \frac{6}{-3} \quad \text{\small Divide both sides by } -3; \text{ reverse}$$
$$\text{\small the direction of the inequality.}$$
$$x < -2$$

The solutions are shown in Figure 2.24.

Figure 2.24

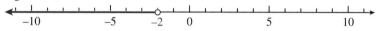

CAUTION

In Example 2, we divided both sides of $-3x > 6$ by -3. We must remember to reverse the direction of the inequality when we divide (or multiply) by a negative number. An answer of $x > -2$ would be incorrect.

Compound Inequalities

An inequality in which the value of the variable expression is bounded from above and from below is called a *compound inequality*. Here is an example of a compound inequality:

$$-3 < 2x - 5 \le 6$$

To solve a compound inequality, we must perform on all three "sides" of the inequality the steps needed to isolate x.

EXAMPLE 3

Solve $-3 < 2x - 5 \le 6$.

Solution

To solve for x, we first add 5 on each side of the inequality symbols.

$$\begin{array}{rcccr} -3 & < & 2x - 5 & \le & 6 \\ +5 & & +5 & & +5 \\ \hline 2 & < & 2x & \le & 11 \end{array}$$

Next, to solve $2 < 2x \le 11$,   we divide each side by 2.

$$\frac{2}{2} < \frac{2x}{2} \le \frac{11}{2}$$
$$1 < x \le \frac{11}{2}$$

The solution consists of all numbers greater than 1 but less than or equal to $\frac{11}{2}$. The graph of the solutions is shown in Figure 2.25.

Figure 2.25

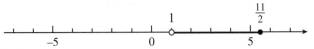

Now try Exercise 4.

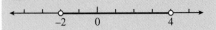
EXERCISE 4
a. Solve $-8 < 4 - 3x < 10$.

b. Graph the solutions on the number line.

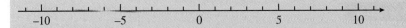

HOMEWORK 2.6

In Problems 1–8:
a. Solve each inequality algebraically.
b. Graph your solutions on the number line.
c. Give at least one value of the variable that is a solution and one value that is not a solution.

1. $x + 10 \leq -5$

2. $x - 4 \geq -13$

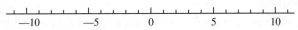

3. $-3y < 15$

4. $5y > -20$

5. $\dfrac{x}{3} \le 4$

6. $\dfrac{x}{-4} > 8$

7. $-8t \ge -60$

8. $-6t \le -76$

In Problems 9–14:
a. **Use the graph to solve each inequality.**
b. **Indicate your solutions along the *x*-axis of the graph.**
c. **Solve the inequality algebraically.**

9. $-2x \ge 8$

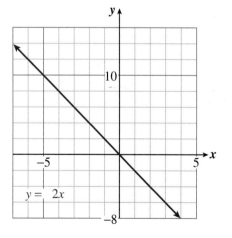

10. $3x > -6$

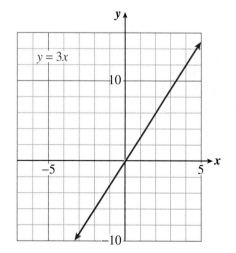

11. $3x + 5 \leq -4$

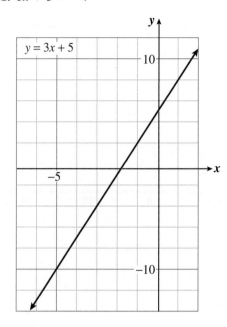

12. $10 > -4x - 6$

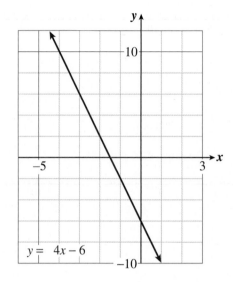

13. $7 - \dfrac{2x}{3} > 1$

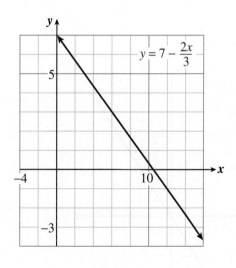

14. $\dfrac{2x - 7}{-5} \leq 3$

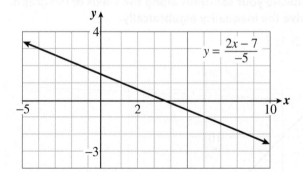

In Problems 15–20, solve each inequality algebraically, and graph the solutions on the number lines.

15. $2x + 3 > 7$

16. $3x - 4 > 11$

17. $-3x + 2 \leq 11$

18. $-4x - 3 \geq -11$

19. $-3 > \dfrac{2x}{3} + 1$

20. $4 < \dfrac{-3x}{4} - 2$

In Problems 21–24, write and graph an equation relating the variables. Then answer each question by solving an equation or inequality. Finally, use your graph to verify your answers.

21. Today, the high temperature was 56°F. If the high temperature is decreasing by 4°F per day, write an equation for the high temperature, *T*, after *d* days. Graph your equation.

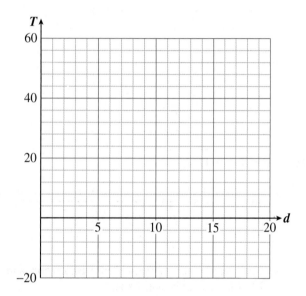

  a. When will the high temperature reach freezing (32°F)?

  b. On which days will the high temperature be below −12°F?

22. Yusuf is now $750 in debt, but he deposits $50 a week into his savings account. Write an equation for Yusuf's net worth, *N*, after *w* weeks. Graph your equation.

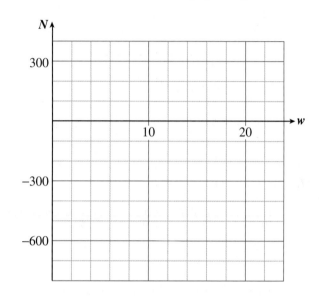

  a. When will Yusuf's net worth be −$100?

  b. When will Yusuf's net worth be over $200?

23. Francine has been scuba diving at a depth of 200 feet, and now she is beginning to ascend at a rate of 15 feet per minute. Write an equation for Francine's elevation, *h*, after *m* minutes. Graph your equation.

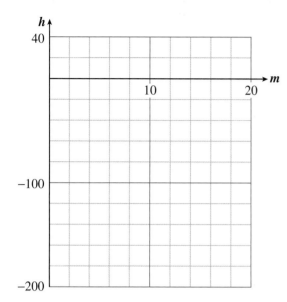

a. When will Francine's elevation be less than −20 feet?

b. When will Francine reach the surface?

24. Briarwood School has an annual budget of $300,000, and it must spend $1500 per year on each student. Write an equation for the amount in Briarwood's account, *B*, if the school enrolls *n* students. Graph your equation.

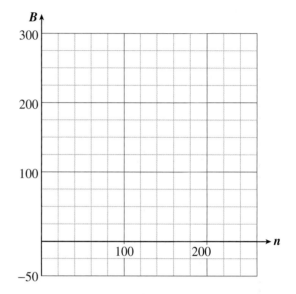

a. How many students are enrolled if Briarwood is in debt (its account is overdrawn)?

b. How many students are enrolled if Briarwood's account balance is exactly −$18,000?

Solve each compound inequality in Problems 25–30, and graph the solutions on the number lines.

25. $-3 \leq 3x \leq 12$

26. $-16 < 4x < 8$

27. $23 > 9 - 2b \geq 13$

28. $-6 > 4 - 5b > -21$

29. $-8 \leq \dfrac{5w + 3}{4} < -3$

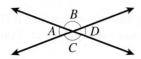

30. $2 < \dfrac{6w - 2}{5} \leq 8$

Use the following facts about angles to answer Problems 31–40.

(1) When two lines intersect, they form two pairs of equal angles called **vertical angles**. For example, angles A and D are one pair of vertical angles, and angles B and C are another pair of vertical angles. *Vertical angles are equal.*

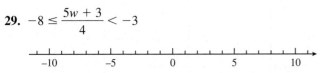

(2) When two lines ℓ_1 and ℓ_2 are both intersected (or "cut") by a third line ℓ_3, four pairs of angles called **corresponding angles** are formed. Angles A and A', B and B', C and C', D and D' are the four pairs of corresponding angles. *If ℓ_1 and ℓ_2 are parallel lines, the corresponding angles are equal.*

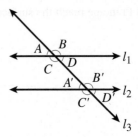

Use the figure below for Problems 31–40. The lines L_1 and L_2 are parallel.

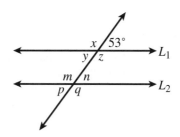

31. Which angle is a vertical angle for z?

32. Which angle is a corresponding angle for z?

33. Which angle(s) are supplementary to z? (Supplementary angles are reviewed in the Homework Problems for Section 1.6.)

In Problems 34–40, find the measure of each angle.

34. y **35.** n **36.** p **37.** z

38. x **39.** m **40.** q

2.7 Intercepts of a Line

Reading

Around 1950, people began cutting down the world's rain forests to clear land for agriculture. In 1970, there were about 9.8 million square kilometers of rain forest left, and by 1990 that figure had been reduced to 8.2 million square kilometers. These two data points are shown in Figure 2.26. On this graph, the horizontal axis displays the number of years since 1950, x, and the vertical axis shows the amount of rain forest remaining, y (in millions of square kilometers). If people continue to cut the rain forest at the same rate, how long will it be before we have completely cleared all the remaining forest?

Figure 2.26   Amount of remaining rain forest

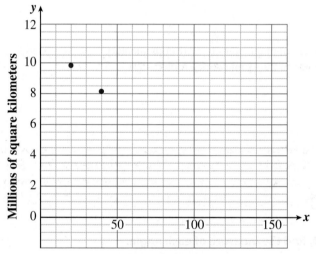

Years after 1950

If we continue to clear the rain forests at the same rate, the graph showing the number of square kilometers left in each year will be a straight line. Draw a straight line through the two points in Figure 2.26. Make sure your line is long enough to cross both axes. The point where the line crosses the x-axis is called the **x-intercept** of the graph. This point has y-coordinate 0, meaning that there is no rain forest left. The x-coordinate of the point gives the number of years after 1950 when this will occur.

If you follow the line back to the vertical axis, you can see how many square kilometers of rain forest were present initially—that is, in 1950, before serious clearing began. The point where the line crosses the y-axis is called the **y-intercept** of the graph. The y-coordinate of this point gives the amount of rain forest present originally, in millions of square kilometers.

Now try Exercise 1.

Skills Review

1. What is the change in a diver's elevation when he ascends from -82 feet to -57 feet?

2. On his third day on the quiz show *That's the Breaks*, Bob's total running score changed from 1200 to -400. What was Bob's score for the third day?

EXERCISE 1

a. How many square kilometers of rain forest were present in 1950?

b. If we continue to clear the rain forest at the same rate, when will it be completely demolished?

3. Nina lost 2 points on her algebra test every time she forgot a minus sign. If she did this on six problems, how did she affect her final score?

4. If stock prices dropped an average of 2 points per day for 14 days, what was the total change in the average price of stocks?

5. Ceci changed her bank balance from −$38 to $184 by depositing her paycheck. How much was her paycheck?

6. In a cryogenics experiment, the temperature must be decreased from 15°F to −280°F in 4 minutes. What is the temperature change per minute?

Answers: *1.* 25 feet   *2.* −1600   *3.* She lost 12 points.   *4.* −28 points   *5.* $222   *6.* −73.75°F per minute

Lesson

Finding the Intercepts of a Graph

It is easy to recognize the intercepts of a line on a good graph. Figure 2.27 shows the graph of the line

$$y = \frac{1}{2}x + 4$$

We can see that its x-intercept is $(-8, 0)$ and its y-intercept is $(0, 4)$. If we have an equation for the line, we can also find the intercepts of its graph algebraically.

Note that because the y-intercept of a graph lies on the y-axis, its x-coordinate is always 0. Thus, to find the y-intercept, we can substitute 0 for x in the equation and solve for y. For the equation graphed in Figure 2.27,

$$y = \frac{1}{2}(0) + 4 = 4$$

Thus the y-intercept is $(0, 4)$, as expected.

Similarly, because the x-intercept lies on the x-axis, its y-coordinate is always zero. To find the x-intercept, we can substitute 0 for y and solve the equation for x. For our example,

$$0 = \frac{1}{2}x + 4 \qquad \text{Subtract 4 from both sides.}$$

$$-4 = \frac{1}{2}x \qquad \text{Multiply both sides by 2.}$$

$$-8 = x$$

Thus the x-intercept is $(-8, 0)$, as expected.

We summarize the methods just described as follows:

To Find the x-intercept of a Graph:

Substitute 0 for y in the equation and solve for x.

To Find the y-intercept of a Graph:

Substitute 0 for x in the equation and solve for y.

Now try Exercise 2.

Figure 2.27

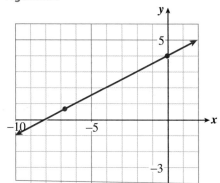

EXERCISE 2
Find the x- and y-intercepts of the graph of $3x - 2y = 12$.

The Intercept Method of Graphing

Once we know the *x*- and *y*-intercepts, we can use them to draw the graph of a linear equation. Instead of choosing several different values of *x* to find points on the graph, we find the two intercepts. We set up a table that looks like this:

x	y
0	
	0

By finding the missing values for *x* and *y*, we are finding the intercepts of the graph. For the equation in Exercise 2, $3x - 2y = 12$, the completed table looks like this:

x	y
0	−6
4	0

Figure 2.28

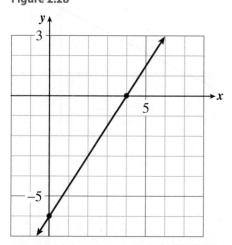

We plot the two points $(0, -6)$ and $(4, 0)$ and draw a line through them to obtain the graph shown in Figure 2.28.

It is still a good idea to find a third point as a check. If we choose $x = 2$, we find

$$3(2) - 2y = 12 \qquad \text{Simplify the left side.}$$
$$6 - 2y = 12 \qquad \text{Subtract 6.}$$
$$-2y = 6 \qquad \text{Divide by } -2.$$
$$y = -3$$

Thus $(2, -3)$ is another point on the graph. You should check to see that this point does lie on the graph in Figure 2.28.

The method described above is a second method for graphing lines, called the **intercept method**. (At this stage, you might want to review the point-plotting method you learned in Section 2.4.)

To Graph a Linear Equation Using the Intercept Method:

1. Use a table to record the *x*- and *y*-intercepts of the graph.
2. Plot the two intercepts and draw the line through them.
3. Find a third point on the graph as a check. (Choose any convenient value for *x* and solve the equation for *y*.)

x	y
0	
	0

EXAMPLE 1

Use the intercept method to graph the equation $3x + 2y = 7$.

Solution

First, we find the *x*- and *y*-intercepts of the graph. To find the *y*-intercept, we substitute 0 for *x* and solve for *y*. To find the *x*-intercept, we substitute 0 for *y* and solve for *x*.

$$3(0) + 2y = 7$$
$$2y = 7$$
$$y = \frac{7}{2} = 3\frac{1}{2}$$

x	*y*
0	$3\frac{1}{2}$
$2\frac{1}{3}$	0

$$3x + 2(0) = 7$$
$$3x = 7$$
$$x = \frac{7}{3} = 2\frac{1}{3}$$

The *y*-intercept is the point $\left(0, 3\frac{1}{2}\right)$, and the *x*-intercept is the point $\left(2\frac{1}{3}, 0\right)$. We plot these points and connect them with a straight line to obtain the graph shown in Figure 2.29. Finally, we find a third point as a check. We choose   *x* = 1   and solve for *y*.

$$3(1) + 2y = 7 \quad \text{Subtract 3 from both sides.}$$
$$2y = 4 \quad \text{Divide by 2.}$$
$$y = 2$$

You should check that the point (1, 2) lies on the graph in Figure 2.29.

Now try Exercise 3.

Interpreting the Intercepts

As you saw in the reading assignment, the intercepts of a graph can often give valuable information about a problem. The intercepts are significant because they often represent starting or ending values for a particular variable. For instance, Example 2 of Section 2.4 considered the temperature in Nome, Alaska on a particular day.

EXAMPLE 2

The temperature in Nome was −12°F at noon and has been rising at a rate of 2°F per hour all day. A graph of the temperature is shown in Figure 2.30. Find the intercepts of this graph, and interpret their meaning in the context of the problem situation.

Solution

In Section 2.4, we wrote an equation for the temperature, *T*, at time *h*, where *h* represented the number of hours after noon. The equation was

$$T = -12 + 2h$$

To find the *T*-intercept, we set   *h* = 0   and solve for *T*.

$$T = -12 + 2(0) = -12$$

The *T*-intercept is (0, −12). This point tells us that when   *h* = 0,   *T* = −12. In other words, the temperature at noon was −12°F. To find the *h*-intercept, we set *T* = 0   and solve for *h*.

$$0 = -12 + 2h \quad \text{Add 12 to both sides.}$$
$$12 = 2h \quad\quad\quad \text{Divide both sides by 2.}$$
$$6 = h$$

Figure 2.29

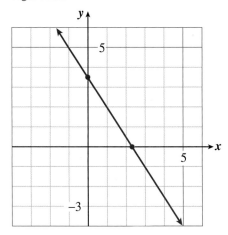

EXERCISE 3

Graph the equation

$$2x = 5y - 10$$

by the intercept method.

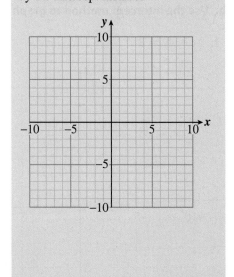

Figure 2.30

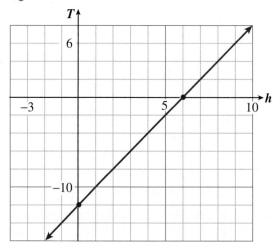

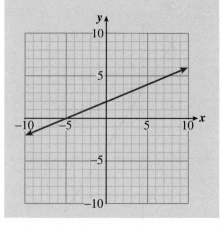

ANSWERS TO 2.7 EXERCISES

1a. Approximately 11.5 million square kilometers

1b. In about $x = 140$, or 2090

2. $(4, 0)$ and $(0, -6)$

3. The intercepts are $(-5, 0)$ and $(0, 2)$.

The h-intercept is the point $(6, 0)$. This tells us that when $h = 6$, $T = 0$, or that the temperature will reach 0°F 6 hours after noon, at 6 P.M.

HOMEWORK 2.7

In Problems 1–12:
a. Find the *x*- and *y*-intercepts of each line.
b. Use the intercept method to graph the line.

1. $2x + 4y = 8$

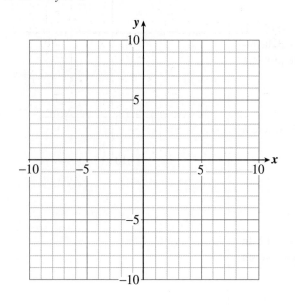

2. $2x - 3y = 6$

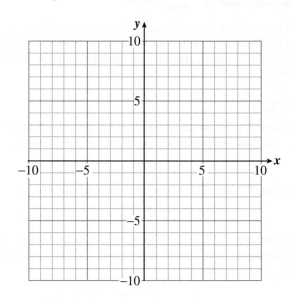

3. $x + 2y + 10 = 0$

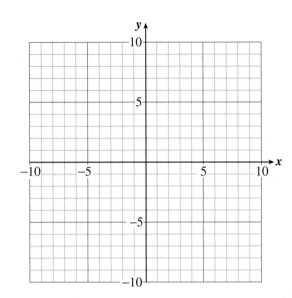

4. $5y + 2x = -15$

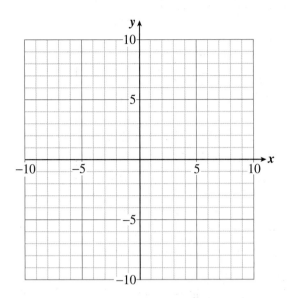

5. $2x = 14 + 7y$

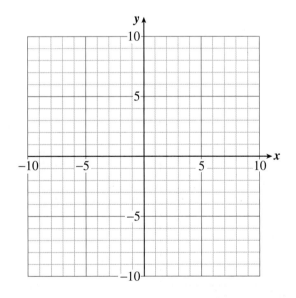

6. $3y - 9 = 2x$

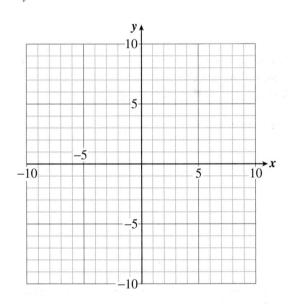

7. $y = -4x + 8$

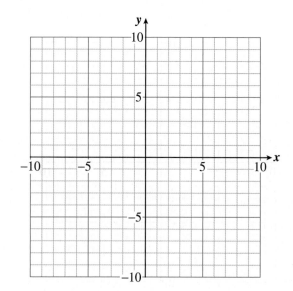

8. $y = -2x - 5$

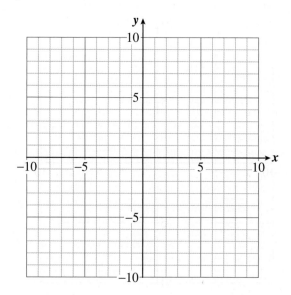

9. $\dfrac{x}{2} + \dfrac{y}{3} = 1$

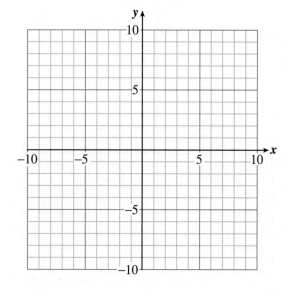

10. $\dfrac{x}{4} - \dfrac{y}{6} = 1$

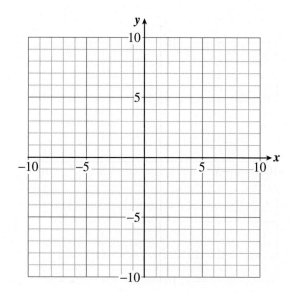

11. $3x - 2y = 120$

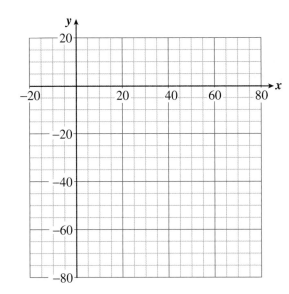

12. $4x + 5y = -400$

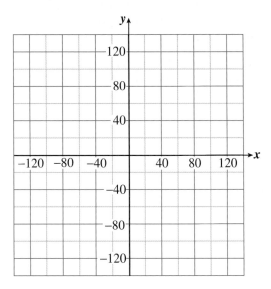

13. During spring break, Francine took the train to San Francisco and then bicycled home. The graph shows Francine's distance d from home after she had biked for h hours.

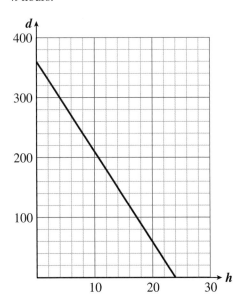

a. How far is it from San Francisco to Francine's home?

b. How many hours did it take Francine to get home?

14. Robin opened a yogurt smoothie shop near campus. The graph shows Robin's profit P after selling s smoothies.

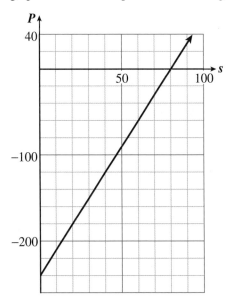

a. How much money did Robin invest to start her shop?

b. How many smoothies did Robin have to sell to break even?

In Problems 15–20:
a. Find the intercepts of each linear equation.
b. Use the intercept method to graph the line.
c. Explain what the intercepts mean in terms of the problem situation.

15. The amount of home heating oil (in gallons) in the Olsons' tank is given by $G = 200 - 15w$,   where w is the number of weeks since they turned on the furnace.

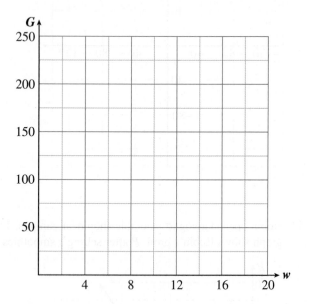

16. Serda's score on her driving test is computed by the equation $S = 120 - 4n$,   where n is the number of wrong answers she gives.

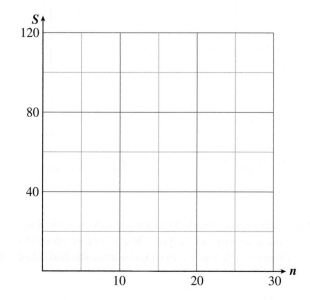

17. Dana joined a savings plan some weeks ago. Her account balance is growing each week according to the formula $B = 225 + 25w$, where $w = 0$ represents this week.

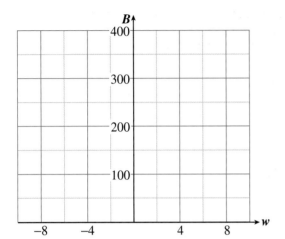

18. Greg is monitoring the growth of a new variety of string beans. The height of the vine each day is given in inches by $h = 18 + 3d$, where $d = 0$ represents today.

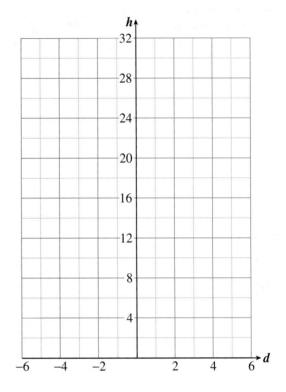

19. Delbert bought some equipment and went into the dog-grooming business. He finds that his profit is increasing according to the equation $P = -600 + 40d$, where d is the number of dogs he has groomed.

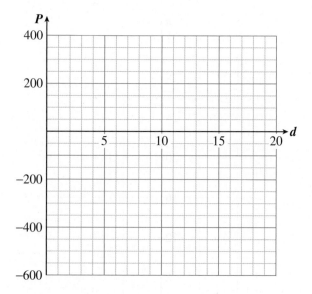

20. Cliff's score was negative at the end of the first round of *College Quiz*, but in *Double Quiz* his score improved according to the equation $S = -400 + 20q$, where q is the number of questions he answered correctly.

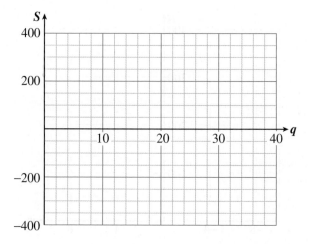

Match each equation in Problems 21–26 with one of the graphs. (More than one equation may describe the same graph.)

21. $2x + 3y = 12$

22. $2x - 3y = 12$

23. $3x - 2y = 12$

24. $-3y - 2x = 12$

25. $\dfrac{x}{6} - \dfrac{y}{4} = 1$

26. $\dfrac{x}{4} - \dfrac{y}{6} = 1$

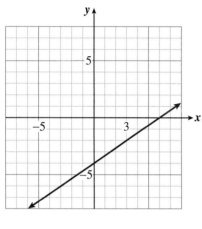

(a)

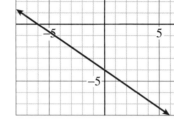

(b)

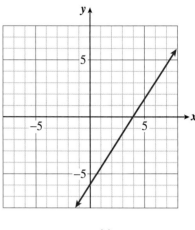

(c)

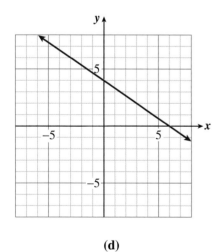

(d)

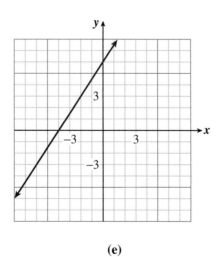

(e)

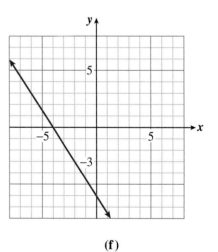

(f)

27. a. At what point does the graph of the equation
   $2x - 3y = 25$ cross the x-axis?

   b. At what point does the graph of the equation
   $1.4x + 3.6y = -18$ cross the y-axis?

28. Explain why the x-coordinate of the y-intercept is 0 and the y-coordinate of the x-intercept is 0.

29. The x-intercept of a line is positive, and its y-intercept is negative. Is the line increasing or decreasing? Sketch a possible example of such a line.

30. The x-intercept of a line is positive, and its y-intercept is also positive. Is the line increasing or decreasing? Sketch a possible example of such a line.

Solve each pair of equations in Problems 31–36. In part (b) of each problem, your answer will involve the constant k.

31. a. $-2x = 6$ **b.** $-2x = k$

32. a. $3x - 8 = 7$ **b.** $3x - 8 = k$

33. a. $15 - 4x = 3$ **b.** $15 - 4x = k$

34. a. $6x + 5 = 2$ **b.** $6x + k = 2$

35. a. $9 + 3x = -1$ **b.** $9 + kx = -1$

36. a. $2 - 5x = -7$ **b.** $2 - kx = -7$

Problems 37–42 focus on properties of angles. You may wish to review the concepts of complementary, supple-mentary, vertical, and corresponding angles. (See pages 76 and 208.) For Problems 37 and 38, find the measure of each angle shown in the figure.

37.

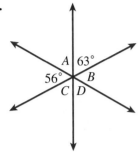

a. A
b. B
c. C
d. D

38.

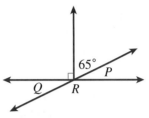

a. P
b. Q
c. R

39. Find the measure of each angle. The lines labeled ℓ_1 and ℓ_2 are parallel.

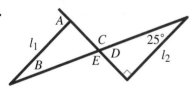

a. v
b. w
c. x
d. y
e. z

In Problems 40–42, find the measure of the indicated angles. The lines labeled l_1 and l_2 are parallel.

40.

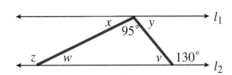

a. B
b. D
c. A
d. C
e. E

41.

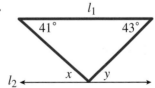

a. x
b. y

42.

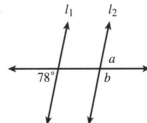

a. a
b. b

2.8 | Like Terms

Reading

In this section, we explore algebraic techniques that will enable you to solve more complicated applied problems. For example, consider the equation

$$2x + 7 = 4x - 3$$

In this equation, *two* terms contain the variable. In order to solve equations like this one, you must be able to add and subtract expressions involving variables.

Equivalent Expressions

When we solve an equation such as $2 + 3x = 17$, we find a value for the variable that makes the equation true. The solution to this equation is 5, so the expressions $2 + 3x$ and 17 are equal if $x = 5$. If we use any other value for x (say, $x = 9$ or $x = -1$), then $2 + 3x$ does *not* equal 17.

It is possible for two algebraic expressions to be equal no matter what value is used for the variable. For instance, the expressions $2x + 3x$ and $5x$ will always give the same value when the same number is substituted for x in each expression. For example,

$$\text{If } x = 3, \quad \text{then} \quad 2x + 3x = 2(3) + 3(3)$$
$$= 6 + 9 = 15$$
$$\text{and} \qquad 5x = 5(3) = 15$$

Both expressions equal 15 when $x = 3$. Verify that $2x + 3x$ and $5x$ are equal when $x = -4$.

$$2x + 3x = \underline{\qquad}$$
$$5x = \underline{\qquad}$$

We say that two algebraic expressions are **equivalent** if they give the same number for *all* values of the variable. Thus $2x + 3x$ and $5x$ are equivalent expressions, and we can replace one by the other if doing so will simplify a problem.

It is important to realize that two expressions that give the same number for *some* values of the variable are not necessarily equivalent.

Try Exercises 1 and 2.

EXERCISE 1

a. Show that the expressions $6 + 2x$ and $8x$ are equal if $x = 1$.

$$6 + 2x = \underline{\qquad}$$
$$8x = \underline{\qquad}$$

b. Show that the expressions $6 + 2x$ and $8x$ are *not* equal if $x = 2$.

$$6 + 2x = \underline{\qquad}$$
$$8x = \underline{\qquad}$$

c. Are the expressions $6 + 2x$ and $8x$ equivalent?

EXERCISE 2

By choosing a value for x, show that the expressions $2 + 3x$ and $5x$ are not equivalent. (You can use $x = 6$ if you like, or some other value.)

Skills Review

Simplify each expression by following the order of operations. (See Sections 1.8 and 2.3.)

1. $7(-3) - 4(-5)$
2. $12 - 3(-2) + 9$
3. $-2(9) + (-5)(-4) - 8(3)$
4. $11 - 6[8 - 5(3)]$
5. $2(-3) - 5(-3) - 3(-3)$
6. $5(-2)(-3) - 2[-4(3) - 1]$
7. $\dfrac{-18}{3 - 6} + \dfrac{2 - 9}{-4 - 3}$
8. $\dfrac{8(-9 + 4)}{10} - \dfrac{2 - 4(-5)}{8 - (-3)}$

Answers: *1.* -1 *2.* 27 *3.* -22 *4.* 53 *5.* 18 *6.* 56 *7.* 7 *8.* -6

Lesson

In the reading assignment, you saw that $2x + 3x$ and $5x$ are equivalent expressions. This should make sense if you consider what these expressions mean: $2x$ means two x's added together, and $3x$ means three x's added together. Then,

$$2x + 3x = (x + x) + (x + x + x) = 5x$$

In other words, we can add two x's to three x's and get five x's, just as we can add two dollars to three dollars and get five dollars. Whenever we have two groups of the same quantity, we can combine them into one larger group by adding the number of items in each group. In algebra, this is called *combining like terms*.

Like terms are any terms that are *exactly alike* in their variable factors.

EXAMPLE 1

a. $2x$ and $3x$ are like terms
         $-4a$ and $7a$ are like terms

because their variable factors are identical.

b. $2x$ and $3y$ are not like terms
         $2x$ and 3 are not like terms

because their variable factors are different.

The numerical factor in a term is called the **numerical coefficient**, or just the **coefficient**. In the expression $3xy$, for example, the number 3 is the numerical coefficient. In a term such as xy or b, the numerical coefficient is 1. If a variable is preceded by a negative sign, the coefficient of the term is -1. In other words,

$$-x \quad \text{means} \quad -1 \cdot x$$
$$-a \quad \text{means} \quad -1 \cdot a$$

and so on.

Now try Exercise 3.

Adding and Subtracting Like Terms

From the preceding discussion, we can state the following rule.

To Add Like Terms:

Add the numerical coefficients of the terms. Do not change the variable factors of the terms.

EXAMPLE 2

$$5n + 3n = (5 + 3)n = 8n$$

The sum in Example 2 still has the variable factor n. We do not change the things we are adding, namely, n's; only the *number* of n's changes.

Now try Exercise 4.

EXERCISE 3
Explain the following phrases, and give an example of each.
a. equivalent expressions

b. like terms

c. numerical coefficient

EXERCISE 4
a. $-4y - 3y$
  $= (-4 - 3)y = $ _____

b. $-6st + 9st$
  $= ($ _____ $)st = $ _____

EXERCISE 5

Delbert and Francine collect aluminum cans to recycle. They are paid x dollars for every pound of cans they collect. At the end of 3 weeks, Delbert has collected 23 pounds of aluminum cans, and Francine has collected 47 pounds.

a. Write algebraic expressions for the amount of money Delbert makes and the amount Francine makes.

b. Write and simplify an expression for the total amount of money Delbert and Francine make.

EXERCISE 6

Combine like terms by adding or subtracting.

a. $-2x + 6x - x$

b. $3bc - (-4bc) - 8bc$

In Exercise 4, we replaced the given expression by a simpler but equivalent expression. In other words, $-7y$ is equivalent to $-4y - 3y$, and $3st$ is equivalent to $-6st + 9st$. Replacing an expression by a simpler equivalent one is known as *simplifying* the expression.

Now try Exercise 5.

We subtract like terms in the same way we add like terms.

To Subtract Like Terms:

Subtract the numerical coefficients of the terms. Do not change the variable factors of the terms.

EXAMPLE 3

$$6a - (-8a) = [6 - (-8)]a = 14a$$

Now try Exercise 6.

We cannot add or subtract *unlike* terms. Thus,

$$2x + 3y \qquad \text{cannot be simplified}$$
$$-6x + 4xy \qquad \text{cannot be simplified}$$
$$6 - 2x \qquad \text{cannot be simplified}$$

Many expressions contain both like and unlike terms. In such expressions, we can combine only the like terms.

EXAMPLE 4

Simplify $-4a + 2 - 5 + 8a + 5a$.

Solution

We combine all the a terms and all the constant terms:

$$-4a + 8a + 5a = 9a \qquad \text{and} \qquad 2 - 5 = -3$$

so

$$-4a + 2 - 5 + 8a + 5a = 9a - 3$$

CAUTION

In Example 4, $9a - 3 \neq 6a$,   because $9a$ and -3 are not like terms. $9a - 3$ cannot be simplified.

Now try Exercises 7 and 8.

Removing Parentheses

Recall that when adding signed numbers, we can remove parentheses that follow an addition symbol. For instance,

$$5 + (-3) = 5 - 3$$

Similarly,

$$5x + (-3x) = 5x - 3x$$

which simplifies to $2x$. This idea extends to expressions in which two or more terms are grouped by parentheses.

> Parentheses that are preceded by a plus sign may be omitted; each term within parentheses keeps its original sign.

EXAMPLE 5

$$(2x + 3) + (5x - 7) = 2x + 3 + 5x - 7$$
$$= 7x - 4$$

Now try Exercise 9.

We must be careful when removing parentheses preceded by a minus sign. In the expression

$$(5x - 2) - (7x - 4)$$

the minus sign preceding $(7x - 4)$ applies to *each term* inside the parentheses. Therefore, when we remove the parentheses, we must change the sign of $7x$ from + to − and change the sign of -4 from − to +. Thus

$$(5x - 2) - (7x - 4) = 5x - 2 - 7x + 4 \quad \text{Each sign is changed.}$$
$$= 5x - 7x - 2 + 4$$
$$= -2x + 2$$

> If parentheses are preceded by a minus sign, change the sign of each term within parentheses and then omit the parentheses.

EXAMPLE 6

Simplify $-5b - (3 - 2b) + (4b - 6)$.

Solution

Before combining like terms, we must remove the parentheses. We change the signs of terms inside parentheses preceded by a minus sign; we do *not* change the signs of terms inside parentheses preceded by a plus sign.

$$-5b - (3 - 2b) + (4b - 6) = -5b - 3 + 2b + 4b - 6$$

signs changed
signs not changed

$$= -5b + 2b + 4b - 3 - 6$$
$$= b - 9$$

EXERCISE 7
Simplify $-5u - 6uv + 8uv + 9u$.

EXERCISE 8
One angle of a triangle is three times the smallest angle, and the third angle is 20° greater than the smallest angle. If the smallest angle is x, write an expression for the sum of the three angles in terms of x, and simplify.

EXERCISE 9
Simplify $(32h - 26) + (-3 + 2h)$.

EXERCISE 10

Simplify

$(3a - 3) - 2a - (5 - 2a)$.

EXERCISE 11

The theater group StageLights plans to sell T-shirts to raise money. It will cost the group $5x + 60$ dollars to print x T-shirts, and they will sell the T-shirts at $12 each. Write an expression for the group's profit from selling x T-shirts, and simplify.

CAUTION

In Example 6, the following step would be *incorrect*:

$$-5b - (3 - 2b) + (4b - 6) \rightarrow -5b - 3 - 2b + 4b - 6$$

(Do you see why?)

Now try Exercises 10 and 11.

Solving Equations

We can now solve equations in which two or more terms contain the variable.

EXAMPLE 7

Solve the equation $2x + 7 = 4x - 3$.

Solution

We first get all terms containing the variable on one side of the equation. We subtract $2x$ from both sides of the equation:

$$
\begin{array}{rcl}
2x + 7 = & 4x - 3 \\
-2x & -2x \\
\hline
7 = & 2x - 3
\end{array}
$$

Now the equation looks like those we already know how to solve, and we proceed as usual to isolate the variable.

$$
\begin{array}{rl}
7 = 2x - 3 & \\
+3 \qquad +3 & \text{Add 3 to both sides.} \\
\hline
10 = 2x & \\
\dfrac{10}{2} = \dfrac{2x}{2} & \text{Divide both sides by 2.} \\
5 = x &
\end{array}
$$

The solution is 5. You should check the solution by substituting 5 into the original equation: Does $2(5) + 7 = 4(5) - 3$?

If one side of an equation or inequality contains like terms, we should combine them before beginning to solve.

EXAMPLE 8

Solve $2x - 4x + 14 < 3 + 5x - 10$.

Solution

We begin by combining like terms on each side of the inequality:

$$2x - 4x + 14 < 3 + 5x - 10 \qquad 2x - 4x = -2x \quad \text{and} \quad 3 - 10 = -7$$
$$-2x + 14 < 5x - 7$$

Now we continue solving as usual. We want to get all the terms containing *x* on one side of the inequality and all the constant terms on the other side.

$$-2x + 14 < 5x - 7$$
$$\underline{-5x \qquad\quad -5x}$$ Subtract $5x$ from both sides.
$$-7x + 14 < \quad -7$$
$$\underline{\quad -14 \qquad -14}$$ Subtract 14 from both sides.
$$-7x \quad < \quad -21$$

$$\frac{-7x}{-7} > \frac{-21}{-7}$$ Divide both sides by -7 and reverse the direction of the inequality.
$$x > 3$$

The solution is all *x*-values greater than 3. You should check the solution by substituting one *x*-value greater than 3 and one *x*-value less than 3 into the original inequality; for instance:

$x = 4$: $\quad 2(4) - 4(4) + 14 < 3 + 5(4) - 10$? True: $6 < 13$
$x = 2$: $\quad 2(2) - 4(2) + 14 < 3 + 5(2) - 10$? False: $10 \nless 3$

Now try Exercise 12.

EXERCISE 12
The theater group StageLights of Exercise 11 would like to make $500 from the sale of T-shirts. How many T-shirts must the group sell?

ANSWERS TO 2.8 EXERCISES
1a. 8, 8 **1b.** 10, 16 **1c.** No

2. If $x = 6$, $2 + 3x = 20$ and $5x = 30$.

3. See pages 224–225. **4a.** $-7y$ **4b.** $3st$

5a. $23x, 47x$ **5b.** $23x + 47x = 70x$

6a. $3x$ **6b.** $-bc$

7. $4u + 2uv$

8. $x + 3x + (x + 20)$; $5x + 20$

9. $34h - 29$

10. $3a - 8$

11. $12x - (5x + 60)$; $7x - 60$

12. 80

HOMEWORK 2.8

In Problems 1–6, choose a value for the variable and show that the pairs of expressions are *not* equivalent.

1. $2 + 7x$ and $9x$

2. $8 - 5b$ and $3b$

3. $-(a - 3)$ and $-a - 3$

4. $-(2a + 1)$ and $-2a + 1$

5. $5(x + 3)$ and $5x + 3$

6. $2(5x)$ and $2(5)(2)(x)$

In Problems 7–14, add or subtract like terms.

7. $-6x + 2x$

8. $-8b + 8b$

9. $-7.6a - 5.2a$

10. $-6.1z + 4.3z$

11. $3t - 4t + 2t$

12. $-w - w + w$

13. $-ab + 5ab - (-3ab)$

14. $3bc - (-4bc) - 8bc$

In Problems 15–18, combine like terms.

15. $3 + 4y - (-8y) - 7$

16. $-2 - 4z + 6z - (-8)$

17. $-2st - 2 + 5s - 6st - (-4s)$

18. $3 - 5u - 6uv + 8uv + 9u$

19. a. Evaluate $-3y + 2 + 7y - 6y - 4y - 8$ for $y = 2.5$.

b. Simplify the expression in part (a).

c. Evaluate your answer to part (b) for $y = 2.5$. Check that you got the same answer for part (a).

20. Evaluate each expression for the given values of the variables.

a. $4 - 3a + 6ab - (-8a) - 10 + 9ab + 2a$ for $a = -2$, $b = 5$

b. $3c - 9 + 2c - 8cd - (-12) + 5cd - 9c$ for $c = -1$, $d = -2$

In Problems 21–26, simplify and combine like terms.

21. $4x + (5x - 2)$

22. $3y - (5y + 1)$

23. $22y - 34 - (16y - 24)$

24. $(14x - 75) - (-3x + 25)$

25. $6a - 5 - 2a - (2a - 5)$

26. $(3a - 2) - 2a - (5 - 2a)$

Write algebraic expressions to answer each question in Problems 27–34.

27. For every smoker in a restaurant, there are three non-smokers.
   a. If there are x smokers, how many nonsmokers are there?

   b. Write an expression for the total number of people in the restaurant.

28. There are eight fewer men than women in a scuba-diving class.
   a. If there are w women, how many men are there?

   b. Write an expression for the total number of people in the class.

29. A tortoise and a hare are having a race. After t seconds, the tortoise has traveled $0.2t$ feet and the hare has traveled $10t$ feet.

   a. Express the distance between them in terms of t.

   b. How far apart are the tortoise and the hare after 10 seconds?

   c. When will they be 147 feet apart?

30. Erik and Christopher left Tampa at the same time. Erik is $300t$ miles due north of Tampa, where t is in hours. Christopher is $48t$ miles due south of Tampa.

   a. Express the distance between them in terms of t.

   b. How far apart are Erik and Christopher after 2 hours?

   c. When will they be 1566 miles apart?

31. The cost (in dollars) of producing m stereos is $251m + 1355$, and each stereo sells for \$847.

   a. What profit is made from producing and selling m stereos?

   b. How many stereos must be sold to make a profit of \$16,525?

32. The cost (in dollars) of producing y videotapes is $2.67y + 5625$, and each tape sells for \$19.98.

   a. What profit is made from producing and selling y tapes?

   b. How many tapes must be sold to make a profit of \$1299?

33. Delbert bought a high-definition TV selling for x dollars. The sales tax in his state is 8%.

   a. Write an expression for the tax on Delbert's purchase.

   b. Write an expression for the total price Delbert paid, including tax.

   c. Delbert paid \$928.80 for his TV. What was the price of the TV without the tax?

34. Francine bought a mountain bike on sale. The regular price was x dollars, but Francine got the bike at 15% off.

   a. Write an expression for the discount Francine received.

   b. Write an expression for the price Francine paid.

   c. Francine paid \$782 for her bike, before tax. What was the regular price of the bike?

In Problems 35–46, solve each equation or inequality.

35. $4m - 3 = 2m + 5$

36. $5p + 12 = 2p + 4$

37. $15 - 9t = 33 - 5t$

38. $24 - 8t = 4 - 2t$

39. $-6s = 3s$

40. $4t - 2 = -4t + 2$

41. $3x + 5 > 2x + 3$

42. $16x + 8 \leq 10x - 4$

43. $-8g + 35 = 9g - 13 + g$

44. $0 = 13q + 25 - 17q + 7$

45. $-15y + 5 - 2y - 4 \geq -12y + 21$

46. $-2y + 18y - 30 \geq 2y - 6 - 5y + 14$

In Problems 47–50, write an algebraic expression for the perimeter of each figure.

47.

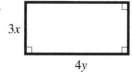

$3x$

$4y$

48.

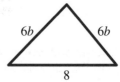

$6b$ $6b$

8

49.

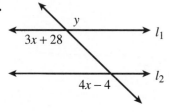

$5a$

$3a$ 3

$5a + 4$

50.

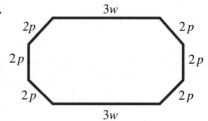

$3w$

$2p$ $2p$

$2p$ $2p$

$2p$ $2p$

$3w$

In Problems 51 and 52, the expressions in the figures give the measures of the angles in degrees. Find the value of each variable. The lines labeled l_1 and l_2 are parallel.

51.

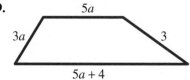

y

$3x + 28$ l_1

$4x - 4$ l_2

a. x
b. y

52.

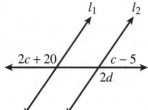

l_1 l_2

$2c + 20$ $c - 5$

$2d$

a. c
b. d

2.9 The Distributive Law

Reading

In this chapter, you have learned skills that will enable you to solve more complex problems using algebra. One type of application involves finding the speed of the current in a river or the speed of the wind.

When a boat travels on a river that has a noticeable current, the current increases the boat's speed when it is traveling downstream (with the current) and decreases its speed when it is traveling upstream (against the current). For example, suppose a boat that moves at 20 miles per hour in still water makes a trip on a river with a 5-mile-per-hour current. On its downstream journey, the boat travels at $20 + 5 = 25$ miles per hour; on the return journey upstream, the boat travels at $20 - 5 = 15$ miles per hour.

EXAMPLE 1

Bridget's motorboat can travel at 30 miles per hour in still water. She travels down the Columbia River to Portland in 2 hours and returns home in 3 hours. What is the speed of the current in the river?

Solution

We let c stand for the speed of the current, and we write an equation we can solve for c. It is helpful to organize the facts of the problem into a table like this one:

	Rate	*Time*	*Distance*
Trip downstream	$30 + c$	2	
Trip upstream	$30 - c$	3	

For the trip downstream, we have *added* the speed of the current to the speed of the boat, and for the trip upstream, we have *subtracted* the speed of the current. We can fill in the last column of the table by applying the formula

$$\text{distance} = \text{rate} \times \text{time:}$$

	Rate	*Time*	*Distance*
Trip downstream	$30 + c$	2	$2(30 + c)$
Trip upstream	$30 - c$	3	$3(30 - c)$

A sketch of the situation is shown in Figure 2.31. Observe that Bridget's boat travels the same distance downstream as it does returning to its starting point upstream. Therefore, the two distances recorded in the table are equal.

$$2(30 + c) = 3(30 - c)$$

To solve this equation, we need to simplify the expressions on each side of the equals sign. We'll use a property called the *distributive law*, which is the subject of the lesson that follows.

Figure 2.31

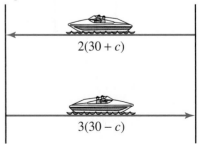

$2(30 + c)$

$3(30 - c)$

Skills Review

Find each percent without using pencil, paper, or calculator. (See Appendix A.9.)

1. 25% of 80   2. 75% of 12   3. 50% of 28   4. 20% of 45

5. 30% of 50   6. 60% of 60   7. $66\frac{2}{3}$% of 36   8. $33\frac{1}{3}$% of 27

Answers: *1.* 20 *2.* 9 *3.* 14 *4.* 9 *5.* 15 *6.* 36 *7.* 24 *8.* 9

Lesson

Products

So far, we have considered sums and differences of algebraic expressions. Now let's take a look at simple products. To simplify a product such as $3(2a)$, we need only multiply the numerical factors:

$$3(2a) = (3 \cdot 2)a = 6a$$

We can do this because of the associative law of multiplication. You can also see why this is reasonable by recalling that multiplication is repeated addition, so

$$3(2a) = \underbrace{2a + 2a + 2a}_{\text{Three terms}} = 6a$$

EXAMPLE 2

$5(3x) = 15x$

Now try Exercise 1.

The Distributive Law

Another useful property for simplifying expressions is the distributive law. First, let's consider a numerical example. Suppose we would like to find the area of the two rooms shown in Figure 2.32. We can do this in two different ways.

Method 1   Think of the rooms as forming one large rectangle 10 feet wide and $12 + 8$ feet long. The area is then given by

$$\text{area} = \text{width} \times \text{length}$$
$$= 10(12 + 8) = 10(20) = 200 \text{ square feet}$$

Method 2   Find the area of each room separately and add the two areas together:

$$10(12) + 10(8) = 120 + 80 = 200 \text{ square feet}$$

Of course, we get the same answer with either method. However, looking at the first step in each calculation, we see that

$$10(12 + 8) = 10(12) + 10(8)$$

These are two ways to compute the expression $10(12 + 8)$.   The first way is the familiar way, which follows the order of operations. The second way *distributes* the multiplication to each term inside parentheses; that is, we multiply each term inside the parentheses by 10.

$$10(12 + 8) = 10(12) + 10(8)$$

This is an example of the distributive law.

EXERCISE 1

Simplify each product.

a. $6(-2b)$

b. $-4(-7w)$

Figure 2.32

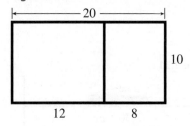

Distributive Law

If a, b, and c are any numbers, then

$$a(b + c) = ab + ac$$

If the terms within parentheses are not like terms, we have no choice but to use the distributive law to simplify the expression.

EXAMPLE 3

Simplify $-2(3x - 1)$.

Solution

Multiply each term inside parentheses by -2:

$$-2(3x - 1) = -2(3x) - 2(-1)$$

$$= -6x + 2$$

Now try Exercises 2 and 3.

Solving Equations and Inequalities

If an equation or inequality contains parentheses, we use the distributive law to simplify each side before we begin to solve.

EXAMPLE 4

Solve $25 - 6x = 3x - 2(4 - x)$.

Solution

We apply the distributive law to the right side of the equation.

$25 - 6x = 3x - 2(4 - x)$	Apply the distributive law.
$25 - 6x = \mathbf{3x - 8 + 2x}$	Combine like terms.
$25 - 6x = 5x - 8$	Subtract $5x$ from both sides.
$25 - 11x = -8$	Subtract 25 from both sides.
$-\ 11x = -33$	Divide both sides by -11.
$x = 3$	The solution is 3.

You should verify each step of the solution, and check the answer.

The techniques you have learned in this chapter are sufficient to solve many linear equations. Although there is no set order in which the steps should always be applied, the following sequence is often appropriate.

EXERCISE 2
Use the distributive law to simplify each expression.
a. $8(3y - 6)$

b. $-3(7 + 5x)$

EXERCISE 3
The length of a rectangle is 3 feet less than twice its width, w.
a. Write an expression for the length of the rectangle in terms of w.

b. Write and simplify an expression for the perimeter of the rectangle in terms of w.

Steps for Solving Linear Equations

1. Use the distributive law to remove any parentheses.
2. Combine like terms on each side of the equation.
3. By adding or subtracting the same quantity on both sides of the equation, get all the variable terms on one side and all the constant terms on the other.
4. Divide both sides by the coefficient of the variable to obtain an equation of the form $x = a$.

Now try Exercises 4 and 5.

EXERCISE 4
Suppose the perimeter of the rectangle in Exercise 3 is 36 feet. Write and solve an equation to find the dimensions of the rectangle.

EXERCISE 5
Solve the equation in Example 1 in this section. What is the speed of the current in the Columbia River?

ANSWERS TO 2.9 EXERCISES
1a. $-12b$ **1b.** $28w$

2a. $24y - 48$ **2b.** $-21 - 15x$

3a. $2w - 3$ **3b.** $6w - 6$

4. 7 feet by 11 feet

5. 6 miles/hour

HOMEWORK 2.9

In Problems 1–4, simplify each product. For which product in each pair must you use the distributive law?

1. a. $8(4c)$ **b.** $8(4 + c)$ **2. a.** $-6(3 + m)$ **b.** $-6(3m)$

3. a. $2(-8 - t)$ **b.** $2(-8t)$ **4. a.** $-3(-12z)$ **b.** $-3(-12 - z)$

In Problems 5–12, use the distributive law to remove parentheses. For some of these problems, you will use the distributive law in the form $(b + c)a = ba + ca$.

5. $5(2y - 3)$ **6.** $(3y - 6)(8)$

7. $-2(4x + 8)$ **8.** $-4(4 - 5a)$

9. $-(5b - 3)$ **10.** $-(-8b - 5)$

11. $(-6 + 2t)(-6)$ **12.** $(3y + 3)(-7)$

In Problems 13–18, simplify and combine like terms.

13. $-6(x + 1) + 2x$ **14.** $4x - 9(2 - 3x)$ **15.** $5 - 2(4x - 9) + 9x$

16. $7(y - 5) + 3(5 - y)$ **17.** $-4(3 + 2z) + 2z - 3(2z + 1)$ **18.** $-3(2t - 5) - 6t - 5(t + 2)$

19. a. Evaluate $(2x + 7) - 4(4x - 2) - (-2x + 3)$ for $x = -3$.

b. Simplify the expression in part (a) by combining like terms.

c. Evaluate your answer to part (b) for $x = -3$. Check that you got the same answer for part (a.)

20. a. Evaluate $6 - 2(6a - 3) - 8 - 4(1 + 3a)$ for $a = 1.8$.

b. Simplify the expression in part (a) by combining like terms.

c. Evaluate your answer to part (b) for $a = 1.8$. Check that you got the same answer for part (a).

21. Simplify each expression if possible. Then evaluate for $x = 3$, $y = 9$.
   a. $2(xy)$ **b.** $2(x + y)$

   c. $2 - xy$ **d.** $-2xy$

22. Simplify each expression if possible. Then evaluate for $x = 4$.
   a. $3 - 5 + x$ **b.** $3(-5 + x)$

   c. $3(-5x)$ **d.** $3 - 5x$

23. Which of the following is a correct application of the distributive law?
   a. $5(3a) = 15a$ **b.** $5(3 + a) = 15 + a$
   c. $5(3a) = 15(5a)$ **d.** $5(3 + a) = 15 + 5a$

24. Which of the following is a correct formula for the perimeter of a rectangle?
   a. $P = 2l + 2w$ **b.** $P = 2(l + w)$

In Problems 25–36, write an algebraic expression to answer each question.

25. An apple and a glass of milk contain a total of 260 calories.
   a. If an apple contains *a* calories, how many calories are in a glass of milk?

   b. Write an expression for the number of calories in two apples and three glasses of milk.

   c. If two apples and three glasses of milk contain 660 calories, find the number of calories in an apple.

26. Last year, Delbert and Francine together made $35,000.
   a. If Delbert's salary was *y*, what was Francine's salary?

   b. This year, Delbert got a 5% raise and Francine got a 6% raise. Write an expression for their total increase in income.

   c. This year, Delbert and Francine together made $36,950. What was Delbert's salary last year?

27. The length of a rectangular vegetable garden is 6 yards more than twice its width.
   a. Write an expression for the perimeter of the garden in terms of its width.

   b. Ann bought 42 yards of fence to enclose the garden. What are the garden's dimensions?

28. A window in Clyde's attic is shaped like an isosceles triangle. One of the sides of the triangle is 8 inches more than one-third of its base.
   a. Write an expression for the perimeter of the window in terms of its base.

   b. Clyde needs 61 inches of weather-stripping to border the window. What is the base of the triangle?

29. Melody sold 47 tickets to a charity concert. Reserved seats cost $10 each, and open seating was $6 per ticket. Let x represent the number of reserved-seat tickets Melody sold. Write an expression in terms of x for each of the following:

a. The number of open-seating tickets Melody sold

b. The amount of money Melody collected for reserved-seat tickets

c. The amount of money Melody collected for open-seating tickets

d. The total amount of money Melody collected

30. A tour director wants to buy 30 seats on a flight to Tahiti. First-class tickets cost $480, and coach tickets cost $350. Let x represent the number of coach tickets he buys. Write an expression in terms of x for each of the following:

a. The number of first-class tickets he buys

b. The total cost of the coach tickets

c. The total cost of the first-class tickets

d. The total cost of all 30 tickets

31. Melody of Problem 29 collected $330 from the sale of concert tickets. How many reserved-seat tickets did she sell?

32. The tour director in Problem 30 spent $12,060 on tickets to Tahiti. How many coach tickets did he buy?

33. A paddleboat journey down the Mississippi takes 5 hours, and the return journey upstream takes 9 hours. The current in the river flows at 8 miles per hour. What is the speed of the paddleboat in still water?

34. A Jetair plane flies from Chicago to Los Angeles in 6 hours against the prevailing wind and returns in 5 hours with the wind behind it. If the jet flies at 330 miles per hour in still air, what is the speed of the wind?

35. Chester can row 9 miles per hour in still water. He rows for 5 hours upstream from his house to an island and returns the next day in 4 hours.
  a. What is the speed of the current in the river?

  b. How far is it from Chester's house to the island?

36. A small plane flew from Fresno to Fullerton with a tailwind of 8 miles per hour in 6 hours. The plane returned against a headwind of 2 miles per hour in 8 hours.
  a. What was the speed of the plane in still air?

  b. How far is it from Fresno to Fullerton?

In Problems 37–46, solve each equation or inequality.

37. $6(3y - 4) = -60$

38. $-3(2x + 1) - 4 = -1$

39. $5w - 64 = -2(3w - 1)$

40. $3(7 + 2d) = 30 + 7(d - 1)$

41. $-22c + 5(3c + 4) = 20 + 8c$

42. $4(2x - 4) - 8(1 - 3x) = 40$

43. $4 - 3(2t - 4) > -2(4 - 3t)$

44. $2(3h - 6) < 5 - (h - 4)$

45. $0.25(x + 3) - 0.45(x - 3) = 0.30$

46. $0.12x + 0.08(8000 - x) = 840$

In Problems 47–50, find an algebraic expression for the area of each rectangle. Use the distributive law to write the expression in two ways.

47.

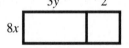

48.

49.

50.

2.10 Line of Best Fit

Reading

In Section 2.7, we used two data points for the years 1970 and 1990 to estimate the size of the rain forest at other times. By following the line through the two given points, we predicted when the rain forest would be completely depleted, and we estimated the size of the rain forest in 1950. These estimates are called **extrapolations** because they are predictions for values outside the range of the data given.

We can also use the line to estimate the size of the rain forest in 1980 (or any other year between the two points). Such an estimate is called an **interpolation**. For example, the point (30, 9) is on the line in Figure 2.33, so we estimate that there were 9 million square kilometers of rain forest in 1980. Interpolation is usually more dependable than extrapolation, which can give nonsensical results. For example, if we use extrapolation to predict the size of the rain forest in the year 2100, when $x = 150$, we obtain -0.6 million square kilometers. This is not a reasonable prediction, because there cannot be a negative amount of rain forest on the earth. Depletion of the rain forest cannot continue indefinitely at its current rate.

Figure 2.33

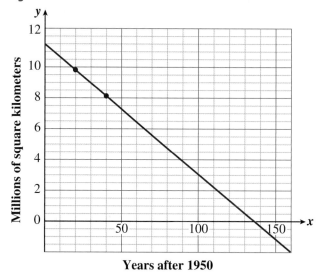

Years after 1950

The process of using a line to predict values is called **linear regression**, and the line itself is called a **regression line**. When actual measurements are involved, no line can provide precisely accurate predictions, but businesspeople and scientists routinely use linear regression to make estimates by interpolation and extrapolation. In this section, we estimate a *line of best fit* to use as a regression line.

Now try Exercise 1.

Skills Review

Use the graph to estimate answers to the following questions.

1. Find the y-coordinate of the point where the x-coordinate is 3.

2. What are the x-coordinates of the points where $y = 3$?

3. What is the x-intercept?

4. What is the y-intercept?

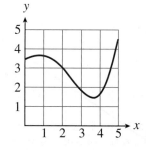

Answers: **1.** 1.8   **2.** 2 and 4.6   **3.** There is no x-intercept.   **4.** 3.4

EXERCISE 1
Use Figure 2.33 to estimate each value. Are you using interpolation or extrapolation in each case?

a. How many millions of square kilometers of rain forest existed in 1960?

b. How many millions of square kilometers of rain forest existed in 1975?

c. How many millions of square kilometers of rain forest existed in 1995?

d. When were there 8.6 million square kilometers of rain forest?

Lesson

Scatterplots

Can we predict a person's height on the basis of his or her shoe size? Try the following experiment. Find the shoe size and height of each person in your class. To convert women's shoe sizes to men's, subtract 1 from each women's shoe size.

Figure 2.34 displays the results of such an experiment. Each point in the graph represents a different individual, and the coordinates of the point give a pair of measurements for that individual. For example, the point labeled *A* has coordinates (11.5, 73.5). Thus point *A* represents a person whose (men's) shoe size is 11.5 and whose height is 73.5 inches. A graph that displays pairs of measurements in this way is called a **scatterplot**.

Figure 2.34

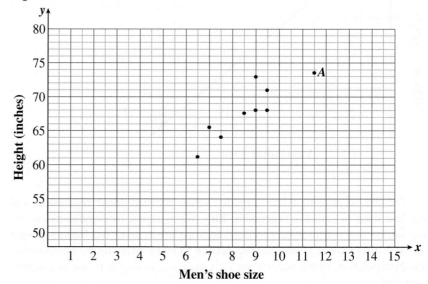

The points on the scatterplot do not all lie on a single line. But even though the points do not line up perfectly, there is a consistent, almost linear trend in the data. The line in Figure 2.35 roughly follows the trend for the scatterplot, and we could use this as a regression line.

Figure 2.35

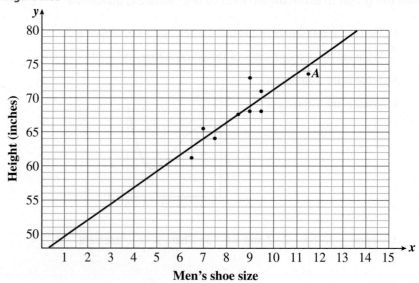

Now try Exercise 2.

Some of the data points in Figure 2.35 lie above the regression line and others lie below the line. This means that some of the individuals represented in the data are taller than predicted by the regression line, and some are shorter. The vertical distance between an actual data point and the regression line represents the *error* in the prediction made by using the regression line.

Now try Exercise 3.

Choosing a Regression Line

The regression line in Figure 2.35 does not make perfect predictions. But it is a good choice for a regression line because the errors in its predictions are all relatively small. A different regression line might reduce some of those errors but make others worse. How can we find the best regression line for a scatterplot? Statisticians use several techniques for finding regression lines, but we will use a graphical approach.

By studying the scatterplot, we try to choose a regression line such that the vertical distances from the data points to the line are as small as possible. For example, Figure 2.36 shows three possible regression lines for the data. Note that there are more data points above line *M* than below it. We can reduce most of the errors by raising line *M* to the position shown by line *L*, so line *L* is a better choice for the regression line. Line *N* has equal numbers of data points on each side, but it does not follow the trend of the data. The errors for line *N* could be reduced by making it steeper.

Figure 2.36

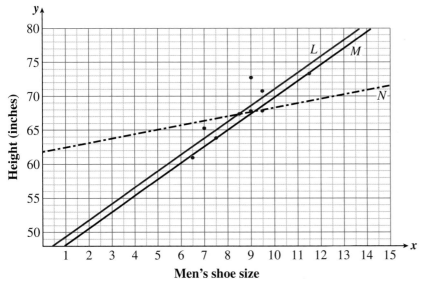

Men's shoe size

EXERCISE 2
What does the regression line in Figure 2.35 predict for the height of a person whose (men's) shoe size is 9?

EXERCISE 3
Is person *A* taller or shorter than predicted by the regression line in Figure 2.35? By how much?

Figure 2.37

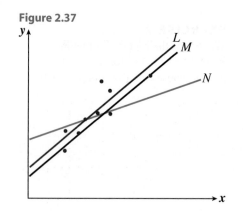

EXAMPLE 1

Determine which of the lines in Figure 2.37 fits the scatterplot best.

Solution

Line *L* is the best fit. Line *M* is very close to most of the points, but it is so far from the other points that the overall fit would be better if the line were shifted upward. Line *N* has equal numbers of points above and below it, but the points would be closer to the line if *N* were steeper.

Of course, when only two data points are available, the line of best fit is the line through those two points.

EXAMPLE 2

Francine's electric car accelerates from 0 to 60 miles per hour in 17.0 seconds. Use a line to predict how long will it take to accelerate from 0 to 55 miles per hour.

Solution

We'll put time on the horizontal axis and speed on the vertical axis. We draw a straight line through the data points (0, 0) and (17, 60). (See Figure 2.38.) On the basis of that line, the car will take about 15.6 seconds to accelerate to 55 miles per hour.

Figure 2.38

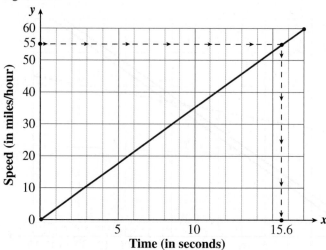

Next, we'll use the height and shoe size data you collected in your class.

Activity 1

1. Create a scatterplot for the data you collected. Use the grid below. (Can you pick out the point that corresponds to you?)

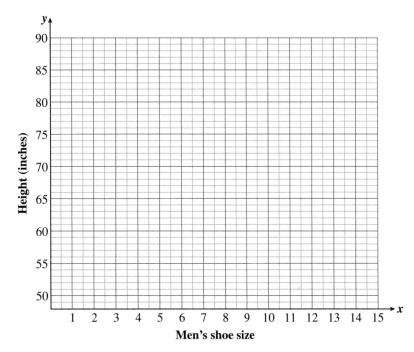

Men's shoe size

2. Draw a regression line on the scatterplot you created. According to that regression line, are you taller or shorter than predicted by your shoe size?

   I am _____ than predicted, by _____ inches.

3. According to your regression line, how tall would you expect a man to be if he wears size 18 shoes? Size 10 shoes?

   Size 18 shoes: height ≈_____ inches

   Size 10 shoes: height ≈_____ inches

ANSWERS TO 2.10 EXERCISES

1a. 10.6 (extrapolation)

1b. 9.4 (interpolation)

1c. 7.8 (extrapolation)

1d. 1985 (interpolation)

2. About 69 inches

3. Shorter by about 1.5 inches

HOMEWORK 2.10

In Problems 1 and 2, which of the lines is the best choice for a regression line for the given scatterplot? Explain your choice.

1. a.
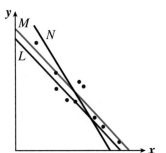

b.
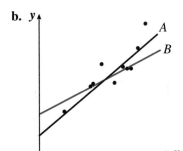

c.
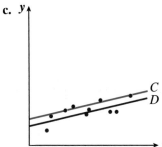

2. a. 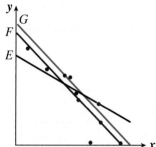   **b.** 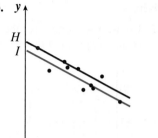   **c.**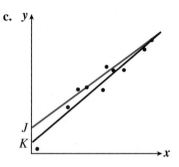

In Problems 3–8, carefully draw a line that fits the given data, and use it to answer the question.

3. A fruit punch recipe that makes 20 servings calls for 12 ounces of orange juice. How much orange juice is needed for 25 servings? (*Hint:* You don't need any orange juice for 0 servings.)

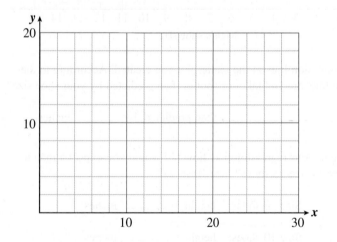

4. Delbert has a rose garden whose area is 24 square feet. He uses nine small pails of compost for his garden. How much should he recommend for his neighbor's garden, which covers an area of 40 square feet? (*Hint:* The neighbor wouldn't need any compost for a garden 0 square feet in area.)

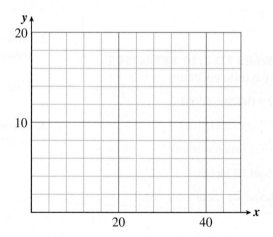

5. Jennie keeps track of the number of gallons of gasoline her car uses and the number of miles she drives between visits to the gas station. Plot the given data.

Gallons	Miles
9.2	274.8
10.7	320.1
10.1	298.2
9.6	285.4
9.7	288.4
9.6	283.1
9.3	275.8

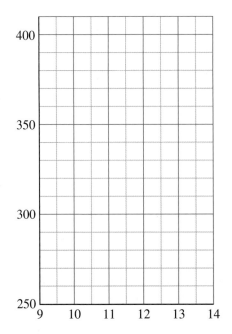

a. How far can Jennie expect to go on a tank of gas if the tank holds 13.2 gallons?

b. How much gasoline will Jennie need to travel 310 miles?

6. Professor Martinez uses an overhead projector to show her math class the display on her calculator. She measures the width of the image on the screen when the projector is at different distances from the screen. Plot the given data.

Distance to screen (cm)	200	210	220	230	240
Width of image (cm)	87.6	91.8	96.4	100.6	104.8

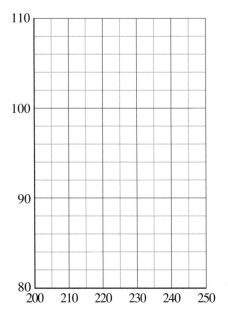

a. Estimate the width of the image if the projector is 250 centimeters from the screen.

b. How far from the screen should Professor Martinez place the projector if she wants the image to be 95 centimeters wide?

7. Maita measured the length of a rubber band when it was supporting various objects of known weight. Plot Maita's data.

Weight (ounces)	5	14.8	14	8
Length (cm)	16.2	27.6	26.7	18.4

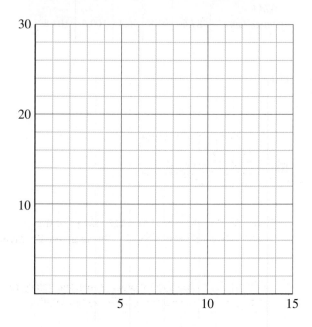

a. Estimate the length of the rubber band when it supports a 12-ounce weight.

b. Estimate the weight of an object that stretches the rubber band to a length of 20 centimeters.

c. What is the physical interpretation of the *y*-intercept of the regression line (if the graph is extended far enough to the left)?

8. Chia-ling measured the volume of a fixed quantity of gas at different temperatures. Plot the data.

Temperature (°C)	20	40	60	80
Volume (liters)	10.3	11	11.7	12.4

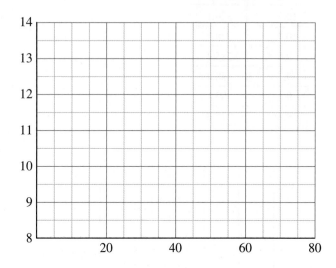

a. Estimate the volume of the gas at a temperature of 37°C.

b. Estimate the temperature needed to bring the gas to a volume of 11.5 liters.

c. What is the physical interpretation of the *y*-intercept of the regression line (if the graph is extended far enough to the left)?

d. What is the physical interpretation of the *x*-intercept of the regression line (if the graph is extended far enough downward)?

9. The scatterplot shows the relationship between the average salaries of men and women college graduates in the years 1970–1990. The average salary for both groups rose each year.

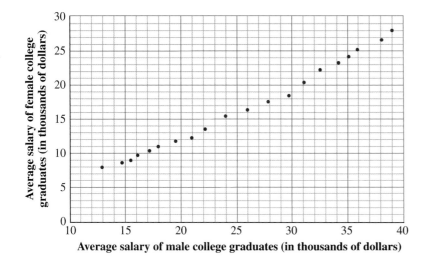

Average salaries of college graduates
Source: U.S. Department of Education, *Digest of Educational Statistics*, 1994.

a. In how many of those years did the women's average salary exceed $25,000? In how many years did the men's average salary exceed $25,000?

b. What were the men earning in the year women's average salary was between $15,000 and $16,000? What were women earning in the year that the men's average salary was between $15,000 and $16,000?

c. Estimate the increase in average salary for the women from 1970 to 1990. Do the same for the men.

d. Estimate the ratio of the women's average salary to the men's average salary in 1970 and in 1990.

10. The scatterplot shows the relationship between the heights of dance partners in a ballroom dance class.

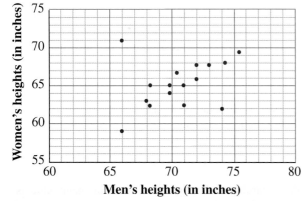

Heights of dancers

a. How many men are 70 inches tall? What are the heights of their partners?

b. What are the heights of the partners of the women who are 65 inches tall?

c. How tall is the shortest woman?

d. How tall is the tallest man?

e. All but two of the points on the scatterplot seem to lie on or near a straight line. What are the heights of the two couples corresponding to these two outlying points?

11. This problem investigates how well a person's hand-span predicts his or her height. Measure the distance from the tip of your thumb to the end of your small finger when your fingers are fully extended. Record your hand-span and your height, in inches, in the table. Obtain the same measurements for four more people.

Span (inches)					
Height (inches)					

a. Plot your data and draw a regression line on the graph.

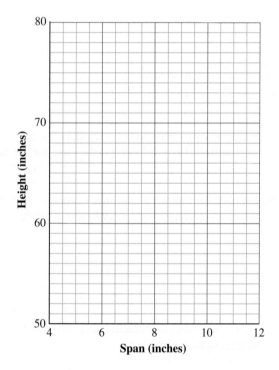

b. Locate and label the point that corresponds to you. Are you taller or shorter than predicted by the regression line? By how much?

c. What does the regression line predict for the height of someone whose hand-span is 10.5 inches?

d. According to the regression line, what is the hand-span of someone who is 58 inches tall?

12. Does the number of course units you take determine the number of books you will need? Poll eight students (including yourself) on how many course units each is taking this semester and how many required books each needs. Record the results in the table.

Units								
Books								

a. Plot your data and draw a regression line on the graph.

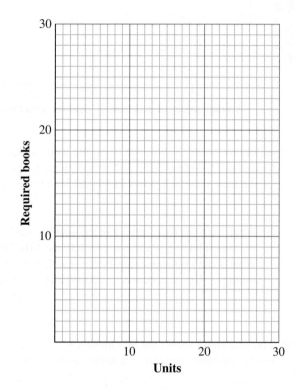

b. Locate and label the point that corresponds to you. Do you have more or fewer required books than predicted by the regression line? By how many?

c. How many required books does the regression line predict for a student carrying 17 course units?

d. According to the regression line, how many units is a student taking if he or she has six required books?

CHAPTER SUMMARY AND REVIEW

Glossary: Write a sentence explaining the meaning of each term.

natural numbers	axes	whole numbers	quadrants
integers	origin	number line	intercepts
opposites	equivalent expressions	absolute value	like terms
order symbols	scatterplot	inequality	regression line

Properties and formulas

1. State three properties that are used in solving inequalities.

2. State the distributive law.

Techniques and procedures

3. Using examples, explain how to add two signed numbers.

4. Using examples, explain how to subtract two signed numbers.

5. Using examples, explain how to multiply or divide two signed numbers.

6. Which quotients involving zero are undefined? Give examples.

7. Refer to the figure and discuss the signs of the coordinates of points in each quadrant.

```
      y▲
  II  |  I
──────┼──────▶ x
  III |  IV
      |
```

8. Explain how to solve an equation by using a graph.

9. Explain how to use the intercept method of graphing a line.

10. Explain how to add or subtract like terms.

11. Explain what a regression line is.

12. What is the difference between interpolation and extrapolation?

Review problems

13. If $x = -3$, evaluate each expression.

  a. $-x$

  b. $|x|$

  c. $-|-x|$

  d. $-(-x)$

14. Graph each set of numbers.

  a. $1.4, 2\frac{1}{2}, -2\frac{1}{3}, -3.5$

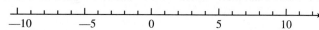

  b. $-15, -5, 10, 25$

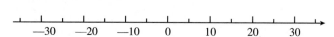

In Problems 15–18, replace the comma by the proper symbol: >, <, or =.

15. $-|-2|, -|-3|$

16. $-2.02, -2.1$

17. $2 - (-5), -7$

18. $-6\left(\dfrac{-1}{3}\right), -2$

Simplify the expressions in Problems 19–26.

19. $28 - 14 - 9 + 15$

20. $11 - 14 + (-24) - (-18)$

21. $12 - [6 - (-2) - 5]$

22. $-2 + [-3 - (-14) + 6]$

23. $5 - (-4)3 - 7(-2)$

24. $5 - (-4)(3 - 7)(-2)$

25. $\dfrac{6(-3) - 8}{-4(-3 - 5)}$

26. $\dfrac{6(-2) - 8(-9)}{5(2 - 7)}$

Evaluate the expressions in Problems 27–30.

27. $2 - ab - 3a$,   for $a = -5, \ b = -4$

28. $(8 - 6xy)xy$,   for $x = -2, \ y = 2$

29. $\dfrac{-3 - y}{4 - x}$, for $x = -1$, $y = 2$

30. $\dfrac{5}{9}(F - 32) + 273$, for $F = -22$

31. If the overnight low in Lone Pine was $-4°$F, what was the temperature after it had warmed by $10°$F?

32. The winter temperature in a city is typically $3°$F warmer than in an adjoining suburb. If the temperature in the suburb is $-7°$F, what is the temperature in the city?

33. Jordan's clothing company is worth \$280,000, and Asher's is worth $-$\$180,000. How much more is Jordan's company worth than Asher's?

34. A certain arsenic compound has a melting point of $-8.5°$C. If the melting point is reduced by $1.1°$C, what is the new melting point?

35. In his first football game, Bo rushes for three consecutive losses of 4 yards each. What is his net yardage for the three plays?

36. In testing a military aircraft's handling at low altitude, the pilot makes a series of flights over a set course, each flight occurring at an altitude 150 feet lower than the previous one. What is the net change in altitude between the first and the sixth flights?

In Problems 37–40, find three solutions for each inequality.

37. $x - 2 < 5$

38. $4x \geq -12$

39. $-9 \leq -3x$

40. $-15 > 5 + x$

41. The temperature in Maple Grove was 18°F at noon, and it has been dropping ever since at a rate of 3°F per hour.

  a. Fill in the table. T stands for the temperature h hours after noon. Negative values of h represent hours before noon.

h	-4	-2	0	1	3	5	8
T							

  b. Write an equation for the temperature, T, after h hours.

  c. Graph your equation on the grid to the right.

  d. What was the temperature at 10 A.M.?

  e. When will the temperature reach $-15°$F?

  f. How much did the temperature drop between 3 P.M. and 9 P.M.?

42. Here is a mystery graph. Its equation is $y = mx + b$. Use the graph to decide whether each statement is true or false.

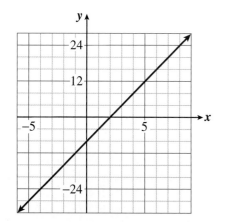

  a. $3m + b = -20$

  b. $2m + b = 0$

  c. $-4m + b > -20$

  d. $m + b < -4$

43. Find the distance between each pair of points.
    a. $(-12, -4)$ and $(-7, -4)$

    b. $(6, -2)$ and $(6, 8)$

44. What is the general form for a linear equation in two variables?

For Problems 45 and 46, make a table of values and graph each equation.

45. $y = -2x + 7$

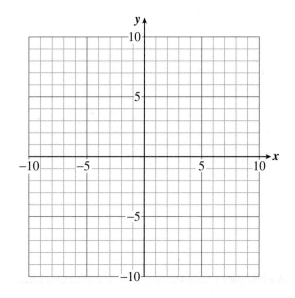

46. $y = \dfrac{4}{3}x - 2$

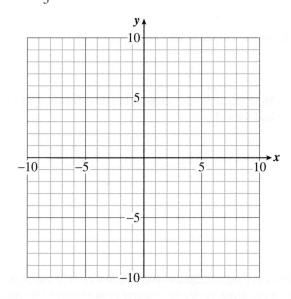

For Problems 47 and 48, use the graph to solve each equation or inequality. Estimate your solutions if necessary.

47. a. $-3x + 9 = 24$

    b. $-3x + 9 \geq 15$

48. a. $24x - 1800 = -1250$

    b. $24x - 1800 < -2500$

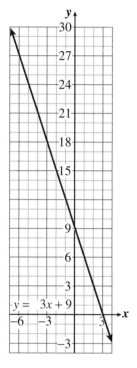

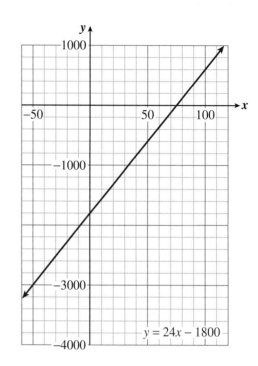

In Problems 49 and 50, simplify each expression. Which product in each pair requires using the distributive law?

49. a. $-5(-6m)$

    b. $-5(6 - m)$

50. a. $9(-3 - w)$

    b. $9(-3w)$

In Problems 51–54, simplify each expression.

51. $(4m + 2n) - (2m - 5n)$

52. $(-5c - 6) + (-11c + 15)$

53. $-7w - 2(4w - 13)$

54. $4(3z - 10) + 5(-z - 6)$

Solve Problems 55–62.

55. $4z - 6 = -10$

56. $3 - 5x = -17$

57. $-1 = \dfrac{5w}{3} + 4$

58. $4 - \dfrac{2z}{5} = 8$

59. $3h - 2 = 5h + 10$

60. $7 - 9w = w + 7$

61. $5p + 10(17 - p) = 2p - 5$

62. $-3(k - 2) - 4(2k + 5) = 10 + 3k$

63. Monica has saved $7800 to live on while she attends college. She spends $600 a month.
   a. Write an equation for the amount, *S*, in Monica's savings account after *m* months.

   b. Find the intercepts of your equation and use them to graph the line.

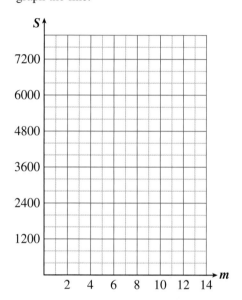

   c. Explain what the intercepts mean in terms of the problem.

   d. Write and solve an equation to answer this question: How long will it be before Monica's savings are reduced to $2400?

64. Ronen is on a rock-climbing expedition. He is climbing out of a deep gorge at a rate of 4 feet per minute, and right now his elevation is −156 feet.
   a. Write an equation for Ronen's elevation, *h*, after *m* minutes.

   b. Find the intercepts of your equation and use them to graph the line.

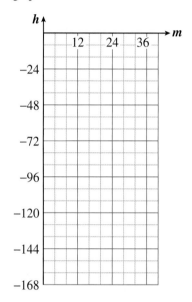

   c. Explain what the intercepts mean in terms of the problem.

   d. Write and solve an equation to answer this question: When will Ronen's elevation be −20 feet?

In Problems 65 and 66, find the *x*- and *y*-intercept of each line, and then graph the line.

65. $6x - 4y = 12$

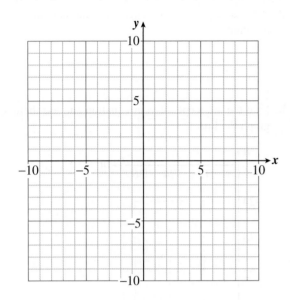

66. $y = 3x - 8$

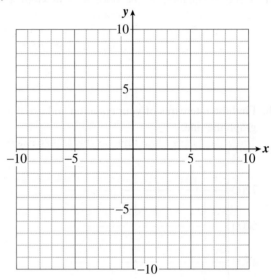

67. The length of a rectangle is three times its width.
   a. If the width of the rectangle is x, what is its length?

   b. Express the perimeter of the rectangle in terms of x.

   c. Suppose the perimeter of the rectangle is 48 centimeters. Find the dimensions of the rectangle.

68. In a city council election, the winner received 132 more votes than her opponent did.
   a. If the winner received y votes, how many did her opponent receive?

   b. Write an expression for the total number of votes cast for the two candidates.

   c. If 12,822 votes were cast, how many did each candidate receive?

69. The principal at Wheaton Elementary school plans to buy 30 computers. Computers with speakers cost $1200 each, and those without speakers cost $800 each. Let x represent the number of computers with speakers. Write expressions in terms of x for each of the following:

a. The number of computers without speakers

b. The total cost of the computers with speakers

c. The total cost of the computers without speakers

d. The total cost of all 30 computers

e. If the principal has $28,800 to spend on the computers, how many of each kind can he buy?

70. The current in the Lazy River flows at 4 miles per hour. A motorboat trip takes $1\frac{1}{2}$ hours upstream; the return trip takes $\frac{1}{2}$ hour. What is the speed of the motorboat in still water?

In Problems 71–76, solve each inequality algebraically, and graph the solutions on the number lines.

71. $2 - 3z \le -7$

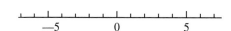

72. $\dfrac{t}{-3} - 1.7 > 2.8$

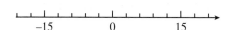

73. $3k - 13 < 5 + 6k$

74. $4(3a - 7) < -18 + 2a$

75. $-9 < 5 - 2n \le -1$

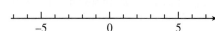

76. $15 \ge -6 + 3m \ge -6$

In Problems 77 and 78, carefully draw a line that fits the given data, and use it to answer the questions.

77. Gokhan was told that the rate at which crickets chirp depends on the air temperature. He records a cricket's chirp rate (in chirps per minute) at various temperatures. Plot the data.

Tempera-ture (°F)	48.3	52.7	60.6	67.1	73.4	75.5
Chirp rate	32	52	84	108	132	142

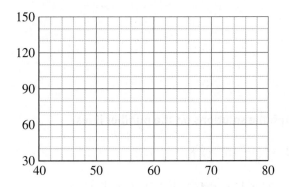

a. Estimate how fast the cricket will chirp at a temperature of 70°F.

b. Estimate the temperature at which the cricket will chirp 100 times per minute.

78. The table shows the total expenditures, in billions of dollars, for all public colleges and universities in the United States for the years 1985–1993. Plot the data.

Year	Expenditures (in billions)
1985	$ 81
1986	$ 86
1987	$ 90
1988	$ 92
1989	$ 96
1990	$102
1991	$106
1992	$108
1993	$111

Source: U.S. Department of Education, *Digest of Educational Statistics*, 1994.

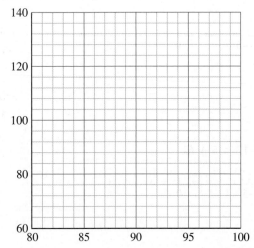

a. Estimate when expenditures were at $74 billion.

b. Use given data to estimate when expenditures will reach $135 billion.

Applications of Linear Equations

3.1 Ratio and Proportion

Reading

Ratios

A **ratio** is a type of quotient that allows us to compare two numerical quantities. For example, saying that the ratio of pupils to computers at a certain elementary school is 10 to 3 means that for every ten pupils in the school there are three computers. We often display a ratio as a fraction because their properties are similar. For the example just mentioned, we can write

$$\frac{\text{number of pupils}}{\text{number of computers}} = \frac{10}{3}$$

Note that this does *not* mean that there are exactly 10 pupils and 3 computers at the school. There could be 40 pupils and 12 computers or 200 pupils and 60 computers (or many other combinations), because

$$\frac{40}{12} = \frac{200}{60} = \frac{10}{3}$$

Try Exercise 1 on page 268.

EXAMPLE 1

In a survey of 100 of his classmates, Greg found that 68 of them supported a strict gun control law and that 32 did not. What is the ratio of survey respondents who support gun control to those who oppose it?

Solution

The ratio of those supporting gun control to those opposing it is 68 to 32, or $\frac{68}{32}$.

This reduces to $\frac{17}{8}$, or 17 to 8.

In Example 1, we could just as easily have computed the ratio of opponents to supporters of the gun control law and obtained the reciprocal of the answer to Example 1: $\frac{32}{68}$, or $\frac{8}{17}$. Either ratio might be useful in applications, so it is important to note which quantity corresponds to the numerator and which to the denominator. When we write the ratio of *a* to *b* as a fraction, *a* is the numerator and *b* is the denominator.

> **The ratio of *a* to *b*** is written $\frac{a}{b}$.

Any ratio can be expressed as a decimal instead of as a common fraction. For example, because $\frac{17}{8}$ is equal to 2.125, we can say that the ratio of supporters of gun control to opponents in Example 1 is 2.125 to 1. We usually express a ratio as a decimal fraction if the numbers being compared are decimal numbers.

EXAMPLE 2

A small circuit board measures 4.2 centimeters long by 2.4 centimeters wide. What is the ratio of its length to its width?

Solution

The ratio of length to width is $\dfrac{4.2 \text{ cm}}{2.4 \text{ cm}}$, or 1.75, as you can verify with your calculator. (Can you express this ratio as a common fraction?)

Rates

A **rate** is a ratio that compares two quantities with different units. You are already familiar with many rates. If you say that your average speed was 50 miles per hour or that apples cost 89 cents per pound, you are using rates. Recall that "per" usually indicates division, so

50 miles per hour	means	$\dfrac{50 \text{ miles}}{1 \text{ hour}}$
89 cents per pound	means	$\dfrac{89 \text{ cents}}{1 \text{ pound}}$

EXAMPLE 3

Sarah traveled 390 miles on 15 gallons of gas. Express her rate of fuel consumption as a ratio and then as a rate.

Solution

As a ratio, Sarah's rate of fuel consumption was $\dfrac{390 \text{ miles}}{15 \text{ gallons}}$. We simplify by dividing the denominator into the numerator, to get $\dfrac{26 \text{ miles}}{1 \text{ gallon}}$, or 26 miles per gallon.

Now try Exercise 2.

Skills Review

Convert each common fraction to a decimal and each decimal to a common fraction. (You may want to review Section A.5 of Appendix A.)

1. $\dfrac{3}{8}$ 2. $\dfrac{7}{5}$ 3. $\dfrac{4}{3}$ 4. $\dfrac{5}{12}$

5. 0.04 6. $0.\overline{6}$ 7. 1.875 8. 2.2

Answers: *1.* 0.375 *2.* 1.4 *3.* $1.\overline{3}$ *4.* $0.41\overline{6}$ *5.* $\dfrac{1}{25}$ *6.* $\dfrac{2}{3}$

7. $\dfrac{15}{8}$ *8.* $\dfrac{11}{5}$

Lesson

Proportions

A **proportion** is a statement that two ratios are equal. In other words, a proportion is a type of equation in which both sides are ratios. For example, the following equations are proportions:

$$\frac{6}{9} = \frac{8}{12} \quad \text{and} \quad \frac{3.2}{8} = \frac{1}{2.5}$$

Like any other equation, a proportion may involve variables. How can we solve a proportion such as this one?

$$\frac{7}{5} = \frac{x}{6}$$

In Section 1.6, you learned that if x is divided by 6, you should multiply both sides of the equation by 6. This has the effect of clearing the fraction. We can solve a proportion more efficiently by clearing both fractions at the same time. To accomplish this, we can multiply both sides of the equation by the LCD (lowest common denominator) of the two denominators involved. In this case, the LCD is $5 \cdot 6 = 30$. Multiplying both sides of the proportion by 30 gives us

$$30\left(\frac{7}{5}\right) = \left(\frac{x}{6}\right)30$$
$$42 = 5x$$

Now we can divide both sides by 5 to get the answer:

$$x = \frac{42}{5} = 8.4$$

There is a shortcut we can use that avoids calculating an LCD. Observe that we can arrive at the equation $42 = 5x$ directly from the original proportion by *cross-multiplying*:

$$\frac{7}{5} \searrow_{\nearrow} = {}_{\nearrow}^{\searrow} \frac{x}{6}$$
$$42 = 5x$$

We can then proceed as before to complete the solution.

The cross-multiplying shortcut is a fundamental property of proportions.

Property of Proportions

If $\dfrac{a}{b} = \dfrac{c}{d}$, then $ad = bc$.

EXAMPLE 4

Solve $\dfrac{7.6}{1.2} = \dfrac{x+3}{2x-1}$.

Solution

We apply the property of proportions and cross-multiply to get

$$7.6(2x - 1) = 1.2(x + 3) \qquad \text{Apply the distributive law.}$$
$$15.2x - 7.6 = 1.2x + 3.6 \qquad \text{Subtract } 1.2x \text{ from both sides;}$$
$$\text{add } 7.6 \text{ to both sides.}$$
$$14x = 11.2$$
$$x = \dfrac{11.2}{14} = 0.8 \qquad \text{Divide both sides by 1.}$$

Check the solution by substituting $x = 0.8$ into the original proportion.

Now try Exercise 3.

EXERCISE 3

Solve $\dfrac{q}{3.2} = \dfrac{1.25}{4}$.

Proportional Variables

Suppose grape juice costs 80 cents per quart. Complete Table 3.1.

Table 3.1

Number of quarts	Total price (in cents)	$\dfrac{\text{Total price}}{\text{Number of quarts}}$
1	80	$\dfrac{80}{1} = 80$
2	160	$\dfrac{160}{2} =$
3	240	
4	320	

You should find that the ratio $\dfrac{\text{total price}}{\text{number of quarts}}$, or price per quart, is the same for each pair of values in the table. This agrees with common sense: The price per quart of grape juice is the same no matter how many quarts you buy.

Two variables are said to be **proportional** if their ratios are always the same. We have just seen that the price of grape juice is proportional to the number of quarts purchased. If we want to know how many quarts can be purchased for $10.00, we can set up a proportion using the ratio *price per quart*.

EXAMPLE 5

Suppose one quart of grape juice costs 80 cents. How much grape juice can you buy for $10.00?

Solution

Let x represent the number of quarts that can be purchased for $10.00, or 1000 cents. Because the variables *total price* and *number of quarts* are proportional, we know that their ratio is constant. Thus the ratio

$$\frac{1000}{x \text{ quarts}} \quad \text{is equal to the ratio} \quad \frac{80 \text{ cents}}{1 \text{ quart}}$$

or

$$\frac{1000}{x} = \frac{80}{1}$$

We solve the proportion by cross-mutiplying to get

$$1000(1) = 80x \qquad \text{Divide both sides by 80.}$$

$$x = \frac{1000}{80} = 12.5$$

Thus we can buy 12.5 quarts of grape juice for $10.00. (Or, if the juice is sold only in quart bottles, then we can buy 12 quarts and have 40 cents left over.)

CAUTION!

You could solve Example 5 by equating two ratios of the form $\frac{\text{number of quarts}}{\text{total price}}$. This approach leads to the proportion $\frac{x}{1000} = \frac{1}{80}$, which has the same solution as before. The important thing is that *both* ratios have the *same units* in their numerators and the *same units* in their denominators. It would *not* be correct to equate $\frac{1000}{x}$ and $\frac{1}{80}$, because the ratios do not have the same units.

EXERCISE 4

Show that the proportions

$$\frac{1000 \text{ cents}}{x \text{ quarts}} = \frac{80 \text{ cents}}{1 \text{ quart}}$$

and

$$\frac{x \text{ quarts}}{1000 \text{ cents}} = \frac{1 \text{ quart}}{80 \text{ cents}}$$

have the same solution, but the equation

$$\frac{1000 \text{ cents}}{x \text{ quarts}} = \frac{1 \text{ quart}}{80 \text{ cents}}$$

has a different solution.

Now try Exercise 4.

EXERCISE 5

If Sarah can drive 390 miles on 15 gallons of gas, how much gas will she need to travel 800 miles? Follow the suggested steps to solve this problem.

Step 1 The variables *miles driven* and *gallons of gas used* are proportional. Make a ratio with these variables:

$$\frac{\text{miles}}{\text{gallons}}$$

$$\left(\text{You could also choose } \frac{\text{gallons}}{\text{miles}}. \right)$$

Step 2 Create a proportion using the information in the problem: One ratio uses 390 miles and 15 gallons; the other uses the unknown quantity, x gallons, and 800 miles. Use the ratio from Step 1 as a guide.

Step 3 Solve the proportion.

To solve a problem involving ratios, begin by writing down a ratio of the units involved. Use this ratio to write a proportion in which both sides have the same units.

EXAMPLE 6

Uncle Ebenezer leaves $500,000 to be divided between his niece Penelope and his nephew Wilfrid in the ratio 7 to 5. How much should each get?

Solution

If Penelope inherits x dollars, then Wilfrid will inherit $500,000 - x$ dollars. The ratio of Penelope's inheritance to Wilfrid's inheritance should be 7 to 5. Thus

$$\frac{x}{500,000 - x} = \frac{7}{5}$$

We solve the proportion by first cross-multiplying.

$5x = 7(500,000 - x)$ Apply the distributive law.

$5x = 3,500,000 - 7x$ Add $7x$ to both sides.

$12x = 3,500,000$ Divide both sides by 12.

$x \approx 291,666.67$

Penelope's share of the legacy is $291,666.67, and Wilfrid's share is

$$500,000 - \$291,666.67 = \$208,333.33$$

HOMEWORK 3.1

1. What is a ratio? What is a proportion?

2. When are two variables proportional?

Write ratios or rates for Problems 3–10.

3. A survey of 300 employees at a large company showed that 125 used public transportation to commute to work. What is the ratio of employees who use public transportation to those who do not?

4. A study of 80 nesting waterfowl at Saltmarsh Refuge showed that the eggs of 56 birds suffered damage from chemical pollutants. What is the ratio of birds with damaged eggs to birds with undamaged eggs?

5. Erica tutored in the math lab for 6.5 hours last week and made $56.68. What is Erica's rate of pay in dollars per hour?

6. In 1985, the average American consumed 63.4 pounds of refined sugar. What was the average American's rate of sugar consumption in ounces per day?

7. The instructions for mixing a rose fertilizer call for $\dfrac{3}{4}$ cup of potash and $1\dfrac{3}{8}$ cups of nitrogen. What is the ratio of nitrogen to potash?

8. A recipe for chocolate frosting calls for $\dfrac{1}{2}$ cup of butter and $1\dfrac{1}{3}$ cups of melted chocolate. What is the ratio of butter to chocolate?

9. An orange contains 0.14 milligram of thiamine and 0.6 milligram of niacin. What is the ratio of niacin to thiamine?

10. The mass of Venus is about 0.81 times the mass of Earth, and the mass of Neptune is about 17.1 times the mass of Earth. What is the ratio of the mass of Venus to the mass of Neptune?

In Problems 11–18, solve each proportion.

11. $\dfrac{x}{16} = \dfrac{9}{24}$

12. $\dfrac{6}{15} = \dfrac{x}{20}$

13. $\dfrac{182}{65} = \dfrac{21}{w}$

14. $\dfrac{16}{w} = \dfrac{204}{119}$

15. $\dfrac{a}{a+2} = \dfrac{2}{3}$

16. $\dfrac{14}{10} = \dfrac{a}{a-2}$

17. $\dfrac{0.3}{0.5} = \dfrac{b+2}{12-b}$

18. $\dfrac{-1.8}{1.2} = \dfrac{b-8}{9+b}$

In Problems 19–24, decide whether the two variables are proportional.

19.

Time	1	2	4	5
Distance	45	90	180	225

20.

Principal	200	500	800	1000
Interest	12	30	48	60

21.

Length	3	4	8	10
Area	9	16	64	100

22.

Length	2	4	6	12
Width	18	9	6	3

23.

Rate	20	40	50	80
Time	40	20	16	10

24.

Width	4	8	16	24
Perimeter	88	96	112	128

Solve Problems 25–36 using proportions.

25. If 3 pounds of coffee makes 225 cups, how many pounds of coffee are needed to make 3000 cups of coffee?

26. It takes the office computer 12 minutes to prepare 30 paychecks. How long will the computer take to prepare 80 paychecks?

27. Gunther's car uses 32 liters of gas to travel 184 kilometers. How many liters will he need to travel 575 kilometers?

28. A pharmacist is mixing a prescription in which 24 milligrams of medication is dissolved in 150 milliliters of solution. How many milligrams of the medication will she need for 820 milliliters of solution?

29. On a map of Fairfield County, $\frac{3}{4}$ centimeter represents a distance of 10 kilometers. If Eastlake and Kenwood are 6 centimeters apart on the map, what is the actual distance between the two towns?

30. On a scale model of Fantasy Valley, $1\frac{1}{2}$ inches represents 50 yards. If the distance from the Water Slide to the Black Hole is 20 inches on the model, what is the actual distance between the two rides?

31. In a survey of 1200 voters, 863 favored the construction of a light rail system. If 8000 people vote, about how many will vote for the light rail system?

32. Park Service rangers release 60 tagged pheasant into a national park already populated with some untagged pheasant. Two months later, the rangers capture 40 pheasant and discover that 13 of them are tagged. Approximately how many pheasant are in the park?

33. If 1 inch equals 2.54 centimeters, what is the length in inches of a wire 35 centimeters long?

34. If 1 kilogram equals 2.2 pounds, what is the weight in kilograms of a suitcase that weighs 63 pounds?

35. Revenue from a state lottery is divided between education and administrative costs in the ratio of 4 to 3. If the lottery revenue this year is $24,000,000, how much money will go to education?

36. The Very Big Corporation divides its profits between new investments and dividends for stockholders in the ratio of 5 to 3. If the corporation made profits of $76,000,000 this year, how much money will go to dividends?

37. The following recipe for Cadet Mess Sloppy Joes comes from the *West Point Officers' Wives Club Cookbook*:

For 1000 servings:

310 pounds ground beef

100 pounds ground turkey

$5\frac{1}{2}$ cans mushrooms

4 pounds green peppers, sliced

18 cans tomato puree

12 cans tomato catsup

1 quart cider vinegar

2 cans diced red peppers

20 pounds onions, diced

$3\frac{1}{2}$ pounds sugar

1 pound garlic powder

3 pounds chili powder

1 quart light Flavor Glow

28 ounces salt and pepper

Brown turkey and beef in oven; place in pots. Braise chopped peppers and diced onions with meat; meat will absorb liquid. Add other ingredients. Stir together. Heat till piping hot. Serve on rolls.

a. What is the ratio of tomato puree to tomato catsup? How many servings of Sloppy Joes can you make if you have only 5 cans of tomato catsup? How many cans of tomato puree will you need?

b. How much chili powder (in ounces) will you need to serve Sloppy Joes to 50 people? How many pounds of ground beef?

c. The cookbook also lists the ingredients needed for 4500 servings. How many cans of mushrooms will be required for 4500 servings? How many pounds of onions?

38. Here is a recipe for Neiman-Marcus in Dallas Chocolate Chip Cookies:

For 112 cookies:

2 cups butter

2 cups sugar

2 cups brown sugar

4 eggs

1 teaspoon vanilla

4 cups flour

5 cups oatmeal, powdered

1 teaspoon salt

2 teaspoons baking powder

1 teaspoon soda

24 ounces chocolate chips

1 8-ounce Hershey bar, grated

3 cups chopped nuts

Cream butter and both sugars together. Add eggs and vanilla; mix together with flour and next four ingredients. Add last three ingredients. Roll into balls and place 2 inches apart on a cookie sheet. Bake for 10 minutes at 350°F.

a. What is the ratio of flour to oatmeal? If you have only 3 cups of flour, how many cookies can you make? How much oatmeal will you need?

b. If you eat 7 cookies, about how many ounces of chocolate chips will you eat?

c. If you would like to make 7 dozen cookies, how many eggs will you need? How many cups of chopped nuts?

3.2 Similarity

Reading

Centuries before any aircraft traveled beyond the surface of the earth, astronomers were able to estimate the distance from the earth to the moon and to the sun, as well as other large distances. They made these indirect measurements with a few insightful assumptions, some geometric facts, and proportions.

In about 200 B.C., the Greek astronomer Eratosthenes estimated the circumference of the earth. He knew that Alexandria was directly north of the ancient city of Syene in what is now Egypt. This meant that as the earth rotated, the sun would reach its highest point in the sky at the same moment over both cities. He also knew that at the moment when the sun was directly overhead in Syene, observers in Alexandria would see the sun at an angle 7.2° below the zenith. This situation is illustrated in Figure 3.1.

Figure 3.1

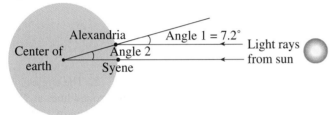

Eratosthenes assumed that the sun was so far away that light rays hitting Syene would be effectively parallel to light rays hitting Alexandria. Therefore, angles 1 and 2 in Figure 3.1 are corresponding angles and must be equal. Eratosthenes concluded that the angle at the earth's center must also be 7.2°.

Try Exercise 1 now.

Now, the fraction of the earth's circumference between Alexandria and Syene is the same as the fraction of a whole circle formed by the angle 7.2°. In other words, we can write a proportion as follows:

$$\frac{\text{distance from Alexandria to Syene}}{\text{circumference of earth}} = \frac{7.2}{360}$$

The Greeks used a unit of distance called the *stade*, and the distance from Alexandria to Syene was 5000 stades. If we let C stand for the circumference of the earth, we can simplify this proportion to

$$\frac{5000}{C} = \frac{7.2}{360}$$

Now try Exercise 2.

Eratosthenes' measurement was astonishingly accurate. The error in his estimate is less than 1% of the earth's actual circumference.

Proportional measurements are often useful in studying geometric objects or diagrams that involve geometric figures. In this section, we use proportions to describe objects with the same shape but different sizes.

EXERCISE 1
What fraction of a whole circle is an angle of 7.2°?

EXERCISE 2
a. Solve the proportion to find the circumference of the earth in stades.

b. If 1 stade is about 157.5 meters, find the circumference of the earth in kilometers.

Skills Review

Explain why the measurements shown in each figure cannot be accurate.

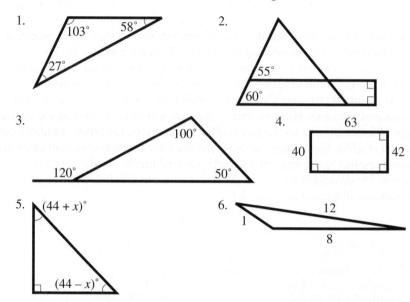

1.

103°   58°
27°

2.

55°
60°

3.

100°
120°   50°

4.

63
40   42

5.

$(44 + x)°$
$(44 - x)°$

6.

12
1
8

Answers: *1.* The sum of the angles is more than 180°.   *2.* The angles marked 55° and 60° should be equal because the two horizontal lines are parallel.   *3.* The angle marked 120° should be equal to the sum of 100° and 50°.   *4.* The opposite sides of the rectangle have different lengths.   *5.* The sum of the angles is less than 180°.   *6.* The triangle inequality is violated: The longest side, 12, is more than the sum of the other two sides.

Lesson

When we enlarge, reduce, or reproduce a photograph, the image in the copy should have the same shape as the original. For example, a copy of a circle should be a circle, and a copy of a square should be a square. If the original object is twice as tall as it is wide, then the copy should also be twice as tall as it is wide. When one geometric figure is an enlargement, a reduction, or a reproduction of another, the two figures are said to be **geometrically similar** (or simply **similar**). In Figure 3.2, (a) and (b) are similar, but (c) is not similar to the other two.

Figure 3.2

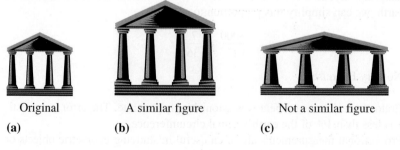

Original   A similar figure   Not a similar figure

(a)   **(b)**   **(c)**

Distances in Similar Figures

How can we tell whether two images of different size have precisely the same shape? Each dimension of one image must be **scaled**—that is, enlarged or reduced—by the same factor to produce the other image. If you measure the two similar buildings in Figure 3.2, you will find that the ratio of their heights is about 3 to 2:

$$\frac{\text{height of larger building}}{\text{height of smaller building}} = \frac{3}{2}$$

The ratio of their widths is also 3 to 2. On the other hand, the building in part (c) is the same height as the original, but it is nearly twice as wide. Its dimensions were not scaled by the same factor, so the ratios of corresponding dimensions are not the same. We can state the following general rule.

Distances in Similar Figures

In similar figures, the ratios of corresponding distances are equal.

Thus, if the distances a and b on one figure correspond to the distances a' and b' on a similar figure, as shown in Figure 3.3, then

$$\frac{a}{a'} = \frac{b}{b'},$$

In general, corresponding distances in similar figures are proportional. In fact, this condition is sufficient to determine whether two *rectangles* are similar: If their lengths and widths are proportional, the rectangles are similar.

Figure 3.3

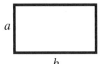
a
b

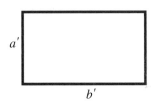
a'
b'

EXAMPLE 1

Which of the following pairs are similar rectangles?

a.
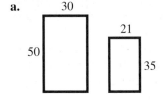
30
21
50
35

b.
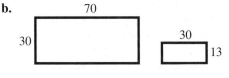
70
30
30
13

c.

5
5
7
7

d.

45
45
20
20

Figure 3.4

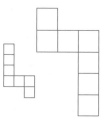

(a)

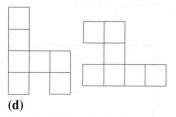

(b)

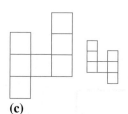

(c)

(d)

Solutions

a. The ratio of the widths of the two rectangles is

$$\frac{\text{width of larger}}{\text{width of smaller}} = \frac{30}{21} = \frac{10}{7}$$

The ratio of their lengths is

$$\frac{\text{length of larger}}{\text{length of smaller}} = \frac{50}{35} = \frac{10}{7}$$

The ratio is the same for both dimensions, so the rectangles are similar.

b. The two rectangles appear to have the same shape, but the ratio of their lengths is $\frac{70}{30} = \frac{7}{3}$, and the ratio of their heights is $\frac{30}{13}$. The rectangles are not similar.

c. Both figures are squares, and any two squares are similar.

d. The lengths and widths of the rectangles are identical, so they are similar.

_____

The orientation of geometric figures makes no difference when we are deciding whether they are similar. For example, in Figure 3.4, the shapes (made from squares) in each pair are geometrically similar.

EXERCISE 3
Which of the following figures are similar to the original?

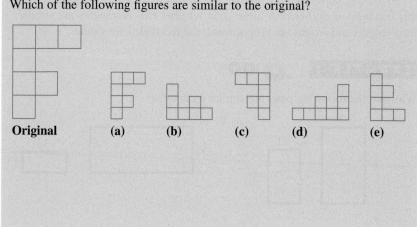

Original   **(a)**   **(b)**   **(c)**   **(d)**   **(e)**

Angles in Similar Figures

We have seen that two rectangles are similar if their dimensions are proportional. However, two figures of another shape may not be similar even if their corresponding lengths are proportional. For example, the parallelograms in Figure 3.5 are not similar, even though the ratio of the sides in each figure is $\frac{1}{2}$.

Figure 3.5

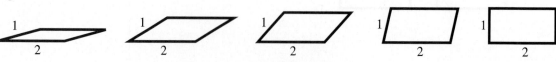

It is easy to see that the parallelograms in Figure 3.5 do not have the same shape. The *angles* between their sides are not equal. Because the angles are different, the figures are not similar.

Angles in Similar Figures

In similar figures, corresponding angles are equal.

Having equal corresponding angles is usually not enough to guarantee that two geometric figures are similar. For example, the angles in every rectangle are all 90°, but not all rectangles are similar. In order for two figures to be similar, they must satisfy *both* of the conditions discussed above.

Similar Geometric Figures

Two geometric figures are similar if and only if

1. The ratios of corresponding distances are equal, and

2. Corresponding angles are equal.

Now try Exercise 4.

Similar Triangles

In general, two geometric figures are similar only if they satisfy the two conditions discussed above: They must have equal corresponding angles and proportional corresponding sides. But triangles present a special case. To decide whether two triangles are similar, it is enough to establish only one of the two conditions for similarity. If one of the conditions is true for two triangles, the other condition is also true. If one of the conditions is *not* true the other condition is also not true.

Similar Triangles

Two triangles are similar if *either* of the following conditions is true:

1. Their corresponding angles are equal.

   or

2. Their corresponding sides are proportional.

EXERCISE 4
The following two figures are similar. Find the missing measurements, assuming that *c* is longer than 6.

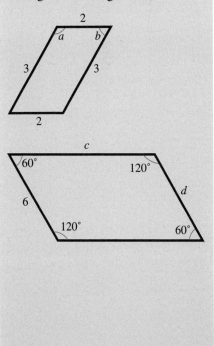

EXAMPLE 2

Which of the following pairs are similar triangles?

a.

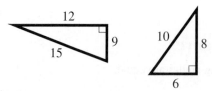

b.

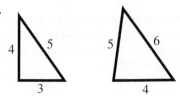

c.

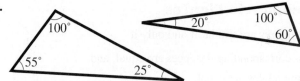

d.

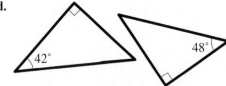

Solutions

a. The ratios of the sides of the larger triangle to the corresponding sides of the smaller triangle are $\frac{15}{10}$, $\frac{12}{8}$, and $\frac{9}{6}$. Because these ratios are equal, the triangles are similar.

b. The ratios of corresponding sides are not equal: The ratio of the longest sides is $\frac{6}{5}$, but the ratio of the smallest sides is $\frac{4}{3}$. The triangles are not similar.

c. The corresponding angles are not equal, so the triangles are not similar.

d. The unmarked angle of the triangle on the left is 48°, and the unmarked angle in the triangle on the right is 42°, so three pairs of angles match. The triangles are similar.

EXERCISE 5
Which of the following pairs are similar triangles? Explain why or why not in each case.

a.

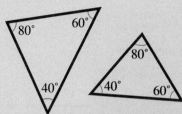

b.

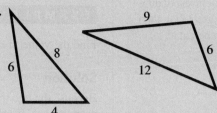

c.

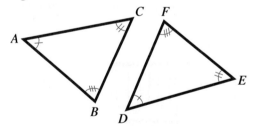

d.

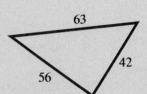

To indicate on a diagram that two angles are equal, we can use hatch marks through the arcs of the angles. Equal angles are given the same number of hatch marks. For instance, in Figure 3.6, the angles at *A* and *D* are equal, as are the angles at *B* and *F*, and the angles at *C* and *E*. Thus the diagram tells us that the two triangles are similar.

Figure 3.6

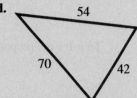

Using Proportions with Similar Triangles

Consider the two right triangles in Figure 3.7. Can we find the unknown lengths *a* and *h* in the larger triangle? First, we note that two pairs of corresponding angles in the triangles are equal: The right angles are equal, and the angles at the lower right vertex of each triangle are equal. But if two pairs of corresponding angles in the triangles are equal, then the third pair must also be equal. This means that the two triangles are similar.

Now try Exercise 6.

Figure 3.7

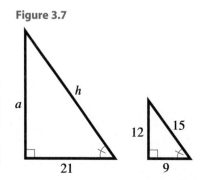

EXERCISE 6

Explain why the following statement is true: "If two pairs of corresponding angles in two triangles are equal, then the third pair must also be equal."

EXERCISE 7

Find the value of h in Figure 3.7.

Once we know that the two triangles in Figure 3.7 are similar, we can use the fact that their corresponding sides are proportional to find a and h.

EXAMPLE 3

Find the value of a in Figure 3.7.

Solution

Because the two triangles are similar, we know that their corresponding sides are proportional. If we form the ratios of the short sides and the medium sides, we obtain the following proportion.

larger triangle:
smaller triangle: $\dfrac{a}{12} = \dfrac{21}{9}$

    ratio of   ratio of
    medium sides   short sides

To solve the proportion, we cross-multiply and obtain

$$9a = 252 \quad \text{Divide both sides by 9.}$$
$$a = 28$$

Now try Exercise 7.

Your answer to Exercise 6 shows that if two *right* triangles have an acute angle with the same measure, then the two right triangles are similar. We can use this fact about right triangles to make indirect measurements, as shown in Example 4.

EXAMPLE 4

You would like to know the height of a certain building. You have a friend hold up a 5-foot pole near the building, and you measure the length of its shadow. The shadow of the pole is 3 feet long. At the same time, the shadow of the building is 12 feet long, as shown in Figure 3.8. Use similar triangles to write a proportion involving the height of the building.

Figure 3.8

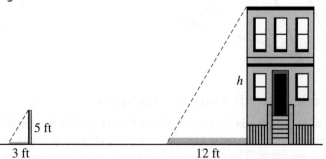

5 ft

h

3 ft   12 ft

Solution

In Figure 3.8, we see two right triangles: One triangle is formed by the building and its shadow, the other by the pole and its shadow. Because the light rays from the sun are parallel, the two angles at the tips of the shadows are equal. Therefore, the two right triangles are similar, and consequently their corresponding sides are proportional. We set the ratio of their heights equal to the ratio of their bases. We write each ratio with the side of the larger triangle in the numerator.

$$\begin{matrix}\text{larger triangle:} \\ \text{smaller triangle:}\end{matrix} \quad \frac{h}{5} = \frac{12}{3}$$

$$\quad\quad\quad\;\; \begin{matrix}\text{ratio of} \\ \text{heights}\end{matrix} \quad \begin{matrix}\text{ratio of} \\ \text{bases}\end{matrix}$$

Now try Exercise 8.

In some applications, the similar triangles may share a side or an angle.

EXAMPLE 5

Identify two similar triangles in Figure 3.9, and write a proportion to find the variable *H*.

Solution

The two triangles overlap, sharing the marked angle, as shown in Figure 3.10. Because each triangle also has a right angle, they are similar. Note that the base of the larger triangle is $24 + 12 = 36$. The ratio of the heights and the ratio of the bases must be equal, so

$$\begin{matrix}\text{larger triangle:} \\ \text{smaller triangle:}\end{matrix} \quad \frac{H}{10} = \frac{36}{24} \quad \text{Cross-multiply.}$$

$$24H = 360 \quad \text{Solve for } H.$$

$$H = 15$$

Figure 3.10

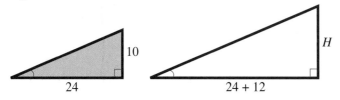

EXERCISE 8

Find the height of the building in Figure 3.8.

Figure 3.9

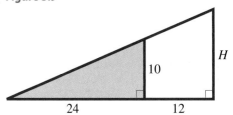

ANSWERS TO 3.2 EXERCISES

1. $\dfrac{1}{50}$

2a. 250,000 stades

2b. 39,375 kilometers

3. (a), (b), (c), and (e)

4. $a = 120°$, $b = 60°$, $c = 9$, $d = 6$

5. *b, c*

6. Because the sum of the angles in each triangle is 180°.

7. $h = 35$

8. 20 feet

9. 15 feet

EXERCISE 9

Heather wants to know the height of a street lamp. She discovers one night that when she is 12 feet from the lamp, her shadow is 6 feet long. Find the height of the street lamp.

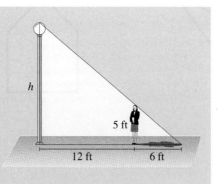

HOMEWORK 3.2

In Problems 1–8, which of the pairs of figures are geometrically similar? If the two figures are similar, give the ratio of lengths on the second figure to lengths on the first figure. If the figures are not similar, explain why not.

1.

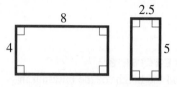

2.

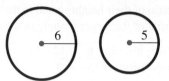

3.

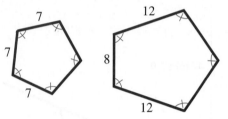

4.

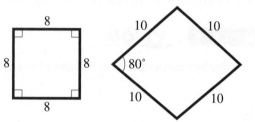

5.

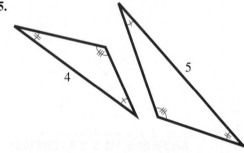

6.

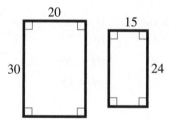

7.

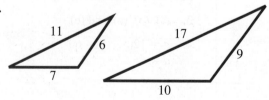

8.

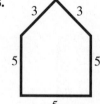

In Problems 9 and 10, determine which of the five figures are similar to the original.

9.

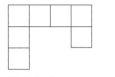

Original

(a)   (b)   (c)   (d)   (e)

10.

Original

(a)   (b)   (c)   (d)   (e)

In Problems 11–16, assume that the figures in each pair are geometrically similar. Solve for the variables. (Figures are not drawn to scale.)

11.

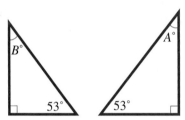

12.

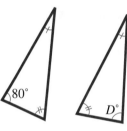

13.

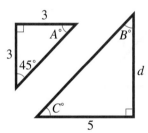

14.

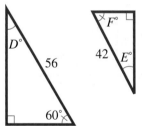

15.

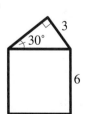

16.

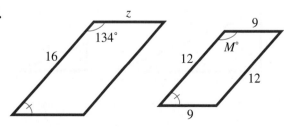

In Problems 17–20, determine whether the triangles in each pair are similar.

17.

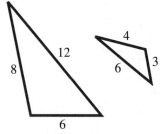

18.

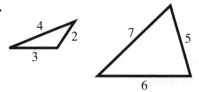

19.

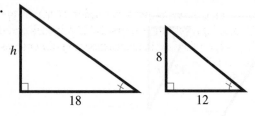

20.

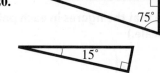

In Problems 21–26, use properties of similar triangles to solve for the variable.

21.

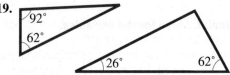

22.

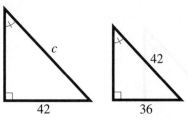

23.

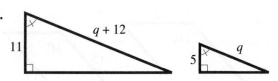

24.

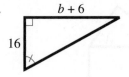

25.

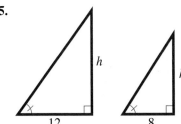

26.

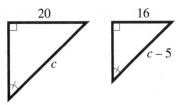

Use properties of geometric similarity to solve Problems 27–32.

27. In making the movie *Titanic*, the director used a 775-foot model of the ill-fated ship. The length of the original vessel was 882 feet 9 inches.
   a. To the nearest percent, what was the scale factor used to build the model?

   b. If the real *Titanic* was 175 feet tall from the keel to the top of the funnels, how tall is the model?

28. When they are planning to build a new bridge, civil engineers sometimes make a scale model of the site in order to study the flow of the water beneath the bridge and the stresses on the proposed structure.
   a. The Straits of Mackinac, between Michigan's Upper and Lower Peninsulas, is 5 miles wide. Engineers studying the Mackinac bridge, the world's longest suspension bridge, build a model in which the straits is 50 feet wide. Express the scale factor as a fraction.

   b. In the model, the suspension portion of the bridge is 7.2 feet long, and the towers are 12.5 inches above the water. What are the corresponding dimensions of the actual bridge?

29. Edo estimates the height of the Washington Monument as follows. He notices that he can see the reflection of the top of the monument at a point in the reflecting pool. He is 35 feet from that point, and that point is 1080 yards from the Washington Monument. From his physics class, Edo knows that the angles marked A and A' are equal. If Edo is 6 feet tall, what is his estimate for the height of the Washington Monument?

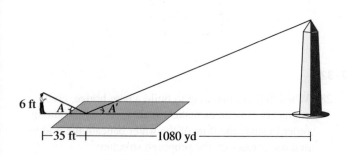

30. Judy is observing the Mr. Freeze roller coaster from a safe distance of 1000 feet. She notices that she can see the reflection of the highest point of the roller coaster in a puddle of water. Judy is 23.5 feet from that point in the puddle. If Judy is 5 feet 3 inches tall, how tall is the roller coaster?

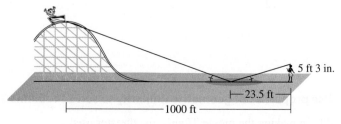

31. In the sixth century B.C., the Greek philosopher and mathematician Thales used similar triangles to measure the distance from the shore to a ship at sea. Two observers on the shore at points A and B sight a ship and measure the angles formed, as shown in figure (a). They then construct a similar triangle with the same angles at A and B, as shown in figure (b), and measure its sides. (This method is called *triangulation*.) Use the lengths given in the figure to find the distance from observer A to the ship.

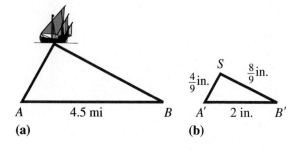

32. The Capilano Suspension Bridge is a free-swinging footbridge that spans a 230-feet-deep gorge in the hills north of Vancouver, British Columbia. Before crossing the bridge, you decide to estimate its length. You walk 100 feet downstream from the entrance to the bridge and sight the far end of the bridge, as shown in figure (a). You then construct a similar right triangle with a 2-centimeter base, as shown in figure (b). You find that the height of your triangle is 8.98 centimeters. How long is the Capilano Suspension Bridge?

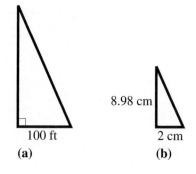

In Problems 33–36, assume the figures in each pair are geometrically similar. Solve for *y* in terms of *x*. (Figures are not drawn to scale.)

33.

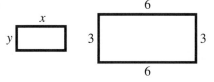

34.

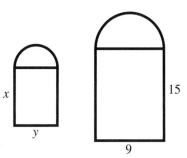

35.

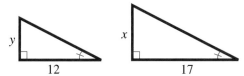

36.

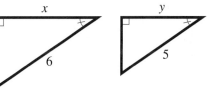

Use properties of similar triangles to solve for the variable in Problems 37–42.

37.

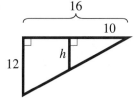

38.

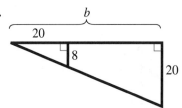

39.

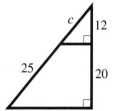

40.

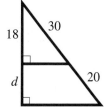

41.

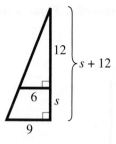

42.

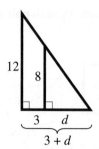

In Problems 43–46, solve for *y* in terms of *x*.

43.

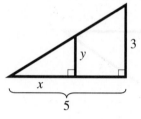

44.

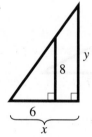

45.

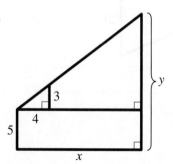

46.

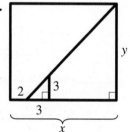

3.3 Direct Variation

Reading

Recall that if two variables are proportional, the ratio of their corresponding values is always the same. In Section 3.1, we saw that the price of grape juice was proportional to the amount purchased. The ratio $\dfrac{\text{total price}}{\text{number of quarts}}$ was always equal to 80. The number 80 is called the **constant of proportionality.**

If we let P stand for the price of the grape juice and q stand for the number of quarts purchased, we have the equation

$$\frac{P}{q} = 80$$

or

$$P = 80q$$

When we graph this equation, we obtain a line that passes through the origin, as shown in Figure 3.11.

In general, whenever two variables x and y are proportional, they are related by the equation

$$y = kx$$

where k is the constant of proportionality. When x and y are related in this way, we say that x and y **vary directly.** In the lesson, you will learn more about direct variation and the graphs of these relationships.

Figure 3.11

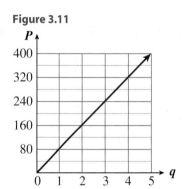

Skills Review

Try to do the following calculations mentally.

1. What is 80% of 120?
2. What is 150% of 36?
3. What percent of 20 is 12?
4. What percent of 16 is 20?
5. $\dfrac{9}{12}$ is the same as $\dfrac{?}{16}$.
6. $\dfrac{36}{30}$ is the same as $\dfrac{?}{20}$.
7. What is $\dfrac{5}{8}$ of 64?
8. What is $\dfrac{8}{3}$ of 27?

Answers: **1.** 96 **2.** 54 **3.** 60% **4.** 125% **5.** 12 **6.** 24 **7.** 40 **8.** 72

Lesson

In the following Activities, we consider the graphs of proportional variables. In each part, complete the table to help you decide whether the variables in each pair are proportional. Then graph the points.

Figure 3.12

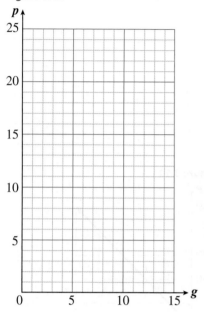

Activity

1. Table 3.2 below shows the price, p, for g gallons of gasoline. Plot the data on the grid in Figure 3.12.

Table 3.2

Gallons	Total price	$\dfrac{Price}{Gallon}$
4	$ 6.00	$\dfrac{6.00}{4} = ?$
6	$ 9.00	$\dfrac{9.00}{6} = ?$
9	$13.50	?
12	$18.00	?
15	$22.50	?

Figure 3.13

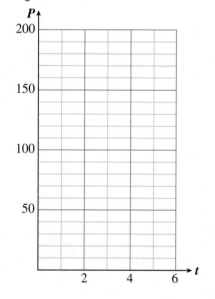

2. Table 3.3 shows the growth in population, P, of a new suburb t years after it was built. Plot the data on the grid in Figure 3.13.

Table 3.3

Years	Population	$\dfrac{People}{Year}$
1	10	$\dfrac{10}{1} = ?$
2	20	$\dfrac{20}{2} = ?$
3	40	?
4	80	?
5	160	?

3. At this point, can you make a conjecture (educated guess) about the graphs of proportional variables? To help you decide whether your conjecture is true, continue with parts 4 and 5 of the Activity.

4. Tuition at Woodrow University is $400 per semester plus $30 per unit.

   a. Write an equation for tuition, T, in terms of the number of units, u.

   b. Use your equation to fill in the second column of Table 3.4.

   c. Graph the equation on the grid in Figure 3.14.

   d. Are the variables proportional? Compute their ratios to decide. (Use the third column of the table.)

Table 3.4

Units	Tuition	$\dfrac{Tuition}{Unit}$
3		
5		
8		
10		
12		

Figure 3.14

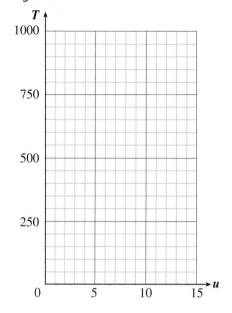

5. Anouk is traveling by train across Alaska at 60 miles per hour.

   a. Write an equation for the distance, D, Anouk travels in terms of hours, h.

   b. Use your equation to fill in the second column of Table 3.5.

   c. Graph the equation on the grid in Figure 3.15.

   d. Are the variables proportional? Compute their ratios to decide. (Use the third column of the table.)

Table 3.5

Units	Distance	$\dfrac{Distance}{Hour}$
3		
5		
8		
10		
12		

Figure 3.15

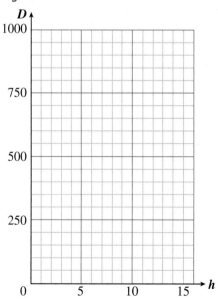

6. Can you revise your conjecture about the graphs of proportional variables so that it applies to the graphs in parts 4 and 5 as well? (*Hint:* Look at the two graphs of proportional variables. What is the *y*-intercept of both graphs?)

HOMEWORK 3.3

Answer the questions in Problems 1–6 using complete sentences.

1. If two variables are proportional, what is true about their ratio?

2. Using your answer to Problem 1, show that if M is proportional to z, then $M = kz$, where k is a constant.

3. Which of the four graphs in the Activity in this section illustrate direct variation?

4. State your revised conjecture from Activity 6 about the graphs of proportional variables.

5. If S varies directly with w, what does the graph of S versus w look like?

6. The circumference, C, of a circle varies directly with its diameter, d. What is the constant of proportionality?

7. Bruce is curious whether his water bill varies directly with the number of units (hundred cubic feet, or HCF) of water he uses. The table shows information from three of his recent water bills.

Month	HCF used	Amount due
March	43	$ 76.90
May	77	$156.51
July	101	$220.17

Does Bruce's water bill vary directly with the amount of water he uses? Support your answer with calculations.

8. An article in the *Los Angeles Times* (June 7, 1997) disputed the theory that lower crime rates are a direct result of a decline in population in the 15- to 24-year-old age bracket, often called the most crime-prone group. The table shows the drop in the crime rate in several California cities from 1995 to 1996 and the drop in the number of young adults.

City	Change in number of 15- to 24-years-olds	Change in crime rate
Los Angeles	−23.2%	−11.6%
San Diego	−19.2%	−4.1%
San Jose	−17.0%	−5.0%

Does the change in the crime rate vary directly with the change in the number of 15- to 24-year-olds? Support your answer with calculations.

For each pair of variables in Problem 9–12:
a. **Find the constant of proportionality.**
b. **Write an equation relating the variables.**
c. **Make a table of values, and graph the equation.**

9. Everett can bicycle 16 miles in 2 hours and 24 miles in 3 hours. The distance d that Everett travels varies directly with the time t that he rides.

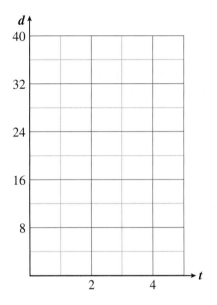

10. Josh made $60 when he worked for 5 hours and $144 when he worked for 12 hours. His paycheck P varies directly with the number of hours h he works.

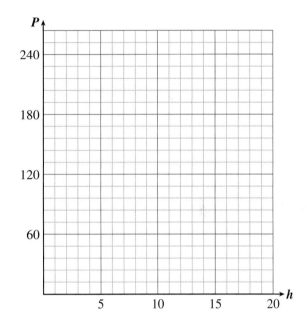

11. The sales tax on $12 is 72 cents, and the sales tax on $15 is 90 cents. The tax t varies directly with the price p.

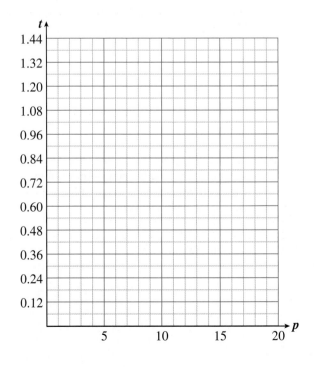

12. A 20-ounce bottle of Kola contains 3 ounces of sugar, and a 64-ounce bottle contains 9.6 ounces of sugar. The amount of sugar s varies directly with the amount of Kola, K.

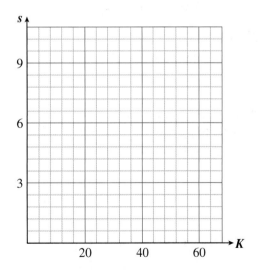

Use the equations you wrote in Problems 9–12 to answer the questions in Problems 13–16.

13. a. Fill in the table using your equation from Problem 9.

Time (hours)	Distance (miles)	Time (hours)	Distance (miles)
2		4	
3		6	
5		10	

   b. What happens to the distance Everett travels when he doubles the time he rides?

14. a. Fill in the table using your equation from Problem 10.

Hours worked	Paycheck	Hours worked	Paycheck
4		12	
6		18	
8		24	

   b. What happens to Josh's paycheck when the number of hours he works triples?

15. a. Fill in the table using your equation from Problem 11.

Price	Sales tax	Price	Sales tax
12		6	
20		10	
30		15	

   b. What happens to the sales tax when the price is cut in half?

16. a. Fill in the table using your equation from Problem 12.

Kola (ounces)	Sugar (ounces)	Kola (ounces)	Sugar (ounces)
20		5	
32		8	
64		16	

   b. What happens to the amount of sugar you consume when you drink only one-quarter as much Kola?

17. Generalize your results from Problems 13–16: If y varies directly with x, and you multiply the value of x by a constant n, what happens to the value of y?

18. The weight of an aquarium full of water varies directly with its volume. Elizabeth would like to replace her 20-cubic-foot aquarium with a new 30-cubic-foot model. If her current aquarium weighs 1248 pounds, how much will the new one weigh?

Does the rule you formulated in Problem 17 hold if the variables are *not* proportional? Use Problems 19 and 20 to help you decide.

19. a. The cost of tuition at Walden College is given by the formula $T = 500 + 40u,$ where u is the number of units you take. Is T proportional to u?

20. a. The temperature in degrees Fahrenheit is given by $F = 1.8C + 32,$ where C is the temperature in degrees Celsius. Is F proportional to C?

b. Fill in the table.

Units	Tuition	Units	Tuition
3		6	
5		10	
8		16	

c. Does doubling the number of units you take double your tuition?

b. Fill in the table.

Temperature (°C)	Temperature (°F)
10	
30	
40	
5	
15	
20	

c. If the Celsius temperature is reduced by half, is the Fahrenheit temperature also reduced by half?

In Problems 21–28, decide whether the two variables are proportional. Try to do the calculations mentally.

21. A 9-inch pizza costs $6, and a 12-inch pizza costs $10.

22. A quart of milk costs $0.90, and a gallon of milk costs $3.20.

23. A commuter train travels 10 miles in 20 minutes and 15 miles in 30 minutes.

24. A 4-inch chocolate bunny weighs 8 ounces, and a 6-inch bunny weighs 27 ounces.

25. It takes $4\frac{1}{2}$ cups of flour to make two loaves of bread and 18 cups to make eight loaves.

26. Alonso traveled 250 miles on 10 gallons of gas and 300 miles on 12 gallons of gas.

27. An 18-foot sailboat sleeps four, and a 32-foot sailboat sleeps six.

28. A shipment of 320 microwave ovens included 12 defective ovens, and a shipment of 240 ovens included 9 defective ones.

In Problems 29–32, use the points in each graph to decide whether y varies directly with x.

29.

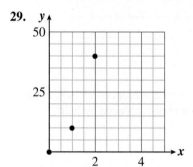

30.

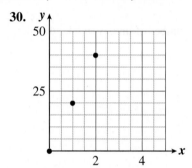

31.

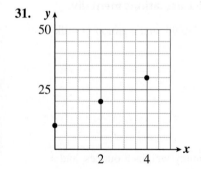

32.

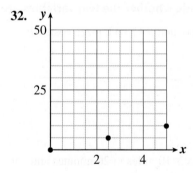

Problems 33–36 give the equations for three examples of direct variation. Complete a table of values for each equation, and graph all three on the grid provided.

33. a. $y = 2x$

34. a. $y = -2x$

b. $z = 3x$

b. $z = -2.5x$

c. $w = 1.5x$

c. $w = -x$

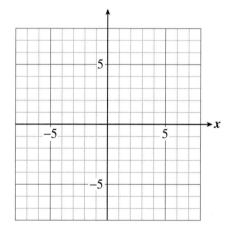

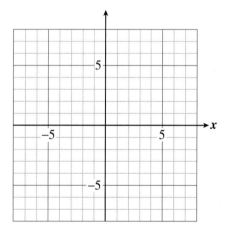

35. a. $y = -\dfrac{1}{3}x$

b. $z = -\dfrac{2}{3}x$

c. $w = -\dfrac{4}{3}x$

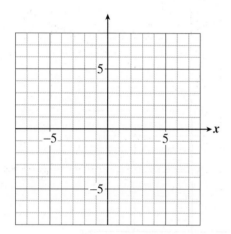

36. a. $y = \dfrac{1}{4}x$

b. $z = \dfrac{3}{4}x$

c. $w = \dfrac{7}{4}x$

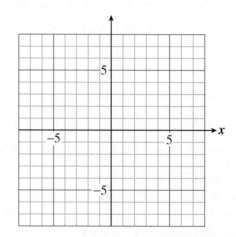

37. What are the *x*- and *y*-intercepts for each graph in Problems 33–36?

38. Can you graph an equation of the form $y = kx$ using the intercept method? If so, explain how; if not, explain why not.

3.4 Slope

Reading

In order to be fired correctly, pieces of a particular kind of pottery must first be cured by raising the temperature of the oven slowly and evenly. Sonia checks the temperature in the oven at 10-minute intervals and records the following data.

Time, x	0	10	20	30	40	50	60
Temperature, y	70	74	78	82	86	90	94

A graph of the data is shown in Figure 3.16.

The temperature in the oven should not be allowed to increase any faster than 0.5 degree per minute. Sonia can calculate the rate at which the temperature is rising by finding the following ratio for two of her data points:

$$\frac{\text{change in temperature}}{\text{change in time}}$$

For example, over the first 10 minutes, the temperature rises from 70 degrees to 74 degrees, so

$$\frac{\text{change in temperature}}{\text{change in time}} = \frac{4 \text{ degrees}}{10 \text{ minutes}}$$

or 0.4 degree per minute. This is less than the maximum rate recommended for curing the pottery.

You can check that over each 10-minute interval, the temperature rises by 4 degrees, so the oven is heating up at an acceptable rate. In order to facilitate such calculations, we use some special notation. The Greek letter Δ (which is read "delta") is often used in mathematics to indicate change. Because we have used the variable x to represent time in minutes and y to represent the temperature in degrees, we can denote the ratio $\frac{\text{change in temperature}}{\text{change in time}}$ by $\frac{\Delta y}{\Delta x}$.

With this notation, we calculate the rate of change of temperature between the data points (20, 78) and (50, 90) as follows:

$$\frac{\Delta y}{\Delta x} = \frac{12 \text{ degrees}}{30 \text{ minutes}} = 0.4 \text{ degree per minute}$$

We can also illustrate this rate of change on the graph of the data, as shown in Figure 3.17. Think of moving from the point (20, 78) to the point (50, 90) by moving first horizontally a distance of $\Delta x = 30$ and then vertically a distance of $\Delta y = 12$.

Calculating the rate of change of one variable with respect to another is so important in applications that the ratio $\frac{\Delta y}{\Delta x}$ is given a name; it is called **slope** and is usually denoted by the letter m. We will most often be concerned with the slope of a line, which we define as follows:

Figure 3.16

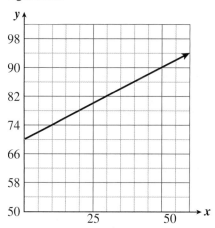

Figure 3.17

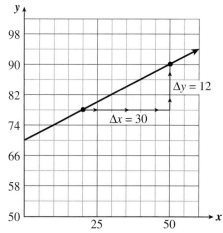

> ## Slope
>
> The slope of a line is defined by the ratio
>
> $$\text{slope} = \frac{\text{change in } y\text{-coordinate}}{\text{change in } x\text{-coordinate}}$$
>
> as we move from one point to another on the line. In symbols,
>
> $$m = \frac{\Delta y}{\Delta x}$$

Now try Exercise 1.

EXERCISE 1

Calculate the slope of the line in Figure 3.16 from the point (10, 74) to the point (30, 82). Illustrate Δy and Δx on the graph.

Skills Review

Write a rate for each of the following situations, including units.

1. Zack's average speed, if he drove 426 miles in 9 hours
2. Zelda's average speed, if she ran 6.6 miles in 55 minutes
3. The rate at which water flows through a pipe, if the pipe fills a 400-gallon storage tank in 20 minutes
4. A baby whale's rate of growth, if it gains 3000 pounds in its first 40 days of life
5. Ernest's rate of pay, if he earns $344 for a 40-hour week
6. Meg's rate of pay, if she charges $90 to type a 40-page paper

Answers: *1.* $47.\overline{3}$ miles per hour   *2.* 0.12 mile per minute   *3.* 20 gallons per minute   *4.* 75 pounds per day   *5.* $8.60 per hour   *6.* $2.25 per page

Lesson

In the following Activity, you will calculate slopes for the graphs in Section 3.3. You should work in groups of three or four so that you can compare your results with those of other students. Recall that to calculate the slope, we choose two points on the graph and compute the ratio

$$\frac{\text{change in vertical coordinate}}{\text{change in horizontal coordinate}}$$

Be sure to include the units with your ratios! Record your answers in your workbook.

Activity

1. Choose two points from the graph in Figure 3.12, which shows the price, p, of gasoline in terms of the number of gallons, g, purchased.

   First point:

   Second point:

   Change in vertical coordinates: $\Delta p =$

   Change in horizontal coordinates: $\Delta g =$

   Slope: $\dfrac{\Delta p}{\Delta g} =$

2. Choose two points from the graph in Figure 3.13, which shows the population, P, of a new suburb t years after it was built.

First point:

Second point:

Change in vertical coordinates: $\Delta P =$

Change in horizontal coordinates: $\Delta t =$

Slope: $\dfrac{\Delta P}{\Delta t} =$

3. Choose two points from the graph in Figure 3.14, which shows the cost, T, of tuition in terms of the number of units, u, taken.

First point:

Second point:

Change in vertical coordinates: $\Delta T =$

Change in horizontal coordinates: $\Delta u =$

Slope: $\dfrac{\Delta T}{\Delta u} =$

4. Choose two points from the graph in Figure 3.15, which shows the distance, d, Anouk travels in terms of hours, h, elapsed.

First point:

Second point:

Change in vertical coordinates: $\Delta d =$

Change in horizontal coordinates: $\Delta h =$

Slope: $\dfrac{\Delta d}{\Delta h} =$

5. a. Did everyone in your group get the same value for the slope of the graph in Figure 3.12?

b. Do you think you will always get the same value for the slope of this graph, no matter which two points you choose?

6. a. What about the graphs in Figures 3.13 through 3.15? Which of these graphs will give different values for the slope, depending on which points you choose?

b. What is different about this graph, compared to the other two graphs?

7. Remember that the slope is a *rate* of change. For each of the four graphs you just considered, answer the following questions.

a. What units is the rate expressed in?

b. What does the slope tell you about the variables involved?

Negative Slopes

All of the slopes you calculated in parts 1–4 of the Activity have positive values because the graphs are increasing. An increasing quantity has a positive rate of change, or slope, and a decreasing quantity has a negative rate of change.

EXAMPLE 1

The value of new office equipment decreases, or depreciates, over time. The graph in Figure 3.18 shows the value, V, in thousands of dollars, of a large photocopying machine t years after it was purchased.

    a. Compute the slope of the line in Figure 3.18.

    b. What does the slope tell you about the value of the machine?

Figure 3.18

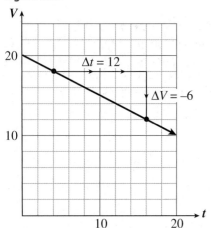

Solution

    a. We choose two points on the graph—say, (4, 18) and (16, 12). We imagine moving from the first point to the second point in two steps, first horizontally and then vertically, as shown in the figure. Each grid line crossing the horizontal axis counts for 2 units, and we move 6 spaces, so $\Delta t = 12$.   To get to the second point, we must move vertically downward, so ΔV is negative. Each grid line crossing the vertical axis also counts for 2 units, so $\Delta V = -6$.   Thus the slope of the line is

$$\frac{\Delta V}{\Delta t} = \frac{-6}{12} = -0.5$$

    b. To interpret the slope, we consider the units of the ratio $\dfrac{\Delta V}{\Delta t}$.   The vertical axis is scaled in thousands of dollars, so the units of ΔV are thousands of dollars, and the units of Δt are years. Accordingly,

$$\frac{\Delta V}{\Delta t} = -0.5 \text{ thousand dollars per year}$$

Thus the slope gives the rate at which the photocopying machine is depreciating as $500 per year.

    When we move downward on the graph in Example 1, ΔV is considered negative. Similarly, when we move to the left horizontally, Δt is considered negative. If we compute the slope in Example 1 by moving from the point (16, 12) to the point (4, 18), as shown in Figure 3.19, then we have

$$\frac{\Delta V}{\Delta t} = \frac{6}{-12} = -0.5$$

This is the same answer as before. In general, it doesn't matter which direction we move along a line to compute its slope; the answer will be the same.

Figure 3.19

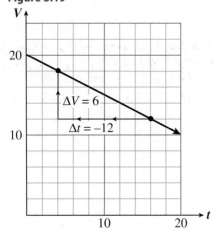

Now try Exercises 2 and 3.

EXAMPLE 2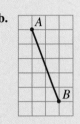

Calculate the slope of the line $y = \dfrac{2}{3}x - 2,$ shown in Figure 3.20, in two ways:

   a. Use the points $P(-3, -4)$ and $Q(3, 0)$
   b. Use the points $R(6, 2)$ and $S(0, -2)$

Figure 3.20

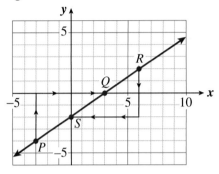

Solution

   a. Starting at point $P(-3, -4)$, we move *up* 4 units in the *y*-direction and *right* 6 units in the *x*-direction to get to point $Q(3, 0)$. Thus

$$m = \frac{\Delta y}{\Delta x} = \frac{4}{6} = \frac{2}{3}$$

   b. Starting at point $R(6, 2)$, we move *down* 4 units in the *y*-direction and *left* 6 units in the *x*-direction to get to point $S(0, -2)$. Thus,

$$m = \frac{\Delta y}{\Delta x} = \frac{-4}{-6} = \frac{2}{3}$$

We get the same value for the slope, $\dfrac{2}{3}$, using either pair of points.

Geometric Meaning of Slope

Suppose we graph two lines with positive slope on the same coordinate system. If we move along the lines from left to right, then the line with the larger slope will be steeper. This makes sense if you think of the slope as a rate of change: The line whose *y*-coordinate is increasing faster with respect to *x* is the steeper line.

EXERCISE 2
Find the slope of each line segment. (Each square counts for 1 unit, horizontally and vertically.)

a.

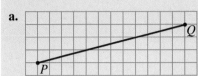

b.

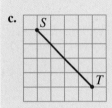

c.

EXERCISE 3
In Exercise 2a, does it matter whether you move from *P* to *Q* or from *Q* to *P* to compute the slope of the line? Verify that you get the same answer if you move in either direction.

You can verify the slope given for each line in Figure 3.21 by computing $\frac{\Delta y}{\Delta x}$. You can also see that for each unit you increase in the *x*-direction, the steepest line increases 2 units in the *y*-direction, the middle line increases 1 unit in the *y*-direction, and the least steep line increases only $\frac{1}{3}$ unit.

Figure 3.21

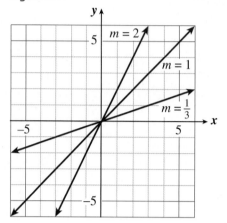

Figure 3.22 shows several lines with negative slopes. These lines slant downward, or decrease, as we move from left to right. The more negative the slope, the more sharply the line decreases. For both increasing and decreasing graphs, the larger the absolute value of the slope, the steeper the graph.

Figure 3.22

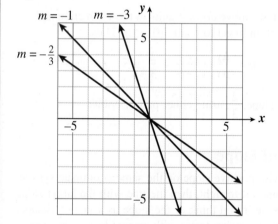

You should note that slopes reveal the relative steepness of two lines only if the lines are graphed on axes with the same scales. Changing the scale on either the *x*-axis or the *y*-axis can greatly alter the appearance of a graph.

EXERCISE 4

a. Which of the two lines shown below appears steeper?

b. Compute the slopes of the two lines. Which has the greater slope?

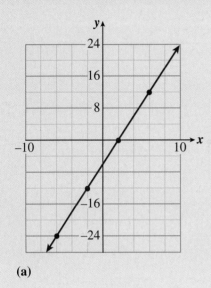

(a)

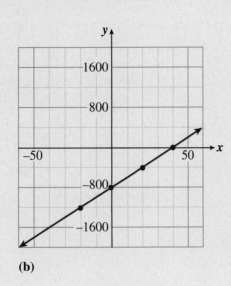

(b)

What about the slopes of horizontal and vertical lines? In the homework problems, you will show that the slope of a horizontal line is zero and that the slope of a vertical line is undefined. We sometimes say that a vertical line has no slope, but a horizontal line *does* have a slope: Its slope is zero.

ANSWERS TO 3.4 EXERCISES

1. 0.4

2a. $\dfrac{1}{4}$

2b. $\dfrac{-5}{2}$

2c. -1

3. No

4a. The line in Figure (a) is steeper.

4b. For the line in Figure (a), $m = 3$; for the line in Figure (b), $m = 20$. The line in Figure (b) has the greater slope.

HOMEWORK 3.4

In Problems 1–4, find the slope of each line segment.

1. a.

b.

2. a.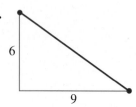

b.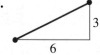

3. a.

b.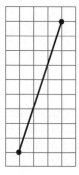

4. a.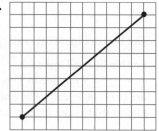

b.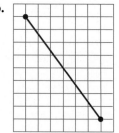

Find the slope of each line in Problems 5–8. Illustrate Δx and Δy on the graph.

5. $x + 2y = 6$

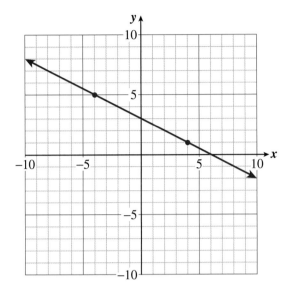

6. $2x - y = 8$

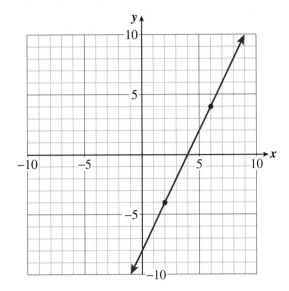

7. $3x - 2y = 0$

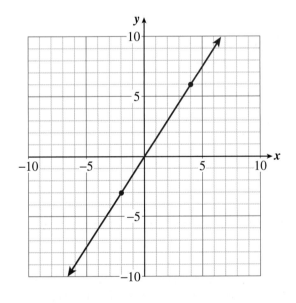

8. $x - 3y = 0$

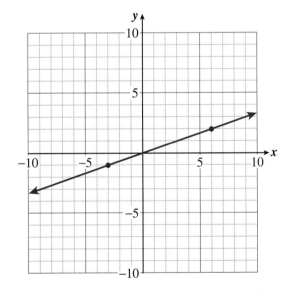

In Problems 9–16:

a. Find the intercepts of each line.

b. Use the intercept method to graph the line.

c. Use the intercepts to calculate the slope of the line.

d. Calculate the slope again using the suggested points on the line.

9. $2x + 3y = 12$
$(-3, 6)$ and $(3, 2)$

10. $3x + 5y = 15$
$(-5, 6)$ and $(10, -3)$

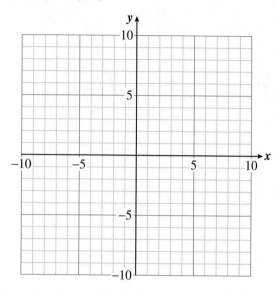

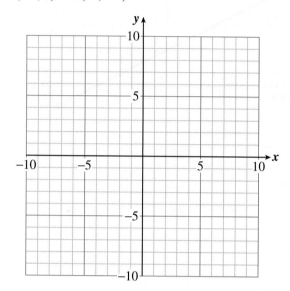

11. $5x - 2y = 10$
$(-2, -10)$ and $(4, 5)$

12. $4x - 3y = 12$
$(-3, -8)$ and $(6, 4)$

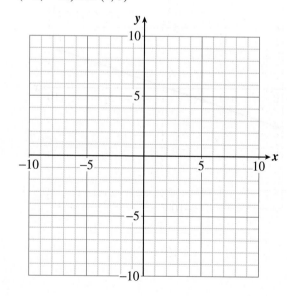

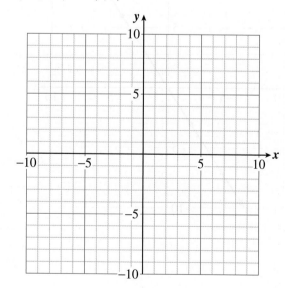

13. $x + y = 5$
$(-3, 8)$ and $(8, -3)$

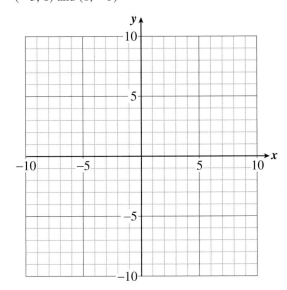

14. $x - y = 4$
$(6, 2)$ and $(-2, -6)$

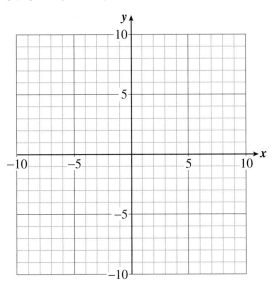

15. $x - 2y = 4$
$(6, 1)$ and $(-4, -4)$

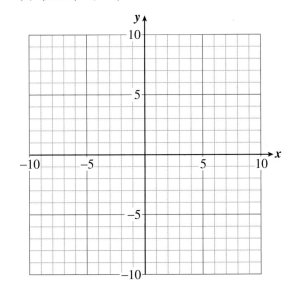

16. $3x + y = 6$
$(3, -3)$ and $(1, 3)$

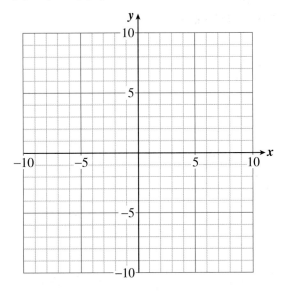

In Problems 17–20, imagine that both lines are graphed on the same coordinate system.

17. Which line is steeper: one with slope $\dfrac{3}{5}$ or one with slope $\dfrac{5}{3}$?

18. Which line is steeper: one with slope -3 or one with slope -5?

19. Which line is decreasing: one with slope $\dfrac{1}{4}$ or one with slope -2?

20. Which line is steeper: one with slope -4 or one with slope 2?

Use the definition of slope to help you answer Problems 21–24.

21. The line shown has slope $\dfrac{5}{2}$. If $\Delta x = 7$, find Δy.

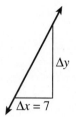

22. The line shown has slope $\dfrac{-2}{3}$. If $\Delta y = 9$, find Δx.

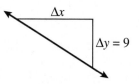

23. The line shown has slope -4. If $\Delta y = -6$, find Δx.

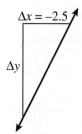

24. The line shown has slope 3. If $\Delta x = -2.5$, find Δy.

In Problems 25–28, draw and label a sketch illustrating each situation. Use a proportion to answer the questions.

25. A sign on a highway says "6% grade, next 3 miles." This means that the slope of the road ahead is $\frac{6}{100}$. How much will you climb in elevation (in feet) as you travel 3 miles horizontally?

26. The slope of a roof is often called the pitch of the roof. The pitch of the roof on Arch's A-frame cottage is 1.6. If it is 12 feet from the center of the floor of the loft to one wall, how far is it from the floor to the peak of the roof?

27. A wheelchair ramp must have a slope of 0.125. If the ramp must reach a door whose base is 2 feet off the ground, how far from the door will the other end of the ramp touch the ground?

28. A staircase has a 70% grade. If the riser on each step is 8.4 inches high, how deep is each step from front to back?

In Problems 29 and 30, choose two points on the graph from the given table and compute the slope of the line.

29.

x	0	2	6	8
y	−30	0	60	90

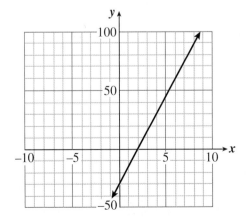

30.

x	−2	0	2	4
y	108	60	12	−36

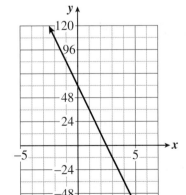

In Problems 31 and 32, graph the line and compute its slope.

31. $y = -12x + 32$

x	-2	0	3	4
y				

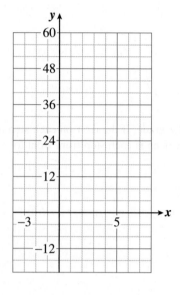

32. $y = \dfrac{1}{2}x + 20$

x	-40	0	20	40
y				

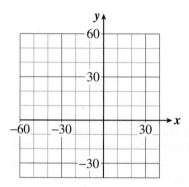

33. A line contains the points $(0, 0)$ and $(3, 2)$. What is its slope?

34. A line contains the points $(0, 0)$ and $(3, -4)$. What is its slope?

35. A line contains the points $(0, 0)$ and $(-30, 50)$. What is its slope?

36. A line contains the point $(0, 0)$ and the point $(-500, -400)$. What is its slope?

Problems 37–40 describe direct variation.
a. Write an equation that relates the variables.
b. Compute the slope of the graph, including units.
c. Interpret the slope as a rate. What does it tell you about the problem situation?

37. Audrey can drive 150 miles on 6 gallons of gas and 225 miles on 9 gallons of gas. Write an equation for the distance d that Audrey can drive on g gallons of gas.

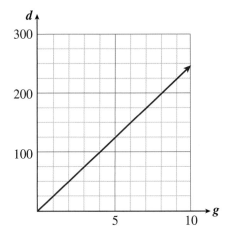

38. Jeremy can type 120 words in 4 minutes and 450 words in 15 minutes. Write an equation for the number of words w that Jeremy can type in t minutes.

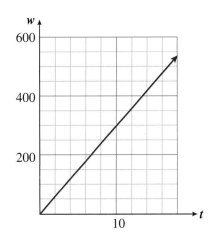

39. The sales tax on a \$15 purchase is 60 cents; the tax on a \$20 purchase is 80 cents. Write an equation for the tax T on a purchase of p dollars.

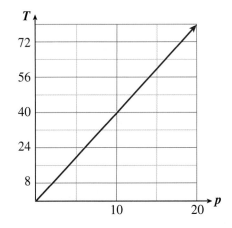

40. Mitra makes \$48 for 3 hours of work and \$240 for 15 hours of work. Write an equation for Mitra's salary S for h hours of work.

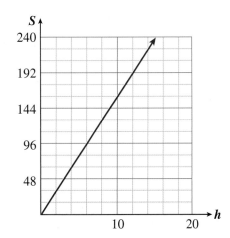

In Problems 41–44, calculate the slope of the line.

41.

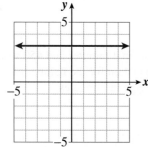

42.

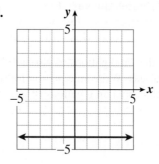

43.

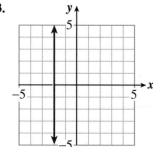

44.

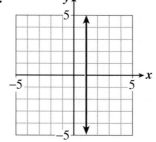

45. a. Explain why $\Delta y = 0$ for any two points on the lines in Problems 41 and 42.

b. Explain why the slope of any horizontal line is zero.

46. a. Explain why $\Delta x = 0$ for any two points on the lines in Problems 43 and 44.

b. Explain why the slope of any vertical line is undefined.

3.5 Slope-Intercept Form

Reading

Recall that a linear equation has the form $Ax + By = C$ and that its graph is a straight line. In Chapter 2, we studied two methods for graphing linear equations: point plotting and the intercept method. In this section, we see how the slope of a line is related to its equation and to its graph.

Skills Review

Simplify each expression. Write your answers in the form $ax + b$.

1. $2x + 3(4x - 2) + 1$

2. $7x - 5(2x + 3) - 2$

3. $\dfrac{3}{2}(4x - 6) - 3x$

4. $\dfrac{2}{3}(6x + 3) + x$

5. $\dfrac{2x + 5}{3} - 1$

6. $\dfrac{3x - 1}{4} + 2$

7. $x + 4 - \dfrac{3 - 5x}{2}$

8. $8 - x - \dfrac{4 + 2x}{3}$

Answers: **1.** $14x - 5$ **2.** $-3x - 17$ **3.** $3x - 9$ **4.** $5x + 2$

5. $\dfrac{2}{3}x + \dfrac{2}{3}$ **6.** $\dfrac{3}{4}x + \dfrac{7}{4}$ **7.** $\dfrac{7}{2}x + \dfrac{5}{2}$ **8.** $\dfrac{-5}{3}x + \dfrac{20}{3}$

In Section 3.4, we considered the following data for the temperature inside an oven used to cure pottery.

Time, x	0	10	20	30	40	50	60
Temperature, y	70	74	78	82	86	90	94

We also computed the slope of the graph of these data, shown in Figure 3.23. The slope is

$$\frac{\text{change in temperature}}{\text{change in time}} = \frac{\Delta y}{\Delta x}$$

$$= \frac{4 \text{ degres}}{10 \text{ minutes}}$$

or

$$m = 0.4 \text{ degree per minute}$$

Figure 3.23

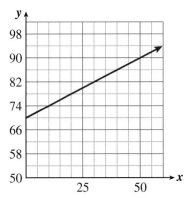

You can see from either the table of data or the graph that the *y*-intercept of the graph is the point (0, 70). This means that the initial temperature inside the drying oven was 70 degrees. In this section, you'll see that knowing these two pieces of information about a graph—its slope and its initial value, or *y*-intercept—enables you not only to sketch the graph but also to find its equation.

Lesson

We begin by considering how different slopes and different *y*-intercepts affect the graph of a line. In the following Activity, you'll compare the graphs of several related equations.

Activity

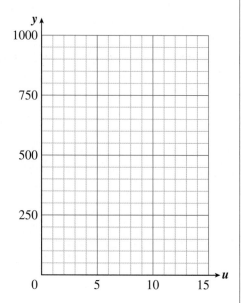

1. **a.** Tuition at Woodrow University is $400 per semester plus $30 per unit. Write an equation for tuition, *W*, in terms of the number of units, *u*, and fill in the second column of the table.

   b. At Xavier College, the tuition, *X*, is $200 per semester plus $30 per unit. Write an equation for *X* and fill in the third column of the table.

   c. At the Yardley Institute, the tuition, *Y*, is $30 per unit. Write an equation for *Y* and fill in the last column of the table.

u	*W*	*X*	*Y*
3			
5			
8			
10			
12			

   d. Graph all three equations on the grid.
   e. Find the slope and the *y*-intercept for each equation.

   f. How are your results from part (e) reflected in the graphs of the equations?

2. a. Anouk is traveling by train across Alaska at 60 miles per hour. Write an equation for the distance, *A*, Anouk travels in terms of hours, *h*, and fill in the second column of the table.

b. Boris is traveling by snowmobile at 30 miles per hour. Write an equation for the distance, *B*, Boris travels and fill in the third column of the table.

c. Chaka is traveling in a small plane at 100 miles per hour. Write an equation for the distance, *C*, Chaka travels and fill in the last column of the table.

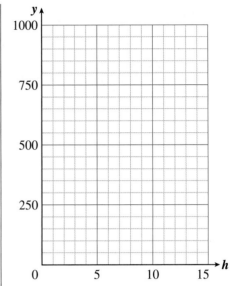

h	*A*	*B*	*C*
3			
5			
8			
10			
16			

d. Graph all three equations on the grid.
e. Find the slope and the *y*-intercept for each equation.

f. How are your results from part (e) reflected in the graphs of the equations?

Slopes and Intercepts

Let's summarize the results of the preceding Activity. In part 1, you graphed three equations:

$$W = 30u + 400$$
$$X = 30u + 200$$
$$Y = 30u$$

All of these equations have the same slope—namely, 30—but different *y*-intercepts. In each case, the *y*-intercept is the same as the constant term in the equation. (This makes sense, because we find the *y*-intercept by setting $u = 0$.)

In part 2, you graphed these three equations:

$$A = 60h$$
$$B = 30h$$
$$C = 100h$$

All of these equations pass through the origin, so their *y*-intercepts are all 0, but each has a different slope. In each case, the slope is the same as the coefficient of

EXERCISE 1

On Memorial Day weekend, Arturo drives from his home to a cabin on Diamond Lake. His distance from Diamond Lake after x hours of driving is given by the equation

$$y = 450 - 50x$$

a. What are the slope and the y-intercept of the graph of this equation?

b. What do the slope and the y-intercept tell you about the problem situation?

h in the equation. This also makes sense if you think about it: If h increases by 1 hour, then A increases by 60 miles, B increases by 30 miles, and C increases by 100 miles.

In general, we see that the coefficients of a linear equation tell us something about its graph. The constant term tells us the y-intercept of the graph, and the coefficient of x (or the independent variable) tells us the slope of the graph. For this reason, we say that the linear equation

$$y = mx + b$$

is written in slope-intercept form.

> A linear equation written in the form
>
> $$y = mx + b$$
>
> is said to be in **slope-intercept form**. The coefficient m is the slope of the graph, and b is the y-intercept.

EXAMPLE 1

The temperature H inside a pottery curing oven is given by the equation

$$H = 70 + 0.4t$$

where t is the number of minutes since the oven was turned on. What does the equation tell us about the temperature?

Solution

The equation can be written in slope-intercept form, $H = mt + b$,   as

$$H = 0.4t + 70$$

The slope of the graph is 0.4, and the y-intercept is 70. This means that the initial temperature in the oven was 70 degrees and that after the heat was turned on, the temperature rose at a rate of 0.4 degree per minute.

Now try Exercise 1.

Slope-Intercept Method for Graphing

We can use the slope-intercept form of a linear equation to graph the equation quickly, without having to plot a lot of points. For example, suppose we would like to graph the equation

$$y = \frac{3}{4}x - 2$$

We note that the slope of the line is $\frac{3}{4}$ and its y-intercept is the point $(0, -2)$. We begin by plotting the y-intercept, as shown in Figure 3.24.

Next, we use the slope to find another point on the line. Recall that the slope, $\frac{\Delta y}{\Delta x} = \frac{3}{4}$, gives the ratio of the change in y-coordinate to the change in x-coordinate as we move from any point on the line to another point on the line. Thus, if we start at the point $(0, -2)$ and move 3 units up (the positive y-direc-

Figure 3.24

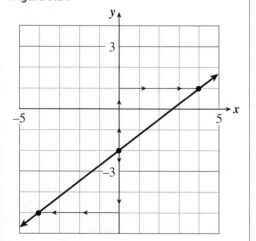

tion), then 4 units right (the positive *x*-direction), we will locate another point on the line. The coordinates of this new point are (4, 1).

To improve the accuracy of our graph, we can find a third point by writing the slope as

$$\frac{\Delta y}{\Delta x} = \frac{-3}{-4}$$

$\left(\text{Note that } \frac{-3}{-4} = \frac{3}{4}.\right)$ Starting again from the *y*-intercept (0, −2), we now move 3 units down and 4 units left and find another point on the graph, (−4, −5). Finally, we draw a line through the three points (−4, −5), (0, −2), and (4, 1) to obtain the graph in Figure 3.24.

The slope-intercept method is the third method we have studied for graphing lines. It can be used to graph any nonvertical line.

To Graph a Line Using the Slope-Intercept Method:

1. Write the equation in the form $y = mx + b$.

2. Plot the *y*-intercept, (0, *b*).

3. Write the slope as a fraction: $m = \dfrac{\Delta y}{\Delta x}$.

4. Use the slope to find a second point on the graph: Starting at the *y*-intercept, move Δy units in the *y*-direction and then Δx units in the *x*-direction.

5. Find a third point by starting from the *y*-intercept and moving $-\Delta y$ units in the *y*-direction and then $-\Delta x$ units in the *x*-direction.

6. Draw a line through the three plotted points.

Now try Exercise 2.

It is not necessary to know the *y*-intercept in order to graph a line. If we know the slope and any point on the line, we can obtain the graph.

EXAMPLE 2

Graph the line of slope $\dfrac{-1}{2}$ that passes through the point (−3, −2).

Solution

First, we plot the given point, (−3, −2). (See Figure 3.25 on page 322.) Then we use the slope to find a second point on the graph. Because the slope is $\dfrac{\Delta y}{\Delta x} = \dfrac{-1}{2}$, we start from the point (−3, −2), and move 1 unit down and then 2 units to the right to find the point (−1, −3) on the line. We can also write the slope as

$$\frac{\Delta y}{\Delta x} = \frac{1}{-2}$$

EXERCISE 2
Graph the equation $y = -3x + 4$ by the slope-intercept method.

$\left(\textit{Hint: } \text{Write the slope as a fraction,}\right.$

$\left. m = \dfrac{\Delta y}{\Delta x} = \dfrac{-3}{1}.\right)$

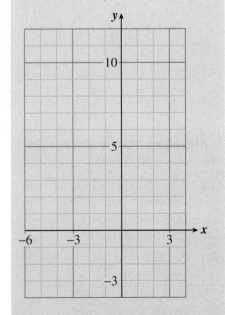

Using this form for the slope, we can obtain a third point by starting at $(-3, -2)$ and moving 1 unit up and then 2 units left to find $(-5, -1)$. Finally, we draw a line through the three points to obtain the graph in Figure 3.25.

Figure 3.25

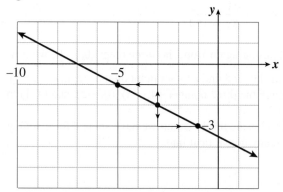

Rewriting an Equation in Slope-Intercept Form

We have seen that the slope-intercept form of a linear equation,

$$y = mx + b$$

gives us useful information about the variables involved: The y-intercept is the initial value of the dependent variable, and the slope is the rate of change. We can also use the slope-intercept method to obtain a sketch of the graph.

   Not all linear equations are given in slope-intercept form. Recall that the general form for a linear equation is

$$Ax + By = C$$

However, we can rewrite the equation of any nonvertical line in slope-intercept form by solving the equation for y in terms of x.

EXAMPLE 3

Find the slope and y-intercept of the graph of $3x - 4y = 8$.

Solution

We rewrite the equation in slope-intercept form by solving for y in terms of x.

$$3x - 4y = 8$$
$$\underline{-3x \qquad\qquad -3x}\qquad\qquad\text{Subtract } 3x \text{ from both sides.}$$
$$-4y = 8 - 3x\qquad\qquad\text{Divide both sides by } -4.$$
$$\frac{-4y}{-4} = \frac{-3x + 8}{-4}\qquad\qquad\text{Divide } -4 \text{ into each term of the right side.}$$
$$y = \frac{-3x}{-4} + \frac{8}{-4}\qquad\qquad\text{Simplify each quotient.}$$
$$y = \frac{3}{4}x - 2$$

The equation is now in slope-intercept form, with $m = \dfrac{3}{4}$ and $b = -2$.

Thus the slope of the graph is $\dfrac{3}{4}$, and the y-intercept is the point $(0, -2)$.

Now try Exercise 3.

EXERCISE 3
Write an equation for the line whose y-intercept is $(0, 4)$ and whose slope is -3.

ANSWERS TO 3.5 EXERCISES

1a. $m = -50, b = 450$

1b. Arturo's distance from the lake decreases at a rate of 50 miles per hour; his home is 450 miles from the lake.

2.

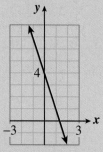

3. $y = -3x + 4$

HOMEWORK 3.5

In Problems 1–8, compare the three equations.
a. Fill in the *y*-values in the table, and graph the lines.
b. Choose two points on each line and compute its slope.

1.   i. $y = 2x - 6$

x	-1	0	1	2	3
y					

ii. $y = 2x + 1$

x	-1	0	1	2	3
y					

iii. $y = 2x + 3$

x	-1	0	1	2	3
y					

2.   i. $y = \dfrac{1}{3}x - 2$

x	-6	-3	0	3	6
y					

ii. $y = \dfrac{1}{3}x$

x	-6	-3	0	3	6
y					

iii. $y = \dfrac{1}{3}x + 4$

x	-6	-3	0	3	6
y					

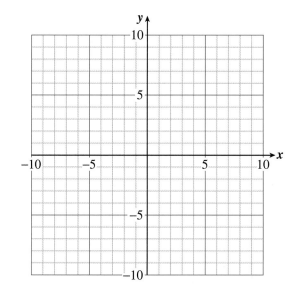

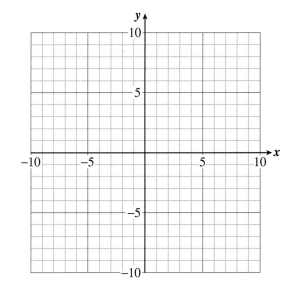

3.   **i.** $y = \dfrac{-3}{2}x - 4$

x	−6	−4	−2	0	2
y					

ii. $y = \dfrac{-3}{2}x + 2$

x	−6	−4	−2	0	2
y					

iii. $y = \dfrac{-3}{2}x + 6$

x	−6	−4	−2	0	2
y					

4.   **i.** $y = -x - 1$

x	−4	−2	0	2	4
y					

ii. $y = -x$

x	−4	−2	0	2	4
y					

iii. $y = -x + 3$

x	−4	−2	0	2	4
y					

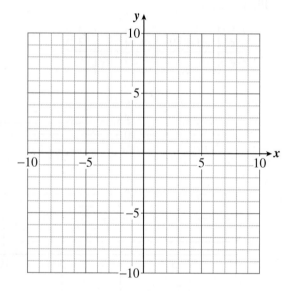

5. i. $y = \frac{1}{4}x + 2$

x	-4	-2	0	2	4
y					

ii. $y = \frac{1}{2}x + 2$

x	-4	-2	0	2	4
y					

iii. $y = x + 2$

x	-4	-2	0	2	4
y					

6. i. $y = 2x + 2$

x	-4	-2	0	2	4
y					

ii. $y = 4x + 2$

x	-4	-2	0	2	4
y					

iii. $y = \frac{5}{2}x + 2$

x	-4	-2	0	2	4
y					

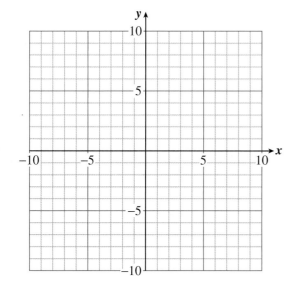

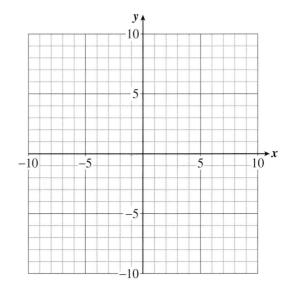

7.   **i.** $y = -3x - 2$

x	-6	-3	0	3	6
y					

ii. $y = -2x - 2$

x	-6	-3	0	3	6
y					

iii. $y = \dfrac{-5}{3}x - 2$

x	-6	-3	0	3	6
y					

8.   **i.** $y = \dfrac{-1}{3}x - 2$

x	-6	-3	0	3	6
y					

ii. $y = \dfrac{-2}{3}x - 2$

x	-6	-3	0	3	6
y					

iii. $y = -x - 2$

x	-6	-3	0	3	6
y					

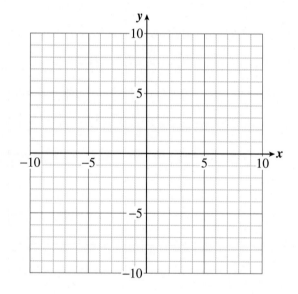

Write each equation in Problems 9–16 in slope-intercept form. State the slope and the *y*-intercept of the graph of each equation.

9. $y = 3x + 4$

10. $y = 3 - 4x$

11. $6x + 3y = 5$

12. $4x + 3y = -2$

13. $2x - 3y = 6$

14. $x - 2y = -7$

15. $5x = 4y$

16. $3y = -5x$

In Problems 17–22:
a. Find the slope and the *y*-intercept of the line.
b. Write an equation for the line.

17.
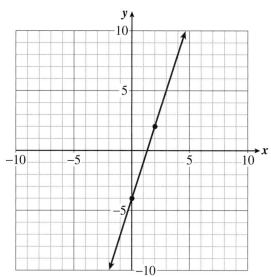

18.
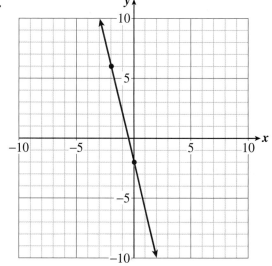

19.
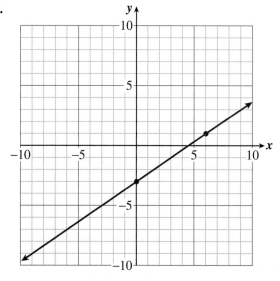

20.
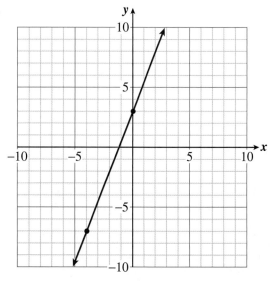

21.

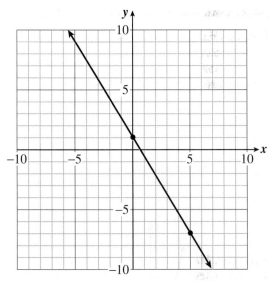

22.

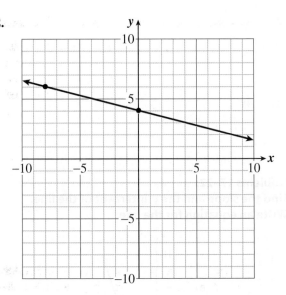

In Problems 23–26:

a. **Find the intercepts of the graph.**
b. **Graph the line.**
c. **Compute the slope of the line.**
d. **Put the equation in slope-intercept form.**

23. $3x + 4y = 12$

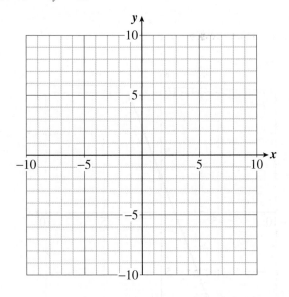

24. $2x - 3y = 6$

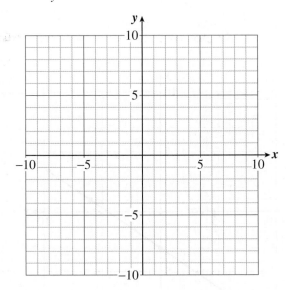

25. $y + 3x - 8 = 0$

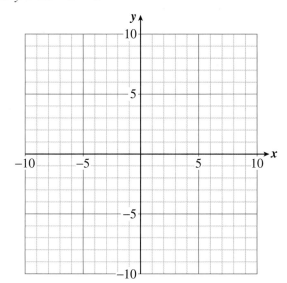

26. $5 - x - 2y = 0$

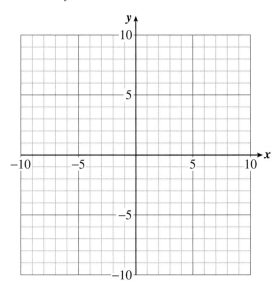

In Problems 27–30:
a. **Write the given equation in slope-intercept form.**
b. **What is the *y*-intercept of each line? What is its slope?**
c. **Use the slope to find four more points on the line.**
d. **Graph the line.**

27. $3x - 5y = 0$

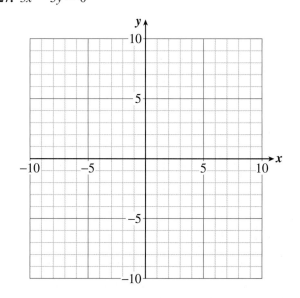

28. $2x + 3y = 0$

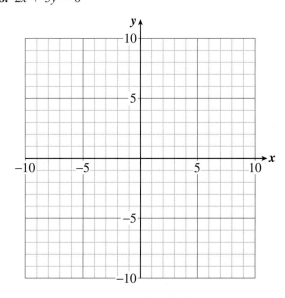

29. $5x + 4y = 0$

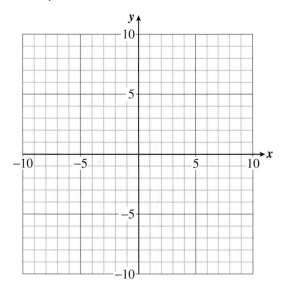

30. $5x - 2y = 0$

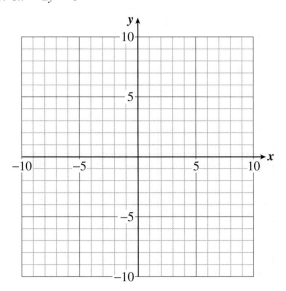

In Problems 31–36, graph the equation using the slope-intercept method.

31. $y = 3 - x$

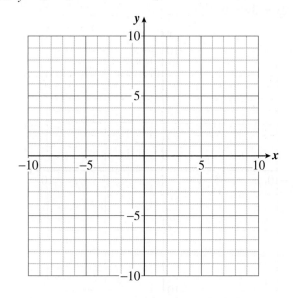

32. $y = x - 3$

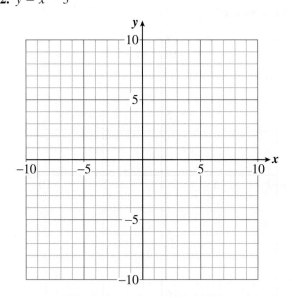

33. $y = 3x - 1$

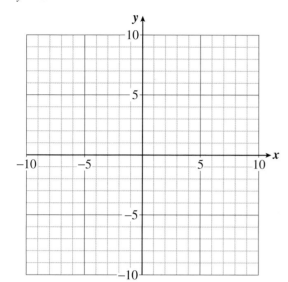

34. $y = -2x + 4$

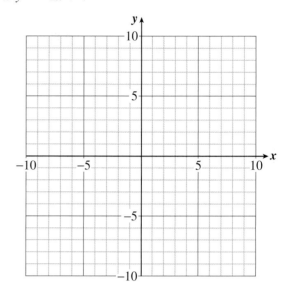

35. $y = \dfrac{3}{4}x + 2$

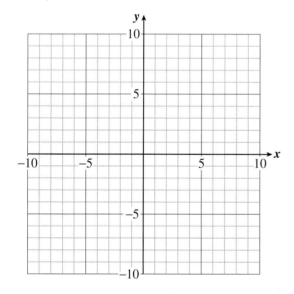

36. $y = \dfrac{-1}{3}x - 3$

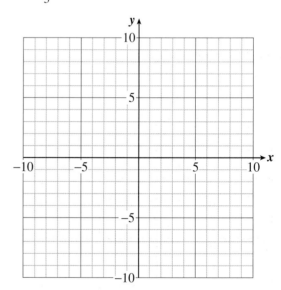

In Problems 37–42, graph the line that has the given slope and passes through the given point.

37. $m = \dfrac{3}{2}$, $(-1, -2)$

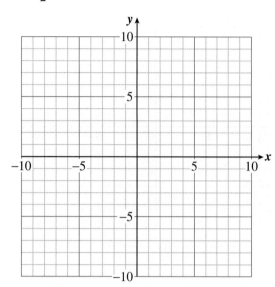

38. $m = \dfrac{1}{4}$, $(-3, 1)$

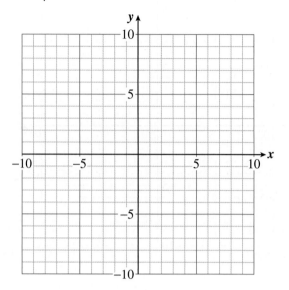

39. $m = -2$, $(-2, 6)$

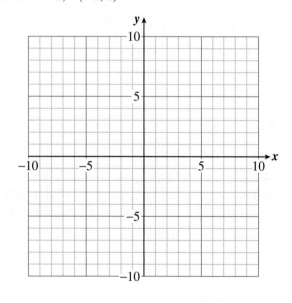

40. $m = -3$, $(-1, 5)$

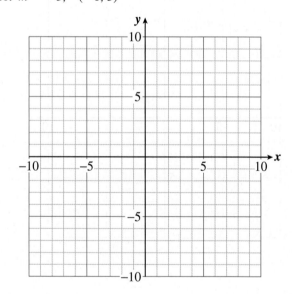

41. $m = \dfrac{-4}{5}$, (3, 4)

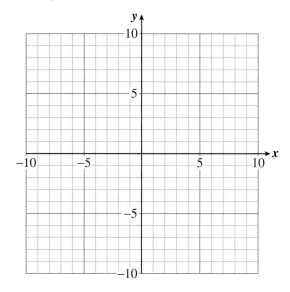

42. $m = \dfrac{-4}{3}$, (1, 2)

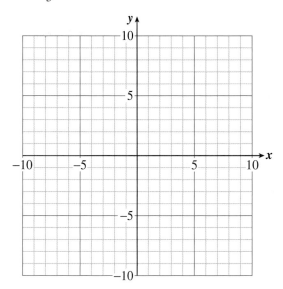

MIDCHAPTER REVIEW

In Problems 1–10, use complete sentences or fill in the blanks to answer the questions.

1. What is the difference between a ratio and a proportion? Give an example.

2. When are two variables proportional?

3. When two figures are geometrically similar, their corresponding angles are _____, and distances in one figure are _____ to corresponding distances in the other.

4. You have determined that one of the acute angles in a right triangle is 27° and that one angle in a larger right triangle is also 27°. Can you conclude that the two right triangles are similar? Explain.

5. If two variables vary directly, what equation do they satisfy?

6. You know that (−4, 1) is one point on a line. Explain how to use the slope to find another point on the line.

7. What is the slope-intercept form of a linear equation? Give an example.

8. Explain the difference between a line whose slope is undefined and a line whose slope is 0.

9. Suppose D varies directly with n and you graph the equation relating them. Then the constant of proportionality is the same as the _____ of the graph, and the y-intercept is equal to _____.

10. A student from a statistics class says that the slope of the line $y = a + bx$ is b. Do you agree or disagree? Give an example with numerical values for a and b.

Write ratios or rates for the situations in Problems 11 and 12.

11. In an anthropology class of 35 students, 14 are men. What is the ratio of men to women in the class?

12. Zach's car went 210 miles on 8.4 gallons of gasoline, and Tasha's car went 204 miles on 8 gallons of gasoline. Which car had the higher mileage per gallon?

Solve each proportion in Problems 13–16.

13. $\dfrac{x}{25} = \dfrac{36}{5}$

14. $\dfrac{105}{y} = \dfrac{15}{17}$

15. $\dfrac{4}{7} = \dfrac{z-6}{z}$

16. $\dfrac{a-3}{a+4} = \dfrac{34.8}{95.7}$

In Problems 17 and 18, decide whether the two variables are proportional.

17.

Time	5	10	15	20
Cost	1.50	2.50	3.50	4.50

18.

Speed	30	40	50	60
Distance	42	56	70	84

Solve Problems 19–22 using proportions.

19. A mason uses 660 bricks to build a wall 7.4 meters long. How many bricks would the mason use to build a wall 9.25 meters long?

20. Cindy wants to cut a 90-inch length of speaker wire into two pieces with lengths in the ratio 6 to 2. How long should each piece be?

21. On a map of Arenac County, 3 centimeters represents 5 miles.
   a. What are the true dimensions of a rectangular township whose dimensions on the map are 6 centimeters by 9 centimeters?

   b. What is the perimeter of the township? What is the perimeter of the corresponding region on the map?

   c. What is the ratio of the perimeter of the actual township to the perimeter of the corresponding region on the map?

   d. What is the area of the township? What is the area of the corresponding region on the map?

   e. What is the ratio of the area of the actual township to the area of the corresponding region on the map?

22. On a map of Euclid County, $\frac{1}{3}$ inch represents 2 miles. Lake Pythagoras is represented on the map by a right triangle with sides 1 inch, $\frac{4}{3}$ inch, and $\frac{5}{3}$ inch.
   a. What are the true dimensions of Lake Pythagoras?

   b. What is the perimeter of Lake Pythagoras? What is the perimeter of the corresponding region on the map?

   c. What is the ratio of the perimeter of the actual lake to the perimeter of the corresponding region on the map?

   d. What is the area of Lake Pythagoras? What is the area of the corresponding region on the map?

   e. What is the ratio of the area of the actual lake to the area of the corresponding region on the map?

Use properties of similar triangles to solve for the variable in Problems 23 and 24.

23.

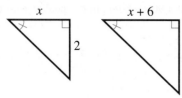

24.

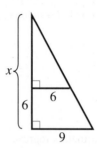

25. The Palm Springs aerial tramway ascends Mt. San Jacinto along a slope of approximately 0.516. In making its ascent, the tram car traverses a horizontal distance of about 11,380 feet. The elevation at the bottom of the tramway is 2643 feet. What is the elevation at the top?

26. The ruins of the Pyramid of Cholula in Guatemala have a square base 1132 feet on each side. The sides of the pyramid rise at a slope of about 0.32. How tall was the pyramid originally?

In Problems 27 and 28:
a. Find the *x*- and *y*-intercepts of the line.
b. Find the slope of the line.
c. Write the equation of the line in slope-intercept form.
d. Sketch the line.

27. $5x - 4y = 20$

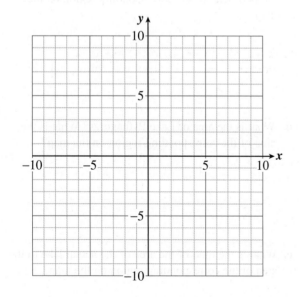

28. $\dfrac{y}{5} - \dfrac{x}{3} = 1$

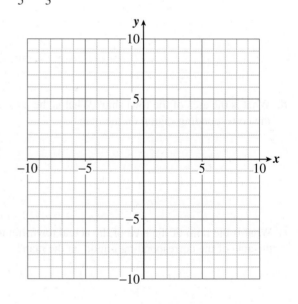

In Problems 29 and 30:
a. Find the slope and the *y*-intercept of the line.
b. Write an equation for the line.

29.

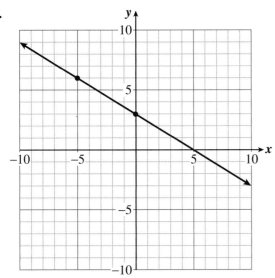

30.

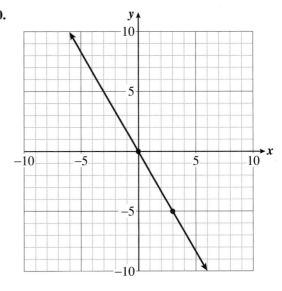

31. The amount of water in a municipal swimming pool, is given in gallons by

$$y = 500{,}000 - 5000h$$

where *h* is the number of hours since the pool custodian started draining the pool for the winter. State the slope and *y*-intercept of the equation, and explain their meaning in terms of the problem situation.

32. This year Francine bought a 3-feet-tall blue spruce sapling for her front yard, and it is supposed to grow about 6 inches per year. Write an equation for the height, *h*, of the tree *t* years from now.

In Problems 33 and 34, graph the line that has the given slope and passes through the given point.

33. $m = -\dfrac{3}{4}$, $(2, -1)$

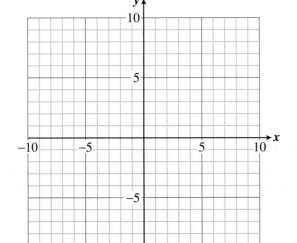

34. $m = \dfrac{1}{3}$, $(0, -3)$

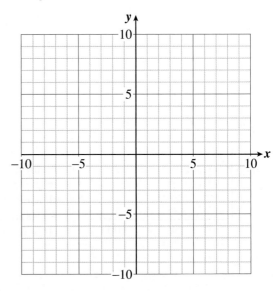

3.6 System of Linear Equations

Reading

The slope-intercept form for a linear equation, $y = mx + b$, gives information about the variables in the equation: The y-intercept, b, tells us the initial value of y, and the slope, m, tells us the rate of change of y with respect to x. In many applications, it is useful to compare two or more linear equations.

EXAMPLE 1

Delbert and Francine are buying appliances for their new home. They have a choice of two different refrigerators: a standard model that sells for $1000 and an energy-efficient model at a price of $1200. The standard model costs $6 per month to run, and the energy-efficient model costs $2 per month. Write a linear equation for the total cost of each refrigerator after t months.

Solution

We let S stand for the total cost of running the standard-model refrigerator for t months, so $S = mt + b$. The initial cost of the standard model is $1000, so $b = 1000$. The cost increases at a rate of $6 per month, so $m = 6$. Thus

$$S = 6t + 1000$$

Similarly, we let E stand for the total cost of running the energy-efficient refrigerator. For this model, the initial cost is $b = 1200$, and the total cost increases at a rate of $2 per month, or $m = 2$. Thus

$$E = 2t + 1200$$

Although the standard model costs less initially, the savings will eventually be offset because this model costs more to run each month. Delbert and Francine want to know when the energy-efficient model will begin to pay for itself—in other words, when the total cost of running the standard model will exceed the total cost of the energy-efficient model. One way to discover when this happens is by trial and error: We can evaluate both S and E for various values of t. The results of such a search are given in Table 3.6.

Figure 3.26

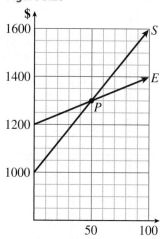

Table 3.6

t	6	12	18	24	30
S	1036	1072	1108	1144	1180
E	1212	1224	1236	1248	1260

After 30 months (or $2\frac{1}{2}$ years), the total cost of the energy-efficient model is still more than that of the standard model, although the gap is closing. A faster way to compare a lot of values is to compare the graphs of the two equations, as shown in Figure 3.26. The graph of S has a smaller initial value than the graph of E (1000 versus 1200), but that the graph of S increases more rapidly because its slope is greater.

The two graphs intersect at the point labeled P in Figure 3.26. At this point, the values of S and E are the same, approximately 1300. The t-coordinate of point P is 50, so we can say that the total cost of operating either the standard model or the energy-efficient model is about \$1300 for the first 50 months. After 50 months of operation (or $4\frac{1}{6}$ years), the total cost of the energy-efficient model will be less than that of the standard model.

It is difficult to read the coordinates of the intersection point P exactly from the graph. It is possible that we have not found the exact moment when the energy-efficient refrigerator begins to be more economical but have only approximated it. Do both refrigerators really have a total cost of \$1300 when $t = 50$? We can check this by substituting the coordinates $(50, 1300)$ into each equation.

Try Exercise 1 now.

When we consider two equations together, as in Exercise 1, we often use the same variables for both equations, like this:

$$y = 6x + 1000$$
$$y = 2x + 1200$$

This pair of equations is called a **system of linear equations**. A **solution** to a system of equations is an ordered pair (x, y) that satisfies each equation in the system. In Exercise 1, you verified that the ordered pair $(50, 1300)$ is a solution to this system of linear equations.

Now try Exercise 2.

Skills Review

Write algebraic expressions to answer each question.

1. How much interest is earned after 2 years on $d + 50$ dollars at an annual interest rate of 6%?

2. How much manganese is there in $8 - z$ grams of an alloy that is 35% manganese?

3. How far will a boat travel in 2 hours at a speed of $r + 3$ miles per hour?

4. How far will a train moving at 40 miles per hour travel in $3 + x$ hours?

5. A small plane has a top airspeed of v miles per hour. How far can the plane travel in 5 hours against a headwind of 15 miles per hour?

6. A fishing boat has a top speed in still water of 26 miles per hour. How far can the boat travel downstream in 3 hours if the speed of the current is w miles per hour?

Answers: **1.** $0.12(d + 50)$ **2.** $0.35(8 - z)$ **3.** $2(r + 3)$ **4.** $40(3 + x)$
5. $5(v - 15)$ **6.** $3(26 + w)$

EXERCISE 1
Verify that the point $(50, 1300)$ is a solution of each of the following equations:

$$S = 6t + 1000$$
$$E = 2t + 1200$$

EXERCISE 2
Decide whether the ordered pair $(2, 3)$ is a solution to the system

$$3x - 4y = -6$$
$$x + 2y = -4$$

Lesson

Solving Systems by Graphing

In the reading assignment, you learned that a solution to a system of equations is an ordered pair that satisfies both equations. Because the solution is a point that lies on both graphs, it is the intersection point of the two lines described by the system. Thus we can solve a system of equations by graphing each equation and looking for the point (or points) where the graphs intersect.

EXAMPLE 2

Robert and Ruth are moving from Los Angeles to Baltimore. Robert sets out in a rental truck at an average speed of 50 miles per hour. Ruth leaves one day later in their car and averages 65 miles per hour. When Ruth left, Robert had already traveled 300 miles. When will Ruth catch up with Robert?

Solution

We let t stand for the number of hours that Ruth has traveled. When she catches up with Robert, they will have traveled the same distance, so we begin by writing equations for the distance each has traveled at time t. Because Ruth travels at 65 miles per hour, an equation for the distance she has traveled is

$$\text{Ruth:} \quad d = 65t$$

Robert's speed is 50 miles per hour, but he has already traveled 300 miles when Ruth starts, so his distance is given by

$$\text{Robert:} \quad d = 300 + 50t$$

Together, the two equations form a system:

$$d = 65t$$
$$d = 300 + 50t$$

Next, we graph both equations on the same axes. The equation for Robert's distance is in slope-intercept form: The slope is 50, and the d-intercept is 300. We plot the d-intercept at $(0, 300)$ and use

$$m = \frac{\Delta d}{\Delta t} = \frac{50}{1} = \frac{250}{5}$$

to find another point on the graph, as shown in Figure 3.27.

To graph the equation for Ruth's distance, it is probably easiest to plot a few points.

Try Exercises 3 and 4.

Figure 3.27

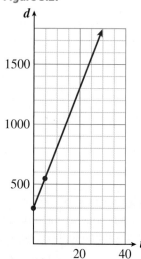

EXERCISE 3

Fill in the table and graph the equation for Ruth's distance, $d = 65t$ on the grid in Figure 3.27.

t	0	5	10
d			

EXERCISE 4

Locate the point where the two graphs intersect, and answer these questions:

a. What are the coordinates of the intersection point?

b. What does the *t*-coordinate of the intersection point tell you about the problem situation?

c. What does the *d*-coordinate of the point tell you?

d. Verify that the intersection point is a solution of both equations of the system developed in Example 2.

You should find that Ruth will catch up with Robert after 20 hours of driving.

Now try Exercise 5.

EXERCISE 5

Solve the system

$$y = x - 3$$
$$y = 2x - 8$$

Use the slope-intercept method to graph each line. Verify that your solution satisfies both equations in the system.

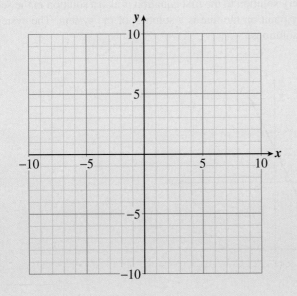

Figure 3.28

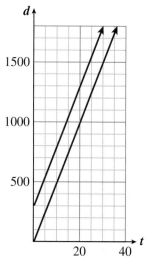

EXERCISE 6

Solve the system

$$3x = 2y + 6$$

$$y = \frac{3}{2}x - 1$$

Use the intercept method to graph the first equation and the slope-intercept method to graph the second equation.

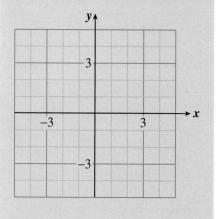

Inconsistent and Dependent Systems

Not every system of equations has a solution. In Example 2, suppose both Ruth and Robert drive at an average speed of 50 miles per hour. Then Ruth will never catch up with Robert. The system of equations that describes this situation is

$$d = 300 + 50t$$

$$d = 50t$$

The graph of this system is shown in Figure 3.28. Note that the two lines are parallel; they will never intersect. Thus there is no point of intersection that lies on both graphs.

If the two equations in a system describe parallel lines, the system has no solution. Such a system is called **inconsistent**.

Now try Exercise 6.

We've examined systems whose graphs intersect and systems whose graphs are parallel. There is a third possibility: Both equations in the system may have the same graph.

EXAMPLE 3

Solve the system

$$y = 2x + 2$$

$$6x - 3y = -6$$

Solution

We graph the first equation by the slope-intercept method:

$$b = 2 \quad \text{and} \quad m = \frac{\Delta y}{\Delta x} = \frac{2}{1}$$

The graph is shown in Figure 3.29. The second equation, which has the form $Ax + By = C$, is easier to graph by the intercept method. You can verify that its intercepts are $(0, 2)$ and $(-1, 0)$.

Thus the graph of the second equation is identical to the graph of the first equation. Every solution to the first equation is also a solution to the second equation, so every point on the line is a solution of the system. The system has infinitely many solutions.

Figure 3.29

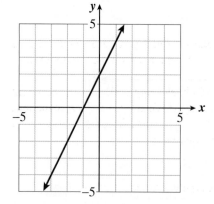

We can verify that the two equations in Example 3 are equivalent by solving the second equation for *y* in terms of *x*:

$$6x - 3y = -6 \qquad \text{Add } 3y \text{ to both sides.}$$
$$\underline{+\ 3y \qquad\quad +\ 3y}$$
$$6x \qquad = -6 + 3y \qquad \text{Add 6 to both sides.}$$
$$\underline{+\ 6 \quad +6}$$
$$6x + 6 = \qquad 3y \qquad \text{Divide both sides by 3.}$$
$$\frac{6x + 6}{3} = \frac{3y}{3} \qquad \text{Divide 3 into each term on the left side.}$$
$$2x + 2 = \qquad y$$

We see that the second equation has the same solutions as the first equation. A system in which the two equations are equivalent, as in Example 3, is called **dependent**. A dependent system has infinitely many solutions: Every point on the graph is a solution to the system.

We have seen three types of systems of linear equations. These are summarized in Figure 3.30. Most of the systems we will study have exactly one solution. Such systems are called **consistent and independent**.

Figure 3.30

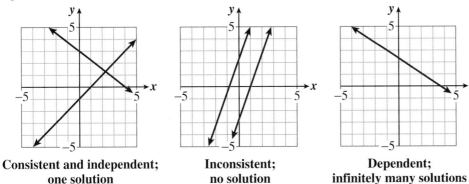

| Consistent and independent; one solution | Inconsistent; no solution | Dependent; infinitely many solutions |

Problem Solving with Systems

In previous sections, we have used single equations to solve problems. Systems of equations can also be useful in problem solving. If there are two unknown quantities in a problem, we can try to write two equations relating those variables.

Activity

Allen has been asked to design a rectangular Plexiglas plate whose perimeter is 28 inches and whose length is three times its width. What should the dimensions of the plate be?

The dimensions of a rectangle are its width and length, so we let *x* represent the width of the plate and *y* represent its length. We must write two equations about the width and length of the plate. We know a formula for the perimeter of a rectangle, $P = 2w + 2l$, so our first equation is

$$2x + 2y = 28$$

We also know that the length is three times the width, or

$$y = 3x$$

These two equations make a system. We'll graph them both on the same grid.

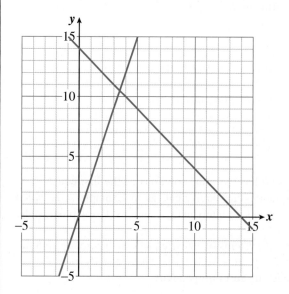

We use the intercept method to graph the first equation, $2x + 2y = 28$:

x	y
0	
	0

We use the slope-intercept method to graph $y = 3x$:

$$b = \underline{\hspace{2cm}} \qquad m = \frac{\Delta y}{\Delta x} = \underline{\hspace{2cm}}$$

The two graphs intersect at approximately (3.5, 10.5). You can verify that $x = 3.5$ and $y = 10.5$ solve the system by checking that these values make *both* equations true. Thus the width of the plate should be 3.5 inches, and its length should be 10.5 inches.

In the preceding Activity, we used two different techniques to graph the linear equations of a system. How can you decide which technique to use for a given equation? Both techniques will work, but it is usually easier to use the slope-intercept method if the equation is given in slope-intercept form, $y = mx + b$. If the equation is given in the form $Ax + By = C$,   it is usually easier to find the x- and y-intercepts of the graph.

You may have found it difficult to read the exact values of the solution, $x = 3.5$ and $y = 10.5$,   from the graph in this Activity. In the next section, we will consider algebraic techniques for solving systems that locate the exact coordinates of the solution without graphing.

ANSWERS TO 3.6 EXERCISES

1. $1300 = 6(50) + 1000$;
     $1300 = 2(50) + 1200$

2.   Not a solution

3.

t	0	5	10
d	0	325	650

4a. $(20, 1300)$

4b. The t-coordinate tells us when Ruth catches up to Robert.

4c. The d-coordinate tells us how far they have traveled.

5. $(5, 2)$

6.   There is no solution.

HOMEWORK 3.6

For Problems 1–8, decide whether each statement is true or false.

1. If an ordered pair satisfies either equation in a system, then it is a solution to the system.

2. If the lines described by the equations in a system have different slopes, then the system is consistent and independent.

3. If two lines have at least two points in common, then they are actually the same line.

4. If the two lines described by the equations in a system are perpendicular, then the system is inconsistent.

5. A system of linear equations may have exactly two solutions.

6. If the lines described by the equations in a system have the same slope, then the system has a unique solution.

7. A system of equations whose graphs are two horizontal lines is inconsistent.

8. If one equation in a system is a constant multiple of the other equation, then the system is dependent.

In Problems 9–12, decide whether the given ordered pair is a solution to the system.

9. $x + 2y = -8$
    $2x - y = 4$
    $(4, -2)$

10. $3x + 2y = -5$
    $x + 3y = -9$
    $(-3, -2)$

11. $x = 5y + 13$
    $2x + 3y = 0$
    $(3, -2)$

12. $4x + 6 = y$
    $2x = 8 + 3y$
    $(1, -2)$

In Problems 13–18, decide which graphing technique, the intercept method or the slope-intercept method, would be easier to use for each equation. Explain why you made the choice you did.

13. $3x - 2y = 18$

14. $y = \dfrac{5x}{3} - 8$

15. $y = \dfrac{-5}{6}x$

16. $\dfrac{x}{8} + \dfrac{y}{3} = 1$

17. $5y - 4x = 0$

18. $y - \dfrac{3}{4}x = -5$

Solve each system in Problems 19–26 by graphing.

19. $y = -3x$

$y = 3x - 6$

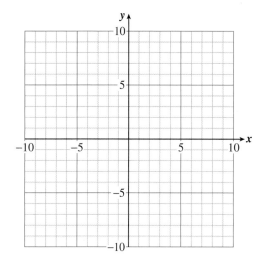

20. $y = \dfrac{1}{2}x$

$y = 2x + 3$

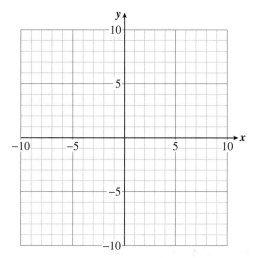

21. $y = \dfrac{3}{4}x - 2$

$2y - 5x = 10$

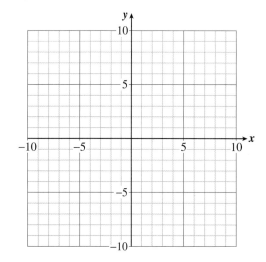

22. $y = \dfrac{2}{3}x + 2$

$2x + y = 10$

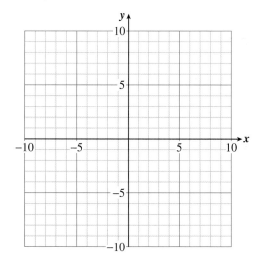

23. $x + y = 4$
  $2x + 2y = 8$

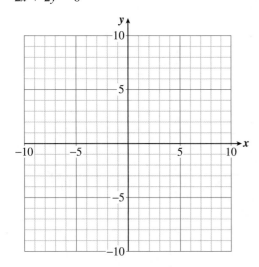

24. $x - 3y = 6$
  $2x - 6y = 6$

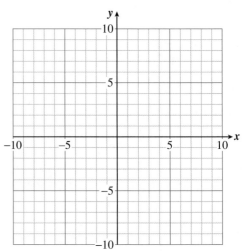

25. $2y = -3x$
  $y = -2x - 1$

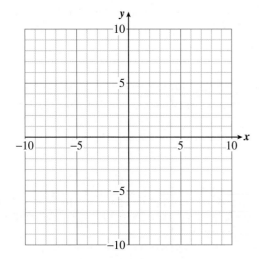

26. $3y = -4x$
  $y = -2x + 2$

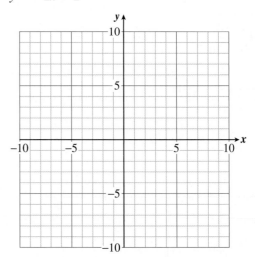

27. Without graphing, explain how you can tell that the following system has no solution.

$$3x = y + 3$$
$$6x - 2y = 12$$

28. Without graphing, explain how you can tell that the following system has infinitely many solutions.

$$-x + 2y = 4$$
$$3x = 6y - 12$$

29. Delbert has accepted a sales job and is offered a choice of two salary plans. Under plan A, he will receive $20,000 a year plus a 3% commission on his sales. Plan B offers a $15,000 annual salary plus a 5% commission.

a. Let x stand for the amount of Delbert's sales in one year. Write equations for his total annual earnings under each plan.

b. Fill in the table, where x is given in thousands of dollars.

x	Earnings under Plan A	Earnings under Plan B
0		
50		
100		
150		
200		
250		
300		
350		
400		

c. Graph both equations on the grid. (Both axes are scaled in thousands of dollars.)

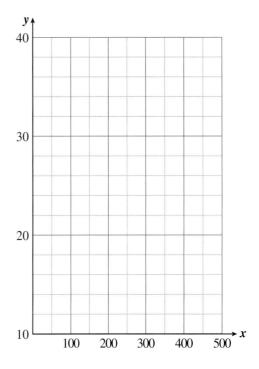

d. What amount of sales gives Delbert equal earnings under either plan?

e. Under what circumstances should Delbert prefer plan A?

f. If Delbert chooses plan B, how much must he sell in order to make more than $30,000 a year? What if he chooses plan A?

g. Verify your answers to part (f) by writing and solving inequalities.

30. Francine wants to join a health club and has narrowed
her choices down to two clubs. Sportshaus charges an
initial fee of $500 and $10 per month thereafter. Fitness
First has an initial fee of $50 and charges $25 per
month.

   a. Let *x* stand for the number of months Francine uses
the health club. Write equations for the total cost of
each health club for *x* months.

   b. Fill in the table for the total cost of each club.

x	Sportshaus Total Cost	Fitness First Total Cost
6		
12		
18		
24		
30		
36		
42		
48		

   c. Graph both equations on the grid.

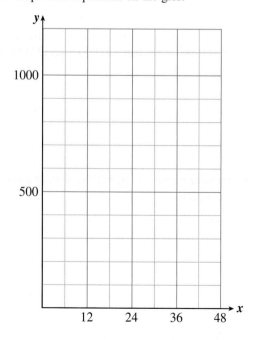

   d. When will the total cost of the two health clubs be
equal?

   e. Under what circumstances should Francine prefer
Fitness First?

   f. If Francine chooses Fitness First, when will her total
cost exceed $650? What if she chooses Sporthaus?

   g. Verify your answers to part (f) by writing and solving inequalities.

31. Orpheus Music plans to manufacture clarinets for schools. The company's start-up costs are $6000, and each clarinet costs $60 to make. The clarinets will sell for $80 each.

a. Let x stand for the number of clarinets Orpheus manufactures. Write equations for the total cost of producing x clarinets and the revenue earned from selling x clarinets.

b. Fill in the table.

x	Cost	Revenue
0		
50		
100		
150		
200		
250		
300		
350		
400		

c. Graph both equations on the grid. (The vertical axis is scaled in thousands of dollars.)

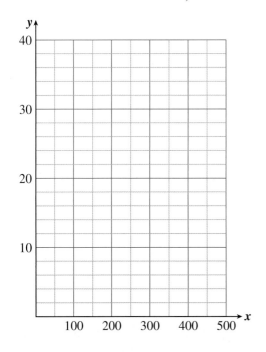

d. How many clarinets must Orpheus sell for its revenue to exceed $20,000?

e. How many clarinets must Orpheus sell in order to break even?

f. What is the company's profit if it sells 500 clarinets? 200 clarinets?

g. Illustrate your answers to part (f) on your graph.

32. When sailing upstream in a canal or in a river that has rapids, ships must sometimes negotiate locks that raise them to a higher water level. Suppose your ship is in one of the lower locks on a canal, at an elevation of 20 feet. The next lock is at an elevation of 50 feet. Water begins to flow from the higher lock to the lower one, raising your ship by 1 foot per minute and simultaneously lowering the water level in the next lock by 1.5 feet per minute.

a. Let t stand for the number of minutes the water has been flowing. Write equations for the water level in each lock after t minutes.

b. Fill in the table.

t	Lower lock water level	Upper lock water level
0		
2		
4		
6		
8		
10		

c. Graph both equations on the grid.

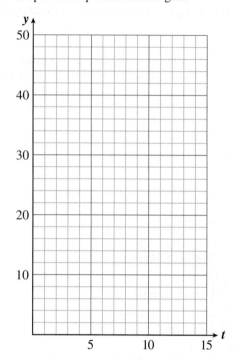

d. When will the water levels in the two locks be 10 feet apart?

e. When will the water levels in the two locks be the same?

f. Write an equation you can use to verify your answer to part (e), and solve it.

In Problems 33–36, write a system of equations in two variables to model the situation, and solve the system graphically. Verify your solution algebraically.

33. A bouquet of four roses and eight carnations costs $14. A bouquet of six roses and nine carnations costs $18. How much does one rose cost? One carnation?

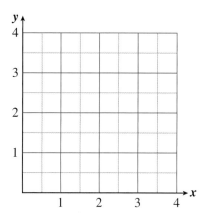

34. Frank's history test consisted of true-false and completion questions. Frank got 16 true-false and 8 completion questions correct, and he earned a score of 80. Frank's friend Fred got 12 true-false and 9 completion questions correct, and his score was 72. How many points was each true-false question worth? Each completion question?

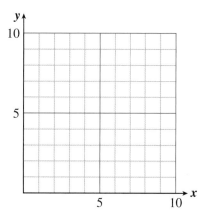

35. The vertex angle of an isosceles triangle is 15° less than each base angle. Find the measure of each angle of the triangle. (*Hint:* Recall that the sum of the angles in any triangle is 180°.)

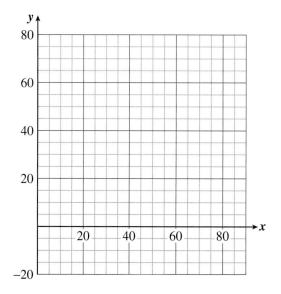

36. Last year, Dan and Barbara together made $35,000. This year Barbara doubled her salary, but Dan took a $5000 pay cut, and their joint earnings were $50,000. How much did each make last year?

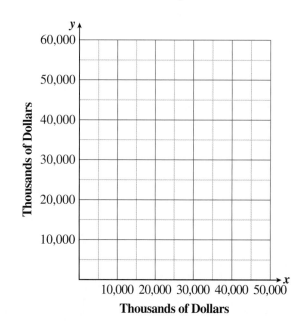

3.7 Algebraic Solution of Systems

Reading

In this section, we discuss how to solve systems of linear equations using algebra. In Example 1 of Section 3.6, we compared the costs of operating two different refrigerators, a standard model and an energy-efficient model. We wrote the following system of equations for this problem, where x is the number of months the refrigerator is in use:

$$y = 6x + 1000$$
$$y = 2x + 1200$$

We were interested in when the total costs for the two refrigerators would be equal. Thus we looked for a point on the graphs where the two y-values, which represent the costs, are equal. This point is the intersection point of the graphs, as shown in Figure 3.31. The x-coordinate of the intersection point, 50, shows that the operating costs for the two refrigerators are equal after 50 months.

In addition to the graphical method for solving a system of equations, there is an algebraic method. We are looking for the x-value of the point where the two y-values are equal. Therefore, we can set the two expressions for y equal, and we'll have an equation in x to solve:

$$6x + 1000 = 2x + 1200 \quad \text{Subtract } 2x \text{ from both sides.}$$
$$4x + 1000 = 1200 \qquad\qquad \text{Subtract 1000 from both sides.}$$
$$4x = 200 \qquad\qquad\quad \text{Divide both sides by 4.}$$
$$x = 50$$

We find that $x = 50$, which is the same answer we got by graphing.

This algebraic method, setting the two expressions for y equal to each other and solving for x, is a special case of the substitution method, which we'll discuss in the lesson.

Now try Exercise 1.

Figure 3.31

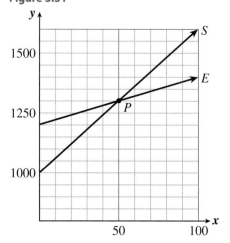

EXERCISE 1
Solve the following system algebraically:

$$d = 65t$$
$$d = 300 + 50t$$

(This is the system from Example 2 in Section 3.6.)

Skills Review

Solve for the indicated variable.

1. $2q + p = 10$ for p
2. $2l + 2w = 18$ for l
3. $3a + 9 = -6b$ for a
4. $2c = 2d + 22$ for d
5. $5r - 4s = 24$ for s
6. $2m = 11 - 3n$ for n

Answers: **1.** $p = 10 - 2q$ **2.** $l = 9 - w$ **3.** $a = -3 - 2b$

4. $d = c - 11$ **5.** $s = \dfrac{5r - 24}{4}$ **6.** $n = \dfrac{11 - 2m}{3}$

Lesson

Substitution Method

In the reading assignment, we solved a system of equations by equating the two expressions for y. You can think of this procedure as substituting the expression for y from one equation into the other equation, like this:

$$y = \underbrace{6x + 1000}$$

Substitute into second equation.

$$y = 2x + 1200$$

$$6x + 1000 = 2x + 1200$$

This method works well here because both equations in the system are already in the form $y = mx + b$. Sometimes we must solve one of the equations for y in terms of x before we can substitute.

EXAMPLE 1

Use substitution to solve the following system:

$$y + 3x = 1$$
$$2y + 5x = 5$$

Solution

We must first solve one of the equations for y in terms of x (or x in terms of y). It will be easiest to solve the first equation for y. We subtract $3x$ from both sides of the equation.

$$y = 1 - 3x$$

We now substitute the expression $1 - 3x$ for y in the other equation.

$$y = \underbrace{1 - 3x}$$

$$2y + 5x = 5$$

$$2(1 - 3x) + 5x = 5$$

This gives us an equation in one variable, which we can solve as usual. We begin by simplifying the left side.

$2(1 - 3x) + 5x = 5$	Apply the distributive law.
$2 - 6x + 5x = 5$	Combine like terms.
$2 - x = 5$	Subtract 2 from both sides.
$-x = 3$	Divide both sides by -1.
$x = -3$	

This is the x-coordinate of the solution point. To find the y-coordinate, we can substitute $x = -3$ into either of the original equations. It is easiest to use the equation we solved for y—namely, $y = 1 - 3x$.

$$y = 1 - 3(-3) = 10 \quad \text{Substitute } -3 \text{ for } x.$$

Thus the solution to the system is $x = -3$, $y = 10$, or the point $(-3, 10)$. You should check that these values satisfy both of the original equations in the system.

Here is a summary of the steps for solving a system of equations by substitution.

To Solve a System by Substitution:

1. Choose one of the variables in one of the equations. (It is easiest to choose a variable whose coefficient is 1 or -1.) Solve the equation for that variable.
2. Substitute the result of Step 1 into the other equation. This gives an equation in one variable.
3. Solve the equation obtained in Step 2. This gives the solution value for one of the variables.
4. Substitute this value into the result of Step 1 to find the solution value of the other variable.

EXERCISE 2
Follow the suggested steps to solve the following system by substitution.

$$3y - 2x = 3$$
$$x - 2y = -2$$

Step 1 Solve the second equation for x in terms of y.

Step 2 Substitute your expression for x into the first equation.

Step 3 Solve the equation you got in Step 2.

Step 4 You now have the solution value for y. Substitute that value into your result from Step 1 to find the solution value for x.

As always, you should check that your solution values satisfy *both* equations in the system.

Now try Exercise 2.

Elimination Method

A second algebraic method for solving systems of equations is called **elimination**. As with the substitution method, we try to obtain an equation in a single variable, but we do it by eliminating one of the variables in the system. In order to use the elimination method, we must first put both equations into the form $Ax + By = C$.

EXAMPLE 2

Solve this system:

$$5x = 2y + 21$$
$$2y = 19 - 3x$$

Solution

First, we rewrite each equation in the form $Ax + By = C$.

$$
\begin{array}{ll}
\begin{aligned}
5x &= 2y + 21 \quad \text{Subtract } 2y \\
-2y &= -2y \quad\ \text{ from both sides.} \\
\hline
5x - 2y &= 21
\end{aligned}
&
\begin{aligned}
2y &= 19 - 3x \quad \text{Add } 3x \text{ to both sides.} \\
+3x &= +3x \\
\hline
3x + 2y &= 19
\end{aligned}
\end{array}
$$

We add the two equations by adding the left side of the first equation to the left side of the second equation by combining like terms, and then adding the two right sides together, as follows:

$$
\begin{aligned}
5x - 2y &= 21 \\
3x + 2y &= 19 \\
\hline
8x &= 40
\end{aligned}
$$

Note that the *y*-terms canceled, or were eliminated. We are left with an equation in *x* that is easy to solve:

$$8x = 40 \quad \text{Divide both sides by 8.}$$
$$x = 5$$

We are not finished yet, because we must still find the solution value of *y*. We can substitute the value for *x* into either of the original equations and solve for *y*. Let's use the second equation, $3x + 2y = 19$.

$$3(5) + 2y = 19 \quad \text{Subtract 15 from both sides.}$$
$$2y = 4 \quad \text{Divide by 2.}$$
$$y = 2$$

Thus our solution is the point $(5, 2)$.

You may have noticed that this method worked because the coefficients of *y* in the two equations were opposites, 2 and -2. This caused the *y*-terms to cancel out when we added the two equations together. What if the coefficients of neither *x* nor *y* are opposites? Then we must multiply one or both of the equations in the system by a suitable constant. Consider the system

$$4x + 3y = 7$$
$$3x + y = -1$$

We can choose to eliminate either the *x*-terms or the *y*-terms in a system. For this example, it will be more efficient to eliminate the *y*-terms. If we multiply each term of the second equation by -3, then the coefficients of *y* will be opposites.

$$-3(3x + y = -1) \longrightarrow -9x - 3y = 3$$

Because we are applying the multiplication property of equality, we must multiply *each term* by -3, not just the *y*-term. We can now replace the second equation by its new version to obtain this system:

$$4x + 3y = 7$$
$$-9x - 3y = 3$$

Now try Exercise 3.

Sometimes it is necessary to multiply *both* equations by suitable constants in order to eliminate one of the variables.

EXAMPLE 3

Use elimination to solve this system:

$$5x - 2y = 22$$
$$2x - 5y = 13$$

Solution

This time we choose to eliminate the *x*-terms. We must arrange things so that the coefficients of the *x*-terms are opposites, so we look for the smallest integer into which both 2 and 5 divide evenly. (This number is called the **least common multiple**, or **LCM**, of 2 and 5.) The LCM of 2 and 5 is 10. We want one of the coefficients of *x* to be 10 and the other to be -10.

EXERCISE 3

Follow the suggested steps to solve this system:

$$4x + 3y = 7$$
$$-9x - 3y = 3$$

Step 1 Add the equations together.

Step 2 Solve the resulting equation for *x*.

Step 3 Substitute your value for *x* into either original equation to find the solution value for *y*.

Step 4 Check that your solution satisfies both original equations.

To accomplish this, we'll multiply the first equation by 2 and the second equation by -5.

$$2(5x - 2y = 22) \longrightarrow \quad 10x - 4y = 44$$
$$-5(2x - 5y = 13) \longrightarrow -10x + 25y = -65$$

Adding these new equations eliminates the x-term and yields an equation in y:

$$10x - 4y = 44$$
$$\underline{-10x + 25y = -65}$$
$$21y = -21$$

Solve for y to find $y = -1$.   Finally, substitute $y = -1$ into the first equation and solve for x.

$$5x - 2(-1) = 22$$
$$5x + 2 = 22 \quad \text{Subtract 2 from both sides.}$$
$$5x = 20 \quad \text{Divide both side by 5.}$$
$$x = 4$$

The solution to the system is the point $(4, -1)$.

EXERCISE 4
Follow the suggested steps to solve the following system by elimination.

$$3x = 2y + 13$$
$$3y - 15 = -7x$$

Step 1 Write each equation in the form $Ax + By = C$.

Step 2 Eliminate the y-terms: Multiply each equation by an appropriate constant.

Step 3 Add the new equations, and solve the result for x.

Step 4 Substitute your value for x into the second equation, and solve for y.

Here are the steps for solving a system by elimination.

To Solve a System by Elimination:

1. Write each equation in the form $Ax + By = C$.
2. Decide which variable to eliminate. If necessary, multiply each equation by an appropriate constant so that the coefficients of that variable are opposites.
3. Add the equations from Step 2, and solve for the remaining variable.
4. Substitute the value found in Step 3 into one of the original equations, and solve for the other variable.

Now try Exercise 4.

Inconsistent and Dependent Systems

Recall that a system of equations whose graphs are parallel lines has no solution and is called inconsistent. If the two equations in a system have the same graph, then every point on the graph is a solution and the system is called dependent. The elimination method reveals whether a system is inconsistent or dependent.

EXAMPLE 4

Solve each system by elimination.

a. $3x - y = 4$
$-6x + 2y = 3$

b. $x - 2y = 3$
$2x - 4y = 6$

Solution

a. To eliminate the y-terms, we multiply the first equation by 2 and add the two equations.

$$6x - 2y = 8$$
$$\underline{-6x + 2y = 3}$$
$$0 = 11$$

Both variables are eliminated, and we are left with the false statement $0 = 11$. There are no values of x or y that will make this equation true, so the system has no solutions. You can see in Figure 3.32 that the system is inconsistent.

b. To eliminate the x-terms, we multiply the first equation by -2 and add the two equations.

$$-2x + 4y = -6$$
$$\underline{2x - 4y = 6}$$
$$0 = 0$$

We are left with the true but unhelpful equation $0 = 0$. The two equations are in fact equivalent (one is a constant multiple of the other), so the system is dependent. The graph of both equations is shown in Figure 3.33.

We summarize the observations from Example 5 as follows:

Figure 3.32

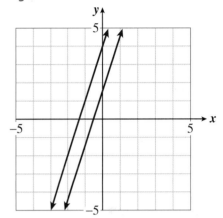

Figure 3.33

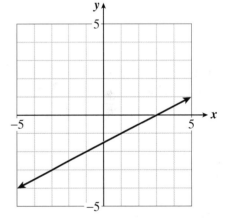

When Using Elimination to Solve a System:

1. If combining the two equations of a system results in an equation of the form

$$0x + 0y = k \qquad (k \neq 0)$$

then the system is inconsistent.

2. If combining the two equations of a system results in an equation of the form

$$0x + 0y = 0$$

then the system is dependent.

ANSWERS TO 3.7 EXERCISES

1. $t = 20, d = 1300$

2. $(0, 1)$

3. $(-2, 5)$

4. $(3, -2)$

HOMEWORK 3.7

In Problems 1–12, solve each system by the substitution method.

1. $y = 2x$
  $3x + y = 10$

2. $y = 3x$
  $x + 2y = 14$

3. $2a + 3b = 4$
  $b = 2a + 4$

4. $3a - 2b = 10$
  $b = a - 5$

5. $2x - 3y = 4$
  $x + 1 = 3y$

6. $2z - w = 5$
  $2w = 1 - 7z$

7. $2a + b = 16$
  $4a - 4 = -3b$

8. $6x - 2y = -3$
  $4x + y = 5$

9. $8r + s = 4$
  $2r + 7s = -8$

10. $2l + 2w = 24$
  $7w = 5l$

11. $36a = 6 - 3b$
  $2b = 3a - 5$

12. $2y + 10 = 18x$
  $50x - 20y = 9$

In Problems 13–16, write a system of equations to model the problem situation, and then solve the system by substitution.

13. A hamburger and a chocolate shake together contain 1030 calories. Two shakes and three hamburgers contain 2710 calories. How many calories are there in one hamburger? In one chocolate shake?

14. In a recent election, the winning candidate received 50 votes more than her opponent. If a total of 4376 votes were cast for the two candidates, how many votes did each receive?

15. Darryl is in charge of buying furniture for a new restaurant. Chairs cost $175, and tables cost $250. Darryl's budget allows $11,400 for tables and chairs, and he plans to buy four chairs for each table. How many chairs can he buy? How many tables?

16. The Fairview Little League sold 300 tickets to the championship game to raise money. Adults' tickets cost $3.50, and children's tickets cost $1.50. If Little Leaguers collected $750 from ticket sales, how many of each kind of ticket did they sell?

In Problems 17–28, solve each system by elimination.

17. $x + y = 5$
$x - y = 1$

18. $7x - 3y = -10$
$x + 3y = 2$

19. $2a + b = 3$
$a + b = 2$

20. $4u - 4v = -4$
$4u - 3v = -3$

21. $3x + 2y = 7$
$x + y = 3$

22. $a = 4b - 8$
$3a + 2b = -3$

23. $3a = 1 + 5b$
$2b = 14 + 6a$

24. $3u = 5v - 5$
$6u = -1 + 9v$

25. $2x + 3y = -1$
    $3x + 5y = -2$

26. $2x = 3y - 1$
    $4y = 24 - 3x$

27. $5z = 1 - 3w$
    $5w = 2 - 7z$

28. $7n + 1 = 9p$
    $3n + 6 = -11p$

In Problems 29–32, write a system of equations to model each problem situation, and then solve the system by elimination.

29. The perimeter of a rectangle is 42 meters, and its width is 13 meters less than its length. Find the dimensions of the rectangle.

30. Two suitcases together weigh 45 kilograms. One of the suitcases weighs 6 kilograms less than twice the other. How much does each weigh?

31. Three pounds of bacon and 2 pounds of coffee cost $17.80. Two pounds of bacon and 5 pounds of coffee cost $32.40. How much does 1 pound of bacon cost? 1 pound of coffee?

32. Nabil and Needa visited a local orchard where customers can pick their own fruit. Nabil picked 11 pounds of apples and 5 pounds of pears for a total charge of $10.85, and Needa picked 9 pounds of apples and 6 pounds of pears for a total charge of $10.50. How much does 1 pound of each kind of fruit cost?

Decide whether each system in Problems 33–38 is inconsistent, dependent, or consistent and independent.

33. $3x - 6y = 6$
    $x - 2y = 3$

34. $3x + 2y = -6$
    $6x + 4y = -12$

35. $8a = 6 + 12b$
$\quad 4 = 6a - 9b$

36. $6a = 2b - 8$
$\quad 10 = 5b - 15a$

37. $3p = 1 + 7q$
$\quad 21q = 9p - 3$

38. $4p - 2 = -5q$
$\quad -15q = 12p - 10$

Which method, elimination or substitution, would be easier to apply to each system in Problems 39–42?

39. $2u + 3v = -8$
$\quad 2u - 3v = -6$

40. $y = 3x - 1$
$\quad 5x - 2y = 9$

41. $17a - b = 0$
$\quad 11a + 3b = 4$

42. $7w - 2z = 5$
$\quad 3w + 4z = -6$

3.8 Applications of Systems of Equations

Reading

In this section, we explore some applications of systems of equations. We begin by considering numerical examples of the kinds of applications we'll study.

Interest

Recall that the formula for calculating interest is

$$I = Prt$$

where I is the interest earned after t years if you invest a principal of P dollars in an account that earns simple interest at an annual rate r. If you make two or more investments simultaneously, then the total interest you earn is the sum of the interest earned on each of the investments.

EXERCISE 1

Harvey deposited $1200 in two accounts. He put $700 in a savings account that pays 6% annual interest, and he put the rest in his credit union account, which pays 7% annual interest. How much interest will Harvey's investments earn in 2 years?

Step 1 Calculate the interest earned by each account.

Step 2 Add the earnings from the two investments.

Now consider a similar example with variables.

EXERCISE 2

You have $5000 to invest for 1 year. You want to put part of the money into bonds that pay 7% interest and the rest of the money into stocks, which involve some risk but will pay 12% if the investment is successful.
a. Fill in the table.

Amount invested in stocks	Amount invested in bonds	Interest from stocks	Interest from bonds	Total interest
$ 500				
$1000				
$3200				
$4000				

Suppose you decide to invest x dollars in the stocks and y dollars in the bonds. Write algebraic expressions for each of the following:
b. Sum of amounts invested

c. Interest earned on the stocks; interest earned on the bonds

d. Total interest earned

Mixtures

Suppose a pharmacist has on hand 20 ounces of a certain drug at 40% strength, but she needs a small quantity of the drug at 75% strength for a prescription. She decides to add a pure form of the drug to the 40% solution. How much should she add to make a mixture of 75% strength?

To solve problems like this one, we need to review some properties of percents. We will need the percent formula,

$$P = rW$$

where P stands for the part obtained when we take r percent of a whole amount, W.

EXERCISE 3

You have two jars of marbles. The first contains 40 marbles, of which 10 are red, and the second contains 60 marbles, of which 30 are red.

a. What percent of the marbles in the first jar are red?

What percent of the marbles in the second jar are red?

You pour both jars of marbles into a larger jar and mix them together.

40 marbles,   60 marbles,
  10 red   30 red

b. How many marbles total are in the larger jar?

How many red marbles are in the larger jar?

What percent of marbles in the larger jar are red?

c. Can you add the percent of red marbles in each of the first two jars to get the percent of red marbles in the mixture?

In Exercise 3, you should find that adding the percent of red marbles in the first two jars does *not* give the percent of red marbles in the mixture. That is,

$$25\% + 50\% \neq 40\%$$

In general, we cannot add percents unless they are percents of the same whole amount.

EXERCISE 4

In a local city council election, your favored candidate, Justine Honest, ran in a small district with two precincts. Ms. Honest won 30% of the 500 votes cast in precinct 1 and 70% of the 300 votes cast in precinct 2. Did candidate Honest win a majority (more than 50%) of the votes in her district?

Step 1 Fill in the first two rows of the table, using information from the problem and the formula $P = rW$.

	Total votes (W)	Percent of votes for honest (r)	Number of votes for honest (P)
Precinct 1			
Precinct 2			
Entire district			

Step 2 Add vertically to complete the *W* and *P* columns of the table.

Step 3 Fill in the last entry in the table to answer the question in the problem. Use the formula $P = rW$ again.

Skills Review

For part (b) of each exercise, write an algebraic expression in two variables.

1. a. How much interest will you earn in 1 year if you invest $2400 in a T-bill that pays $7\frac{1}{2}\%$ interest and $800 in a savings account that pays 4.8% interest?

    b. How much will you earn if you invest *x* dollars in the T-bill and *y* dollars in the savings account?

2. a. How far will you travel if you jog for 40 minutes at 9 miles per hour and then jog for 30 minutes at 5 miles per hour?

    b. How far will you travel if you jog for 40 minutes at *x* miles per hour and then jog for 30 minutes at *y* miles per hour?

3. a. How long will it take you to drive 180 miles on a highway at an average
      speed of 60 miles per hour and then 30 miles on a gravel road at an aver-
      age speed of 40 miles per hour?
   b. How long will it take you if you drive on the highway at x miles per hour
      and on the gravel road at y miles per hour?
4. a. How much nitrogen is in a mixture consisting of 10 pounds of fertilizer
      that is 6% nitrogen and 4 pounds of fertilizer that is 60% nitrogen?
   b. How much nitrogen is in a mixture of x pounds of the first fertilizer and y
      pounds of the second fertilizer?

Answers: **1. a.** $218.40 **b.** $0.075x + 0.048y$ **2. a.** 8.5 miles
b. $\frac{2}{3}x + \frac{1}{2}y$ **3. a.** 3.75 hours **b.** $\frac{180}{x} + \frac{30}{y}$ **4. a.** 3 pounds
b. $0.06x + 0.60y$

Lesson

For each of the following applications, we will write a system of two equations
in two variables.

Interest

Let's consider interest problems that involve two investments.

EXAMPLE 1

Mort invested money in two accounts, a savings plan that pays 8% interest and a
mutual fund that pays 7% interest. He put twice as much money in the savings
plan as in the mutual fund. At the end of the year, Mort's total interest income
was $345. How much did he invest in each account?

Solution

The two unknown quantities are the amounts Mort invested in his two accounts.

    Amount invested in savings: x
    Amount invested in mutual fund: y

Do not confuse the amount Mort invested in each account (the principal) with the
amount of interest he earned! Making a table is a good way to keep the amounts
straight.

	Principal	Interest rate	Interest
Savings plan	x	0.08	$0.08x$
Mutual fund	y	0.07	$0.07y$

We need to write two equations for the problem, one about the principal and one
about the interest. From the statement "He put twice as much money in the savings plan as in the mutual fund," we obtain this equation:

$$\left(\begin{array}{c}\text{amount invested}\\\text{in savings plan}\end{array}\right) = 2 \cdot \left(\begin{array}{c}\text{amount invested}\\\text{in mutual fund}\end{array}\right)$$

Equation about principal: $x = 2y$

From the statement "Mort's total interest income was $345," we obtain the equation

$$\left(\begin{array}{c} \text{interest from} \\ \text{savings plan} \end{array} \right) + \left(\begin{array}{c} \text{interest from} \\ \text{mutual fund} \end{array} \right) = \text{total interest}$$

Equation about interest: $0.08x$ $+$ $0.07y$ $= 345$

This gives us a system of equations:

$$x = 2y$$
$$0.08x + 0.07y = 345$$

Because the first equation is already solved for x in terms of y, we'll use the substitution method to solve the system. We substitute $2y$ for x in the second equation to get

$$0.08(2y) + 0.07y = 345$$

Now we solve for y.

$$0.16y + 0.07y = 345 \quad \text{Combine like terms.}$$
$$0.23y = 345 \quad \text{Divide both sides by 0.23.}$$
$$y = 1500$$

Finally, we substitute $y = 1500$ into the first equation to find a value for x.

$$x = 2y = 2(1500) = 3000$$

Thus Mort invested $3000 in the savings plan and $1500 in the mutual fund.

Follow the suggested steps to complete Exercise 5.

EXERCISE 5

Jerry invested a total of $2000, part of it at 4% interest and the remainder at 9%. His yearly income from the 9% investment is $37 more than his yearly income from the 4% investment. How much did he invest at each rate?

Step 1 Choose variables for the unknown quantities, and fill in the table.

	Principal	*Interest rate*	*Interest*
First investment			
Second investment			

Step 2 Write two equations, one about the principal and one about the interest.

Step 3 Solve the system. (Which method seems easiest?)

Mixtures

In the reading assignment, we considered mixture problems involving discrete objects, such as marbles or votes. The same principles apply to mixtures of liquids.

EXAMPLE 2

A chemist wants to produce 45 milliliters of a 40% solution of acetic acid by mixing a 20% solution with a 50% solution. How many milliliters of each should he use?

Solution

We let x represent the number of milliliters of the 20% solution the chemist needs and y the number of milliliters of the 50% solution (see Figure 3.34).

Figure 3.34

x milliliters y milliliters   45 milliliters

We use a table like the one in Exercise 4. The first two columns of the table contain the variables and information given in the problem: the number of milliliters of each solution and its strength as a percent. We fill in the last column of the table by using the formula $P = rW$. The entries in this last column give the *amount* of the important ingredient (in this case, milliliters of acid) in each component solution and in the mixture.

	Number of milliliters (W)	*Percent of acid (r)*	*Amount of acid (P)*
20% solution	x	0.20	$0.20x$
50% solution	y	0.50	$0.50y$
Mixture	45	0.40	$0.40(45)$

We write two equations about the mixture problem. The first equation is about the total number of milliliters mixed together. The chemist must mix x milliliters of one solution with y milliliters of another solution and end up with 45 milliliters of the mixture, so

Total number of milliliters: $x + y = 45$

The second equation uses the fact that the acid in the mixture can come only from the acid in each of the two original solutions. We used the last column of the table to calculate how much acid is in each component, and we add these quantities to get the amount of acid in the mixture.

Amount of acid: $0.20x + 0.50y = 0.40(45)$

These two equation make up a system:

$$x + y = 45$$
$$0.20x + 0.50y = 18$$

Once we have completed the table, it is easy to obtain these equations; we can simply add down the W and P columns of the table. (Note that we cannot add down the middle column, because percents don't add!) To solve our system, we first multiply the second equation by 100 to clear the decimals. This gives us

$$x + y = 45$$
$$20x + 50y = 1800$$

We'll solve this system by elimination. We multiply the first equation by -20 and add the equations together.

$$
\begin{array}{r}
-20x - 20y = -900 \\
\underline{20x + 50y = 1800} \\
30y = 900
\end{array}
$$

Solving for y gives $y = 30$. We substitute $y = 30$ into the first equation to find

$$x + 30 = 45$$

or $x = 15$. Thus the chemist needs 15 milliliters of the 20% solution and 30 milliliters of the 50% solution for the mixture.

You will find mixture problems easy to solve if you complete an appropriate table first. The rows of the table represent the two components and the final mixture, and the columns are used to calculate the amount of the important ingredient in each. The table below is a sample that you can customize for the specifics of a particular problem.

	Total amount (W)	Percent of important ingredient (r)	Amount of important ingredient (P)
First component			
Second component			
Mixture			

EXERCISE 6

Polls conducted by Senator Quagmire's campaign manager show that the senator can win 60% of the rural vote in his state but only 45% of the urban vote. If 800,000 citizens from urban areas vote, how many voters from rural areas must come to the polls in order for the senator to win 50% of the vote?

Step 1 Let x represent the number of rural voters and y the total number of voters. Fill in the table.

	Number of voters (W)	Percent of voters for quagmire (r)	Number of voters for quagmire (P)
Rural			
Urban			
Total			

Step 2 Add down the first and third columns to write a system of equations.

Step 3 Solve your system and answer the question in the problem.

Motion

We can also use systems of equations to solve problems involving motion at a constant speed. We need the formula

$$D = RT$$

EXAMPLE 3

Geologists can calculate the distance from a seismograph to the epicenter of an earthquake by timing the arrival of what are known as P and S waves. They know that P waves travel at about 5.4 miles per second and that S waves travel at 3.0 miles per second. If the P waves arrive 3 minutes before the S waves, how far away is the epicenter of the quake?

Solution

We let x represent the distance from the seismograph to the epicenter. The time it takes for the waves to arrive is also unknown, so we let y be the travel time for the P waves, in seconds. The travel time for the S waves is then $y + 180$ seconds. We'll organize all this information in a table.

	Rate (R)	Time (T)	Distance (D)
P waves	5.4	y	x
S waves	3.0	$y + 180$	x

We can now write two equations, one for the P waves and one for the S waves, using the formula $RT = D$.

$$5.4y = x$$
$$3.0(y + 180) = x$$

We'll solve the system by substitution. We substitute $5.4y$ for x in the second equation and then solve for y.

$$3.0(y + 180) = 5.4y \quad \text{Apply the distributive law.}$$
$$3y + 540 = 5.4y \quad \text{Subtract } 3y \text{ from both sides.}$$
$$540 = 2.4y \quad \text{Divide both sides by 2.4.}$$
$$225 = y$$

Thus it takes the P waves 225 seconds to arrive at the seismograph. To solve for the distance x, substitute $y = 225$ into the first equation to find

$$x = 5.4\,(225) = 1215$$

The epicenter is located 1215 miles from the seismograph.

EXERCISE 7

A steamboat requires 3 hours to travel 24 miles upstream on a river and 2 hours for the return trip downstream. Find the speed of the current and the speed of the steamboat in still water.

Step 1 Choose variables to express the speed of the current and the speed of the steamboat in still water.

	Rate (R)	Time (T)	Distance (D)
Upstream			
Downstream			

Step 2 Fill in the table.

Step 3 Write two equations about the steamboat.

Step 4 Solve your system and answer the questions in the problem.

ANSWERS TO 3.8 EXERCISES
1. $154

2b. $x + y = \$5000$

2c. $0.12x, 0.07y$

2d. $0.12x + 0.07y$

3a.   25%; 50%

3b.   100; 40; 40%

3c.   No

4.   No, she received only 45% of the votes.

5. $900 at 9%; $1100 at 4%

6.   400,000 voters

7.   Speed of the current is 2 miles per hour; speed of the steamboat in still water is 10 miles per hour.

HOMEWORK 3.8

Use systems of equations to solve Problems 1–6 about interest.

1. Goodlife Insurance Company has $150,000 in fees from its clients to invest. The company's treasurer puts part of the money into bonds that pay 6.5% annual interest and the rest into a mutual fund that pays 11.8% annual interest.
   a. Assign variables to the amounts of money deposited into the two accounts, and write an equation about the sum of the deposits.

   b. Write expressions for the interest earned on each account after 1 year.

   c. Goodlife earned $12,930 interest in 1 year. Write an equation to express this fact.

   d. Solve your system of equations to find out how much Goodlife invested in each account.

2. Setsuko has saved $18,000 and plans to invest it in two accounts: a savings account that pays 5.2% annual interest and a stock that promises a 9% annual return.
   a. Assign variables to the amounts of money Setsuko invests in the two accounts, and write an equation about their sum.

   b. Write expressions for the interest earned on each investment after 1 year.

   c. Setsuko earned $1430 on her investments in 1 year. Write an equation to express this fact.

   d. Solve your system of equations to find out how much Setsuko invested in each place.

3. Mario borrowed $30,000 from two banks to open a print shop. The first bank charges 12% annual interest, and the second charges 15% annual interest. Mario's total annual interest payment on the two loans is $3750. How much did he borrow at each rate?

   a. Assign variables to the unknown quantities, and make a table like the one in Example 1 of this section.

   b. Write two equations for the problem, one about the principal and one about the interest.

   c. Solve your system and answer the question.

4. Rani earns an annual income of $1060 on two investments in stocks. Her stock in Solar Enterprises earns 11% annually, and her Compucorp stock earns 10%. Rani invested $6200 more in Solar Enterprises than in Compucorp. How much did she invest in each company?

   a. Assign variables to the unknown quantities, and make a table like the one in Example 1 of this section.

   b. Write two equations for the problem, one about the principal and one about the interest.

   c. Solve your system and answer the question.

5. Stefan borrowed twice as much for his car loan, at 7% interest, as for his student loan, at 4% interest. The annual interest on the car loan is $500 more than the interest on the student loan. How much did Stefan borrow for each loan?

6. Rhoda invested $34,000 in federal bonds that pay 8% annual interest and city bonds that pay 9% interest. The yearly income on each of the investments is the same. How much did Rhoda invest in each type of bond?

Solve Problems 7–10 about mixtures.

7. The chemistry department has 80 students, of whom 35% are women. The physics department has 60 students, of whom 15% are women.
   a. How many chemistry students are women? How many physics students are women?

   b. How many students are there in both departments? How many of them are women?

   c. What percent of the students in chemistry and physics are women?

8. Joe Cleat came to bat 125 times during the first half of the baseball season and got a hit 16% of the time. During the second half of the season, Joe came to bat 75 times and got a hit 40% of the time.
   a. How many hits did Joe get in the first half of the season? How many hits did he get in the second half of the season?

   b. How many times did Joe come to bat all season? How many hits did he get?

   c. What percent of Joe's at-bats resulted in a hit?

9. Pipette, a French chemistry student, has 30 milliliters of a 50% solution of an acid. She wants to reduce the strength by adding 12 milliliters of a 15% solution of the same acid.

  a. How much acid is in the 30 milliliters of 50% solution? How much acid is in the 12 milliliters of 15% solution?

  b. How much acid is in the mixture? How many milliliters of the mixture are there?

  c. What percent of the mixture is acid?

  d. Fill in the table below with your answers to parts (a)–(c).

	Number of milliliters (W)	Percent acid (r)	Amount of acid (P)
50% solution			
15% solution			
Mixture			

10. An auto mechanic keeps supplies of antifreeze solution (mixed with water) in two strengths. The first solution is 20% antifreeze, and the second solution is 30% antifreeze.

  a. How much antifreeze is in 5 quarts of the 20% solution? How much antifreeze is in 10 quarts of the 30% solution?

  b. The mechanic mixes 5 quarts of the 20% solution with 10 quarts of the 30% solution. How much antifreeze is in the mixture? How many quarts of the mixture does he have?

  c. What percent of the mixture is antifreeze?

  d. Fill in the table with your answers from parts (a)–(c).

	Number of quarts (W)	Percent antifreeze (r)	Amount of antifreeze (P)
20% solution			
30% solution			
Mixture			

Use a system of equations to solve Problems 11–14 about mixtures.

11. A pet store owner wants to mix a 12% salt water solution and a 30% salt water solution to obtain 45 liters of a 24% solution. How many liters of each ingredient does he need?

    a. Choose variables, and make a table for the problem.

    b. Use your table to write two equations about the mixture.

    c. Solve your system and answer the question.

12. How many quarts of pure pigment must be mixed into a batch of paint that is 30% pigment to produce 24 quarts of paint that is 65% pigment?

    a. Choose variables, and make a table for the problem.

    b. Use your table to write two equations about the mixture.

    c. Solve your system and answer the question.

13. A newspaper poll of 400 people found that 58% were in favor of a recycling program. It also found that 50% of the men polled and 70% of the women favored the program. How many women were polled?

14. The Clean Air Watchdogs tested 1200 new cars from two manufacturers and found that 80% met emission standards. Major Motors claims that 88% of its cars meet the standards, and Ace Autos claims that 76% of its cars meet the standards. If these claims are true, how many of each company's cars did the Watchdogs test?

Solve Problems 15–18 about motion.

15. Delbert and Francine leave Cedar Rapids at the same time and drive in opposite directions for 6 hours.
    a. Choose variables for Delbert's and Francine's speeds and fill in the table.

	Rate (R)	Time (T)	Distance (D)
Delbert			
Francine			

    b. Make a sketch showing Cedar Rapids, Delbert, and Francine. Label your sketch with the distance that each has traveled.

    c. Francine drives 5 miles per hour slower than Delbert does. After 6 hours, they are 570 miles apart. Write two equations about the problem.

    d. Solve your system to find Delbert's and Francine's speeds.

16. Rose and Iris leave their homes in different towns and drive toward each other. After 3 hours they meet and have lunch.
    a. Choose variables for Rose's and Iris's speeds and fill in the table.

	Rate (R)	Time (T)	Distance (D)
Rose			
Iris			

    b. Make a sketch showing Rose's and Iris's hometowns and their lunch location. Label your sketch with the distance each travels.

    c. Rose drives 10 miles per hour faster than Iris does. They live 240 miles apart. Write two equations about the problem.

    d. Solve your system to find Rose's and Iris's speeds.

17. Bonnie left Dallas and drove north at 40 miles per hour. Three hours later Clyde headed north from Dallas on the same road at 70 miles per hour until he caught up with Bonnie.
   a. Make a sketch showing Dallas, Bonnie, and Clyde.

   b. Complete the table.

	Rate	Time	Distance
Bonnie		*t*	*d*
Clyde			

   c. Use your table to write two equations about the problem.

   d. Solve your system. How long did Bonnie drive before Clyde caught up to her? How far had she driven?

18. Steven bicycled uphill into the mountains at 12 miles per hour and then returned downhill at 20 miles per hour. His return trip took 2 hours less than his trip out.
   a. What can you say about the distances Steven bicycled uphill and downhill?

   b. Complete the table.

	Rate	Time	Distance
Uphill		*t*	*d*
Downhill			

   c. Use your table to write two equations about the problem.

   d. Solve your system. How long did Steven bicycle uphill? How far did he go?

Use a system of equations to solve Problems 19–22 about motion.

19. A yacht leaves San Diego and heads south, traveling at 25 miles per hour. Six hours later, a Coast Guard cutter leaves San Diego traveling at 40 miles per hour and pursues the yacht. How long will it take the cutter to catch the yacht? How far will the boats have traveled?

20. Kristin conducts sight-seeing tours from a blimp, which travels at 50 miles per hour in still air. She flies north along the lake shore and then returns, for a 2-hour round trip. Today, there is a 20-mile-per-hour wind from the north. How long should Kristin fly north before she turns around? How far will she go on the northbound journey?

21. Byron and Ada conduct sight-seeing tours of New England by bicycle. Byron leads the tour group at an average speed of 10 miles per hour, while Ada goes ahead at a speed of 12 miles per hour to prepare lunch.

Ada arrives at the lunch stop 40 minutes $\left(\dfrac{2}{3}\text{ of an hour}\right)$ before the tour group. How far did the group bicycle that morning?

22. Irma flies her plane with the wind and makes a 450-mile trip in 3 hours. The return trip against the wind takes 5 hours. Find the speed of the wind and the speed of the plane in still air.

Solve Problems 23 and 24 about mixtures using a single equation in one variable.

23. In Julio's history class, the final grade is computed by adding 60% of the test average to 40% of the term paper grade. Julio's test average is 72.
   a. Write an equation for Julio's final grade, g, in terms of his term paper grade, t.

   b. What will Julio's final grade be if his term paper grade is 65? What if his term paper grade is 80?

   c. Use the graph to estimate what grade Julio must make on the term paper in order to earn a final grade of 80 in the class.

   d. Use the equation from part (a) to verify your answer algebraically.

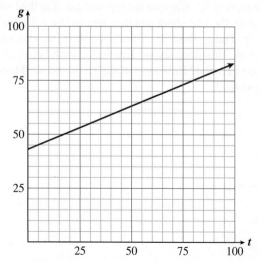

24. In Nicola's physics class, the final grade is computed by adding 70% of the test average to 30% of the final exam grade. Nicola's test average is 84.
   a. Write an equation for Nicola's final grade, g, in terms of her final exam grade, t.

   b. What will Nicola's final grade be if she makes 70 on the final exam? What if she makes 100?

   c. Use the graph to estimate what grade Nicola must make on the final exam to earn a final grade of 75 in the class.

   d. Use the equation from part (a) to verify your answer algebraically.

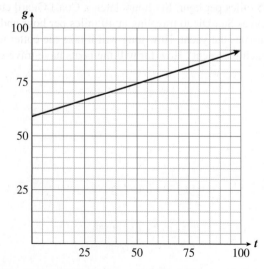

3.9 Point-Slope Form

Reading

A New Formula for Slope

In Section 2.4, you learned how to compute the distance between two points that lie on the same horizontal or vertical line. For example, the distance between points $A(6, 5)$ and $B(6, -7)$ in Figure 3.35 is

$$\Delta y = 5 - (-7) = 12$$

Figure 3.35

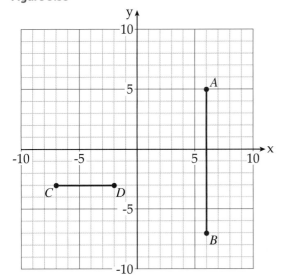

The distance between points $C(-7, -3)$ and $D(-2, -3)$ is

$$\Delta x = -2 - (-7) = 5$$

When we compute slope, the direction in which we move along the line makes a difference. Thus we need to use **directed distance**, which can be either positive or negative, to compute slopes.

If we move from C to D in Figure 3.35, we have moved in the positive x-direction, so the directed distance is

$$\Delta x = -2 - (-7) = 5$$

If we move from D to C, we have moved in the negative x-direction, so the directed distance is

$$\Delta x = -7 - (-2) = -5$$

In either case we *subtract the initial coordinate from the final coordinate.*

In our work so far, we have computed the slope of a line by finding Δy and Δx on a graph. Now we will develop a formula for slope that we can use instead of a graph.

EXAMPLE 1

Compute the slope of the line segment joining P and R in Figure 3.36 in two ways:

    a. Find Δy and Δx using the graph.

    b. Find Δy and Δx using coordinates.

Figure 3.36

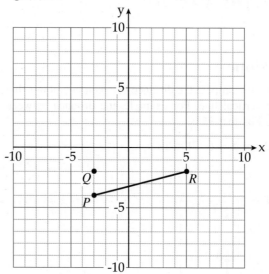

Solution

a. As we move from P to Q, we move up 2 squares on the graph, so

$$\Delta y = 2$$

As we move from Q to R, we move 8 squares to the right, so

$$\Delta x = 8$$

Thus the slope of the line is

$$m = \frac{\Delta y}{\Delta x} = \frac{2}{8} = \frac{1}{4}$$

b. First, we write the coordinates of P and Q:

$$P(-3, -4)$$
$$Q(-3, -2)$$

We compute the directed distance from P to Q:

$$\Delta y = \underset{\text{final} - \text{initial}}{-2 - (-4)} = 2$$

Then we write the coordinates of Q and R:

$$Q(-3, -2)$$
$$R(5, -2)$$

We compute the directed distance from Q to R:

$$\Delta x = \underset{\text{final} - \text{initial}}{5 - (-3)} = 8$$

We get the same value for the slope as in part (a):

$$m = \frac{\Delta y}{\Delta x} = \frac{2}{8} = \frac{1}{4}$$

EXERCISE 1

Compute the slope of the line segment joining *A* and *C* in two ways.

a. Find Δy and Δx using the graph.

$\Delta y = \underline{\hspace{4cm}}$

$\Delta x = \underline{\hspace{4cm}}$

$m = \dfrac{\Delta y}{\Delta x} = \underline{\hspace{4cm}}$

b. Find Δy and Δx using coordinates.

Coordinates of *A*: $\underline{\hspace{3cm}}$

Coordinates of *C*: $\underline{\hspace{3cm}}$

$\Delta y = \underline{\hspace{3cm}}$

    final − initial

$\Delta x = \underline{\hspace{3cm}}$

    final − initial

$m = \dfrac{\Delta y}{\Delta x} = \underline{\hspace{4cm}}$

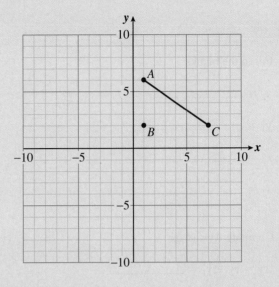

Do you get the same answers for parts (a) and (b)? You should!

Subscript Notation

When computing the slope of a line joining two points, we must be careful to sub-tract their *x*-coordinates and their *y*-coordinates in the same order, final minus ini-tial. To make this easier, we use a notation called a **subscript**. For example, in Figure 3.37, suppose we call *D* the first point and *F* the second point. Then we denote the coordinates of *D* by (x_1, y_1) and the coordinates of *F* by (x_2, y_2), so that

$$x_1 = -6 \quad \text{and} \quad y_1 = -2$$
$$x_2 = 9 \quad \text{and} \quad y_2 = 4$$

Figure 3.37

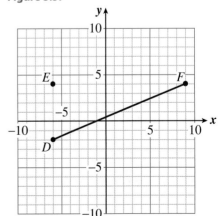

The subscript 1 on x_1, for instance, has nothing to do with the *value* of the coor-dinate; it merely identifies this coordinate as the *x*-coordinate of the *first* point.

Now we can describe the formula for slope in an organized way. First, we ob-serve that the coordinates of point *E* are $(-6, 4)$, or (x_1, y_2). From Figure 3.37, we can see that

$$\Delta y = y_2 - y_1 = 4 - (-2) = 6$$

and

$$\Delta x = x_2 - x_1 = 9 - (-6) = 15$$

Thus the slope of the line segment joining *D* and *F* is

$$m = \frac{y_2 - y_1}{x_2 - x_1} = \frac{6}{15} = \frac{2}{5}$$

EXERCISE 2

Follow the steps to compute the slope of the line segment joining H and K.

Step 1 Let H be the first point and K the second point. Write down the coordinates of each.

Step 2 Fill in the blanks:

$y_2 =$ _____

$y_1 =$ _____

$x_2 =$ _____

$x_1 =$ _____

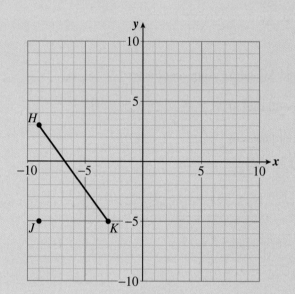

Step 3 Compute Δy and Δx.

$\Delta y = y_2 - y_1 =$ _____

$\Delta x = x_2 - x_1 =$ _____

Step 4 Compute the slope.

$m = \dfrac{y_2 - y_1}{x_2 - x_1} =$ _____

Illustrate Δy and Δx on the graph. Is your value for the slope reasonable?

The method we have described gives us a new formula for computing slope, often called the **two-point formula**:

The Two-Point Formula for Slope

The slope of the line joining points $P_1(x_1, y_1)$ and $P_2(x_2, y_2)$ is

$$m = \frac{y_2 - y_1}{x_2 - x_1} \qquad \text{if} \qquad x_2 \neq x_1$$

We don't need a graph of the line in order to use this formula, just the coordinates of two points on the line.

EXAMPLE 2

Compute the slope of the line joining the points $(-6, 2)$ and $(3, -1)$.

Solution

It doesn't matter which point is P_1 and which is P_2; we'll denote the point $(-6, 2)$ as P_1. Then $(x_1, y_1) = (-6, 2)$ and $(x_2, y_2) = (3, -1)$. Thus

$$m = \frac{y_2 - y_1}{x_2 - x_1}$$

$$= \frac{-1 - 2}{3 - (-6)} = \frac{-3}{9} = \frac{-1}{3}$$

In Example 2, we could have reversed the order of *both* subtractions to get the same answer as before:

$$m = \frac{y_1 - y_2}{x_1 - x_2}$$

$$= \frac{2 - (-1)}{-6 - 3} = \frac{3}{-9} = \frac{-1}{3}$$

The order of the points does not matter, as long as we are consistent when computing Δy and Δx.

Now try Exercise 3.

Skills Review

Cross-multiply to solve each proportion for *y* in terms of *x*.

1. $\dfrac{y}{x} = \dfrac{-5}{2}$

2. $\dfrac{y - 3}{4} = \dfrac{x}{2}$

3. $\dfrac{x + 1}{5} = \dfrac{y - 1}{3}$

4. $-2 = \dfrac{y + 6}{x}$

5. $\dfrac{y + 2}{x - 5} = \dfrac{3}{4}$

6. $\dfrac{-1}{3} = \dfrac{4 - y}{1 - x}$

Answers: **1.** $y = \dfrac{-5}{2}x$ **2.** $y = 2x + 3$ **3.** $y = \dfrac{3}{5}x + \dfrac{8}{5}$

4. $y = -2x - 6$ **5.** $y = \dfrac{3}{4}x - \dfrac{23}{4}$ **6.** $y = \dfrac{-1}{3}x + \dfrac{13}{3}$

Lesson

Finding Points on a Line

In Section 3.5, you learned that if you know one point on a line and its slope, you can graph the line using the slope-intercept method. We can also use the slope to find the coordinates of other points on the line.

Suppose we know that a line has slope $m = \dfrac{3}{2}$ and passes through the point $(x_1, y_1) = (1, 2)$. Follow the steps given in the Activity.

EXERCISE 3
Compute the slope of the line joining the points $(-4, -7)$ and $(2, -3)$.

Activity

1. Plot the point (1, 2) on the grid.

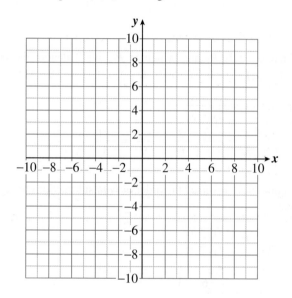

2. Use Δy and Δx to move to another point on the graph, as you would when using the slope-intercept method of graphing.

3. What are the coordinates of the new point, (x_2, y_2)? Note that you can find the new coordinates as follows:

$$x_2 = x_1 + \Delta x = \underline{\hspace{2cm}}$$
$$y_2 = y_1 + \Delta y = \underline{\hspace{2cm}}$$

4. Now, suppose you want to find a third point on the graph by starting at (1, 2) and moving 4 units in the x-direction—that is, $\Delta x = 4$.

Step 1 How far should you move in the y-direction? In other words, what is Δy? You can find Δy because you know that the *ratio* $\dfrac{\Delta y}{\Delta x}$ must always be $\dfrac{3}{2}$. You can solve the following proportion:

$$\frac{\Delta y}{4} = \frac{3}{2}$$

to find that $\Delta y = 6$. (You should check this.)

Step 2 Use this information to plot the new point on the graph and to find its new coordinates:

$$x_3 = x_1 + \Delta x = \underline{\hspace{2cm}}$$
$$y_3 = y_1 + \Delta y = \underline{\hspace{2cm}}$$

5. Use the two steps described in part 4 to find the coodinates of yet another point on the graph by moving $\Delta x = -6$ from the initial point, (1, 2).

Step 1 To find Δy, solve the proportion

$$\frac{\Delta y}{-6} = \frac{3}{2}$$

Step 2 To find the coordinates of the new point, compute

$$x_4 = x_1 + \Delta x = \underline{\hspace{2cm}}$$
$$y_4 = y_1 + \Delta y = \underline{\hspace{2cm}}$$

6. Write the coordinates of the three new points you found:

$$(x_2, y_2) =$$
$$(x_3, y_3) =$$
$$(x_4, y_4) =$$

If you use any one of these three points, along with the original point $(1, 2)$, to compute the slope, you should get $\frac{3}{2}$. Check that this is true:

$$\frac{\Delta y}{\Delta x} = \frac{y_2 - 2}{x_2 - 1} =$$
$$\frac{\Delta y}{\Delta x} = \frac{y_3 - 2}{x_3 - 1} =$$
$$\frac{\Delta y}{\Delta x} = \frac{y_4 - 2}{x_4 - 1} =$$

In fact, *any point (x, y) on the line must satisfy the following equation*:

$$\frac{y - 2}{x - 1} = \frac{3}{2}$$

because the slope between any two points is always $\frac{3}{2}$ for this line.

7. Use the equation $\dfrac{y - 2}{x - 1} = \dfrac{3}{2}$ to decide whether the following points lie on the line:

   a. $(-3, -4)$ **b.** $(-5, -3)$ **c.** $(6, 4)$

8. Graph the line on the grid and check visually whether the three points in part 7 are on the line.

The Activity shows that if you know the slope m of a line and any one point (x_1, y_1) on the line, you can use the formula for slope to write an equation that any other point on the line must satisfy:

$$\frac{y - y_1}{x - x_1} = m$$

This is, in fact, an equation for the line.

EXERCISE 4

Consider a line that has slope -2 and passes through the point $(1, 3)$.

a. Graph the line on the grid.

b. The slope of the line is $m = -2$, so

$$\frac{\Delta y}{\Delta x} = \frac{-2}{1}$$

The given point is $(x_1, y_1) = (1, 3)$. Use the formula

$$\frac{\Delta y}{\Delta x} = \frac{y - y_1}{x - x_1}$$

to write an equation for the line.

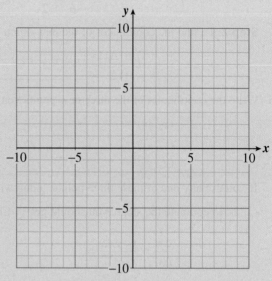

c. Simplify the equation in part (b) to write the equation of the line in the form $y = mx + b$. What is the y-intercept of the line? Verify your answer using the graph you created.

EXERCISE 5

Find an equation for the line that has slope $\dfrac{-1}{2}$ and passes through the point $(-3, -2)$.

Step 1 Use the formula

$$\frac{y - y_1}{x - x_1} = m.$$

Step 2 Cross-multiply to simplify the equation.

Step 3 Solve for y.

In Example 1 in the reading assignment, we knew two points on the line and we used them to calculate the slope, m. In Exercise 4, you used the slope formula in a new way: to find the equation of a line. In this case, you knew the slope and one point and used them to write the equation of the line. The only difference in the formula is that instead of using a particular point, (x_2, y_2) on the line, you obtained an equation for *any* point (x, y) on the line.

We can simplify the formula by clearing the denominator:

$$(x - x_1)\frac{y - y_1}{x - x_1} = m(x - x_1) \quad \text{Multiply both sides by } (x - x_1).$$

to get

$$y - y_1 = m(x - x_1)$$

This version of the formula is called the **point-slope formula** for linear equations.

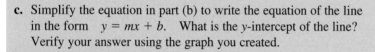

Point-Slope Formula

To find an equation for the line of slope m passing through the point (x_1, y_1), use the point-slope formula:

$$\frac{y - y_1}{x - x_1} = m$$

or

$$y - y_1 = m(x - x_1)$$

Now try Exercise 5.

ANSWERS TO 3.9 EXERCISES

1. $m = \dfrac{-2}{3}$

2. $m = \dfrac{-4}{3}$

3. $m = \dfrac{2}{3}$

4a.

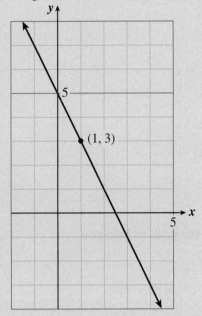

4b. $\dfrac{-2}{1} = \dfrac{y - 3}{x - 1}$

4c: $y = -2x + 5;\ (0, 5)$

5: $y = \dfrac{-1}{2}x - \dfrac{7}{2}$

HOMEWORK 3.9 www

Simplify the expressions in Problems 1–6.

1. $\dfrac{10 - 2}{2 - 8}$

2. $\dfrac{-6 - (-12)}{3 - (-5)}$

3. $\dfrac{-5 - (-5)}{2 - 9}$

4. $-3(-4 - 2) - 6$

5. $\dfrac{3}{2}(4 - 7) + \dfrac{1}{2}$

6. $\dfrac{5}{3}(5 - 8) + 3$

In Problems 7–12, compute the slope of the line joining the given points.

7. $(5, 2), (8, 7)$

8. $(2, 4), (6, 1)$

9. $(3, -2), (0, 1)$

10. $(-3, -4), (-7, 1)$

11. $(6, -2), (-3, -3)$

12. $(3, 4), (-2, 4)$

In Problems 13–16, you are given the slope of a line and one point on it. Use the grids to help you find the missing coordinates of the other points on the line.

13. $m = 2, \quad (1, -1)$
    a. $(0, ?)$ **b.** $(?, -7)$

14. $m = -3, \quad (-2, 4)$
    a. $(-1, ?)$ **b.** $(?, -5)$

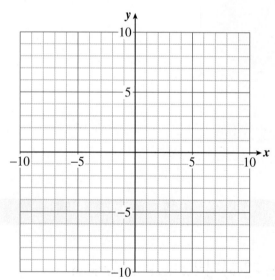
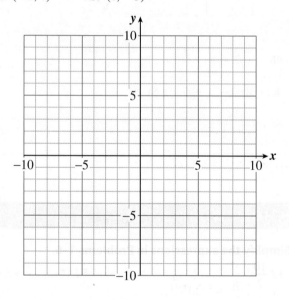

15. $m = \dfrac{-1}{2}$, $(2, -6)$

    a. $(6, ?)$ **b.** $(?, -3)$

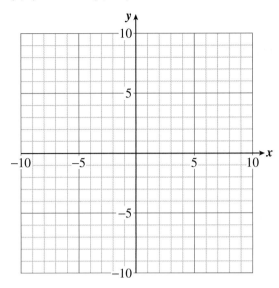

16. $m = \dfrac{2}{3}$, $(2, -3)$

    a. $(8, ?)$ **b.** $(?, -9)$

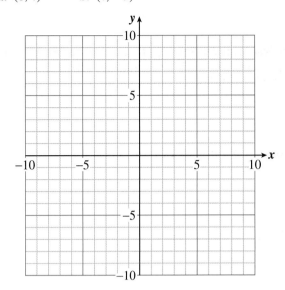

In Problems 17–22, find an equation for the line that has the given slope and passes through the given point. Write your equation in slope-intercept form.

17. $m = -2$, $(-3, 4)$

18. $m = 3$, $(-1, 1)$

19. $m = \dfrac{1}{2}$, $(4, -3)$

20. $m = \dfrac{7}{4}$, $(3, -2)$

21. $m = \dfrac{-2}{3}$, $(-6, 2)$

22. $m = \dfrac{-3}{4}$, $(-3, 1)$

In Problems 23–28, give the slope of each line and the coordinates of one point on the line without doing any calculations.

23. $y = \dfrac{3}{5}x - 7$

24. $y = 6 - x$

25. $y - 2 = 3(x + 5)$

26. $\dfrac{y + 8}{x - 1} = \dfrac{-7}{4}$

27. $\dfrac{y}{x} = \dfrac{4}{5}$

28. $y = -16$

In Problems 29–34:
a. Find the x- and y-intercepts of each line.
b. Use the intercepts to compute the slope of the line.

29. $\dfrac{x}{5} + \dfrac{y}{7} = 1$

30. $\dfrac{x}{8} - \dfrac{y}{2} = 1$

31. $\dfrac{-x}{2.4} + \dfrac{y}{1.6} = 1$

32. $\dfrac{x}{4.5} + \dfrac{y}{7.1} = 1$

33. $3x + \dfrac{2}{7}y = 1$

34. $\dfrac{5}{6}x + 8y = 1$

3.10 Using the Point-Slope Form

EXERCISE 1

Calculate the slopes of the two parallel lines. (Denote the slope of the first line as m_1 and the slope of the second line as m_2.)

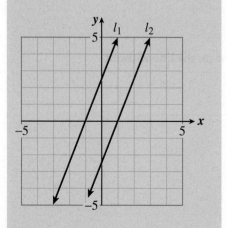

Reading

In this section, we see how to use the point-slope form in some applications. We also consider the equations of horizontal and vertical lines and investigate some properties of parallel and perpendicular lines.

Parallel and Perpendicular Lines

Lines that lie in the same plane but never intersect are called **parallel lines**. It is easy to understand that parallel lines have the same slope.

   Try Exercise 1.

   Lines that intersect at right angles are called **perpendicular lines**. It is a little harder to see the relationship between the slopes of perpendicular lines.

   Now try Exercise 2.

   In Exercise 2, note that the product of m_1 and m_2 is -1. That is,

$$m_1 m_2 = \dfrac{1}{2}(-2) = -1$$

This relationship holds for any pair of perpendicular lines. We summarize the results of Exercises 1 and 2 as follows:

Parallel and Perpendicular Lines

1. Two lines are **parallel** if their slopes are equal—that is, if

$$m_1 = m_2$$

   or if both lines are vertical.

2. Two lines are **perpendicular** if the product of their slopes is -1—that is, if

$$m_1 m_2 = -1$$

   or if one of the lines is horizontal and one is vertical.

Another way to state the condition for perpendicular lines is

$$m_2 = \frac{-1}{m_1}$$

Because of this relationship, we often say that the slope of one perpendicular line is the *negative reciprocal* of the slope of the other.

EXAMPLE 1

Decide whether the following lines are parallel, perpendicular, or neither:

$$2x + 3y = 6 \quad \text{and} \quad 3x - 2y = 6$$

Solution

We could graph the lines, but we can't be sure from a graph if two lines are exactly parallel or exactly perpendicular. A more accurate way to settle the question is to find the slope of each line. To do this, we write each equation in slope-intercept form—that is, we solve for y.

$2x + 3y = 6$ Subtract $2x$ from both sides.
$\quad 3y = -2x + 6$ Divide by 3.
$\quad\quad y = \dfrac{-2}{3}x + 2$

$3x - 2y = 6$ Subtract $3x$ from both sides.
$\quad -2y = -3x + 6$ Divide by -2.
$\quad\quad y = \dfrac{3}{2}x - 3$

The slope of the first line is $m_1 = \dfrac{-2}{3}$, and the slope of the second line is $m_2 = \dfrac{3}{2}$. The slopes are not equal, so the lines are not parallel. The product of the slopes is

$$m_1 m_2 = \left(\frac{-2}{3}\right)\left(\frac{3}{2}\right) = -1$$

so the lines are perpendicular.

Now try Exercise 3.

EXERCISE 2
Calculate the slopes of the perpendicular lines.

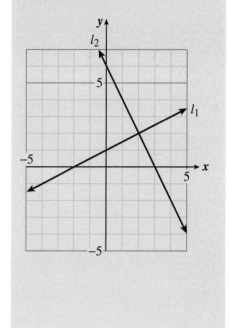

EXERCISE 3
a. What is the slope of a line that is parallel to $x + 4y = 2$?

b. What is the slope of a line that is perpendicular to $x + 4y = 2$?

Skills Review

Find the negative reciprocal of each number.

1. $\dfrac{3}{4}$ 2. $\dfrac{-1}{3}$ 3. -6 4. $\dfrac{11}{8}$

5. $1\dfrac{2}{3}$ 6. $-2\dfrac{1}{2}$ 7. -3.2 8. 0.625

Answers: *1.* $\dfrac{-4}{3}$ *2.* 3 *3.* $\dfrac{1}{6}$ *4.* $\dfrac{-8}{11}$ *5.* $\dfrac{-3}{5}$ *6.* $\dfrac{2}{5}$ *7.* 0.3125

8. -1.6

Lesson

Equations of Horizontal and Vertical Lines

In Section 3.4, you learned that the slope of a horizontal line is 0. We can use the point-slope form to find the equation of a horizontal line that passes through a particular point.

EXAMPLE 2

Find the equation of the horizontal line that passes through (5, 3).

Solution

We substitute $m = 0$ and $(x_1, y_1) = (5, 3)$ into the point-slope formula, $y - y_1 = m(x - x_1)$.

$$y - 3 = 0(x - 5)$$
$$y - 3 = 0$$
$$y = 3$$

The equation $y = 3$ is actually the slope-intercept form for this line, with $m = 0$ and $b = 3$. The fact that x does not appear in the equation means that $y = 3$ for every point on the line, no matter what the value of x is. (See Figure 3.38.)

What about the equation of a vertical line? The slope of a vertical line is undefined; a vertical line does not have a slope. We cannot use the point-slope form to find the equation of a vertical line. However, we can use what we know about horizontal lines from Example 2. Look at the graph of the vertical line in Figure 3.39. Every point on the line has x-coordinate -2, no matter what its y-coordinate is. An equation for this line is $x = -2$. The fact that y does not appear in the equation means that the value of x does not depend on y; it is constant.

The y-intercept of the horizontal line $y = 3$ is $(0, 3)$; the line has no x-intercept. The x-intercept of the vertical line $x = -2$ is $(-2, 0)$; the line has no y-intercept.

Figure 3.38

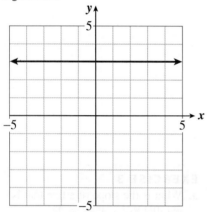

Figure 3.39

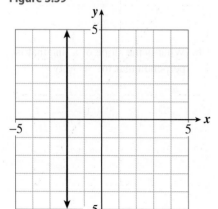

We summarize our results as follows:

Horizontal and Vertical Lines

1. The equation of the **horizontal line** passing through $(0, b)$ is

$$y = b$$

2. The equation of the **vertical line** passing through $(a, 0)$ is

$$x = a$$

Now try Exercise 4.

The Line Through Two Points

How many lines pass through two given points? There is only one. Using the point-slope formula, we can find its equation.

EXAMPLE 3

Find an equation for the line that passes through $(2, -1)$ and $(-1, 3)$.

Solution

We will solve this problem in two steps: We will find the slope of the line and then use the point-slope formula.

Step 1 Let $(x_1, y_1) = (2, -1)$ and $(x_2, y_2) = (-1, 3)$. Using the slope formula, we find that

$$m = \frac{y_2 - y_1}{x_2 - x_1}$$

$$= \frac{3 - (-1)}{-1 - 2} = \frac{4}{-3} = \frac{-4}{3}$$

Step 2 Apply the point-slope formula with $m = \dfrac{-4}{3}$ and

$(x_1, y_1) = (2, -1)$. (We can use either point to find the equation of the line.) Then

$$\frac{y - y_1}{x - x_1} = m$$

becomes

$$\frac{y - (-1)}{x - 2} = \frac{-4}{3}$$

We cross-multiply to find

$$3(y + 1) = -4(x - 2) \quad \text{Apply the distributive law.}$$
$$3y + 3 = -4x + 8 \quad \text{Subtract 3 from both sides.}$$
$$3y = -4x + 5 \quad \text{Divide both sides by 3.}$$
$$y = \frac{-4}{3}x + \frac{5}{3}$$

The graph of the line is shown in Figure 3.40.

Now try Exercise 5.

EXERCISE 4
a. Find an equation for the vertical line passing through $(-4, -1)$.

b. Find an equation for the horizontal line passing through $(-4, -1)$.

Figure 3.40

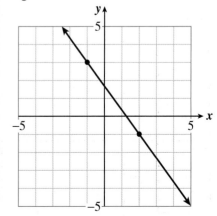

EXERCISE 5

In Section 2.7, we considered some data about the clearing of the world's rain forests. In 1970 there were about 9.8 million square kilometers of rain forest left; in 1990 there were about 8.2 million square kilometers.

a. Use these data points to find a linear equation for the number of million square kilometers, y, of rain forest left x years after 1950.

b. If we continue to clear the rain forest at the same rate, when will it be completely destroyed? How does your answer compare to the estimate we made in Section 2.7?

Step 1 From the information given, write the coordinates of two points on the line.

Step 2 Find the slope of the line.

Step 3 Use the point-slope formula to find an equation for the line.

Step 4 To answer part (b), find the x-intercept of the line.

Equations from Graphs

The easiest way to find the equation of a line from its graph is to identify the slope and the *y*-intercept of the line. Then we can write the equation in the form $y = mx + b$. However, if the *y*-intercept is difficult to read from the graph, it may be more accurate to choose another point and then use the point-slope formula to derive the equation.

EXAMPLE 4

Find an equation for the line in Figure 3.41.

Solution

The intercepts are difficult to estimate accurately, so we look for two points that lie on grid lines. In this case, we choose $(-6, -1)$ and $(1, 3)$. First, we use these two points to calculate the slope.

$$m = \frac{y_2 - y_1}{x_2 - x_1} = \frac{3 - (-1)}{1 - (-6)} = \frac{4}{7}$$

Next we use the point $(-6, -1)$ to write the point-slope formula for the line.

$$\frac{y - (-1)}{x - (-6)} = \frac{4}{7}$$

Finally, we solve for *y* to obtain the slope-intercept form.

$$\frac{y + 1}{x + 6} = \frac{4}{7} \qquad \text{Cross-multiply.}$$

$$7(y + 1) = 4(x + 6) \qquad \text{Apply the distributive law.}$$

$$7y + 7 = 4x + 24 \qquad \text{Subtract 7 from both sides.}$$

$$7y = 4x + 17 \qquad \text{Divide both sides by 7.}$$

$$y = \frac{4}{7}x + \frac{17}{7}$$

Now try Exercise 6.

Figure 3.41

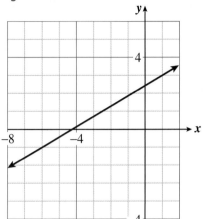

ANSWERS TO 3.10 EXERCISES

1. $m_1 = m_2 = 3$

2. $m_1 = \dfrac{1}{2}$, $m_2 = -2$

3a. $m = \dfrac{-1}{4}$

3b. $m = 4$

4a. $x = -4$

4b. $y = -1$

5a. $y = -0.08x + 11.4$

5b. 2092

6b. $y \approx 2.4x + 47.2$

6c. 66.4 inches

6d. $9\dfrac{1}{2}$

EXERCISE 6

In Section 2.10, we estimated a regression line for data relating shoe size and height. The figure shows that regression line for the data.

a. Find (approximate) coordinates of two widely spaced points on the line. (None of the data points lies exactly on the line, so you should not choose any of the data points.)

b. Use the point-slope formula to find the equation of the line through the two points you chose.

c. Use the equation to estimate the height of a person whose shoe size is 8.

d. Use the equation to predict the shoe size of a person who is 70 inches (5 feet 10 inches) tall.

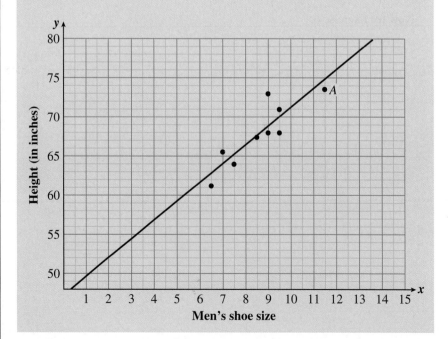

HOMEWORK 3.10

Determine whether the lines in Problems 1–6 are parallel, perpendicular, or neither.

1. $3x - 4y = 2$
$8y - 6x = 6$

2. $5x + 3y = 1$
$3y - 5x = 5$

3. $2x = 4 - 5y$
$2y = 4x - 5$

4. $y = 4x + 3$
$x = 3 - 4y$

5. $y + 3x - 2 = 0$
$x - 3y + 2 = 0$

6. $2x + 6y + 5 = 0$
$3y + x - 5 = 0$

In Problems 7–14, give the equation of the line described.

7. Horizontal, passing through $(6, -5)$

8. Vertical, passing through $(5, -7)$

9. m undefined, passing through $(2, 1)$

10. $m = 0$, passing through $(-3, -1)$

11. Parallel to $x = 4$, passing through $(-8, 3)$

12. Perpendicular to $y = -2$, passing through $(6, 1)$

13. The x-axis

14. The y-axis

15. a. What is the slope of the line $y = -3x + 2$?

  b. What is the slope of a line parallel to $y = -3x + 2$?

  c. A line is parallel to $y = -3x + 2$ and passes through $(1, 3)$. What is its equation?

16. a. What is the slope of the line $y = 2x + 1$?

  b. What is the slope of a line parallel to $y = 2x + 1$?

  c. A line is parallel to $y = 2x + 1$ and passes through $(-2, 1)$. What is its equation?

17. a. What is the slope of the line $y = -2x - 4$?

  b. What is the slope of a line perpendicular to $y = -2x - 4$?

  c. A line is perpendicular to $y = -2x - 4$ and passes through $(4, 2)$. What is its equation?

18. a. What is the slope of the line $y = -3x - 2$?

  b. What is the slope of a line perpendicular to $y = -3x - 2$?

  c. A line is perpendicular to $y = -3x - 2$ and passes through $(1, 3)$. What is its equation?

19. The vertices of a triangle are points $B(-1, 8)$, $C(4, -2)$, and $D(10, 1)$.

   a. Sketch the triangle on the grid.

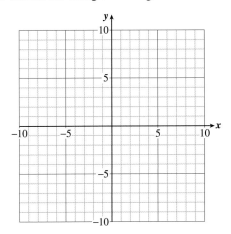

   b. Show that the triangle is a right triangle. (Show that two of the sides are perpendicular.)

20. The vertices of a quadrilateral are $P(-4, -2)$, $Q(6, 2)$, $R(3, 10)$, and $S(-2, 8)$.

   a. Sketch the quadrilateral on the grid.

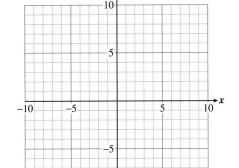

   b. Show that the quadrilateral is a trapezoid. (Show that two of the sides are parallel.)

21. Three of the vertices of a parallelogram $WXYZ$ are $W(2, 5)$, $X(6, 1)$ and $Y(9, 4)$.

   a. Sketch the parallelogram on the grid.

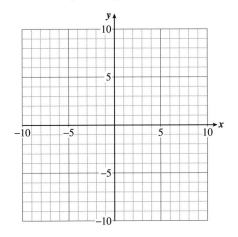

   b. Find the coordinates of the fourth vertex, Z.

   c. Show that both pairs of opposite sides are parallel.

22. The line segment from $F(-5, 3)$ to $G(3, 9)$ forms one leg of a right triangle. The hypotenuse is horizontal. Find the coordinates of the third vertex, H, of the triangle. (There are two possible solutions.)

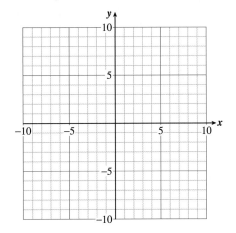

In Problems 23–28, find an equation for the line passing through the given points. Write your answer in slope-intercept form.

23. $(-2, 4), (1, 7)$

24. $(1, 3), (0, 7)$

25. $(3, 5), (-3, -5)$

26. $(2, 4), (-3, 4)$

27. $(6, 4), (-2, 5)$

28. $(5, -2), (-1, 3)$

In Problems 29–32, write a linear equation relating the variables. Then use the equation to answer the questions.

29. a. Francine is driving into the mountains and stopping periodically to record the temperature at various altitudes. The temperature at an altitude of 3200 feet is 77°F, and the temperature at 8000 feet is 65°F. What is the temperature, y, at an altitude of x feet?

b. What is the slope, including units? What does the slope tell you about the problem situation?

c. What will the temperature be at 10,000 feet?

d. What was the temperature at sea level?

30. a. Delbert bought a condominium 5 years ago. One year after he bought it, it was worth $62,500, and last year it was worth $58,000. Delbert would like to know what the value, y, of the condominium will be x years after it was purchased.

b. What is the slope, including units? What does the slope tell you about the problem situation?

c. When will the condominium be worth $46,000?

d. What did Delbert pay for the condominium?

31. a. Envirotech is marketing a new line of clothes dryers that use microwaves rather than hot air to dry clothes. It cost the company $45,000 to produce the first 100 dryers. With the production of 180 dryers, the total cost was up to $61,000. What is the total cost, *y*, of producing *x* dryers?

b. What is the slope, including units? What does the slope tell you about the problem?

c. Envirotech has budgeted $100,000 for manufacturing microwave clothes dryers this year. How many can the company produce?

d. How much did Envirotech invest in development before it made the first dryer?

32. a. There was a lot of snow in the mountains last winter, and now a thaw has started. The water level in the river has been rising since Sunday at noon. At 9 A.M. on Monday, the water reached the 10-foot mark on the levee, and by 5 P.M., it was at 10 feet 4 inches. What will the water level, *y*, be in inches after the water rises for *x* hours?

b. What is the slope, including units? What does the slope tell you about the problem?

c. The levee is 13 feet high. If the water continues to rise, when will flooding start?

d. What was the water level before the thaw?

In Problems 33–36, find the equation of the line shown in the graph.

33.

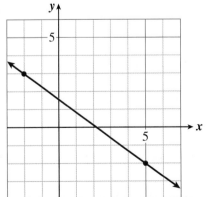

34.

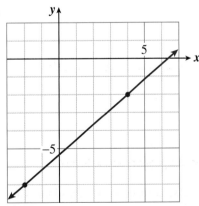

35.

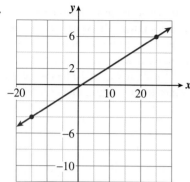

36.

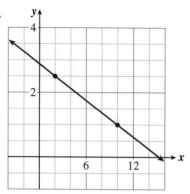

In Problems 37–40, plot the data on the grid provided. Carefully draw a line that fits the data. Find the equation of the line, and use the equation to answer each question.

37. The heights and shadow lengths of several objects were measured at the same time of day, and the results were recorded. Plot the data.

Height (inches)	10	20	30	40	50
Shadow length (inches)	3.8	7.3	11.5	14.6	18.6

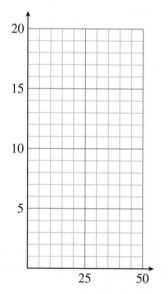

a. Estimate the length of the shadow of an object 17 inches tall.

b. A maple tree casts a shadow 84 inches long. How tall is the tree?

38. Irene measured the height of a burning candle at different times and recorded the data. Plot Irene's data.

Time (minutes)	Height (centimeters)
0	30
5	29.1
10	28.7
15	28.1
20	27.3
25	26.6
30	26.1

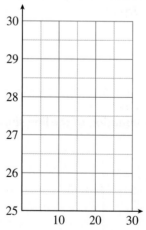

a. How tall can Irene expect the candle to be after 35 minutes?

b. How many minutes should Irene expect the candle to last?

39. As Percy raises the temperature of a gas in a container, the pressure inside the container changes. Plot Percy's data.

Temperature (°C)	20	40	60	80
Pressure (atmospheres)	1.01	1.08	1.15	1.22

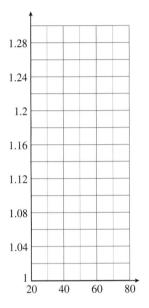

a. Estimate the pressure of the gas at a temperature of 37°C.

b. Estimate the temperature needed to bring the gas to a pressure of 1.5 atmospheres.

c. What is the physical interpretation of the y-intercept of the regression line (if the graph is extended far enough to the left)?

d. What is the physical interpretation of the x-intercept of the regression line (if the graph is extended far enough downward)?

40. Francine has designed a fuel-efficient automobile that runs on a combination of gasoline and battery power. She records how far she can drive on various amounts of gasoline. Plot Francine's data.

a. Estimate the distance that the automobile can travel on 0.25 gallon of gasoline.

Gasoline (gallons)	0.1	0.2	0.3	0.4
Distance (miles)	6.6	15.5	23.9	33.1

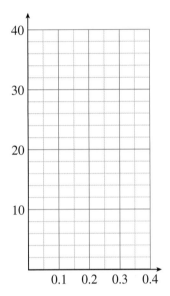

b. Estimate the amount of gasoline needed to drive 50 miles.

c. What is the physical interpretation of the x-intercept of the regression line (if the graph is extended far enough downward)?

41. Over the history of the Olympic Games, the winning times for both men and women in the 100-meter dash have gradually decreased. The table lists the winning times (in seconds) from the 1928 games, the first in which women ran the race, through the 1996 games. Answer the following questions about the data.

a. Has the men's winning time been faster than the women's winning time in each Olympic year?

b. What was the difference in the men's and women's winning times in 1928?

c. What was the difference in the men's and women's winning times in 1996?

d. In how many Olympics after 1928 did the winning man break the previous record in the 100-meter dash? In how many Olympics did the winning woman break the previous record?

e. By how many seconds did the men's winning time improve from 1928 to 1996? By how many seconds did the women's time improve in that same period?

Men's times				*Women's times*			
Year	*Time*	*Year*	*Time*	*Year*	*Time*	*Year*	*Time*
1928	10.8	1968	9.95	1928	12.2	1968	11.0
1932	10.3	1972	10.14	1932	11.9	1972	11.07
1936	10.3	1976	10.06	1936	11.5	1976	11.08
1948	10.3	1980	10.25	1948	11.9	1980	11.6
1952	10.4	1984	9.99	1952	11.5	1984	10.97
1956	10.5	1988	9.92	1956	11.5	1988	10.54
1960	10.2	1992	9.96	1960	11.0	1992	10.82
1964	10.0	1996	9.84	1964	11.4	1996	10.94

Figures (a) and (b) show plots of the data, each with a line of best fit. Note that we have set $x = 0$ as 1928 and that the winning times are in seconds.

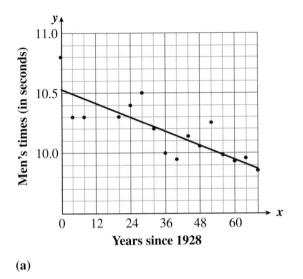

(a)

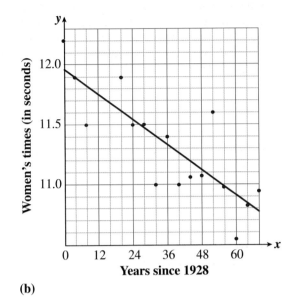

(b)

f. Find approximate equations for both of the lines.

g. Which of the two lines has the steeper slope?

h. Predict the winning time in the men's 100-meter dash in 2004 (when $x = 76$).

i. Predict the winning time in the women's 100-meter dash in 2004.

j. If both lines were drawn on the same axes, where would they intersect? What is the practical meaning of the intersection point?

k. How do you know that extrapolation along either of these lines will eventually lead to nonsensical conclusions? (*Hint:* Interpret the *x*-intercepts of the lines.)

42. According to the *Digest of Educational Statistics 1994*, the average salaries of men and women college graduates increased each year from 1970 to 1990. The table shows the salary data.

a. Has the average salary for men been greater than the average salary for women in each year?

b. What was the difference in the men's and women's average salaries in 1970?

c. What was the difference in the men's and women's average salaries in 1990?

d. When did the men's average salary first exceed $20,000? The women's average salary?

e. By how much did the men's average salary increase from 1970 to 1990? By how much did the women's average salary increase?

Men's salaries				Women's salaries			
Year	Salary	Year	Salary	Year	Salary	Year	Salary
1970	13,264	1981	26,394	1970	8156	1981	16,322
1971	13,730	1982	28,030	1971	8451	1982	17,405
1972	14,879	1983	29,892	1972	8736	1983	18,452
1973	15,503	1984	31,487	1973	9057	1984	20,257
1974	16,240	1985	32,822	1974	9523	1985	21,389
1975	17,477	1986	34,391	1975	10,349	1986	22,412
1976	18,236	1987	35,327	1976	11,010	1987	23,399
1977	19,603	1988	36,434	1977	11,605	1988	25,187
1978	20,941	1989	38,565	1978	12,347	1989	26,709
1979	22,406	1990	39,238	1979	13,441	1990	28,017
1980	24,311			1980	15,143		

Figures (a) and (b) show plots of the data, each with a line of best fit.

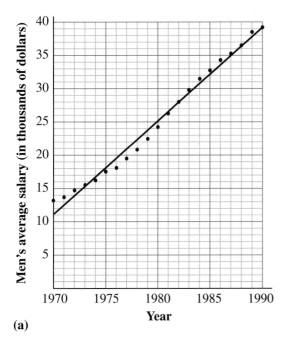

(a)

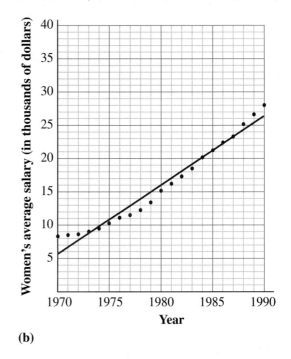

(b)

f. Find approximate equations for both of the lines.

i. Predict the average salary of female college graduates in the year 2004.

g. Which of the two lines has the steeper slope?

j. If both lines were drawn on the same axes, where would they intersect? What is the practical meaning of the intersection point?

h. Predict the average salary of male college graduates in the year 2004.

k. How do you know that extrapolation along either of these lines will eventually lead to nonsensical conclusions? (*Hint:* Interpret the *x*-intercepts of the lines.)

CHAPTER SUMMARY AND REVIEW

Glossary: Write a sentence explaining the meaning of each term.

ratio	solution of a system	inconsistent system	subscript
proportion	direct variation	dependent system	parallel lines
proportional	slope	directed distance	perpendicular lines
constant of proportionality	system of equations		

Properties and Formulas

1. State the fundamental property of proportions.

2. If two quantities vary directly, what equation relates them?

3. Give a formula for slope using the Δ notation.

4. State the slope-intercept formula, and explain the meaning of the coefficients.

5. State the two-point formula for slope.

6. State the point-slope formula for a line.

7. What is true about the slopes of parallel lines?

8. What is true about the slopes of perpendicular lines?

9. What does the equation of a vertical line look like? What is the slope of a vertical line?

10. What does the equation of a horizontal line look like? What is the slope of a horizontal line?

Techniques and Procedures

11. Explain how to tell whether two variables are proportional.

12. Explain the slope-intercept method for graphing a line.

13. What is the easiest way to find the slope of a line from its equation?

14. How can you find the solution to a system of linear equations by graphing?

15. Name two algebraic methods for solving a system of linear equations.

16. Suppose you are using the elimination method to solve a system of linear equations. How can you tell whether the system is dependent or inconsistent?

17. How would you label the columns when making a table for a problem about interest?

18. How would you label the columns when making a table for a problem about a mixture?

19. How would you label the columns when making a table for a problem about motion?

20. How can you find the equation of a line when you know the slope of the line and one point on the line?

Review Problems

21. Decide whether the two variables are proportional.

a.

D	0.5	1	2	4
V	0.25	2	16	64

b.

s	0.2	0.8	1.6	2.5
M	0.08	0.32	0.64	1

Solve the proportions in Problems 22 and 23.

22. $\dfrac{16}{q} = \dfrac{52}{86.125}$

23. $\dfrac{2x-1}{x+3} = \dfrac{3}{2}$

24. At Van's Hardware, 36 metric frimbles cost \$4.86. How many metric frimbles can you buy for \$6.75?

25. Bob's weekly diet includes 70 grams of fat, 200 grams of protein, and 1800 grams of carbohydrates. Jenny's weekly diet includes 112 grams of fat. How many grams of protein and carbohydrates should she consume so that her diet has the same proportions of the three types of nutrients as Bob's diet?

26. Terence has to divide a 100-pound sack of concrete into two portions in the ratio of 7 to 5. How many pounds of concrete should he put into each portion?

In Problems 27 and 28, find the unknown length in each figure.

27.

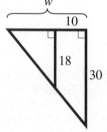

28.

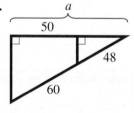

29. Ubi stands 12 feet from a lamppost and casts a shadow 9 feet long. If Ubi is 6 feet tall, how tall is the lamppost?

30. Rizalinda is making a science fiction video that includes a scene involving a giant calculator. She wants the giant calculator to be proportional to a real one and as wide as she is tall, namely 160 centimeters. If her real calculator is 8 centimeters wide and 17 centimeters tall, how tall must the giant calculator be?

31. Simon joined a co-op that provides access to the Internet. His user bill, B, varies directly with the number of hours, h, that he is logged on. Last month, he was logged on for 28 hours, and his bill was $21.
  a. Write an equation expressing B in terms of h.

  b. This month Simon's bill was $39. How many hours did he log on to the Internet?

  c. Graph the equation you wrote in part (a).

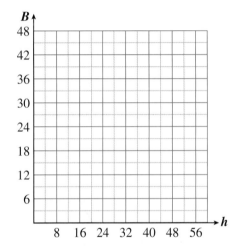

32. A small orchard that is 150 yards on each side contains 100 apple trees. A larger orchard that is 600 yards on each side contains 1600 trees. Does the number of trees vary directly with the length of the orchard?

33. The number of defective computer chips in a shipment varies directly with the size of the shipment. Last month, Vantage Electronics found 12 defective chips in its order. If the company increases its order by 25%, how many defective ones should it expect to receive?

34. Forensic scientists can estimate the height, H, of a person from the length of the femur (thigh bone) using the formula $H = 0.61 + 0.02L$, where L is the length of the femur. Does a person's height vary directly with the length of the femur?

In Problems 35 and 36:
a. Find the intercepts of each line.
b. Use the intercepts to graph the line.
c. Use the intercepts to find the slope of the line.

35. $6x - 3y = 18$

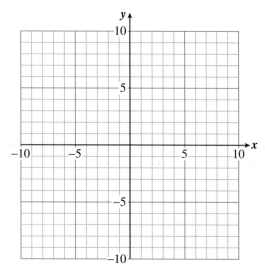

36. $4y + 9x = 36$

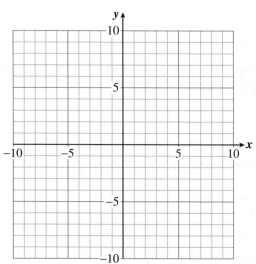

In Problems 37 and 38, use the graph to find the slope of each line, and illustrate Δx and Δy on the graph.

37. $3x + 2y = -7$

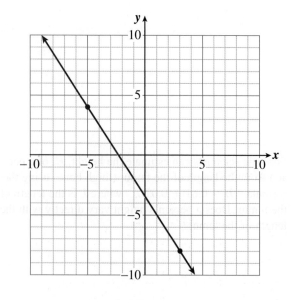

38. $3y = 5x$

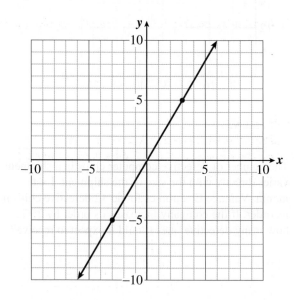

39. Each of the switchbacks on a mountain road is about 3600 feet long. If the grade on the road is 8%, how much elevation is gained on each switchback?

40. This year, because of heavy snowfall last winter, the water in Saginaw Bay is 44 inches higher than usual. If the slope of the beach at the water's edge is $18.\overline{3}\%$, how much narrower is the beach this year than usual?

41. It costs Delbert \$16.80 to fill up the 12-gallon gas tank in his sports car and \$35 to fill the 25-gallon tank in his recreational vehicle.

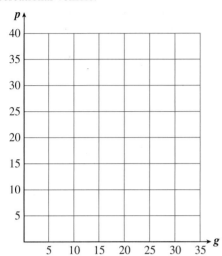

a. Plot price, *p*, on the vertical axis, and gallons, *g*, on the horizontal axis, and compute the slope of the line segment joining the two given points.

b. What does the slope tell you about the problem situation?

42. The floor tile for an 8-foot-by-10-foot kitchen costs \$1200, and for the same tile a 12-foot-by-15-foot kitchen costs \$2700.

a. Verify that the cost, *C*, varies directly with the area, *A*, of the kitchen.

b. Write an equation for *C* in terms of *A*.

c. Interpret the slope in your equation as a rate. What does it tell you about the problem situation?

In Problems 43 and 44:
a. Find the slope and *y*-intercept of the line.
b. Graph the line using the slope-intercept method.

43. $2y + 5x = -10$

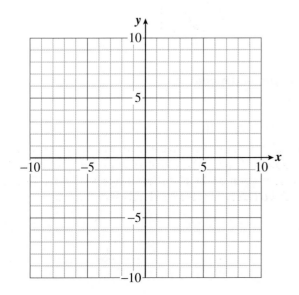

44. $y + x = 0$

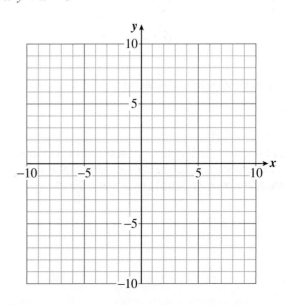

45. Find the slope and the *y*-intercept of the line $y = 4$.

46. The price of a medium sundae at a frozen yogurt shop is given by $y = 1.35 + 0.85t$,   where *t* is the number of toppings you select. Find the slope and the *y*-intercept of the graph, and explain what each means in terms of the problem situation.

In Problems 47 and 48, decide whether the given point is a solution of the given system.

47. $x + y = 8$ $(-2, 10)$
     $x - y = 2$

48. $8x + 3y = 21$ $(-3, 1)$
     $5x = y - 16$

Solve each system in Problems 49 and 50 by graphing.

49. $x + y = 5$
     $2x - y = 4$

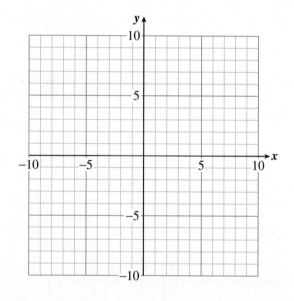

50. $x - y = 7$
     $y = \dfrac{-2}{3}x - 2$

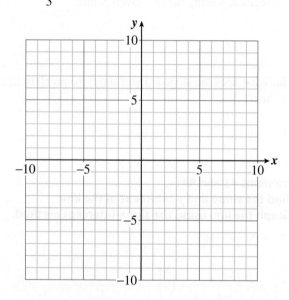

Classify each system in Problems 51 and 52 as consistent, inconsistent, or dependent.

51. a. $y = 3x - 4$
       $y = 3x + 4$

b. $y = \dfrac{2}{5}x - 2$
    $5y = 2x - 10$

52. a. $2y = 3x + 7$

$y = \dfrac{3}{2}x + 7$

b. $3x - 4y = 6$

$4x + 3y = 6$

Solve each system in Problems 53 and 54 by substitution.

53. $y = 2x + 1$
$2x + 3y = -21$

54. $x + 4y = 1$
$2x + 3y = -3$

Solve each system in Problems 55 and 56 by elimination.

55. $2x + 7y = -19$
$5x - 3y = 14$

56. $4x + 3y = -19$
$5x + 15 = -2y$

Solve Problems 57–64 by writing a system of equations.

57. A health food store wants to produce 30 pounds of granola that sells for 80 cents per pound. It plans to mix cereal that sells for 65 cents per pound with dried fruit that sells for 90 cents per pound. How much of each should it use?

58. The perimeter of a rectangle is 50 yards, and its length is 9 yards greater than its width. Find the dimensions of the rectangle.

59. Last year, Veronica made $93 in interest from her two savings accounts, one of which paid 6% interest and the other 9%. Veronica withdrew the $93 interest but left the original principal in each account. This year, the interest rates dropped to 4% and 8%, respectively, and she made $76 interest. How much did Veronica invest in each account?

60. Marvin invested $300 more at 6% than he invested at 8%. His total annual income from his two investments is $242. How much did he invest at each rate?

61. How many pounds of an alloy containing 60% copper must be melted with an alloy containing 20% copper to obtain 8 pounds of an alloy containing 30% copper?

62. Jerry Glove, a teammate of Joe Cleat (whom we met in Homework Problem 8 in Section 3.8), came to bat only 20 times in the first half of the season and got a hit 15% of the time. During the second half of the season, Jerry came to bat 140 times and got a hit 35% of the time.

  a. How many hits did Jerry get in the first half of the season? How many hits did he get in the second half?

  b. How many times did Jerry come to bat all season? How many hits did he get?

  c. What percent of Jerry's at-bats resulted in a hit?

  d. Who had a better batting average in the first half of the season, Joe or Jerry? Who had a better batting average in the second half? Who had the better batting average overall?

63. Alida and Steve are moving to San Diego. Alida is driving their car, and Steve is driving a rental truck. They start together, but Alida drives twice as fast as Steve. After 3 hours, they are 93 miles apart. How fast is each traveling?

64. Jake rides for the Pony Express, covering his portion of the route in 6 hours and then returning home, riding 8 miles per hour slower, in 9 hours. How far does Jake ride?

In Problems 65 and 66, find the slope of the line described.

65. Passing through the points $(1, -2)$ and $(-7, 3)$

66. Perpendicular to the line $y = \dfrac{1}{3}x - 2.$

In Problems 67–72, find an equation for the line described. Write your answer in slope-intercept form if possible.

67. With x-intercept $(5, 0)$ and y-intercept $(0, -1)$

68. Having slope $-\dfrac{1}{4}$ and passing through the point $(3, -5)$

69. Parallel to the y-axis and passing through the point $(-4, 2)$

70. Horizontal and passing through the point $(-3, 8)$

71. Passing through the points $(-3, 1)$ and $(-1, -1)$

72. Perpendicular to the line $2x + 3y = 1$ and passing through the point $(-4, 1)$

73. Graph the line that has slope $\dfrac{-5}{3}$ and passes through the point $(-2, 1)$

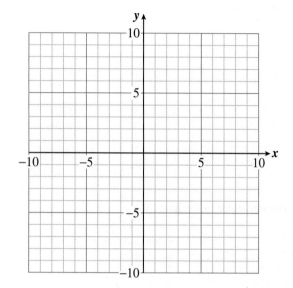

74. Which pair of lines are parallel, which are perpendicular, and which are neither?

a. $5x + 4y = 3$
$3y - 5x = 4$

b. $5x + 3y = 1$
$5y - 3x = 6$

c. $4y - 5x = 1$
$2y - \dfrac{5}{2}x = 2$

Quadratic Equations

4.1 Exponents and Formulas

Reading

Figure 4.1

In this reading assignment, we review some basic terminology about exponents. It is not unusual for the same factor to occur more than once in a product. A simple example is the formula for the volume of a cube with edge s, as shown in Figure 4.1:

$$V = s \cdot s \cdot s$$

We can write such products with a notation called an exponent. An **exponent** is a number that appears above and to the right of a given factor and tells how many times that factor occurs in the expression. The factor to which the exponent applies is called the **base**, and the product is called a **power** of the base. For example,

$$\overset{\text{exponent}}{2^5} = \underbrace{2 \cdot 2 \cdot 2 \cdot 2 \cdot 2}_{\text{5 factors of 2}} = 32$$

base

The notation 2^5 is read as "the fifth power of 2" or as "2 raised to the fifth power," or simply as "2 to the fifth." In general,

$$a^n = a \cdot a \cdot a \cdot \ldots \cdot a$$

(n factors of a, where n is a positive integer)

Try Exercise 1 on page 422.

Because the exponents 2 and 3 are used frequently, they have special names:

5^2 means $5 \cdot 5$ and is read "5 *squared*" or "the square of 5"

5^3 means $5 \cdot 5 \cdot 5$ and is read "5 *cubed*" or "the cube of 5"

These names come from the formulas for the area of a square with side s and the volume of a cube with edge s:

$$A = s^2$$
$$V = s^3$$

Now try Exercise 2.

423

EXERCISE 1

Compute the following powers. (For part d, find out how to use your calculator to compute powers.)

a. 6^2 **b.** 3^4

c. $\left(\dfrac{2}{3}\right)^3$ **d.** 1.4^6

EXERCISE 2

a. Find the area of a square whose side is $2\dfrac{1}{2}$ centimeters long.

b. Find the volume of a cube whose base is the square in part (a).

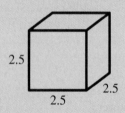

2.5

2.5

2.5

Powers of Negative Numbers

To indicate that a negative number is raised to a power, we must enclose the negative number in parentheses. For example, to indicate "the square of -5," we write

$$(-5)^2 = (-5)(-5) = 25$$

If the negative number is not enclosed in parentheses, the exponent applies *only* to the positive number, and the negative sign is applied *after* the power has been computed. Compare the two expressions

$$(-3)^2 = (-3)(-3) = 9 \quad \text{\small Exponent applies to } (-3).$$

and

$$-3^2 = -(3 \cdot 3) = -9 \quad \text{\small Exponent applies only to 3.}$$

The first expression tells us to square -3 (multiply -3 by itself), and the second expression tells us to square 3 and then make the result negative. Note how the placement of parentheses changes the meaning of the expressions in Exercise 3. (Only one of the answers is positive. Which is it?)

EXERCISE 3

Compute each power.

a. -4^2 **b.** $(-4)^2$

c. $-(4)^2$ **d.** (-4^2)

In Exercise 4, you will compute successive powers of -2.

EXERCISE 4

Compute each power of -2.

a. $(-2)^1$ **b.** $(-2)^2$ **c.** $(-2)^3$

d. $(-2)^4$ **e.** $(-2)^5$

f. The sign of the power is _____ if the exponent is _____.

Do you see a pattern developing? What is the connection between the exponent and the sign of the answer?

Skills Review

Follow the order of operations to simplify each expression. See Lessons 1.8 and 2.3.

1. $-2[-3(-5) - 8(4)]$

2. $[-8 + 6(-4)(-3)][5 - (-2)]$

3. $\frac{4}{3}(-6)(6 - 9)(6 - 9)$

4. $\frac{3}{8}(-7 - 5)(-4 - 4)$

5. $-2.4(-3) + (8 - 4.5)(-7.2)$

6. $-9.6 - 3.2(-8 - 2.4)(-3)$

Answers: **1.** 34 **2.** 448 **3.** -72 **4.** 36 **5.** -18 **6.** -109.44

Lesson

Repeated products of variables can also be written with exponents. Thus, for example,

$$x^4 = x \cdot x \cdot x \cdot x$$

Compare the expression above with this one:

$$4x = x + x + x + x$$

An *exponent* on a variable indicates *repeated multiplication*, whereas a *coefficient* in front of a variable indicates *repeated addition*. The expressions x^4 and $4x$ are not equivalent; in other words, they are not equal for all values of x. (Can you think of one value of x for which they *are* equal?)

Note that we do not usually write x^1, because x^1 is simply the same as x.

Now try Exercise 5.

Order of Operations

In the reading assignment, we compared two expressions involving exponents, -3^2 and $(-3)^2$. We saw that $-3^2 \neq (-3)^2$. Here is a similar problem that includes variables: Compare the expressions $2x^3$ and $(2x)^3$. In the first expression, only the x should be cubed:

$$2x^3 \quad \text{means} \quad 2xxx \quad \text{\small Exponent applies to the base, } x, \text{ only.}$$

In the second expression, the quantity $2x$ is cubed:

$$(2x)^3 \quad \text{means} \quad (2x)(2x)(2x) \quad \text{\small Exponent applies to the base, } 2x.$$

The two expressions are not equivalent, as you can see by evaluating each for, say, $x = 5$:

$$2x^3 = 2(5^3) = 2(125) = 250$$
$$(2x)^3 = (2 \cdot 5)^3 = 10^3 = 1000$$

In general, *an exponent applies only to its base*, and not to any other factors in the product. If you want an exponent to apply to more than one factor, you must enclose those factors in parentheses.

Now try Exercise 6.

We are ready to state a rule about evaluating expressions with exponents. Observe that to evaluate $2x^3$, we computed the power x^3 first and then the product, $2 \cdot x^3$. To evaluate $(2x)^3$, we first performed the multiplication inside parentheses, $2 \cdot x$, and then computed the power. In general, we compute powers before multiplications, but after operations within parentheses. We revise the order of operations, introduced in Section 1.8, to include exponents.

EXERCISE 5

a. Write each expression as either a repeated addition or a repeated multiplication.

$a^5 = $ _____

$5a = $ _____

b. Evaluate each expression in part (a) for $a = 3$.

$a^5 = $ _____

$5a = $ _____

EXERCISE 6

Compare the expressions by writing them without exponents.

a. $3ab^4$ and $3(ab)^4$

b. $a + 5b^2$ and $(a + 5b)^2$

Order of Operations

1. Perform any operations inside parentheses or above or below a fraction bar.
2. Compute all indicated powers.
3. Perform all multiplications and divisions in the order in which they occur from left to right.
4. Perform additions and subtractions in order from left to right.

EXAMPLE 1

Simplify each expression.

    a. $9 - 5 \cdot 2^3$ **b.** $-2(8 - 3 \cdot 4)^2$

Solution

    a. $9 - 5 \cdot 2^3$ Compute the power first: $2^3 = 8$.

        $= 9 - 5 \cdot 8$ Multiply before subtracting.

        $= 9 - 40 = -31$

    b. $-2(8 - 3 \cdot 4)^2$ Simplify within parentheses first.

        $= -2(8 - 12)^2$

        $= -2(-4)^2$ Compute the power; then multiply.

        $= -2(16) = -32$

It is especially important to follow the order of operations when evaluating an expression. If you are substituting a negative number for the variable, it may be helpful to enclose the negative number in parentheses.

EXAMPLE 2

Evaluate each expression for $x = -6$.

    a. x^2 **b.** $-x^2$ **c.** $2 - x^2$ **d.** $(2x)^2$

Solution

    a. We replace x by (-6) and then square. This means that -6 gets squared, *not just* 6. Thus

$$x^2 = (-6)^2 = (-6)(-6) = 36$$

    The parentheses are essential for this calculation. It would be *incorrect* to write $x^2 = -6^2 = -36$.

    b. In this expression, only x is squared, and then the negative sign is applied to the result.

$$-x^2 = -(-6)^2 = -36$$

    c. Replace x by (-6) to get

$$2 - x^2 = 2 - (-6)^2$$ Compute the power first.

$$= 2 - 36 = -34$$

d. When we replace x by (-6), for clarity we change the existing parentheses to brackets.

$$(2x)^2 = [2(-6)]^2 \qquad \text{Multiply inside brackets first.}$$
$$= [-12]^2 = 144$$

Now try Exercise 7.

EXERCISE 7
Evaluate each expression for $t = -3$.
a. $-3t^2 - 3$

b. $-t^3 - 3t$

Formulas

Many useful formulas involve exponents. Figure 4.2 shows several solid figures, along with formulas for their volumes and surface areas. **Volume** is the amount of space contained within a three-dimensional object, and it is measured in *cubic units*, such as cubic feet or cubic centimeters. **Surface area** is the sum of the areas of all the faces or surfaces that contain a solid, and it is measured in *square units*. (You will find a more extensive list of common formulas from geometry in the front of this book.)

Figure 4.2

Sphere	Cylinder	Cone	Pyramid

 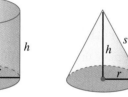 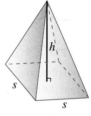

$V = \frac{4}{3}\pi r^3$ $\qquad$ $V = \pi r^2 h$ $\qquad$ $V = \frac{1}{3}\pi r^2 h$ $\qquad$ $V = \frac{1}{3}s^2 h$

$S = 4\pi r^2$ $\qquad$ $S = 2\pi r^2 + 2\pi rh$ $\qquad$ $S = \pi r^2 + \pi rs$

EXAMPLE 3

The Red Deer Pub and Microbrewery has a spherical copper tank in which beer is brewed. If the tank is 4 feet in diameter, how much beer does it hold?

Solution

The formula for the volume of a sphere is $V = \frac{4}{3}\pi r^3$, where r is the radius of the sphere. The tank has a diameter of 4 feet, so its radius is 2 feet. We substitute $r = 2$ into the formula and simplify. (If your calculator does not have a key for π, you can use the approximation $\pi \approx 3.14$.)

$$V = \frac{4}{3}\pi(2)^3 \qquad \text{Compute the power first.}$$

$$= \frac{4\pi(8)}{3} \qquad \text{Multiply by } 4\pi, \text{ and divide by 3.}$$

$$= 33.51\ldots$$

The tank holds approximately 33.5 cubic feet of beer, or about 251 gallons.

If a formula is not given in the most convenient form for a particular problem, you may need to solve an equation after substituting values for the variables.

EXERCISE 8

If the height of a box is the same as its width, as shown in the figure, its surface area is given by the formula

$$S = 2w^2 + 4lw$$

(Do you see how to obtain this formula, using the definition of surface area?)

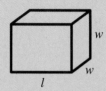

a. Francine would like to wrap a package whose length is 4 feet and whose width and height are both 1.5 feet. She has 24 square feet of wrapping paper. Is that enough paper?

b. What is the longest box whose width and height are 1.5 feet that Francine can wrap? (Assume that the paper is just the right shape and that the edges of the paper do not overlap!)

EXAMPLE 4

The formula $A = P + Prt$ gives the amount, A, of money (principal plus interest) in an account after t years, if a principal of P dollars is invested at simple annual interest rate r. How long will it take an investment of \$5000 to grow to \$7000 in an account that pays $6\frac{1}{2}\%$ simple annual interest?

Solution

We substitute 7000 for A, 5000 for P, and 0.065 for r in the formula and then solve the resulting equation for t.

$$7000 = 5000 + 5000(0.065)t \quad \text{Subtract 5000 from both sides.}$$
$$2000 = 5000(0.065)t \quad \text{Simplify the right side.}$$
$$2000 = 325t \quad \text{Divide both sides by 325.}$$
$$6.15 \approx t$$

Note that the answer is *not* 6.15 years! Because interest is paid only once a year, the investor must wait until the end of the *seventh* year before the account contains at least \$7000.

Now try Exercise 8.

Solving for One Variable

If we plan to use a formula more than once with different values for the variables, it may be faster in the long run to solve for one of the variables in terms of the others. We did something like this when we put a linear equation into slope-intercept form: We solved for y in terms of x.

EXAMPLE 5

Solve the formula $A = P + Prt$ for t.

Solution

For this problem, we treat t as the unknown and treat all the other variables as if they were constants. We begin by isolating the term containing t on one side of the equation:

$$A = P + Prt \quad \text{Subtract } P \text{ from both sides.}$$
$$A - P = P + Prt - P$$
$$A - P = Prt \quad \text{Divide both sides by } Pr.$$
$$\frac{A - P}{Pr} = \frac{Prt}{Pr}$$
$$\frac{A - P}{Pr} = t$$

We now have a new formula for t: $\quad t = \dfrac{A - P}{Pr}$

You can use the formula for t that we found in Example 5 to see how long it will take an investment to grow to various amounts. Try Exercise 9.

EXERCISE 9

a. Use the formula developed in Example 5 to write an equation expressing t in terms of A if you invest \$2000 at 5% annual interest.

b. Fill in the table.

Amount (dollars)	2500	3000	5000	6000	7500
Time (years)					

c. Graph your equation. Is the equation linear?

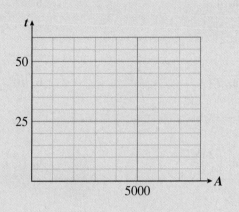

ANSWERS TO 4.1 EXERCISES

1a. 36 **1b.** 81 **1c.** $\dfrac{8}{27}$ **1d.** 7.529536

2a. 6.25 square centimeters **2b.** 15.625 cubic centimeters

3a. -16 **3b.** 16 **3c.** -16 **3d.** -16

4a. -2 **4b.** 4 **4c.** -8 **4d.** 16 **4e.** -32
4f. Negative, odd; or: Positive, even

5a. $a^5 = a \cdot a \cdot a \cdot a \cdot a$; $5a = a + a + a + a + a$

5b. $a^5 = 243$; $5a = 15$

6a. $3abbbb$; $3abababab$ **6b.** $a + 5bb$; $(a + 5b)(a + 5b)$

7a. $-3t^2 - 3 = -30$ **7b.** $-t^3 - 3t = 36$

8a. No **8b.** 3.25 feet

9a. $t = \dfrac{A - 2000}{100}$ **9b.** 5, 10, 30, 40, 55 **9c.** The equation is linear.

10. $l = \dfrac{S - 2w^2}{4w}$

EXERCISE 10

Solve the formula $S = 2w^2 + 4lw$ for l.

HOMEWORK 4.1

Compute the powers in Problems 1–4.

1. a. 4^3 **b.** 5^3 **c.** 5^4

2. a. 2^4 **b.** 10^6 **c.** 8^3

3. a. $\left(\dfrac{2}{3}\right)^4$ **b.** $\left(\dfrac{4}{5}\right)^3$ **c.** $\left(\dfrac{11}{9}\right)^2$

4. a. $\left(\dfrac{12}{7}\right)^2$ **b.** $\left(\dfrac{6}{5}\right)^3$ **c.** $\left(\dfrac{1}{9}\right)^3$

Use a calculator to compute the powers in Problems 5 and 6. Round your answers to the nearest hundredth.

5. a. $(3.1)^3$ **b.** $(2.6)^4$ **c.** $(0.8)^4$

6. a. $(0.3)^3$ **b.** $(-6.2)^5$ **c.** $(-4.8)^6$

Simplify each expression in Problems 7–10.

7. a. -5^2 **b.** -5^3

    c. $(-5)^2$ **d.** $(-5)^3$

8. a. $-6 \cdot 2^3$ **b.** $(-6 \cdot 2)^3$

    c. $6^3 - 2^3$ **d.** $6(-2)^3$

9. a. $-(-2)^2$ **b.** $-(-2)^3$

c. $-2^3 - 2^2$ **d.** $-(2^3 - 2)^2$

10. a. $-2 - 4^2$ **b.** $-2(-4)^2$

c. $-(2 - 4)^2$ **d.** $-2(-4^2)$

Translate each phrase in Problems 11–16 into an algebraic expression, and then simplify it. Round to two decimal places if necessary.

11. The square of the sum of 3 and 4   **12.** The sum of the square of 3 and the square of 4

13. 5 more than 2.3 to the third power   **14.** 1 less than the fifth power of 0.8

15. 25% of the cube of 6   **16.** The square of 40% of 5

17. Evaluate each expression for $x = -2$.
    a. $5x^3$ **b.** $5x^2$

18. Evaluate each expression for $w = -9$.
    a. $(2w)^2$ **b.** $36 - (2w)^2$

    c. $5 - x^2$ **d.** $5 - x^3$ **c.** $-2(4 - w)^2$ **d.** $2(4 - w)^3$

19. Evaluate for $a = -3$ and $b = -4$.
   a. ab^3 **b.** $a - b^3$

   c. $(a - b^2)^2$ **d.** $ab(a^2 - b^2)$

20. Evaluate for $h = -2$ and $g = -5$.
   a. $h^2 - 2hg + g^2$ **b.** $(h - g)^2$

   c. $h^2 - g^2$ **d.** $(h - g)(h + g)$

Simplify each expression in Problems 21–28.

21. a. $x + x + x$

   b. $x \cdot x \cdot x$

22. a. $w \cdot w \cdot w \cdot w$

   b. $w + w + w + w$

23. a. $2a \cdot 2a$

   b. $2a + 2a$

24. a. $5b + 5b$

   b. $5b \cdot 5b$

25. a. $-q - q - q$

   b. $-q(-q)(-q)$

26. a. $-z(-z)$

   b. $-z - z$

27. a. $-3m - 3m$

   b. $(-3m)(-3m)$

28. a. $-4p(-4p)(-4p)$

   b. $-4p - 4p - 4p$

For Problems 29–32, recall that two algebraic expressions are said to be *equivalent* if they are equal for every value of their variables.

29. a. Explain why the expressions $3z^2$ and $(3z)^2$ are not equivalent.

   b. Explain why the expressions $(3z)^2$ and $9z^2$ are equivalent.

30. a. Explain why the expressions $-u^2$ and $(-u)^2$ are not equivalent.

   b. Explain why the expressions $(-u)^2$ and u^2 are equivalent.

31. a. Find two values of x for which $2x = x^2$.

   b. Find four values of x for which $2x \neq x^2$.

   c. Is $2x$ equivalent to x^2?

32. a. Find one value of y for which $y^2 + y^2 = y^4$.

   b. Find two values of y for which $y^2 + y^2 \neq y^4$.

   c. Is $y^2 + y^2$ equivalent to y^4?

In Problems 33–36, write an expression for the shaded area in each combination of squares and circles.

33.

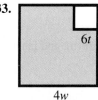

34.

35.

36.

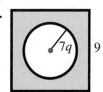

Write an algebraic expression for the volume of each figure in Problems 37–40.

37.

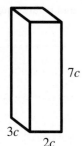

38.

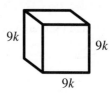

39.

40.

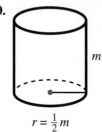

41. The dome on a new planetarium is a hemisphere (half a sphere) of radius 40 feet.
a. How much space is enclosed within the dome?

42. a. A large candle is shaped like a cylinder with a radius of 3 inches and a height of 20 inches. How much wax was used to make the candle?

b. What is the surface area of the dome?

b. You would like to make another candle with a radius of 2 inches, using 126 cubic inches of wax you have left over from making the first candle. How tall will the candle be?

c. Draw a sketch of the dome, and label its radius.

c. Draw a sketch of your candle in part (b), and label its dimensions.

43. a. How much sheet steel is needed to make a cylindrical can with radius 5 centimeters and height 16 centimeters? (*Hint:* Should you calculate the volume or the surface area of the can?)

b. How long is a cylindrical section of concrete pipe if its diameter is 3 feet and the area of its outer surface is approximately 848.25 square feet?

c. Draw a sketch of the pipe in part (b), and label its dimensions.

44. a. Sometime around 200 B.C., the inhabitants of Teotihuacán in central Mexico built the Sun Pyramid, which is 210 feet high and has a square base 689 feet on a side. What is the volume of the Sun Pyramid?

b. What is the slope of one face of the Sun Pyramid?

c. Draw a sketch of the pyramid, and label its dimensions.

In Problems 45 and 46, use the graph provided to answer the questions.

45. Here is a graph of the equation $y = x^2$.

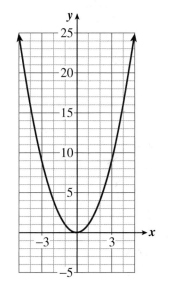

a. Locate the point corresponding to $x = 4$. What is the value of x^2, according to the graph?

b. Locate two points where $y = 9$. What x-values correspond to those points?

c. Use the graph to estimate the value of 2.5^2.

d. Use the graph to estimate the value of $(-3.5)^2$.

e. Use the graph to estimate the solution of the equation $x^2 = 20$. (Did you find two solutions?)

f. Use the graph to estimate the solution of the equation $x^2 = -5$.

46. Here is a graph of the equation $y = x^3$.

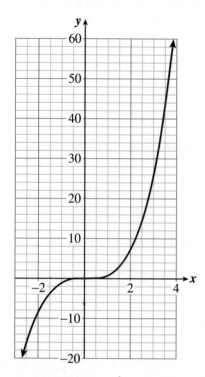

a. Locate the point corresponding to $x = 3$. What is the value of x^3, according to the graph?

b. Locate a point where $y = -8$. What x-value corresponds to this point?

c. Use the graph to estimate the value of 2.5^3.

d. Use the graph to estimate the value of $(-1.5)^3$.

e. Use the graph to estimate the solution of the equation $x^3 = 40$.

f. Use the graph to estimate the solution of the equation $x^3 = -16$.

Solve each formula in Problems 47–56 for the indicated variable.

47. $v = lwh$,   for w

48. $E = IR$,   for I

49. $E = \dfrac{mv^2}{2}$,   for m

50. $V = \pi r^2 h$,   for h

51. $A = \dfrac{h}{2}(b + c)$, for h

52. $F = \dfrac{kmM}{d^2}$, for M

53. $F = \dfrac{9}{5}C + 32$, for C

54. $C = \dfrac{5}{9}(F - 32)$, for F

55. $A = \pi rh + 2\pi r^2$, for h

56. $P = 2l + 2w$, for w

Simplify each expression in Problems 57–62 mentally, without using pencil, paper, or calculator!

57. 24×10^2

58. 8.91×10^5

59. 0.003×10^4

60. $3 \cdot 10 + 9$

61. $2 \cdot 10^2 + 3 \cdot 10 + 4$

62. $3 \cdot 10^3 + 4 \cdot 10^2 + 5 \cdot 10 + 6$

4.2 Square Roots

Reading

Suppose we would like to draw a square whose area is 25 square inches. How long should each side of the square be? Because the formula for the area of a square is $A = s^2$, the side s should satisfy the equation $s^2 = 25$. In other words, we want a number whose square is 25. As you can probably guess, the length of the square should be 5 inches, because $5^2 = 25$. We say that 5 is a *square root* of 25. In general, we have the following definition.

> The number s is called a **square root** of the number b if $s^2 = b$.

You can see that *finding a square root of a number is the opposite of squaring a number.* Thus, for example,

$$5 \text{ is a square root of } 25 \text{ because } 5^2 = 25$$
$$12 \text{ is a square root of } 144 \text{ because } 12^2 = 144$$

5 is not the only square root of 25, because $(-5)^2 = 25$ as well. Thus 25 has two square roots, 5 and -5. In fact, *every positive number has two square roots, one positive and one negative.*

Try Exercise 1 now.

EXERCISE 1

Find two square roots for each number.

a. 225   **b.** $\dfrac{4}{9}$

Radicals

The positive square root of a number is called the **principal square root**. We use the symbol $\sqrt{}$ to denote the positive, or principal, square root. Thus we may write

$$\sqrt{25} = 5 \quad \text{and} \quad \sqrt{144} = 12$$

The symbol $\sqrt{}$ is called the **radical sign**, and the number underneath it is called the **radicand**. Because of this terminology, square roots are often called **radicals**.

What about the other square root, the negative one? If we want to indicate the negative square root of a number, we place a negative sign outside the radical sign, like this:

$$-\sqrt{16} = -4 \quad \text{and} \quad -\sqrt{49} = -7$$

If we want to refer to both square roots, we use the symbol $\pm$, which is read "plus or minus." For example,

$$\pm\sqrt{36} = \pm 6 \quad \text{which means 6 or } -6$$

Note that zero has only one square root: $\sqrt{0} = 0$.

Now try Exercise 2.

We have seen that each positive number has two square roots, and zero has exactly one. What about the square root of a negative number? For example, can we find $\sqrt{-4}$? The answer is no, because the square of any number is positive (or zero). Try this yourself: The only reasonable candidates for $\sqrt{-4}$ are 2 and -2, but

$$2^2 = \underline{}$$
$$(-2)^2 = \underline{}$$

It is true that $2(-2) = -4$, but multiplying a number by its opposite is not squaring the number!

EXERCISE 2

Find each square root.

a. $-\sqrt{81}$

b. $\pm\sqrt{\dfrac{64}{121}}$

Thus $\sqrt{-4}$ is not meaningful for us. In general, *we cannot find the square root of a negative number.* Appendix B introduces a new type of numbers, called *imaginary numbers*, which give meaning to square roots of negative numbers. For now, we say that the square root of a negative number is *undefined.*

Now try Exercise 3.

Skills Review

Solve each equation for x.

1. $3x + 5 = 17$

2. $3x + k = 17$

3. $\dfrac{x}{4} - 9 = -4$

4. $\dfrac{x}{4} - m = -4$

5. $2x - 3 = 4x + 7$

6. $2x - c = 4x + 7$

7. $6x + 1 = 3(2 - x)$

8. $bx + 1 = 3(2 - x)$

Answers: **1.** 4 **2.** $\dfrac{17 - k}{3}$ **3.** 20 **4.** $4(-4 + m)$ **5.** -5

6. $\dfrac{-c - 7}{2}$ **7.** $\dfrac{5}{9}$ **8.** $\dfrac{5}{b + 3}$

Lesson

So far in your study of algebra, you have learned how to solve various kinds of linear equations. In linear equations, the variable cannot have any exponent other than 1; for this reason, such equations are often called *first-degree equations.* In this chapter, we consider *second-degree equations,* or **quadratic equations**. A quadratic equation can be written in the standard form

$$ax^2 + bx + c = 0$$

where a, b, and c are constants and a is not zero. Here are some examples of quadratic equations:

$$2x^2 + 5x - 3 = 0 \qquad 7t - t^2 = 0 \qquad \text{and} \qquad 3a^2 = 16$$

Note that the x^2 term and the x term in a quadratic expression are not like terms, so they cannot be combined.

Now try Exercise 4.

Extraction of Roots

Consider a simple quadratic equation:

$$x^2 = 16$$

This equation tells us that x is a number whose square is 16. Thus x must be a square root of 16. Because 16 has two square roots, 4 and -4, these are the solutions of the equation.

We can think of the solution process in the following way. Since x is squared in the equation, we perform the opposite operation, or take square roots, in order to solve for x.

$$x^2 = 16 \qquad \text{Take square roots of both sides.}$$
$$x = \pm\sqrt{16} \qquad \text{Simplify.}$$
$$x = \pm 4$$

Remember that every positive number has *two* square roots, and consequently a quadratic equation has two solutions. This method for solving quadratic equations is called **extraction of roots**. To apply the method, we isolate the squared term on one side of the equation.

EXAMPLE 1

Solve $3y^2 - 40 = 35$ by extraction of roots.

Solution
We begin by isolating the quadratic term, y^2, and then we extract roots.

$$3y^2 - 40 = 35 \qquad \text{Add 40 to both sides.}$$
$$3y^2 = 75 \qquad \text{Divide both sides by 3.}$$
$$y^2 = 25 \qquad \text{Take square roots of both sides.}$$
$$y = \pm\sqrt{25} \qquad \text{Simplify.}$$
$$y = \pm 5$$

The solutions are 5 and -5.

Now try Exercise 5.

EXERCISE 5
Solve $4a^2 + 25 = 169$.

Irrational Numbers

Now suppose we want to solve the equation

$$x^2 = 6$$

We can use extraction of roots to find

$$x = \pm\sqrt{6}$$

but we cannot simplify $\sqrt{6}$ as either a whole number or any common fraction or decimal fraction. If you use a calculator with an eight-digit display to obtain a decimal value for $\sqrt{6}$, you will find

$$\sqrt{6} \approx 2.4494897$$

However, this is only an approximation, and not the exact value of $\sqrt{6}$. (Try squaring 2.24494897 and you will see that

$$2.4494897^2 = 5.9999998$$

which is not exactly 6, although it is close.) In fact, no matter how many digits your calculator or computer can display, you can never find an exact decimal equivalent for $\sqrt{6}$. $\sqrt{6}$ is an example of what is called an *irrational number*. (You will learn more about irrational numbers in the reading assignment for Section 4.3.)

Now try Exercise 6.

Of course, even though many radicals are irrational numbers, some radicals, such as $\sqrt{16} = 4$ and $\sqrt{\dfrac{9}{25}} = \dfrac{3}{5}$,   represent integers or fractions. You should learn to recognize *perfect squares*, which are integers such as 9, 16, and 25 whose square roots are integers.

EXERCISE 6
Use your calculator to find two approximate solutions to the equation

$$2t^2 = 26$$

Round your answers to three decimal places.

Order of Operations: Expressions Containing Radicals

When we evaluate algebraic expressions that involve radicals, we must follow the order of operations as usual. Square roots occupy the same position as exponents in the hierarchy of operations: They are computed after parentheses but before multiplication.

EXAMPLE 2

Find an approximation to three decimal places for $8 - 2\sqrt{7}$.

Solution

(You may be able to enter the expression into your calculator just as it is written. If not, you must follow the order of operations.) The given expression has two terms, 8 and $-2\sqrt{7}$; the second term is the product of $\sqrt{7}$ and -2. We should *not* begin by subtracting 2 from 8, because multiplication precedes subtraction. First, we find an approximation for $\sqrt{7}$:

$$\sqrt{7} \approx 2.6457513$$

(Do *not* round off your approximations at any intermediate steps in the problem or you will lose accuracy at each step! You should be able to work directly with the value on your calculator's display.)

Next, we multiply our approximation by -2 to find

$$-2\sqrt{7} \approx -5.2915026$$

Finally, we add 8 to get

$$8 - 2\sqrt{7} \approx 2.7084974$$

Rounding to three decimal places gives 2.708.

A radical symbol acts like parentheses to group operations. *You should perform any operations that appear under a radical before evaluating the root.*

Try Exercise 7 now.

We can update the order of operations given in Section 4.1 by modifying Steps 1 and 2 to include radicals.

> ## Order of Operations
>
> 1. Perform any operations inside parentheses, under a radical, or above or below a fraction bar.
> 2. Compute all indicated powers and roots.
> 3. Perform all multiplications and divisions in the order in which they occur from left to right.
> 4. Perform additions and subtractions in order from left to right.

Cube Roots

Imagine a cube whose volume is 64 cubic inches. What is the length, c, of one edge of this cube? Because the volume of a cube is given by the formula $V = c^3$, we must solve the following equation to find the edge of the cube.

$$c^3 = 64$$

EXERCISE 7

a. Simplify $\sqrt{6^2 - 4(3)}$.

b. Approximate your answer for part (a) to three decimal places.

c. Explain why the following is *incorrect*:

$$\sqrt{6^2 - 4(3)} \rightarrow \sqrt{6^2} - \sqrt{4(3)}$$
$$= \sqrt{36} - \sqrt{12}$$
$$\approx 6 - 3.464 = 2.536$$

In other words, we are looking for a number c whose cube is 64. With a little trial and error, we soon discover that $c = 4$. The number c is called the *cube root* of 64 and is denoted by $\sqrt[3]{64}$. In general,

> The number c is called a **cube root** of the number b if $c^3 = b$.

Try Exercise 8 now.

Recall that every positive number has two square roots and that negative numbers do not have square roots. The situation is different with cube roots: Every number has exactly one cube root. The cube root of a positive number is positive, and the cube root of a negative number is negative. When you study complex numbers, you will learn more about square and cube roots.

Like square roots, some cube roots are irrational numbers and some are not. Cube roots are treated the same as square roots in the order of operations. We can also define fourth roots, fifth roots, and so on; we will not study such roots here.

EXERCISE 8

Find each cube root. (Use a scientific calculator for part d.)

a. $\sqrt[3]{8}$

b. $\sqrt[3]{-125}$

c. $\sqrt[3]{-1}$

d. $\sqrt[3]{50}$

ANSWERS TO 4.2 EXERCISES

1a. $15, -15$ **1b.** $\dfrac{2}{3}, \dfrac{-2}{3}$

2a. -9 **2b.** $\pm\dfrac{8}{11}$

3a. 8 **3b.** Undefined **3c.** -8 **3d.** ± 8

4. a and d

5. $a = \pm 6$

6. $t = \pm 3.606$

7a. $\sqrt{24}$ **7b.** 4.899

7c. Operations under a radical must be performed first:
$\sqrt{6^2 - 4(3)} = \sqrt{36 - 12} = \sqrt{24}$.

8a. 2 **8b.** -5 **8c.** -1 **8d.** $3.684\ldots$

HOMEWORK 4.2

1. a. If $p = \sqrt{d}$, then $d =$ _____.

b. If $v^2 = k$, then $v =$ _____.

2. a. What is a principal square root?

b. What is a radicand?

3. a. Explain why the square root of a negative number is undefined.

   b. If $x > 0$, explain the difference between $\sqrt{-x}$ and $-\sqrt{x}$.

4. a. Make a list of the squares of all the integers from 1 to 20. These are the first 20 perfect squares. Now make a list of the square roots of these perfect squares.

   b. Make a list of the cubes of all the integers from 1 to 10. These are the first ten perfect cubes. Now make a list of the cube roots of these perfect cubes.

Simplify each expression in Problems 5–10. Do not use a calculator!

5. a. $4 - 2\sqrt{64}$

   b. $\dfrac{4 - \sqrt{64}}{2}$

6. a. $\dfrac{\sqrt{225}}{5} - \dfrac{\sqrt{49}}{2}$

   b. $\dfrac{36}{6 + \sqrt{36}}$

7. a. $\sqrt{9 - 4(-18)}$

   b. $\sqrt{\dfrac{4(50) - 56}{16}}$

8. a. $-2(3\sqrt{16} - \sqrt{3(27)})$

   b. $10 + 2(3 - \sqrt{169})$

9. a. $5\sqrt[3]{8} - \dfrac{\sqrt[3]{64}}{8}$

b. $\dfrac{3 + \sqrt[3]{-729}}{6 - \sqrt[3]{-27}}$

10. a. $6 - 3\sqrt[3]{27} - 7(5)$

b. $\dfrac{8 - 2\sqrt[3]{-125}}{6}$

11. What is the difference between a *linear* equation and a *quadratic* equation? Give an example of each.

12. Find the mistake in the following solution, and correct it.

$$4x^2 - 1 = 35$$
$$4x^2 = 36$$
$$4x = \pm 6$$
$$x = \pm\dfrac{3}{2}$$

In Problems 13–22, solve each quadratic equation by extraction of roots. Round your answer, if necessary, to three places.

13. $x^2 = 121$

14. $w^2 - 125 = 0$

15. $98 = 2a^2$

16. $108 - 3b^2 = 0$

17. $0 = 3n^2 - 15$

18. $0 = 24 - 4t^2$

19. $400 + \dfrac{k^2}{6} = 625$

20. $144 + \dfrac{h^2}{9} = 169$

21. $55 - 3z^2 = 7$

22. $12 - 5v^2 = 2$

For each expression in Problems 23–34, give a decimal approximation to the nearest thousandth.

23. $5\sqrt{3}$

24. $-6\sqrt{5}$

25. $\dfrac{-2}{3}\sqrt{21}$

26. $\dfrac{3}{5}\sqrt{76}$

27. $-3 + 2\sqrt{6}$

28. $5 - 3\sqrt{7}$

29. $2 + 6\sqrt[3]{-25}$

30. $-1 - 3\sqrt[3]{120}$

31. $\dfrac{8 - 2\sqrt{2}}{4}$

32. $\dfrac{6 + 9\sqrt{3}}{3}$

33. $3\sqrt{3} - 3\sqrt{5}$

34. $2(\sqrt{10} - 5) + \sqrt{6}$

35. The famous waffle cone at Zanner's Ice Cream holds 500 cubic centimeters (a little over 1 pint) of ice cream. The cone is about 16 centimeters tall. What is the radius of the waffle cone at its top?

36. A cylindrical syringe holds 100 cubic centimeters of fluid. If the syringe is 10 centimeters long, what is its radius?

37. The largest monument ever constructed is the Quetzalcoatl pyramid southeast of Mexico City. It is 177 feet tall, and its volume is approximately 116,500,000 cubic feet. Approximately how long is one side of its square base?

38. The existence of the asteroid Ceres was predicted in 1781, and the asteroid was actually discovered in 1801 with the aid of calculations by the mathematician Carl Gauss. The asteroid is approximately spherical, and its volume is approximately 641,431,000 cubic kilometers. What is the radius of Ceres?

In Problems 39–44, one of the two statements is true for all values of x, and the other is not. By trying the values of x suggested in the tables, decide which statement is true.

39. $x + x = 2x$
$x + x = x^2$

x	$x + x$	$2x$	x^2
3			
5			
-4			
-1			

40. $x \cdot x = 2x$
$x \cdot x = x^2$

x	$x \cdot x$	$2x$	x^2
4			
6			
-3			
-1			

41. $x^2 + x^2 = x^4$
$x^2 + x^2 = 2x^2$

x	$x^2 + x^2$	x^4	$2x^2$
2			
3			
-2			
-1			

42. $2x^2 + 3x^2 = 5x^2$
$2x^2 + 3x^2 = 5x^4$

x	$2x^2 + 3x^2$	$5x^2$	$5x^4$
1			
2			
-3			
-2			

43. $x + x^2 = x^3$
$x \cdot x^2 = x^3$

x	$x + x^2$	$x \cdot x^2$	x^3
1			
4			
-3			
-1			

44. $2x + x^2 = 3x^2$
$2x \cdot x^2 = 2x^3$

x	$2x + x^2$	$2x \cdot x^2$	$3x^2$	$2x^3$
2				
3				
-2				
-1				

In Problems 45–56, simplify each expression if possible. Identify any expressions that cannot be simplified.

45. $5a^2 - 7a^2$

46. $-4b^2 + 9b^2$

47. $3t - 2t^2$

48. $8v^2 - 4v$

49. $-m^2 - m^2$

50. $q^2 - q^2$

51. $3k(4k)$

52. $3k + 4k$

53. $3k + 4k^2$

54. $3k(4k^2)$

55. $3k^2 + 4k^2$

56. $3k^2 - 4k^2$

Use the definition of square root to simplify each expression in Problems 57–62. Do not use a calculator.

57. $\sqrt{5}(\sqrt{5})$

58. $\sqrt{29}(\sqrt{29})$

59. $\sqrt{x}(\sqrt{x})$

60. $\sqrt{n}(\sqrt{n})$

61. $\dfrac{6}{\sqrt{6}}$

62. $\dfrac{2}{\sqrt{2}}$

In Problems 63–66, find an expression for the surface area of each box. (*Hint:* Each box has six faces: top and bottom, front and back, left and right. Find the area of each face; then add the areas.

63.

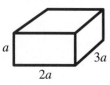

a

$3a$

$2a$

64.

$4b$

$\dfrac{b}{2}$ b

65.

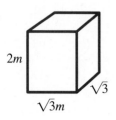

$2m$

$\sqrt{3}$

$\sqrt{3}m$

66.

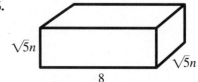

$\sqrt{5}n$

$\sqrt{5}n$

8

Solve Problems 67–74 by extraction of roots. Round your answers to the nearest thousandth.

67. $3g^2 - 54 = 0$

68. $5d^2 - 100 = 0$

69. $2.4m^2 = 126$

70. $3.5k^2 = 9.8$

71. $2x^2 - 200 = x^2 + 25$

72. $s^2 - 90 = 8 - s^2$

73. $3t^2 - 16 = 16t^2$

74. $8b^2 - 12 = 5b^2 + 24$

In Problems 75–78, solve each formula for the indicated variable.

75. $A = 4\pi r^2$,   for r

76. $s = \dfrac{1}{2}gt^2$,   for t

77. $V = \dfrac{1}{3}\pi r^2 h$,   for r

78. $F = \dfrac{km}{d^2}$,   for d

4.3 Nonlinear Graphs

Reading

In Section 4.2, we saw that $\sqrt{6}$ is an irrational number, and we used a calculator to find that $\sqrt{6} \approx 2.4494897$. The term *irrational* has nothing to do with being unreasonable or illogical; it comes from the word *ratio*.

A **rational number** is one that can be expressed as a quotient (or ratio) of two integers, where the denominator is not zero. Thus any fraction such as the following is a rational number:

$$\frac{2}{3}, \qquad \frac{-4}{7}, \qquad \text{or} \qquad \frac{15}{8}$$

Integers are also rational numbers, because any integer can be written as a fraction with a denominator of 1. $\left(\text{For example, } 6 = \dfrac{6}{1}.\right)$ All of the numbers we have encountered in earlier chapters are rational numbers.

When we write a rational number in decimal form, one of two things happens:

1. The decimal representation *terminates*, or ends. For example,

$$\frac{3}{4} = 0.75$$

2. The decimal representation *repeats a pattern*. For example,

$$\frac{4}{11} = 0.363636\ldots = 0.\overline{36}$$

Try Exercise 1 now.

An **irrational number** is one that *cannot* be expressed as a quotient of two integers. There is no terminating decimal fraction that gives the exact value of $\sqrt{6}$, or of any other irrational number. The decimal representation of an irrational number never ends and does not repeat any pattern. The best we can do is round off the decimal form and give an approximate value. Nonetheless, an irrational number still has a precise location on a number line, just as a rational number

EXERCISE 1
Find the decimal form for each rational number. Does the decimal form terminate?

a. $\dfrac{2}{3}$

b. $\dfrac{5}{2}$

c. $\dfrac{13}{27}$

d. $\dfrac{962}{2000}$

does. Figure 4.3 shows the locations of several rational and irrational numbers on a number line.

Figure 4.3

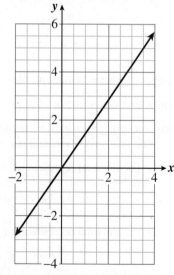

$$-\sqrt{10} \qquad \frac{2}{3} \quad \sqrt{5} \qquad \frac{15}{4} \qquad \sqrt{36}$$

Each point on a number line corresponds to either a rational number or an irrational number, and all such numbers fill up the number line completely. The rational and irrational numbers together make up the **real numbers**, and the number line is sometimes called the **real line**.

It is important that you understand the distinction between an exact value and an approximation:

$\sqrt{6}$ indicates the *exact value* of the square root of 6

2.4494897   is an *approximation* to the square root of 6

The same remarks can be made about rational numbers. We often use an approximation for a rational number if its decimal equivalent is a repeating decimal. For example,

$\dfrac{2}{3}$ indicates the *exact value* of 2 divided by 3

0.666667   is an *approximation* to $\dfrac{2}{3}$

Now try Exercise 2.

EXERCISE 2

Give a decimal equivalent for each radical, and round your answer to three decimal places if necessary. Identify the radical as rational or irrational.

a. $\sqrt{1}$ **b.** $\sqrt{2}$

c. $\sqrt{3}$ **d.** $\sqrt{4}$

e. $\sqrt{5}$ **f.** $\sqrt{6}$

g. $\sqrt{7}$ **h.** $\sqrt{8}$

i. $\sqrt{9}$ **j.** $\sqrt{10}$

Skills Review

The figure shows the graph of $y = \sqrt{2}x$.

1. Use the graph to estimate the solution to $\sqrt{2}x = 2$.

2. Solve $\sqrt{2}x = 2$ algebraically.

3. Use the graph to estimate the value of $3\sqrt{2}$.

4. Find y when $x = 3$ algebraically.

5. What is the slope of the graph?

Answers: *1.* $x = 1.4$
2. $x = \sqrt{2} \approx 1.414$ *3.* 4.25
4. $y = 3\sqrt{2} \approx 4.243$ *5.* $m = \sqrt{2} \approx 1.414$

Lesson

Recall that a linear equation is one that can be written in the form $y = mx + b$. In the following Activities, you will investigate some nonlinear graphs—those whose equations involve exponents or roots. For each activity, use the table of values and the graph of the equation to answer the questions.

Activity

1. At the Custom Pizza Shop, you can buy special smoked chicken pizza in any size you like. The cost, C, of the pizza, in dollars, is given by the equation

$$C = \frac{1}{4}r^2$$

where r is the radius of the pizza. Fill in the table of values and graph the equation.

a. What is the cost of a pizza with a radius of 3 inches? Locate this point on your graph.

b. Use your graph to find out how big a pizza you can buy for $16.

c. If you can spend at most $36 on pizza, what sizes can you buy? Mark all of these sizes on the horizontal axis.

r	C
0	
1	
2	
4	
6	
9	
10	
14	

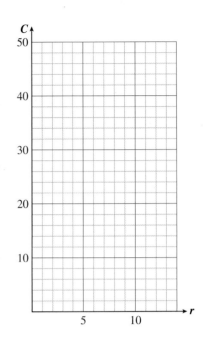

2. While hiking, Delbert drops a stone off the edge of a 400-foot cliff. After it falls for t seconds, the stone's height h above the base of the cliff is given in feet by

$$h = 400 - 16t^2$$

Fill in the table of values, and graph the equation.

a. What is the height of the stone 3 seconds after being dropped? Locate this point on your graph.

b. Use your graph to find out how long the stone will take to hit the ground.

c. On the horizontal axis, mark all of the times when the stone is more than 300 feet high.

t	h
0	
1	
2	
3	
4	
5	

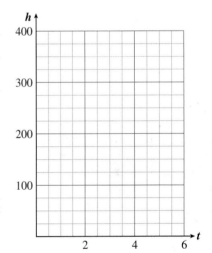

3. The area of a circle of radius r is given by the formula

$$A = \pi r^2$$

a. Rewrite the formula using a decimal approximation for π, rounded to two decimal places.

b. Fill in the table of values, rounding your answers to one decimal place. Then graph the equation.

c. Use your graph to estimate the area of a circle whose radius is 2.25 inches. Then find the area algebraically, using the formula.

d. Use your graph to estimate the radius of a circle whose area is 30 square inches. Then find the radius algebraically, using the formula.

r	A
0	
0.5	
1	
1.5	
2	
2.5	
3	
3.5	
4	

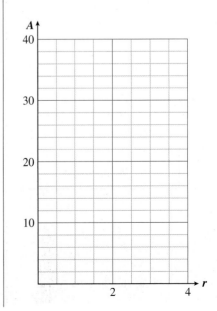

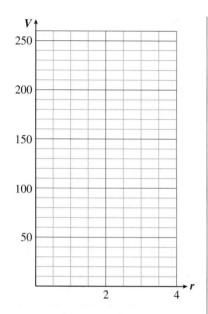

4. The volume of a sphere of radius r is given by the formula

$$V = \frac{4}{3}\pi r^3$$

a. Rewrite the formula using a decimal approximation for $\frac{4}{3}\pi$, rounded to two decimal places.

b. Fill in the table of values, rounding the values of V to one decimal place. Then graph the equation.

c. Use your graph to estimate the volume of a sphere whose radius is 2.25 inches. Then find the volume algebraically, using the formula.

d. Use your graph to estimate the radius of a sphere whose volume is 14 cubic inches. Then find the radius algebraically, using the formula.

r	V
0	
0.5	
1	
1.5	
2	
2.5	
3	
3.5	
4	

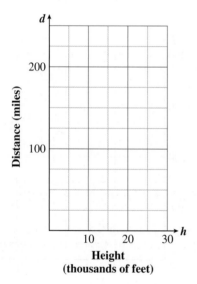

Height
(thousands of feet)

5. If you are flying in an airplane at an altitude of h feet on a clear day, you can see for a distance of d miles, where d is given by

$$d = 1.22\sqrt{h}$$

a. Fill in the table of values. Then graph the equation.

b. How far can you see from an altitude of 10,000 feet? Locate this point on your graph.

c. What is the minimum altitude from which you can see 100 miles? Locate this point on your graph.

d. On the horizontal axis, mark all altitudes from which you can see farther than 175 miles.

h	d
2000	
5000	
10,000	
15,000	
20,000	
25,000	
30,000	

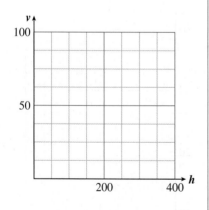

6. If an object falls from a height of h meters, its velocity, v, when it strikes the ground is given in meters per second by

$$v = 4.4\sqrt{h}$$

a. Fill in the table of values. Then graph the equation.

b. If a penny falls off the Washington Monument, which is 170 meters high, what will its velocity be when it hits the ground? Locate this point on your graph.

c. A rock dropped from the Royal Gorge bridge in Colorado will hit the water below at a velocity of 80 meters per second. How high is the bridge? Locate this point on your graph.

d. On the horizontal axis, mark all heights at which the velocity of the penny in part (b) is less than 70 meters per second.

h	v
50	
100	
150	
200	
250	
300	
350	
400	

ANSWERS TO 4.3 EXERCISES

1a. $0.\overline{6}$; no

1b. 2.5; yes

1c. $0.\overline{481}$; no

1d. 0.481; yes

2a. 1

2b. 1.414

2c. 1.732

2d. 2

2e. 2.236

2f. 2.449

2g. 2.646

2h. 2.828

2i. 3

2j. 3.162; $\sqrt{1}$, $\sqrt{4}$, and $\sqrt{9}$ are rational; all the other numbers are irrational.

HOMEWORK 4.3

1. What is a rational number? An irrational number? Give examples of each.

2. How can you tell whether a decimal represents a rational number or an irrational number? Give examples.

3. Give three examples of square roots that represent rational numbers and three examples of square roots that represent irrational numbers.

4. Give three examples of cube roots that represent rational numbers and three examples of cube roots that represent irrational numbers.

Determine whether the statements in Problems 5–10 are true or false.

5. If a number is irrational, it cannot be an integer.

6. Every real number is either rational or irrational.

7. Irrational numbers do not have an exact location on the number line.

8. It is not possible to find an exact decimal equivalent for an irrational number.

9. The number 2.8 is only an approximation for $\sqrt{8}$; the exact value is 2.828427125.

10. The number $\sqrt{17}$ appears somewhere between 16 and 18 on the number line.

Identify each number in Problems 11–22 as rational or irrational.

11. $\sqrt{6}$

12. $\dfrac{-5}{3}$

13. $\sqrt{16}$

14. $\sqrt{\dfrac{5}{9}}$

15. 6.008

16. $3.\overline{23}$

17. $\sqrt{250}$

18. $\dfrac{\sqrt{3}}{2}$

19. $\dfrac{\sqrt{81}}{4}$

20. $\sqrt[3]{16}$

21. $2 + \sqrt{5}$

22. $7 - \sqrt{169}$

For Problems 23–30, first review Section 2.6 on solving inequalities graphically. Graph each equation by plotting points. Then use your graph to solve the equations and inequalities given. (You may have to estimate the solutions.)

23. $y = x^2$

    a. $x^2 = 12$ **b.** $x^2 \leq 6$

24. $y = x^3$

    a. $x^3 = -15$ **b.** $x^3 > -2$

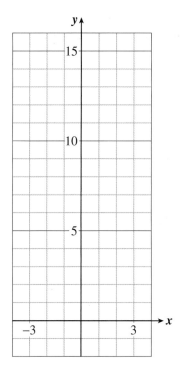

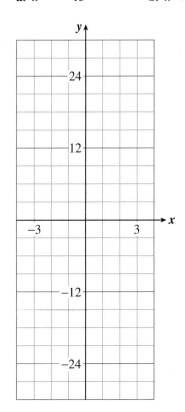

25. $y = \sqrt{x}$

    a. $\sqrt{x} = 2.5$ **b.** $1 < \sqrt{x} \leq 3$

26. $y = \sqrt[3]{x}$

    a. $\sqrt[3]{x} = -2.5$ **b.** $2 \geq \sqrt[3]{x} \geq -2$

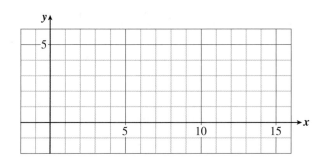

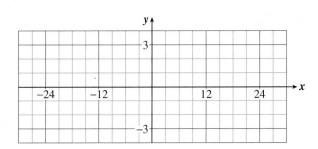

27. $y = 4 - x^2$

   a. $4 - x^2 = -5$ **b.** $4 - x^2 < 0$

28. $y = (x + 3)^2$

   a. $(x + 3)^2 = -1$ **b.** $(x + 3)^2 \leq 8$

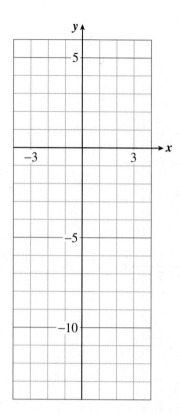
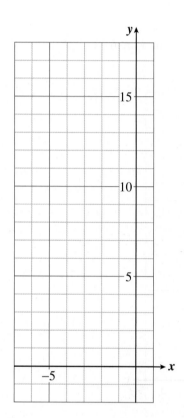

29. $y = \sqrt{x + 2}$

   a. $\sqrt{x + 2} = 3$ **b.** $\sqrt{x + 2} > 2$

30. $y = \sqrt{2 - x}$

   a. $\sqrt{2 - x} = 3$ **b.** $\sqrt{2 - x} > 2$

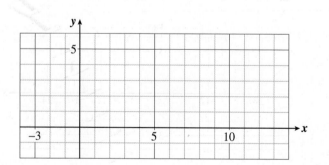
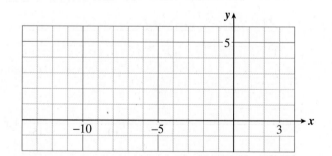

31. a. Sketch three rectangles of different sizes but whose length is always twice the width.

b. If the width of such a rectangle is w inches, write an equation for its area, A, in terms of w.

c. Make a table of values for A in terms of w, and graph your equation. (Use your table of values to help you choose scales for the axes.)

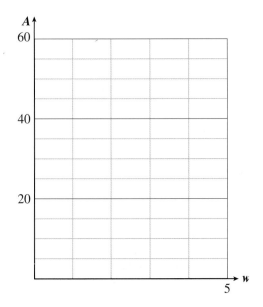

d. If the area of one of these rectangles is 48 square inches, find its width. Locate the point corresponding to this rectangle on your graph.

32. a. Sketch three equilateral triangles of different sizes.

b. The area, A, of an equilateral triangle is given by

$$A = \frac{\sqrt{3}}{4} s^2$$

where s is the length of one side of the triangle. Rewrite the formula for the area, using an approximation to three decimal places.

c. Make a table of values for A in terms of s, and graph your equation. (Use your table of values to help you choose scales for the axes.)

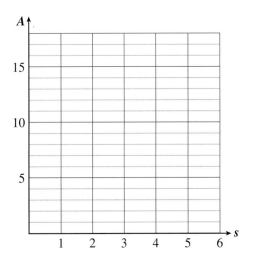

d. If the area of a particular equilateral triangle is 15.6 square inches, find the length of its side. Locate the point corresponding to this triangle on your graph.

33. Engineers must take many factors into account when building highways. For example, the greatest speed, in miles per hour, at which a car can safely travel around an unbanked curve of radius r feet is given by the formula $v = \sqrt{2.5r}$.

    a. Make a table of values for v in terms of r. Use values of r in multiples of 50 feet. Graph the equation.

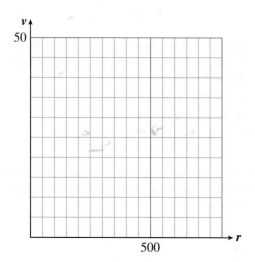

    b. Use your graph to estimate the smallest safe radius for an off ramp from a freeway if cars will be traveling on it at 40 miles per hour.

34. If an object falls freely under the force of gravity, the time it takes to fall d feet is given by the formula

$$t = \frac{\sqrt{d}}{4}$$

    where t is in seconds.

    a. Make a table of values for t in terms of d. Use values of d in multiples of 100 feet. Graph the equation.

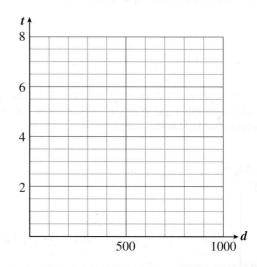

    b. Use your graph to estimate how far an object will fall in 7 seconds.

In Problems 35–40, choose the best approximation for each number. Do not use a calculator.

35. $\sqrt{72}$
    a. 7   **b.** 8   **c.** 36   **d.** 64

36. $\sqrt[3]{120}$
    a. 5   **b.** 10   **c.** 40   **d.** 60

37. $\sqrt{13}$
    a. 6   **b.** 6.5   **c.** 3.5   **d.** 4

38. $\sqrt{19}$
    a. 4   **b.** 4.3   **c.** 4.8   **d.** 5

39. $\sqrt{134}$

    a. 11   **b.** 11.5   **c.** 12   **d.** 15

40. $\sqrt{500}$

    a. 20   **b.** 22   **c.** 134   **d.** 250

41. a. Evaluate $\sqrt{x^2}$ for $x = 3, 5, 8,$ and 12.

    b. Simplify $\sqrt{a^2}$ if $a \geq 0$.

42. a. Evaluate $(\sqrt{x})^2$ for $x = 4, 9, 16,$ and 81.

    b. Simplify $(\sqrt{a})^2$ if $a \geq 0$.

Simplify the expressions in Problems 43–58. Do not use a calculator.

43. $(\sqrt{16})^2$

44. $(\sqrt{25})(\sqrt{25})$

45. $(\sqrt{7})(\sqrt{7})$

46. $(\sqrt{13})^2$

47. $(\sqrt[3]{5})^3$

48. $(\sqrt[3]{4})(\sqrt[3]{4})(\sqrt[3]{4})$

49. $(\sqrt[3]{9})(\sqrt[3]{9})(\sqrt[3]{9})$

50. $(\sqrt[3]{20})^3$

51. $\dfrac{3}{\sqrt{3}}$

52. $\dfrac{10}{\sqrt{10}}$

53. $\dfrac{-11}{\sqrt{11}}$

54. $\dfrac{-15}{\sqrt{15}}$

55. $(\sqrt{2b})(\sqrt{2b})$

56. $(\sqrt{5w})(\sqrt{5w})$

57. $\dfrac{2m}{\sqrt{m}}$

58. $\dfrac{H}{3\sqrt{H}}$

59. As if you were talking to a classmate, explain why you can always simplify $\sqrt{x}\ \sqrt{x}$, as long as x is non-negative.

60. As if you were talking to a classmate, explain why you can always simplify $\dfrac{x}{\sqrt{x}}$, as long as x is positive.

61. What is the smallest positive integer (other than 1) that is both a perfect square and a perfect cube?

62. Find the cube root of 27^2 without using a calculator.

4.4 The Pythagorean Theorem

Reading

Delbert is taking a week-long tour on his trail bike. He is traveling across open country in the West, using a guidebook that describes the route and suggests places to camp each night. This morning he reads the following instructions in the guidebook:

"After breaking camp, ride directly west for 20 miles. You will come to a deep, narrow canyon running north and south. However, there is a dirt road that runs along the east side of the canyon, and precisely where you meet the road, there is a bridge over the canyon. Cross the bridge and continue for 2 more miles to a county park, which is your first rest stop for the day."

Delbert sets out heading west, but his compass is not completely accurate, and he is in fact riding on a course slightly north of west. When he reaches the canyon, his odometer reads 20.5 miles instead of 20 miles, as the guidebook described. "Oh, well," thinks Delbert, who is getting a little tired and is ready for his rest stop, "That's not so far off. If I head south on this dirt road, I should come to the bridge in just a short distance."

Try Exercise 1 now.

Skills Review

Follow the order of operations to simplify. See Lesson 4.1.

1. $3(-5) - 2^3$

2. $6(7 - 4)^2$

3. $-6 - 2 \cdot 4^2$

4. $(5 - 3)^4(3 - 6)^3$

5. $4(4 - 4^2)$

6. $-3(-2)^2 - 5$

7. $-(3 \cdot 2)^2 - 5$

8. $3 - 4(-2)^3(-3)$

9. $\dfrac{-5^2 + 1}{4} + \dfrac{2(-3)^3}{-6}$

10. $\dfrac{2^2(-3^2)}{1 - 3^2} - \dfrac{7^2 - 6^2}{(1 - 3)^2}$

11. $\dfrac{3^3 - 3}{(3 - 5)^3} - \dfrac{2(-2)^3 - 8}{-2^2(8 - 2^2)}$

12. $3 \cdot \dfrac{5^3 - (-10)^2}{3^2 - 3(5)} \cdot \dfrac{2^5 + 4}{3^2 - 2^2}$

Answers: *1.* -23 *2.* 54 *3.* -38 *4.* -432 *5.* -48 *6.* -17
7. -41 *8.* -93 *9.* 3 *10.* $\dfrac{5}{4}$ *11.* $\dfrac{-9}{2}$ *12.* -90

Lesson

Activity

In this Activity we investigate some properties of right triangles. The grid on page 463 is measured in centimeters. Use the grid and follow these steps:

Step 1 First, measure carefully and cut out a strip of paper or cardboard between 12 and 24 centimeters long. (It will be easier if you choose an integer value for the length of the strip.)

Step 2 Use your strip to form the **hypotenuse**, or longest side, of a right triangle on the grid, with the perpendicular legs on the axes.

Step 3 Record the lengths of the legs (and the hypotenuse) in the first three columns of Table 4.1. (We'll get to the other columns in a minute.)

Step 4 Try several different triangles, and record the lengths of the legs.

Table 4.1

Hypotenuse (H)	First leg (FL)	Second leg (SL)	$(FL)^2$	$(SL)^2$	$(FL)^2 + (SL)^2$	$(H)^2$

EXERCISE 1

a. Sketch the situation described in the story. Show Delbert's camp, the canyon and the bridge, Delbert's actual route, and the route he should have taken.

b. Without doing any calculations, guess how far Delbert must ride on the dirt road to reach the bridge.

Repeat Steps 1–4 with a hypotenuse strip of a different length, and record your data in the first three columns of Table 4.2.

Table 4.2

Hypotenuse (H)	First leg (FL)	Second leg (SL)	$(FL)^2$	$(SL)^2$	$(FL)^2 + (SL)^2$	$(H)^2$

Step 5 The last four columns in each table are for recording some calculations. You should be able to interpret what goes in each column:

$(FL)^2$:   Square the length of the first leg.

$(SL)^2$:   Square the length of the second leg.

$(FL)^2 + (SL)^2$:   Add the entries in the previous two columns.

$(H)^2$:   Square the length of the hypotenuse.

Step 6 Compare the entries in the last two columns of each table. Do you notice any pattern? Summarize your observations as a conjecture.

Now try Exercise 2.

The Pythagorean Theorem

The activity revealed a relationship among the sides of a right triangle. This relationship was known to several ancient civilizations, including those in Egypt, Mesopotamia, China, and India, and is called the **Pythagorean theorem** after the Greek mathematician Pythagoras. Note that the relationship holds only for *right* triangles; it is not true for any other kind of triangle.

Pythagorean Theorem

If c stands for the length of the hypotenuse of a right triangle, and the lengths of the two legs are represented by a and b, then

$$a^2 + b^2 = c^2$$

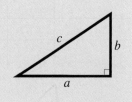

The Pythagorean theorem tells us that if we know the lengths of any two sides of a right triangle, we can find the length of the third side.

EXERCISE 2

Use the grid to model the sketch you made in Exercise 1 for Delbert's trail bike trip. (How long should the hypotenuse strip be?) From the grid, read the distance that Delbert must travel along the dirt road to get to the bridge.

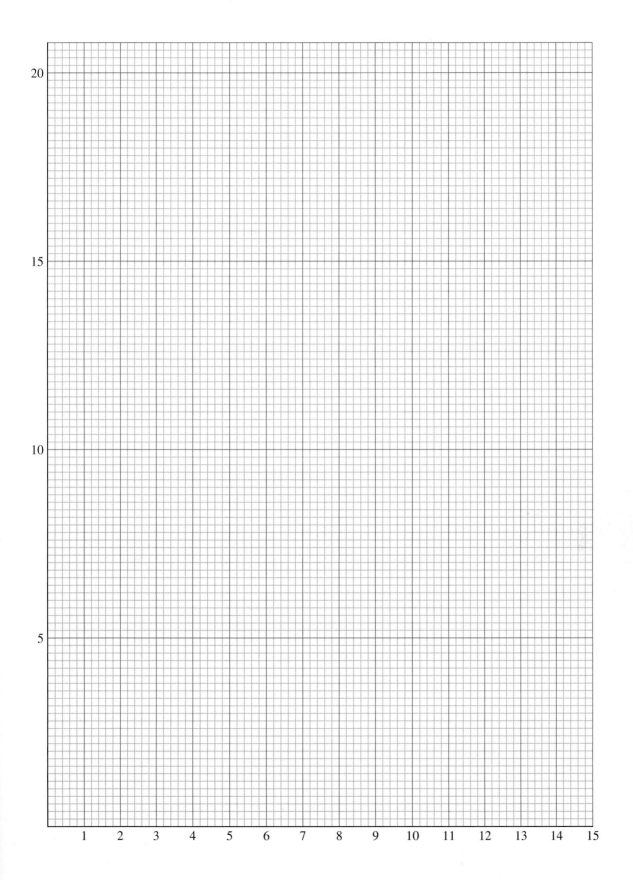

EXAMPLE 1

A 25-foot ladder is placed against a wall so that its foot is 7 feet from the base of the wall. How far up the wall does the ladder reach?

Solution

We make a sketch of the situation, as shown in Figure 4.4, labeling known dimensions. We call the unknown height h. The ladder forms the hypotenuse of a right triangle, so we can apply the Pythagorean theorem, substituting 25 for c, 7 for b, and h for a:

$$a^2 + b^2 = c^2$$
$$h^2 + 7^2 = 25^2$$

Now we solve the equation by extraction of roots:

$$h^2 + 49 = 625 \qquad \text{Subtract 49 from both sides.}$$
$$h^2 = 576 \qquad \text{Extract roots.}$$
$$h = \pm\sqrt{576} \qquad \text{Simplify the radical.}$$
$$h = \pm 24$$

The height must be a positive number, so the solution -24 does not make sense for this problem. The ladder reaches 24 feet up the wall.

Now try Exercise 3.

Figure 4.4

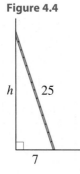

EXERCISE 3

Use the Pythagorean theorem to write an equation that describes the situation in Exercise 1. Solve the equation to find out how far Delbert must ride on the dirt road before reaching the bridge. How does your answer compare with your estimate in Exercise 2?

ANSWERS TO 4.4 EXERCISES

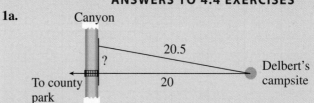

1a.

1b. and 2. About 4.5 miles

3. $d^2 + 20^2 = 20.5^2$; $d = 4.5$ miles

HOMEWORK 4.4

Find the unknown side or sides for each right triangle in Problems 1–10.

1.

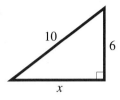

2.

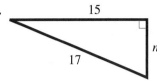

3.

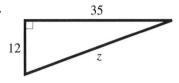

4.

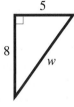

5.

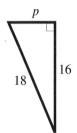

6.

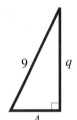

7.

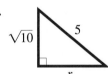

8.

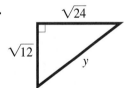

9.

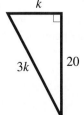

10.

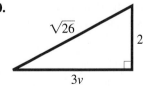

In Problems 11–16, decide whether a triangle with the given sides is a right triangle.

11. 9 inches, 16 inches, 25 inches

12. 12 centimeters, 16 centimeters, 20 centimeters

13. 5 meters, 12 meters, 13 meters

14. 5 feet, 8 feet, 13 feet

15. 5^2 feet, 8^2 feet, 13^2 feet

16. $\sqrt{5}$ feet, $\sqrt{8}$ feet, $\sqrt{13}$ feet

In Problems 17–22, sketch each situation, write an equation to describe the sketch, and solve the equation. Round your answer to two decimal places if necessary.

17. You have a 10-foot ladder, and you'd like to rest the top of the ladder on a windowsill that is 8 feet above the ground. Directly below the window a flower bed extends 5 feet out from the wall, and you don't want the foot of the ladder to crush the flowers. Will the base of the ladder clear the flower bed?

18. A baseball diamond is a square whose sides are 90 feet long. Find the straight-line distance from home plate to second base.

19. How long must a wire be to stretch from the top of a 40-foot telephone pole to a point on the ground 20 feet from the base of the pole?

20. The lighthouse is 5 miles east of Gravelly Point. The marina is 7 miles south of Gravelly Point. How far is it from the lighthouse to the marina?

21. The sun deck at Francine's summer house is 30 feet above the beach. The slope of the stairs from the beach to the deck is 1.2. How long is the handrail for the stairs?

22. The ceiling of the screened porch at Francine's summer house slopes up from front to back. The porch is 10 feet wide from front to back. The front ceiling is 7 feet tall, and the back ceiling is 9 feet tall. If the porch is 20 feet long, what is the area of the ceiling?

23. Akemi wants to know the distance across a small lake, from point A to point B. She chooses the point C so that the line from A to C is perpendicular to the line from A to B. She then measures the distance from A to C as 3 miles and the distance from B to C as 4 miles. How far is it from A to B across the lake?

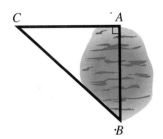

24. Engineers in the Department of Roads want to know the distance through a mountainous region in the path of a proposed highway. The field engineer chooses point C on a line perpendicular to the planned route through the mountains. He then measures the distance from A to C as 2 miles and the distance from B to C as 3 miles. How far is it from A to B through the mountains?

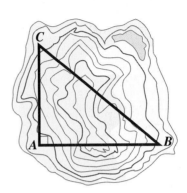

In Problems 25–28, decide whether the two expressions are equivalent by evaluating them for the values of the variables suggested in the tables.

25. Does $(a + b)^2 = a^2 + b^2$?

a	b	$a + b$	$(a + b)^2$	a^2	b^2	$a^2 + b^2$
2	3					
3	4					
1	5					
-2	6					

26. Does $\sqrt{a^2 + b^2} = a + b$?

a	b	$a + b$	a^2	b^2	$a^2 + b^2$	$\sqrt{a^2 + b^2}$
3	4					
2	5					
1	6					
-2	-3					

27. Does $\sqrt{a + b} = \sqrt{a} + \sqrt{b}$?

a	b	$a + b$	$\sqrt{a + b}$	$\sqrt{a}$	$\sqrt{b}$	$\sqrt{a} + \sqrt{b}$
2	7					
4	9					
1	5					
9	16					

28. Does $(\sqrt{a} + \sqrt{b})^2 = a + b$?

a	b	$a + b$	$\sqrt{a}$	$\sqrt{b}$	$\sqrt{a} + \sqrt{b}$	$(\sqrt{a} + \sqrt{b})^2$
4	9					
1	4					
3	5					
6	10					

In Problems 29–34, evaluate each expression for the given values of *x*, and round your answers to three decimal places if necessary.

29. $\sqrt{x^2 - 4}$,   for $x = 3, \sqrt{5}, -2$

30. $(x + \sqrt{3})^2$,   for $x = 2, \sqrt{3}, -1$

31. $\sqrt{x} - \sqrt{x + 3}$,   for $x = 1, 4, 100$

32. $4x - x\sqrt{x}$,   for $x = 4, 3, 1$

33. $2x^2 - 4x$,   for $x = -3, \sqrt{2}, \dfrac{1}{4}$

34. $2x(x - 4)^2$,   for $x = -2, \sqrt{6}, \dfrac{1}{2}$

Put each set of numbers in Problems 35–38 in order from smallest to largest. Try not to use a calculator.

35. $\dfrac{5}{4}, 2, \sqrt{8}, 2.3$

36. $3.5, \sqrt{10}, \dfrac{11}{3}, 3$

37. $2\sqrt{3}, 3, \dfrac{23}{6}, \sqrt{6}$

38. $\sqrt{120}, 10\sqrt{2}, 12, \dfrac{120}{11}$

Solve each equation in Problems 39–44 mentally; do not use a calculator.

39. $t^2 = 10{,}000$

40. $b^2 = \dfrac{4}{9}$

41. $u^2 = \dfrac{2}{98}$

42. $h^2 = 0.04$

43. $n^2 = 0.25$

44. $m^2 = (-17)^2$

In Problems 45–48, give exact answers, not approximations.

45. a. Find the height of an equilateral triangle whose sides are 6 feet long. (*Hint:* The altitude divides the base into two segments of equal length.)

b. Find the area of the triangle.

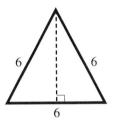

46. a. Find the perimeter of the trapezoid shown if $a = 1$ meter. (*Hint:* Find the lengths of both diagonal sides first.)

b. Find the area of the trapezoid.

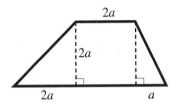

47. a. The pyramid shown has a square base. If $k = 2$ centimeters, find the length of the diagonal of the base.

b. Find the height of the pyramid. (*Hint:* The altitude of the pyramid divides the diagonal of the base into two segments of equal length. Use your answer to part a.)

c. Find the volume of the pyramid. (Refer to Figure 4.2 on page 425.)

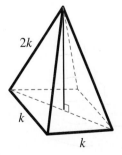

48. a. Suppose you measure the box shown, and you find that the base is 4 inches long and 1 inch wide and the height of the box is 3 inches. Find the diagonal, d, of the base of the box.

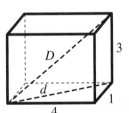

b. Find the diagonal, D, of the box. (*Hint:* D is the hypotenuse of a right triangle. What are its legs?)

49. Use the given areas to find the area of the largest square.

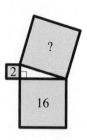

50. Use the given areas to find the area of the largest semi-circle.

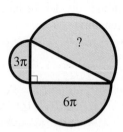

Decide whether the equations in Problems 51–60 are true or false. Explain your answers.

51. $\dfrac{15}{\sqrt{5}} = 3\sqrt{5}$

52. $\dfrac{7}{\sqrt[3]{7}} = \sqrt[3]{7}$

53. $(\sqrt{13})^3 = 13\sqrt{13}$

54. $(\sqrt[3]{3})^2 = 3$

55. $5x + 3x^2 = 8x^3$

56. $5x^2 - 3x = 2x$

57. $(\sqrt{2}m)(\sqrt{2}m) = 2m$

58. $(5\sqrt{x})^2 = 25x$

59. $(\sqrt[3]{b})(2\sqrt[3]{b})(3\sqrt[3]{b}) = 6b$

60. $(\sqrt{a})(\sqrt{a})(\sqrt{a}) = a^3$

4.5 Products of Binomials

Reading

In Section 4.2, we saw that a quadratic equation has the form

$$ax^2 + bx + c = 0$$

So far, most of the quadratic equations we've studied have had the form $ax^2 + c = 0$; that is, they have not had a linear term, bx. You will need some new skills in order to study the more general form of quadratic equations.

In this section, we use the areas of rectangles to investigate products of algebraic expressions. We have already used rectangles to visualize the distributive law. Part (a) of Exercise 1 has been done for you.

EXERCISE 1
Find the area of each rectangle.

Use the distributive law to find the product.

a.

	$3x$	4
5	$5(3x) = 15x$	$5(4) = 20$

$$\text{area} = 5(3x + 4)$$
$$= 5(3x) + 5(4)$$
$$= 15x + 20$$

b.

	x	9
$2x$		

$$\text{area} = 2x(x + 9)$$

c.

	$4b$	7
$3b$		

$$\text{area} = 3b(4b + 7)$$

Now try Exercise 2.

EXERCISE 2
Use the distributive law to find the products.
a. $2a(6a - 5)$

b. $-4v(2v - 3)$

c. $-5x(x^2 - 3x + 2)$

d. $-3xy(4x^2 - 2xy + 2y)$

Skills Review

Find the area and perimeter of each figure.

1.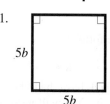

2.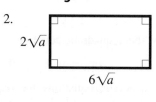

3.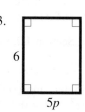

4.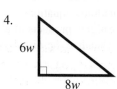

5.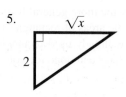

6.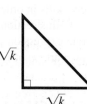

Answers: 1. $25b^2$; $20b$ *2.* $12a$; $16\sqrt{a}$ *3.* $30p$; $12 + 10p$
4. $24w^2$; $24w$ *5.* $\sqrt{x}$; $2 + \sqrt{x} + \sqrt{x+4}$ *6.* $\dfrac{k}{2}$; $2\sqrt{k} + \sqrt{2k}$

Lesson

At this stage, it will be helpful to learn some new terminology. An algebraic expression with only one term, such as $2x^3$, is called a **monomial**. An expression with two terms, such as $x^2 - 16$, is called a **binomial**. An expression with three terms is a **trinomial**. The expression $ax^2 + bx + c$ is thus called a **quadratic trinomial**.

Your goal in this lesson is to understand products in which each factor is a binomial.

Multiplying Binomials

Consider the rectangle shown in Figure 4.5. As you can see, it is divided into four smaller rectangles. You can verify that we get the same answer when we compute its area in two different ways: We can add up the areas of the four smaller rectangles, or we can find the length and width of the entire large rectangle and then find their product.

Sum of Four Smaller Rectangles	*One Large Rectangle*
area $= 8(5) + 8(15) + 4(5) + 4(15)$	area $= (8 + 4)(5 + 15)$
$= 40 + 120 + 20 + 60$	$= (12)(20)$
$= 240$	$= 240$

Figure 4.5

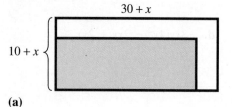

EXAMPLE 1

The City Council plans to install a 10-foot-by-30-foot reflecting pool in front of City Hall. When the cost estimate comes in, the council members realize that they can afford to enlarge the pool. They decide to increase both the length and the width by x feet. Write an equation for the area, A, of the pool in terms of x.

Solution

Look at the drawing of the pool in Figure 4.6(a). Both dimensions of the original pool have been increased by x. The area of the larger pool is thus

$$A = (30 + x)(10 + x)$$

Figure 4.6

30 + x

10 + x

(a)

In Figure 4.6(b), we partition the new pool area into four smaller rectangles and compute the area of each. With this method, we get the following expression for the area of the new pool:

$$A = x^2 + 30x + 10x + 300$$
$$= x^2 + 40x + 300$$

This expression is equivalent to the first expression for area. Thus

$$(30 + x)(10 + x) = x^2 + 40x + 300$$

Now try Exercise 3.

The rectangles used in computing products do not have to be drawn to scale—they are simply tools for visualizing the products. We can use rectangles to represent products involving negative numbers as well. We still follow the rules for adding and multiplying negative numbers.

EXAMPLE 2

a. Use a rectangle to represent the product $(x - 4)(x + 6)$.

b. Write the product as a quadratic trinomial.

Solution

a. We let the first factor represent the width of the rectangle and the second factor its length.

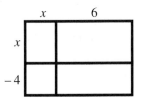

b. We find the area of each smaller rectangle.

	x	6
x	x^2	$6x$
-4	$-4x$	-24

Then we add the areas together.

$$\text{area} = x^2 + 6x - 4x - 24$$
$$= x^2 + 2x - 24$$

Try Exercise 4 now.

The Four Terms in a Binomial Product

Using a rectangle to visualize a product of binomials illustrates how the distributive law operates. In Example 2, we computed the product $(x - 4)(x + 6)$. The top row of the rectangle corresponds to

$$x(x + 6) = x^2 + 6x$$

Figure 4.6

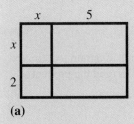

(b)

EXERCISE 3

Write the area of each rectangle in two different ways: as the sum of four smaller areas and as one large rectangle, using the formula

$$\text{area} = \text{length} \times \text{width}.$$

(a)

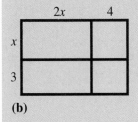

(b)

EXERCISE 4

a. Use a rectangle to represent the product $(3x - 2)(x - 5)$.

b. Write the product as a quadratic trinomial.

EXERCISE 5

Use the rectangle diagram to help you find the linear term in the product

$$(x - 6)(2x + 3)$$

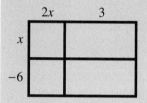

The bottom row corresponds to

$$-4(x + 6) = -4x - 24$$

Thus we multiply each term of the first binomial by each term of the second binomial, resulting in four multiplications in all:

$$(x - 4)(x + 6) = x^2 + 6x - 4x - 24$$

Each term of the product corresponds to the area of one of the four smaller rectangles, as shown in Figure 4.7.

Figure 4.7

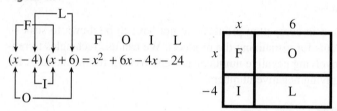

The letters F, O, I, and L indicate how each term of the product arises from the binomial factors:

> **F** stands for the product of the **First** terms in each binomial.
>
> **O** stands for the product of the **Outer** terms.
>
> **I** stands for the product of the **Inner** terms.
>
> **L** stands for the product of the **Last** terms.

Note that the quadratic term in the product comes from the first terms. The linear, or first-degree, term of the product is the sum of the outer and inner terms. The constant term of the product comes from the last terms.

$$(x - 4)(x + 6) = x^2 + 2x - 24$$
$$\quad \textbf{F} \quad \textbf{O+I} \quad \textbf{L}$$

Now try Exercise 5.

In Exercise 6, use the rectangle diagrams to help you compute products involving two variables.

EXERCISE 6

Write each product as a quadratic trinomial in two variables.

a. $(3a - 5b)(3a - b)$

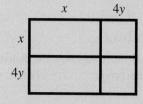

b. $(x + 4y)^2$

HOMEWORK 4.5

In Problems 1–4:
a. Write a product (length × width) for the area of each rectangle.
b. Use the distributive law to compute the product.

1.

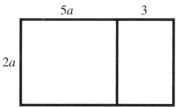

2.

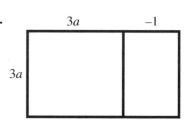

3.

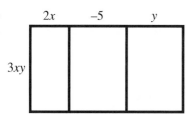

4.

Compute each product in Problems 5–12.

5. $-2b(6b - 2)$

6. $-3b(5b + 3)$

7. $-4x^2(2x + 3y)$

8. $-7y^2(x - 2y)$

9. $(y^3 + 3y - 2)(2y)$

10. $(x^2 - 3x - 4)(4x)$

11. $-xy(2x^2 - xy + 3y^2)$

12. $5ab(-3a + 2ab - b)$

Simplify the expressions in Problems 13–16.

13. $2a(x + 3) - 3a(x - 3)$

14. $4x(3 + 2y) - 3(2x + y)$

15. $2x(3 - x) + 2(x^2 + 1) - 2x$

16. $3y(2y - 5) + 2y^2 - 5(y^2 + 2y)$

In Problems 17–22, fill in the missing parts of each algebraic expression.

17. $4x(2x + ?) = 8x^2 - 12x$

18. $-3x(5 + ?) = -15x + 3x^2$

19. $?(4m - 9) = -12m^2 + 27m$

20. $?(-2p + 7) = -8p^2 + 28p$

21. $6h(? + ?) = 24h - 18h^2$

22. $-2q(? + ?) = 10q^2 - 18q$

In Problems 23 and 24, find each product without a calculator by using the rectangle diagram (not to scale!).

23. 36×42

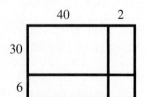

24. 23×27

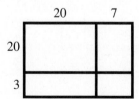

In Problems 25 and 26, make your own rectangle diagram to find the product without a calculator.

25. 82×16

26. 51×34

For Problems 27–30, express the area of each rectangle in two ways:
a. As the sum of four smaller areas
b. As the area of one large rectangle, using the formula   area = length × width

27.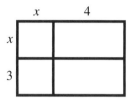

28.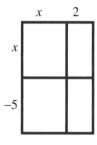

29.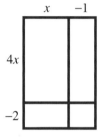

30.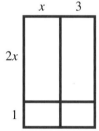

In Problems 31–36:
a. Use a rectangle diagram to represent each product.
b. Write the product as a quadratic trinomial.

31. $(a - 5)(a - 3)$

32. $(y - 3)(y + 1)$

33. $(y + 1)(3y - 2)$

34. $(a + 5)(5a + 2)$

35. $(5x - 2)(4x + 3)$

36. $(2x + 5)(6x - 1)$

In Problems 37–42, use rectangle diagrams to help you multiply the binomials in two variables.

37. $(x + 2y)(x - y)$ **38.** $(x - 3b)(x - b)$

39. $(3s + t)(2s + 3t)$ **40.** $(4s - t)(3s + 2t)$

41. $(2x - a)(x - 3a)$ **42.** $(3x + 4y)(x + 2y)$

In Problems 43–46:
a. Find the linear term in each product.
b. Shade the small rectangles that correspond to the linear term.

43. $(x + 6)(x - 9)$ **44.** $(x - 5)(x + 3)$

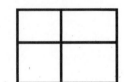   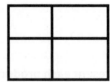

45. $(2x - 5)(x + 4)$ **46.** $(3x - 2)(2x + 1)$

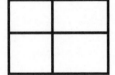   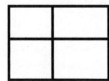

Compute each product in Problems 47–52. Multiply the binomials together, and then multiply the result by the numerical coefficient.

47. $2(3x - 1)(x - 3)$ **48.** $4(x - 2)(2x - 3)$

49. $-3(x + 4)(x - 1)$

50. $-2(x - 5)(x + 2)$

51. $-(4x + 3)(x - 2)$

52. $-(3x - 2)(x + 6)$

Compute each product in Problems 53–58. What do you notice? Explain why this happens.

53. $(x + 3)(x - 3)$

54. $(x - 4)(x + 4)$

55. $(x - 2a)(x + 2a)$

56. $(x + 5z)(x - 5z)$

57. $(3x + 1)(3x - 1)$

58. $(2x - 3)(2x + 3)$

Compute each product in Problems 59–64.

59. $(w + 4)(w + 4)$

60. $(n + 1)(n + 1)$

61. $(z - 6)(z - 6)$

62. $(m - 9)(m - 9)$

63. $(3a - 2c)(3a - 2c)$

64. $(2b + 5c)(2b + 5c)$

In Problems 65 and 66, evaluate the two expressions to decide whether they are equivalent.

65. $(a - b)^2$ and $a^2 - b^2$

a	b	$a - b$	$(a - b)^2$	a^2	b^2	$a^2 - b^2$
5	3					
2	6					
−4	−3					

66. $(a + b)^2$ and $a^2 + b^2$

a	b	$a + b$	$(a + b)^2$	a^2	b^2	$a^2 + b^2$
2	3					
−2	−3					
2	−3					

Write the area of each square in Problems 67–70 in two ways:
a. As the sum of four smaller areas
b. As the area of one large square, using the formula area = (length)^2

67.

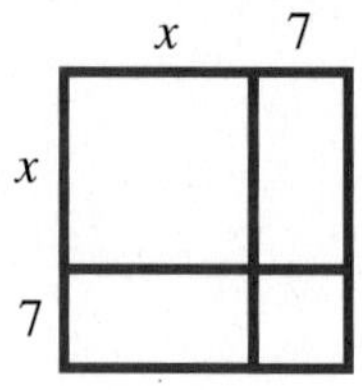

68.

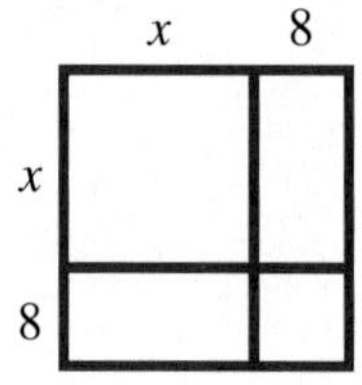

69.

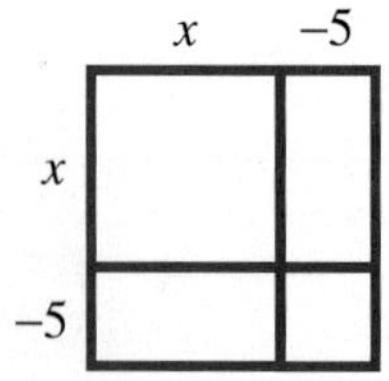

70.

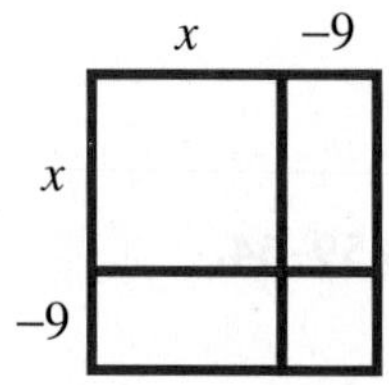

71. Is $(x + 4)^2$ equivalent to $x^2 + 4^2$? Explain why or why not, and give a numerical example to justify your answer.

72. Is $(x - 3)^2$ equivalent to $x^2 - 3^2$? Explain why or why not, and give a numerical example to justify your answer.

Compute each product in Problems 73–78.

73. $(x - 2)^2$

74. $(x + 1)^2$

75. $(2x + 1)^2$

76. $(2x - 3)^2$

77. $(3x - 4y)^2$

78. $(5x + 2y)^2$

In Problems 79–82, use the Pythagorean theorem to express the relationship among the sides of each right triangle.

79.

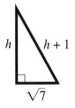

h $h + 1$
$\sqrt{7}$

80.

4
$p - 2$ p

81.

$2x + 1$
$x - 1$
12

82.

$3z - 4$
$z + 1$
$\sqrt{13}$

MIDCHAPTER REVIEW

In Problems 1–10, use complete sentences or fill in the blanks to answer the questions.

1. Explain the difference between -3^2 and $(-3)^2$.

2. If $s^2 = 17$, s is called a _____ of 17.

3. If $s^3 = 17$, s is called a _____ of 17.

4. What is the difference between "the square of 4" and "the square root of 4"?

5. What is the difference between $x^2 = 9$ and $x = \sqrt{9}$?

6. What is the difference between $-\sqrt{25}$ and $\sqrt{-25}$?

7. When asked to state the Pythagorean theorem, a classmate says, "$a^2 + b^2 = c^2$ when the sides of a triangle are a, b, and c." Another classmate claims that the formula does not work for all triangles. Do you agree with either student? What necessary part or parts of the theorem are not mentioned by either of the two?

8. A classmate says that $\sqrt[3]{2}$ must be a rational number because it can be written as $\dfrac{\sqrt[3]{2}}{1}$. Do you agree or disagree? Explain.

9. Give an example of a negative irrational number.

10. What do you get if you square $\sqrt[3]{5}$ and then multiply the result by $\sqrt[3]{5}$? Explain how to get the correct answer without using a calculator.

Simplify each expression in Problems 11–20.

11. a. $3(-7)^2$

b. $3^2(-7^2)$

12. a. $3^2 - 7^2$ **b.** $(3 - 7)^2$

13. a. $\sqrt{13^2 - 5^2}$ **b.** $\sqrt{13^2} - \sqrt{5^2}$

14. a. $\sqrt{13^2}\,\sqrt{5^2}$ **b.** $\sqrt{(13 - 5)^2}$

15. a. $\sqrt{\left(\dfrac{4}{5}\right)^2 + \left(\dfrac{3}{5}\right)^2}$ **b.** $\sqrt{\left(\dfrac{4}{5}\right)^2} + \sqrt{\left(\dfrac{3}{5}\right)^2}$

16. a. $\sqrt{\left(\dfrac{4}{5}\right)^2}\,\sqrt{\left(\dfrac{3}{5}\right)^2}$ **b.** $\sqrt{\left(\dfrac{4}{5} + \dfrac{3}{5}\right)^2}$

17. a. $2 - \sqrt[3]{64}$ **b.** $2\,\sqrt[3]{-64}$

18. a. $(\sqrt[3]{101})^3$ **b.** $\sqrt[3]{7} \cdot \sqrt[3]{7} \cdot \sqrt[3]{7}$

19. a. $(\sqrt[3]{8})^2$ **b.** $\sqrt[3]{8^2}$

20. a. $(\sqrt{9})^3$ **b.** $\sqrt{9^3}$

Identify each expression in Problems 21 and 22 as rational, irrational, or undefined.

21. a. $\sqrt[3]{-5}$

b. $\sqrt{-5^2}$

c. $\sqrt{(-5)^2}$

d. $(\sqrt[3]{-5})^2$

22. a. $-3.1\overline{6}$

b. -3.16

c. $-\sqrt{10}$

d. $-\pi$

In Problems 23–30, evaluate each expression for the given values of the variables. Round to three decimal places any answers that are irrational numbers.

23. a. $5 - a^2, \quad a = 3$

b. $5 - a^2, \quad a = -3$

24. a. $(5 - a)^2, \quad a = 3$

b. $(5 - a)^2, \quad a = -3$

25. a. $-w^2, \quad w = -2$

b. $(-w)^2, \quad w = -2$

26. a. $\sqrt{-w}, \quad w = -2$

b. $\sqrt[3]{-w}, \quad w = -2$

27. a. $\sqrt{x^2 - y^2}, \quad x = 13, \quad y = -5$

b. $(\sqrt{x})^2 + (\sqrt{-y})^2, \quad x = 13, \quad y = -5$

28. a. $\sqrt{b^2 - 4ac}, \quad a = 2, \quad b = -3, \quad c = 1$

b. $\dfrac{-b + \sqrt{b^2 - 4ac}}{2a}, \quad a = 2, \quad b = -3, \quad c = 1$

29. a. $\sqrt[3]{x^3 + y^3 + z^3}, \quad x = 3, \quad y = 4, \quad z = 5$

b. $\sqrt[3]{x^3} + \sqrt[3]{y^3} + \sqrt[3]{z^3}, \quad x = 3, \quad y = 4, \quad z = 5$

30. a. $m^3 + n^3, \quad m = 3, \quad n = 2$

b. $(m + n)(m^2 - mn + n^2), \quad m = 3, \quad n = 2$

In Problems 31 and 32, find the volume and surface area of each box.

31.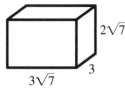
$2\sqrt{7}$
3
$3\sqrt{7}$

32.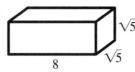
$\sqrt{5}$
$\sqrt{5}$
8

Solve Problems 33–36 by extraction of roots. Round any answers that are irrational numbers to the nearest thousandth.

33. $9x^2 - 4 = 0$

34. $3 + 2t^2 = 9$

35. $5y^2 - 12 = 2y^2$

36. $q^2 - 3 = 5 - q^2$

In Problems 37 and 38, graph the equation by plotting points. Then use your graph to estimate solutions to the equations given in parts (a), (b), and (c).

37. $y = x^3$
  a. $x^3 = 2$
  b. $x^3 = -1.4$
  c. $x = \sqrt[3]{4}$

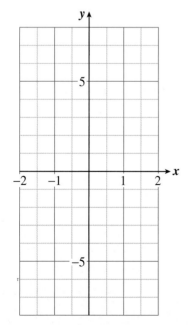

38. $y = x^2$
  a. $x^2 = 5$
  b. $x^2 = -3$
  c. $x = -\sqrt{2}$

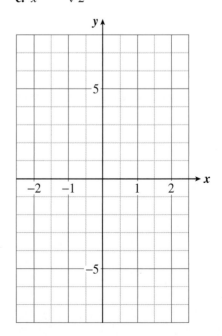

In Problems 39–42, solve the formula for the indicated variable.

39. $s = vt + \dfrac{1}{2}at^2$, for v

40. $A = \dfrac{h}{2}(b + c)$, for c

41. $V = \dfrac{s^2 h}{3}$, for s

42. $A = \dfrac{\pi d^2}{4}$, for d

43. The distance, d, that a penny will fall in t seconds is given, in feet, by

$$d = 16t^2$$

a. If you drop a penny from a height of 144 feet, how long will it take to reach the ground?

b. How far will the penny fall in $1\frac{1}{2}$ seconds?

44. a. How much water would you need to fill a spherical tank of radius 3 feet?

b. A biosphere contains 7000 cubic inches of space. What is the radius of the biosphere?

In Problems 45 and 46, write an algebraic expression for the volume of each figure. (Refer to Figure 4.2 on page 425.)

45.

46.

In Problems 47–50:
a. Find the unknown side or sides of each right triangle.
b. Find the perimeter of the triangle.
c. Find the area of the triangle.

47.

48.

49.

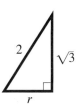

50.

In Problems 51 and 52, draw a sketch of each situation, write an equation to describe the sketch, and solve the equation. Round your answer to the nearest tenth.

51. Francine walks along the edge of a rectangular yard from the northwest corner to the southeast corner. Delbert walks diagonally across the same yard. If the yard is 30 yards long and 20 yards wide, how much farther than Delbert does Francine walk?

52. Rani kayaks due west for 2500 meters and then due south for another 1000 meters. If she then kayaks directly back to her starting point, how long is her total trip?

Compute each product in Problems 53 and 54.

53. $-5x(4 - 3x)$

54. $(3xy - 4x^2 + 2)(-2xy)$

Simplify each expression in Problems 55 and 56.

55. $6a(2a - 1) - (3a^2 - 3a)$

56. $4b - 2(3 - b^2) - b(b - 3)$

Fill in the missing parts of the algebraic expressions in Problems 57 and 58.

57. $-2n(3n + ?) = -6n^2 - 2n$

58. $5b(6 + ?) = 30b - 35b^2$

In Problems 59 and 60, write the area of each rectangle in two ways:
a. As the sum of four smaller areas
b. As the area of one large rectangle, using the formula   area = length × width

59.

$$5x \quad -2$$
$$x$$
$$1$$

60.

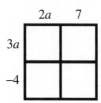

$$2a \quad 7$$
$$3a$$
$$-4$$

Compute each product in Problems 61–66.

61. $(x - 2)(x + 4)$

62. $(2y + 1)(3y - 2)$

63. $(5a + 1)(5a - 1)$

64. $(n - 7)(n - 7)$

65. $(2q + 5)(2q + 5)$

66. $(3c - 8)(3c + 8)$

4.6 Graphing Quadratic Equations

Reading

In this section, we consider two problems that can be solved using quadratic equations. Here is the first problem:

The marketing team at Raingear, Inc. find that the company will sell $150 - 5p$ umbrellas per month if it charges p dollars per umbrella. The marketers would like to know how the price of the umbrellas affects the company's revenue.

To find the total revenue from the sale of a product, we multiply the price of one item by the number of items sold. That is,

revenue = (price per item) · (number of items sold)

Very often, the higher the price of an item, the fewer items a company will sell. This is the case with Raingear's umbrellas: As the price, p, increases, the number of umbrellas the company sells, $150 - 5p$, decreases.

Now try Exercise 1.

The equation you wrote for part (b) of Exercise 1 is a quadratic equation. You have not yet learned an algebraic technique to solve this particular quadratic equation. (Do you see why extraction of roots will not work?) However, you can use trial and error to find the solution. Try Exercise 2.

EXERCISE 1

a. Use the formula for revenue to write an equation that expresses Raingear's monthly revenue from umbrellas, R, in terms of p. Simplify your equation by applying the distributive law.

b. Write an equation to find the price Raingear should charge for each umbrella in order to make $1000 in revenue each month.

EXERCISE 2

a. Use the table to look for a solution to the equation you wrote in part (b) of Exercise 1.

Price per umbrella p	Number of umbrellas sold $150 - 5p$	Revenue $R =$
2		
4		
6		
8		
10		
12		
14		

b. Will the revenue continue to increase as the price per umbrella increases? Why or why not?

Skills Review

For the following problems:
a. Solve each equation.
b. Write the equation in the form $ax + b = 0.$
c. Graph the equation $y = ax + b.$
d. Find the *x*-intercept of your graph. Compare this value with your answer to part (a).

1. $2x + 5 = 11$

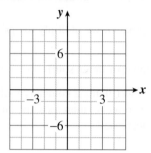

2. $2x - 3 = 5x + 9$

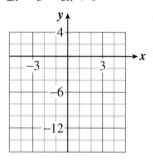

3. $0.7x + 0.2(100 - x) = 0.3(100)$

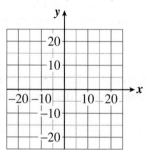

4. $4(7 - 4x) = -2(6x - 5) - 6$

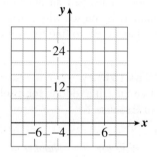

Answers: **1.** $x = 3$ **2.** $x = -4$ **3.** $x = 20$ **4.** $x = 6$

Lesson

For Exercise 1 in the reading assignment, you wrote a quadratic equation about the total revenue Raingear will earn from selling umbrellas. A good way to study such problems is to consider the graph of the equation.

Activity 1

a. Extend the table of values from Exercise 2 by completing the table.

Price per umbrella *p*	Number of umbrellas sold $150 - 5p$	Revenue $R =$
11		
13		
15		
17		
18		
20		
22		
24		
28		
30		

b. Use the values in the tables on pages 487 and 488 to draw a smooth graph of the equation for revenue, *R*, in terms of *p*.

c. Locate the point on your graph that corresponds to a monthly revenue of $1000.

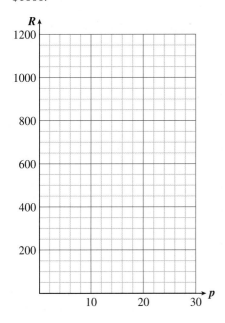

d. Find another point on the graph that corresponds to $1000 monthly revenue. What price per umbrella produces this revenue? How many umbrellas will be sold at this price?

e. What is the maximum monthly revenue that Raingear can earn from umbrellas?

f. What price per umbrella should the company charge in order to earn the maximum revenue?

The graph you drew is called a **parabola**. Its shape is characteristic of the graphs of all quadratic equations

$$y = ax^2 + bx + c, \qquad \text{where} \qquad a \neq 0$$

The most basic parabola is the graph of the equation $y = x^2$, shown in Figure 4.8. (You can use the equation to verify the values shown in the table.)

x	−4	−3	−2	−1	0	1	2	3	4
y	16	9	4	1	0	1	4	9	16

Figure 4.8

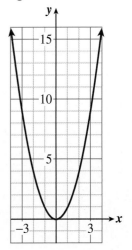

EXERCISE 3

Complete the table of values, and graph each of the quadratic equations. Compare each graph to the basic parabola shown in Figure 4.8.

a. $y = x^2 + 4$

x	-3	-2	-1	0	1	2	3
y							

b. $y = x^2 - 4$

x	-3	-2	-1	0	1	2	3
y							

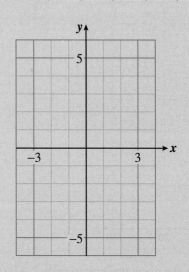

c. $y = 4 - x^2$

x	-3	-2	-1	0	1	2	3
y							

d. $y = (x - 4)^2$

x	1	2	3	4	5	6	7
y							

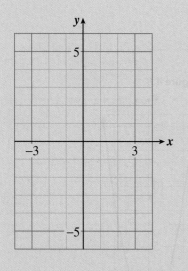

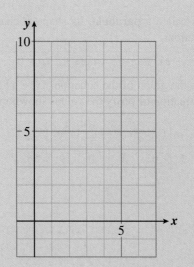

Each graph in Exercise 3 is a smooth curve that bends upward or downward. The high or low point of a parabola is called its **vertex**. For example, the vertex of the basic parabola $y = x^2$ is the point (0, 0).

Try Exercises 4 and 5 now.

Here is a second problem that involves a quadratic equation and its graph:

Activity 2

Delbert is standing at the edge of a 360-foot cliff. He throws his algebra book up and away from the cliff with a velocity of 36 feet per second. The height h of his book above the ground at the base of the cliff after t seconds, in feet, is given by the formula

$$h = -16t^2 + 36t + 360$$

a. Fill in the table, and then graph the equation. (Use your calculator to evaluate the expression for h.)

t	h
0	
0.5	
1	
1.5	
2	
2.5	
3	
4	
5	
6	

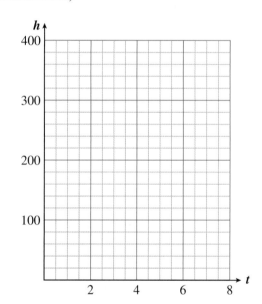

b. What is the height of the book after 1 second? After 3 seconds?

c. Use your graph to find the approximate time when the book is 200 feet high.

d. When does the book hit the ground?

e. Write an algebraic equation to find out when the book hits the ground.

f. Verify that your answer to part (d) is a solution for the equation you wrote in part (e).

It would be convenient to solve the equation you wrote in part (e) algebraically—that is, without using a graph. First, note this important fact: The solution of the equation $-16t^2 + 36t + 360 = 0$ occurs at the point where the graph crosses the t-axis. In other words, it is a t-intercept of the graph. (This makes sense, because the h-coordinate of the point is zero.) There is an important connection between the graph of a quadratic equation and the solutions of the equation. You will explore this connection in the homework problems.

HOMEWORK 4.6

For Problems 1–4, make a table of values, and graph each pair of parabolas on the grid provided.

1. $y = x^2 + 1$
$y = x^2 - 3$

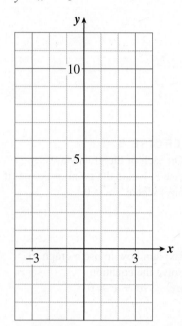

2. $y = x^2 - 1$
$y = x^2 + 2$

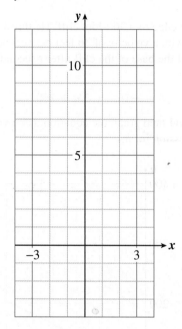

3. $y = -x^2 - 2$
$y = 5 - x^2$

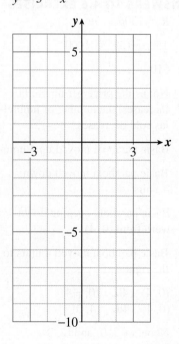

4. $y = -x^2 + 3$
$y = 9 - x^2$

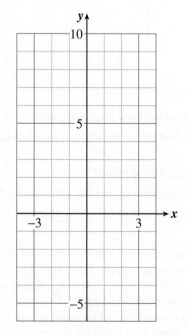

5. For each graph in Problems 1–4, give the coordinates of the vertex.

6. Use your graphs from Problems 1–4 to help you answer the following questions. (In these questions, k is a positive constant.)

   a. What is the vertex of the graph of $y = x^2 + k$?

   b. What is the vertex of the graph of $y = x^2 - k$?

For Problems 7–10, make a table of values, and graph each pair of parabolas on the grid provided.

7. $y = (x + 2)^2$
   $y = (x - 1)^2$

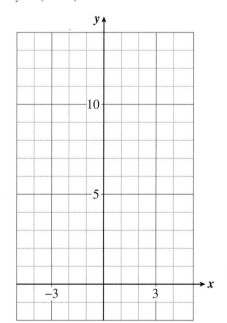

8. $y = (x - 2)^2$
   $y = (x + 3)^2$

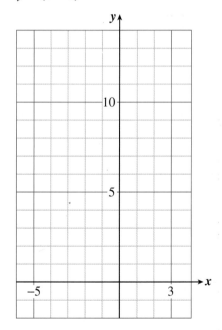

9. $y = -(x + 1)^2$
$y = -(x - 4)^2$

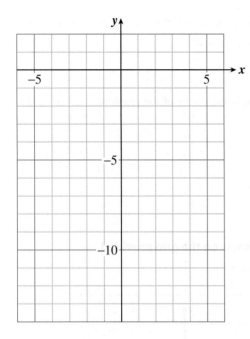

10. $y = -(x - 3)^2$
$y = -(x + 4)^2$

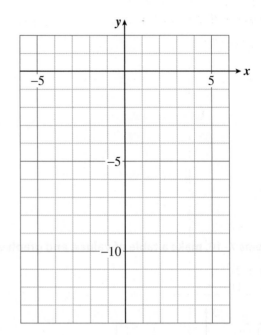

11. For each graph in Problems 7–10, give the coordinates of the vertex.

12. Use your graphs from Problems 7–10 to help you answer the following questions. (In these questions, k is a positive constant.)

   a. What is the vertex of the graph of $y = (x + k)^2$?

   b. What is the vertex of the graph of $y = (x - k)^2$?

In Problems 13–16:
a. Graph each quadratic equation.
b. What is the vertex of the parabola?
c. Write the equation in the form $y = ax^2 + bx + c.$

13. $y = 2 + (x - 3)^2$

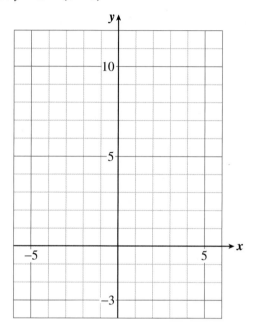

14. $y = -3 + (x + 4)^2$

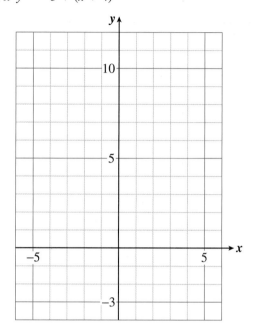

15. $y = 4 - (x + 1)^2$

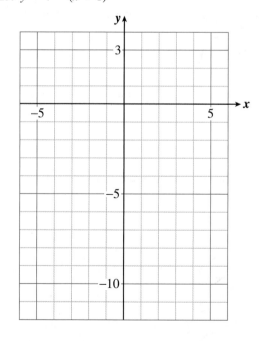

16. $y = -1 - (x - 2)^2$

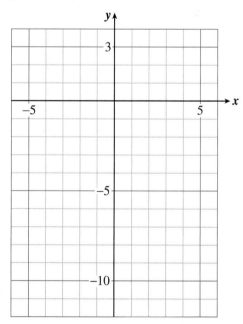

Graph each equation in Problems 17–24 on the grid provided, using the *x*-values in the table. Then answer the questions.

17. $y = x^2 + 2x - 8$

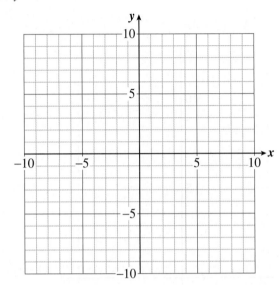

18. $y = x^2 - x - 6$

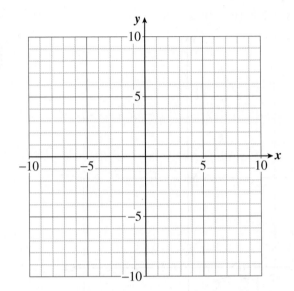

a. Use your graph to find the solutions of the equation $x^2 + 2x - 8 = 0$.

b. Verify algebraically that your answers to part (a) are really solutions.

c. Compute the product $(x + 4)(x - 2)$.

d. What is the value of the expression $(x + 4)(x - 2)$ when $x = -4$? When $x = 2$?

a. Use your graph to find the solutions of the equation $x^2 - x - 6 = 0$.

b. Verify algebraically that your answers to part (a) are really solutions.

c. Compute the product $(x - 3)(x + 2)$.

d. What is the value of the expression $(x - 3)(x + 2)$ when $x = 3$? When $x = -2$?

19. $y = -x^2 + 6x$

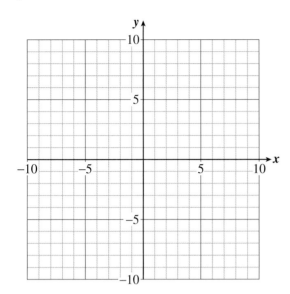

a. Use your graph to find the solutions of the equation $-x^2 + 6x = 0$.

b. Verify algebraically that your answers to part (a) are really solutions.

c. Compute the product $-x(x - 6)$.

d. What is the value of the expression $-x(x - 6)$ when $x = 0$? When $x = 6$?

20. $y = x^2 + 4x$

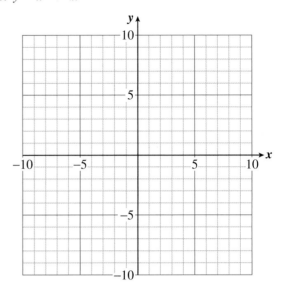

a. Use your graph to find the solutions of the equation $x^2 + 4x = 0$.

b. Verify algebraically that your answers to part (a) are really solutions.

c. Compute the product $x(x + 4)$.

d. What is the value of the expression $x(x + 4)$ when $x = 0$? When $x = -4$?

21. $y = x^2 - 6x + 9$

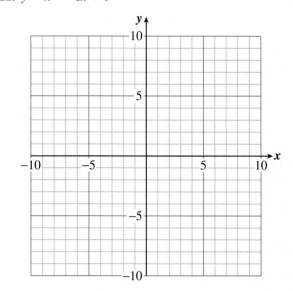

22. $y = 9 - x^2$

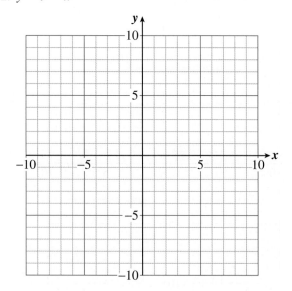

a. Use your graph to find the solutions of the equation $x^2 - 6x + 9 = 0$.

b. Verify algebraically that your answers to part (a) are really solutions.

c. Compute the product $(x - 3)^2$.

d. What is the value of the expression $(x - 3)^2$ when $x = 3$?

a. Use your graph to find the solutions of the equation $9 - x^2 = 0$.

b. Verify algebraically that your answers to part (a) are really solutions.

c. Compute the product $(3 + x)(3 - x)$.

d. What is the value of $(3 + x)(3 - x)$ when $x = 3$? When $x = -3$?

23. $y = -2x^2 + 12x - 10$

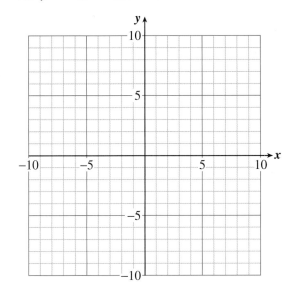

24. $y = \frac{1}{2}x^2 + x - 12$

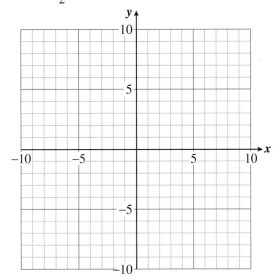

a. Use your graph to find the solutions of the equation $-2x^2 + 12x - 10 = 0$.

b. Verify algebraically that your answers to part (a) are really solutions.

c. Compute the product $-2(x - 1)(x - 5)$.

d. What is the value of the expression $-2(x - 1)(x - 5)$ when $x = 1$? When $x = 5$?

a. Use your graph to find the solutions of the equation $\frac{1}{2}x^2 + x - 12 = 0$.

b. Verify algebraically that your answers to part (a) are really solutions.

c. Compute the product $\frac{1}{2}(x + 6)(x - 4)$.

d. What is the value of the expression $\frac{1}{2}(x + 6)(x - 4)$ when $x = -6$? When $x = 4$?

25. The expression in part (c) of Problems 17–24 is called the *factored form* of the quadratic equation. Explain why the factored form is useful for finding the solutions of a quadratic equation.

26. Explain why the solutions of the quadratic equation $ax^2 + bx + c = 0$ are also the *x*-intercepts of the graph of $y = ax^2 + bx + c$.

27. The bridge over the Rushing River at Marionville is 48 feet high. Francine stands on the bridge and tosses a rock up into the air off the edge of the bridge. The height h of the rock above the water t seconds later, in feet, is given by

$$h = 48 + 32t - 16t^2$$

a. Complete the table of values.

t	0	0.5	1	1.5	2	2.5	3
h							

b. Sketch a graph of the equation.

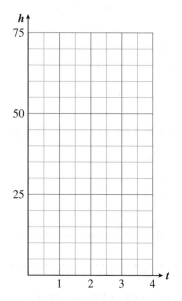

c. Estimate the height of the rock after 1.75 seconds. Verify your answer algebraically.

d. When is the rock about 40 feet above the water?

e. Write an equation for the question in part (d).

f. How long is the rock more than 60 feet high?

g. After reaching its highest point, how long does the rock fall before it hits the water?

28. A rocket in a fireworks display is shot from ground level with an initial speed of 96 feet per second. Its height t seconds later, in feet, is given by

$$h = 96t - 16t^2$$

a. Complete the table of values.

t	0	1	2	3	4	5	6
h							

b. Sketch the graph of the equation.

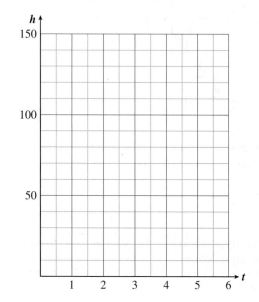

c. Estimate the height of the rocket after 1.75 seconds. Verify your answer algebraically.

d. Find two times when the rocket is at a height of 140 feet.

e. Write an equation for the question in part (d).

f. How long does it take the rocket to reach its highest point?

g. How long does it take the rocket to fall from its highest point back to earth?

29. Kitchenware Appliances finds that the cost of producing *x* toasters per week, in dollars, is given by

$$C = 0.1x^2 - 8x + 400$$

a. Complete the table of values.

x	10	20	30	40	50	60	70	80	90	100
C										

b. Sketch the graph of the equation.

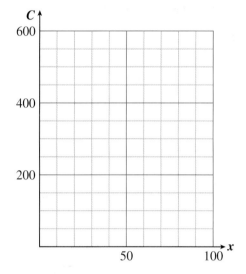

c. How much will it cost to produce 25 toasters per week?

d. How many toasters can be produced for $330? (There are two answers.)

e. Write an equation for the question in part (d).

f. How many toasters should Kitchenware produce if it wants to minimize its costs?

g. How many toasters can be produced if costs must be kept under $500?

30. Mitra plans to build a rectangular pen for her rabbits in the back yard. She has 48 feet of wire fence. If she builds a pen *w* feet wide, the area of the pen, in square feet, is given by

$$A = 24w - w^2$$

a. Complete the table of values.

w	0	4	8	12	16	20	24
A							

b. Sketch a graph of the equation.

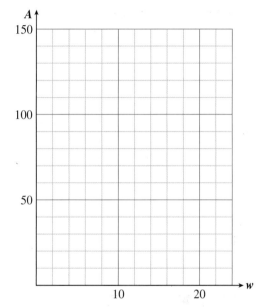

c. If the pen is 10 feet wide, what is its area?

d. If the area of the pen is 120 square feet, what is its width? (There are two answers.)

e. Write an equation for the question in part (d).

f. What is the largest area that Mitra can enclose with her fence?

g. What are the dimensions of the largest rabbit pen Mitra can build?

4.7 Solving Quadratic Equations

Reading

In Section 4.6, we found a quadratic equation for the revenue that Raingear, Inc. earns by selling umbrellas at a price of p dollars each:

$$R = p(150 - 5p) = 150p - 5p^2$$

The graph of this equation is shown in Figure 4.9.

If Raingear charges $p = 0$ dollars for each umbrella, the company will not earn any revenue, so $R = 0$. You can see this fact in the graph, which passes through the point $p = 0$, $R = 0$. But there is another point on the graph where $R = 0$; it has p-coordinate 30. If Raingear charges $30 per umbrella, the company will not earn any revenue. (Why not?)

Figure 4.9

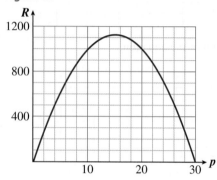

The Zero Factor Principle

Let's look again at the factored form of the equation for revenue,

$$R = p(150 - 5p)$$

The first factor is p, and when $p = 0$,

$$R = 0(150 - 5 \cdot 0) = 0$$

so the revenue equals zero as well. The second factor, $150 - 5p$, equals zero when $p = 30$. If Raingear charges $30 per umbrella, its revenue is

$$R = 30(150 - 5 \cdot 30) = 0$$

(Recall that $150 - 5p$ represents the number of umbrellas that Raingear will sell at a price of p dollars each. Apparently, no one will buy a $30 umbrella.)

Both of these values, $p = 0$ and $p = 30$, are solutions of the quadratic equation

$$p(150 - 5p) = 0$$

In fact, a quadratic equation only has a solution when one of its factors is equal to zero. This is a consequence of the following fact about numbers.

Zero Factor Principle

If the product of two numbers is zero, then one (or both) of the numbers must be zero. Using symbols,

If $ab = 0$, then either $a = 0$ or $b = 0$

The zero factor principle also applies to algebraic expressions: If the product of two factors is zero, then one of the factors must be zero.

EXAMPLE 1

    a. Compute the product $(x + 5)(x - 4)$.

    b. Solve the equation $x^2 + x - 20 = 0$.

Solution

    a. We use the FOIL method to apply the distributive law to the product:

$$(x + 5)(x - 4) = x^2 - 4x + 5x - 20$$
$$= x^2 + x - 20$$

    b. From part (a), we see that $x^2 + x - 20$ can also be written in factored form as $(x + 5)(x - 4)$. Thus the equation $x^2 + x - 20 = 0$ is equivalent to

$$(x + 5)(x - 4) = 0$$

We can apply the zero factor principle: If the product is zero, then one of the factors must be zero. Thus either

$$x + 5 = 0 \quad \text{or} \quad x - 4 = 0$$

This means that either $x = -5$ or $x = 4$. These are the solutions of the equation, as we verify below.

Check: We substitute each value into the original equation.

$$x = -5: \quad (-5)^2 + (-5) - 20 = 25 - 5 - 20 = 0$$
$$x = 4: \quad (4)^2 + (4) - 20 = 16 + 4 - 20 = 0$$

Now try Exercise 1.

Skills Review

Mentally evaluate each expression for the given values of the variable. Do not use pencil, paper, or calculator.

   1. $2x(x - 3)$, $x = 0, 1, 2, 3$ 2. $(x + 1)(x - 6)$, $x = -1, 3, 6, 9$

   3. $2(x + 2)(x + 4)$, $x = -4, -2, 0, 2$

   4. $3n(2n - 1)$, $n = -1, 0, \frac{1}{2}, 1$

Answers: *1.* $0, -4, -4, 0$ *2.* $0, -12, 0, 30$ *3.* $0, 0, 16, 48$
4. $9, 0, 0, 3$

Lesson

If we want to use the zero factor principle to solve quadratic equations, we must be able to write quadratic expressions in factored form. This process is called

EXERCISE 1
Find the solutions of each quadratic equation.
a. $(x - 2)(x + 5) = 0$

b. $y(y - 4) = 0$

factoring, and it is the reverse of multiplying factors together. Here are some examples that compare multiplying and factoring:

Multiplying	*Factoring*
$2 \cdot 3 = 6$	$6 = 2 \cdot 3$
$3(2x - 5) = 6x - 15$	$6x - 15 = 3(2x - 5)$
$(x - 5)(x + 2) = x^2 - 3x - 10$	$x^2 - 3x - 10 = (x - 5)(x + 2)$

Common Factors

The first type of factoring we consider uses the distributive law. For example, the expression $6x - 15$ has two terms, $6x$ and -15, and each term is divisible by 3:

$$6x = 3 \cdot 2x \quad \text{and} \quad -15 = 3(-5)$$

We say that 3 is a **common factor** for the expression $6x - 15$. Using the distributive law, we can write

$$6x - 15 = 3 \cdot 2x - 3 \cdot 5 = 3(2x - 5)$$

We have factored out a common factor from $6x - 15$.

EXAMPLE 2

Factor: $36t^2 - 63t$.

Solution

We look for the largest common factor for the two terms. Both coefficients are divisible by 9, and the largest power of t that divides evenly into both terms is t. We factor out a common factor of $9t$—that is, we write the expression as

$$36t^2 - 63t = 9t(? - ?)$$

You may be able to see at a glance what the missing factors are, but if not, you can divide each term of the original expression by the common factor, as follows:

$$\frac{36t^2}{9t} = 4t \quad \text{and} \quad \frac{63t}{9t} = 7$$

Substitute these factors for the question marks to obtain the factored form,

$$36t^2 - 63t = 9t(4t - 7)$$

You can always check your factorization by computing the product; it should be the same as the original expression.

Now try Exercise 2.

Solving Quadratic Equations by Factoring

Now, let's see how to solve a quadratic equation using factoring. First, we must arrange the terms so that one side of the equation is zero; the zero factor principle applies only to zero products. A quadratic equation is said to be in *standard form* when it is written as

$$ax^2 + bx + c = 0$$

(where b or c could be zero).

EXAMPLE 3

Solve: $2x^2 = 4x$.

Solution

First, we write the equation in standard form by subtracting $4x$ from both sides.

$$2x^2 - 4x = 0$$

Next, we factor the left side of the equation. We can take out a common factor of $2x$.

$$2x(x - 2) = 0$$

Now, we apply the zero factor principle: In order for the product to be zero, one of the two factors must be zero. We set each factor equal to zero, and solve for x.

$$2x = 0 \quad \text{or} \quad x - 2 = 0$$
$$x = 0 \quad \text{or} \quad x = 2$$

Thus the two solutions are $x = 0$ and $x = 2$. You should check that both of these values satisfy the original equation.

_____

In general, we use the following steps.

To Solve a Quadratic Equation by Factoring:

1. Write the equation in standard form,

$$ax^2 + bx + c = 0 \quad (a \neq 0)$$

2. Factor the left side of the equation.

3. Apply the zero factor principle; that is, set each factor equal to zero.

4. Solve each equation to obtain two solutions.

Now try Exercise 3.

Application to Graphing

In the reading assignment for this section, we saw that the solutions of the equation

$$150p - 5p^2 = 0$$

namely, $p = 0$ and $p = 30$, are also the horizontal intercepts of the graph of

$$R = 150p - 5p^2$$

In general, the solutions of the equation

$$ax^2 + bx + c = 0$$

are also the x-intercepts of the graph of

$$y = ax^2 + bx + c$$

Try Exercise 4 now.

EXERCISE 3
Solve each quadratic equation.
a. $2x^2 + 8x = 0$

b. $-3n^2 + 18n = 0$

EXERCISE 4
a. Find the x-intercepts of the graph of $y = 2x^2 + 8x$.

b. Sketch the graph.

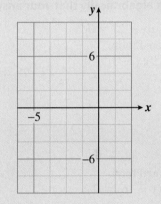

ANSWERS TO 4.7 EXERCISES
1a. $x = 2, -5$ **1b.** $y = 0, 4$

2a. $2x(x + 4)$

2b. $3n(-n + 6)$ or $-3n(n - 6)$

3a. $x = 0, -4$ **3b.** $n = 0, 6$

4a. $(0, 0), (-4, 0)$

4b.

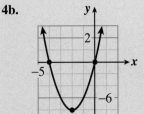

HOMEWORK 4.7

In Problems 1–6, apply the zero factor principle to find the solutions.

1. $(x + 1)(x - 4) = 0$

2. $0 = (w - 6)(w - 7)$

3. $0 = p(p + 7)$

4. $-2t(t - 2) = 0$

5. $(2v + 3)(4v - 1) = 0$

6. $(3u - 5)(2u + 1) = 0$

In Problems 7–10:
a. Verify that the factored form for each expression is correct.
b. Use the factored form to find the solutions of the original equation.
c. Check algebraically that your answers to part (b) are solutions of the original equation.

7. $x^2 - 6x - 27 = 0$
   Factored form:
   $(x - 9)(x + 3) = 0$

8. $x^2 - 13x + 42 = 0$
   Factored form:
   $(x - 6)(x - 7) = 0$

9. $-3x^2 + 48x = 0$
   Factored form:
   $-3x(x - 16) = 0$

10. $x^2 - 16x + 64 = 0$
   Factored form:
   $(x - 8)^2 = 0$

Factor the expressions in Problems 11–16 by finding the largest common factor.

11. $6a^2 - 8a$

12. $15b^2 - 5b$

13. $-18v^2 - 6v$

14. $7w + 3w^2$

15. $4h - 9h^2$

16. $-9n - 12n^2$

Factor the right side of each formula in Problems 17–22.

17. $d = k - kat$

18. $P = 2l + 2w$

19. $A = 2rh - \pi r^2$

20. $A = \frac{1}{2}bh + \frac{1}{2}h^2$

21. $V = \pi r^2 h - \frac{1}{3}s^2 h$

22. $S = 2\pi R^2 + 2\pi R H$

Solve each equation in Problems 23–28.

23. $10x^2 - 15x = 0$

24. $-24x^2 - 16x = 0$

25. $20x^2 = x$

26. $-9x = 81x^2$

27. $0 = 144x + 3x^2$

28. $0 = 140x - 4x^2$

What is wrong with the solutions to the quadratic equations in Problems 29 and 30?

29. $x^2 - 6x + 8 = 0$
$$x^2 = 6x - 8$$
$$x = \frac{6x - 8}{x}$$

30. $x^2 + 2x - 15 = 0$
$$2x = 15 - x^2$$
$$x = \frac{15 - x^2}{2}$$

31. The revenue, in dollars, that you can earn by making and selling n jade bracelets is given by

$$R = -2n^2 + 80n$$

a. Factor the expression for R.

b. Complete the table and sketch the graph.

n	R
0	
5	
10	
20	
30	
40	

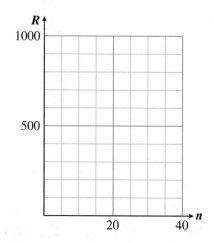

c. What is the maximum revenue that you can earn? How many bracelets must you sell in order to earn that revenue?

32. The revenue, in dollars, that you can earn by making and selling g glasses of lemonade is given by

$$R = -0.10g^2 + 4.80g$$

a. Factor the expression for R.

b. Complete the table and sketch the graph.

g	R
0	
6	
12	
18	
24	
30	
36	
42	
48	

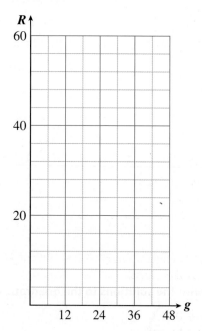

c. What is the maximum revenue that you can earn? How many glasses of lemonade must you sell in order to earn that revenue?

33. The height, in feet, of a football t seconds after it is kicked from the ground is given by

$$h = -16t^2 + 80t$$

a. Factor the expression for h.

b. Complete the table, and sketch the graph.

t	h
0	
1	
2	
2.5	
3	
4	
5	

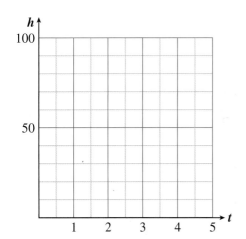

c. What is the maximum height of the football? When does it reach this height?

34. The height, in centimeters, of a golf ball t seconds after it is hit is given by

$$h = -490t^2 + 1470t$$

a. Factor the expression for h.

b. Complete the table, and sketch the graph.

t	h
0	
0.5	
1	
1.5	
2	
2.5	
3	

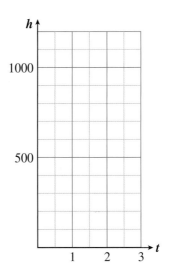

c. What is the maximum height of the golf ball? When does it reach this height?

Without graphing, find the *x*-intercepts of the graph of each equation in Problems 35–40.

35. $y = (x - 3)(2x + 5)$

36. $y = (3x - 4)(x + 2)$

37. $y = 2x^2 - 6x$

38. $y = 3x^2 + 12x$

39. $y = 8x - 3x^2$

40. $y = -6x - 4x^2$

In Problems 41–46:
a. Compute each product.
b. Illustrate each product as the area of a rectangle.

41. $(y + 4)(y + 2)$

42. $(b - 5)(b - 4)$

43. $(w - 6)(w + 3)$

44. $(h + 8)(h - 5)$

45. $(t + 5)(2t + 3)$

46. $(6v + 3)(3v - 2)$

4.8 Factoring Quadratic Trinomials

Reading

In Section 4.7, you learned how to factor quadratic expressions of the form $ax^2 + bx$. In this section, we consider quadratic trinomials of the form $x^2 + bx + c$. Many of these can be factored into the product of two binomials. For example,

$$x^2 + 7x + 12 = (x + 3)(x + 4)$$

How can we come up with the factors of the quadratic trinomial? The product is illustrated in Figure 4.10.

$(x + 3)(x + 4) = x^2 + 4x + 3x + 12$

$\qquad\qquad\quad$ F $\quad$ $\overbrace{\text{O + I}}$ $\quad$ L

$\qquad\qquad = x^2 +\quad 7x\quad + 12$

Figure 4.10

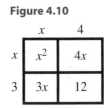

We can make several observations about this product.

1. The quadratic term, x^2, is the product of the **F**irst terms in the two binomials.

2. The constant term, 12, is the product of the **L**ast terms in the binomials.

3. The middle term, $7x$, is the sum of two terms: the product of the **O**utside terms of the binomials and the product of the **I**nside terms.

In general, if a quadratic trinomial is the product of two binomials, $(x + p)(x + q)$, then

$$(x + p)(x + q) = x^2 + qx + px + pq$$
$$= x^2 + (p + q)x + pq$$

We see that the constant term of the trinomial is the product pq and that the linear term of the trinomial has as its coefficient the sum $p + q$. Thus, in order to factor the trinomial, we must look for two numbers p and q whose product is the constant term and whose sum gives the coefficient of the middle term.

Skills Review

For the following problems:
a. Find the linear term of each product mentally.
b. Find the constant term.

1. $(x + 1)(x + 3)$ $\qquad$ 2. $(a - 6)(a - 3)$ $\qquad$ 3. $(p + 7)(p - 4)$

4. $(q - 8)(q + 2)$ $\qquad$ 5. $(3w + 4)(w + 2)$ $\qquad$ 6. $(2z - 5)(3z + 2)$

Answers: 1. a. $4x$ *b.* 3 $\quad$ *2. a.* $-9a$ *b.* 18 $\quad$ *3. a.* $3p$ *b.* -28
4. a. $-6q$ *b.* -16 $\quad$ *5. a.* $10w$ *b.* 8 $\quad$ *6. a.* $-11z$ *b.* -10

Lesson

Factoring Quadratic Trinomials of the Form $x^2 + bx + c$

Now, we'll use what we discussed in the reading assignment to factor a quadratic trinomial into a binomial product, $(x + p)(x + q)$.

EXAMPLE 1

Factor $t^2 + 7t + 12$ into a product of two binomials.

$$t^2 + 7t + 12 = (\underline{\quad} + \underline{\quad})(\underline{\quad} + \underline{\quad})$$

Solution

We apply our observations from the reading assignment to help us fill in the blanks.

1. The product of the **First** terms must be t^2, so we place t's in the First spots:

$$t^2 + 7t + 12 = (t + \underline{\quad})(t + \underline{\quad})$$

2. The product of the **Last** terms is 12, so we should fill in the blanks with two numbers p and q whose product is 12. There are three possibilities:

$$1 \text{ and } 12, \qquad 2 \text{ and } 6, \qquad \text{or} \qquad 3 \text{ and } 4$$

We use the middle term of the trinomial to decide which possibility is correct.

3. The sum of **Outside** and **Inside** is the middle term, $7t$. We can check each possibility above to see which gives the correct middle term:

$$(t + 1)(t + 12) \quad \text{O} + \text{I} = 12t - t = 13t$$
$$(t + 2)(t + 6) \quad \text{O} + \text{I} = 6t + 2t = 8t$$
$$(t + 3)(t + 4) \quad \text{O} + \text{I} = 4t + 3t = 7t$$

We see that the last possibility gives the correct middle term, so the factorization is

$$t^2 + 7t + 12 = (t + 3)(t + 4)$$

We could also write the factored form of the trinomial in Example 1 as $(t + 4)(t + 3)$;   the order of the factors does not affect the product. Aside from rearranging the factors, there is only one correct factorization for a quadratic trinomial.

Try Exercise 1 now.

Factoring Quadratic Trinomials of the Form $x^2 - bx + c$

What if the quadratic trinomial we want to factor has one or more negative coefficients? We first consider the case in which the middle term is negative (but the quadratic and constant terms are positive).

EXAMPLE 2

Factor $x^2 - 12x + 20$ into a product of two binomials.

Solution

As in Example 1, we look for a product of the form

$$x^2 - 12x + 20 = (x + p)(x + q)$$

We must find two numbers p and q that satisfy two conditions:

1. Their product, pq, is the **Last** term and so must equal 20.
2. Their sum, $p + q$, must equal -12 (because
   O + **I** $= qx + px = -12x$).

EXERCISE 1
Factor each trinomial.
a. $x^2 + 8x + 15$

b. $y^2 + 14y + 49$

These two conditions tell us that p and q must both be negative.
We start by listing all pairs of negative factors of 20:

$$-1 \text{ and } -20, \qquad -2 \text{ and } -10, \qquad -4 \text{ and } -5$$

Then we check each possibility to see which one gives the correct middle term:

$$(x - 1)(x - 20) \quad \text{O} + \text{I} = -20x - x = -21x$$
$$(x - 2)(x - 10) \quad \text{O} + \text{I} = -10x - 2x = -12x$$
$$(x - 4)(x - 5) \quad \text{O} + \text{I} = -5x - 4x = -9x$$

The second possibility gives the correct middle term, so the factorization is

$$x^2 - 12x + 20 = (x - 2)(x - 10)$$

Now try Exercise 2.

Factoring Quadratic Trinomials of the form $x^2 + bx - c$ or $x^2 - bx - c$

Finally, we consider the case in which the constant term is negative; the middle term can be either positive or negative.

EXAMPLE 3

Factor $x^2 + 2x - 15$.

Solution

We look for a factorization of the form

$$x^2 + 2x - 15 = (x + p)(x + q)$$

This time the product pq of the two unknown numbers must be negative, -15. This means that p and q must have opposite signs, one positive and one negative. How do we know which is positive and which is negative? The easiest way to resolve this problem is to guess! If we choose the wrong one, we can easily fix it later.

There are only two ways to factor 15, either 1 times 15 or 3 times 5. We will just *guess* that the second factor is negative, and check   O + I   for each possibility:

$$(x + 1)(x - 15) \quad \text{O} + \text{I} = -15x + x = -14x$$
$$(x + 3)(x - 5) \quad \text{O} + \text{I} = -5x + 3x = -2x$$

The second possibility gives a middle term of $-2x$. This is not quite correct, because the middle term we want is $2x$. We can fix this by changing the signs on the factors of 15: Instead of using $+3$ and -5, we'll change to -3 and $+5$. You can check that the correct factorization is

$$x^2 + 2x - 15 = (x - 3)(x + 5)$$

Now try Exercise 3.

The sign patterns we have discovered for factoring quadratic trinomials are summarized below. The order of the terms is very important. For these strategies to work, the trinomial must be written in descending powers of the variable; that is, the quadratic term must come first, then the linear term, and finally the constant term.

$$\underset{\text{quadratic term}}{x^2} \qquad \underset{\text{linear term}}{+ bx} \qquad \underset{\text{constant term}}{+ c}$$

EXERCISE 2
Factor each trinomial.
a. $m^2 - 10m + 24$

b. $m^2 - 11m + 24$

EXERCISE 3
Factor each trinomial.
a. $t^2 + 8t - 48$

b. $t^2 - 8t - 48$

Sign Patterns for Factoring Quadratic Trinomials

Assume that b, c, p, and q are positive integers. Then

1. $x^2 + bx + c = (x + p)(x + q)$

If all the coefficients of the trinomial are positive, then both p and q are positive.

2. $x^2 - bx + c = (x - p)(x - q)$

If the middle term of the trinomial is negative and the other two terms are positive, then p and q are both negative.

3. $x^2 \pm bx - c = (x + p)(x - q)$

If the constant term of the trinomial is negative, then p and q have opposite signs.

Using the Zero Factor Principle to Solve Quadratic Equations

We can now use the zero factor principle to solve many quadratic equations of the form $x^2 + bx + c = 0$.

EXAMPLE 4

a. Solve $x^2 + 3x = 18$.

b. Solve $2x^2 + 6x = 36$.

Solution

a. Recall the steps given in Section 4.6 for solving quadratic equations. The right side must be equal to zero if we want to apply the zero factor principle, so we begin by subtracting 18 from both sides:

$$x^2 + 3x - 18 = 0 \qquad \text{Factor the left side.}$$

$$(x + 6)(x - 3) = 0 \qquad \begin{array}{l}\text{Apply the zero factor principle:}\\ \text{Set each factor equal to zero.}\end{array}$$

$$(x + 6) = 0 \qquad x - 3 = 0$$

$$x = -6 \qquad x = 3 \quad \text{Solve each equation.}$$

The solutions are -6 and 3. You should check that each of these solutions satisfies the original equation.

b. You may have noticed that each term of the equation in part (b) is twice the corresponding term of the equation in part (a). We can factor out a common factor of 2 from the left side of the standard form of this equation:

$$2x^2 + 6x - 36 = 0 \quad \text{Factor out 2 from the left side.}$$

$$2(x^2 + 3x - 18) = 0$$

Then we can divide both sides of the equation by 2:

$$\frac{2(x^2 + 3x - 18)}{2} = \frac{0}{2}$$

$$x^2 + 3x - 18 = 0$$

This new equation is the same as the equation in part (a). Thus the solutions of the equation in part (b) are the same as the solutions in part (a), namely −6 and 3.

Now try Exercise 4.

Graphing Parabolas

Recall that the *x*-intercepts of the graph of

$$y = ax^2 + bx + c$$

are given by the solutions of the equation

$$ax^2 + bx + c = 0$$

EXAMPLE 5

Find the *x*-intercepts of the graph of $y = x^2 + 4x - 5$.

Solution

The *x*-intercepts are the points where $y = 0$,   so we substitute $y = 0$ into the equation and solve for *x*:

$$x^2 + 4x - 5 = 0 \qquad \text{Factor the left side.}$$
$$(x + 5)(x - 1) = 0 \qquad \text{Apply the zero factor principle.}$$
$$x + 5 = 0 \quad x - 1 = 0 \quad \text{Solve each equation.}$$
$$x = -5 \qquad x = 1$$

The *x*-intercepts are the points $(-5, 0)$ and $(1, 0)$.

All the parabolas we'll study are symmetric about a vertical line (called the **axis of symmetry**) that passes through the vertex. Consider the two examples shown in Figures 4.11 and 4.12. Because of this symmetry, the *x*-intercepts are located at equal distances on either side of the axis of symmetry. Or we can say that the *x*-coordinate of the vertex is exactly halfway between the two *x*-intercepts.

Figure 4.11

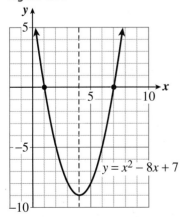

$y = x^2 - 8x + 7$

Figure 4.12

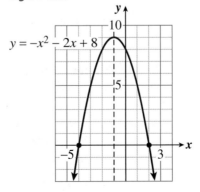

$y = -x^2 - 2x + 8$

Thus we can locate the vertex of a parabola by taking the average of its *x*-intercepts. For the graph in Figure 4.11, the *x*-coordinate of the vertex is

$$x = \frac{1 + 7}{2} = 4$$

EXERCISE 4
Solve each quadratic equation.
a. $a^2 - 13a + 30 = 0$

b. $u^2 - 6u = 16$

c. $3x^2 + 3 = 6x$

d. $9x^2 - 18x = 0$

EXERCISE 5

Find the vertex of the parabola in Figure 4.12, whose equation is $y = -x^2 - 2x + 8.$ The x-intercepts are $(-4, 0)$ and $(2, 0)$.

EXERCISE 6

a. Find the vertex and the y-intercept of the parabola in Example 5, $y = x^2 + 4x - 5.$

b. Sketch a graph of the equation.

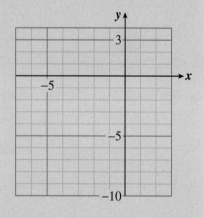

To find the y-coordinate of the vertex, we evaluate the equation of the parabola for $x = 4.$

$$y = 4^2 - 8(4) + 7$$
$$= 16 - 32 + 7 = -9$$

The vertex of the parabola in Figure 4.11 is the point $(4, -9)$.

Now try Exercise 5.

Once we have located the intercepts and the vertex of a parabola, we can sketch a very good graph without needing to plot a lot of points. Don't forget that you can also find the y-intercept easily by substituting $x = 0$ into the equation.

Now try Exercise 6.

ANSWERS TO 4.8 EXERCISES

1a. $(x + 5)(x + 3)$

1b. $(y + 7)(y + 7)$

2a. $(m - 6)(m - 4)$

2b. $(m - 8)(m - 3)$

3a. $(t + 12)(t - 4)$

3b. $(t - 12)(t + 4)$

4a. $a = 3, a = 10$

4b. $u = 8, u = -2$

4c. $x = 1, x = 1$

4d. $x = 0, x = 2$

5. $(-1, 9)$

6a.   Vertex: $(-2, -9)$; y-intercept: $(0, -5)$

6b.
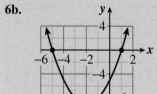

HOMEWORK 4.8

In Problems 1–6, each rectangle diagram represents a product as an area.
a. Fill in the missing expressions.
b. Write the area of the rectangle as a product of binomials; then compute the product.

1.

	x	?
x	x^2	$6x$
?	$5x$	30

2.

	x	?
x	x^2	$-2x$
?	$6x$	-12

3.

	x	-3
x	x^2	$-3x$
?	?	27

4.

	x	?
x	x^2	?
6	$6x$	-36

5.

	x	?
?	x^2	$2x$
?	?	16

6.

	x	?
?	x^2	?
?	$-7x$	-21

Factor each quadratic trinomial in Problems 7–18.

7. $n^2 + 10n + 16$

8. $y^2 + 20y + 100$

9. $h^2 + 26h + 48$

10. $w^2 + 17w + 72$

11. $a^2 - 8a + 12$

12. $y^2 - 3y + 2$

13. $t^2 - 15t + 36$

14. $p^2 - 19p + 48$

15. $x^2 - 3x - 10$

16. $y^2 + 2y - 8$

17. $a^2 + 8a - 20$

18. $b^2 - 4b - 12$

Factor each trinomial in Problems 19–30, if possible. If the trinomial cannot be factored, say so.

19. $x^2 - 17x + 30$

20. $x^2 - 18x + 45$

21. $x^2 + 4x + 2$

22. $x^2 + 3x + 1$

23. $y^2 - 44y - 45$

24. $y^2 - 14y - 51$

25. $t^2 - 9t - 20$

26. $u^2 + 8u - 15$

27. $q^2 - 5q - 6$

28. $m^2 + m - 6$

29. $n^2 - 5n + 6$

30. $p^2 + p + 6$

Factor each expression in Problems 31–36 as a product of binomials. (See Problems 53–58 of Homework 4.5 for a hint.)

31. $x^2 - 9$

32. $z^2 - 64$

33. $4 - w^2$

34. $1 - g^2$

35. $-121 + b^2$

36. $-225 + a^2$

Solve each quadratic equation in Problems 37–50.

37. $x^2 + 3x - 10 = 0$

38. $y^2 - 4y - 12 = 0$

39. $t^2 + t = 42$

40. $t^2 + 4t = 21$

41. $2x^2 - 10x = 12$

42. $3y^2 + 6y = 45$

43. $0 = n^2 - 14n + 49$

44. $x^2 + 3x = 0$

45. $5q^2 = 10q$

46. $a^2 - 36 = 0$

47. $x(x - 4) = 21$

48. $z(z + 2) = 3z + 2$

49. $(x - 2)(x + 1) = 4$

50. $(x + 3)^2 = 2x + 14$

In Problems 51–56:
a. **Find the *x*-intercepts and the *y*-intercept of the graph.**
b. **Find the vertex of the graph.**
c. **Sketch the graph.**

51. $y = x^2 + 2x$

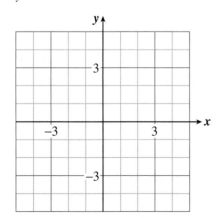

52. $y = x^2 - 6x$

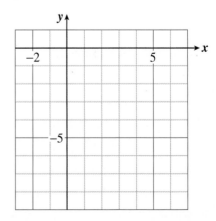

53. $y = x^2 - 2x - 3$

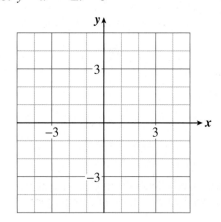

54. $y = x^2 + 4x - 12$

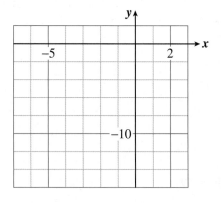

55. $y = 9 - x^2$

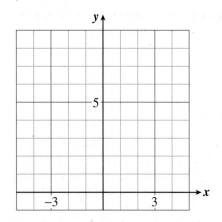

56. $y = 16 - x^2$

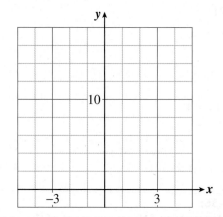

57. An architect is designing rectangular offices, which will be 3 yards longer than they are wide to allow for storage space. Let w represent the width of one office.

    a. Draw a sketch of the floor of an office, and label its dimensions.

    b. Write an equation for the area, A, of one office in terms of its width.

    c. The main office should have an area of 54 square yards. Write an equation to express this requirement.

    d. Solve your equation. You should get two solutions. Which one makes sense for this application?

58. An electronics engineer is designing a computer monitor with a rectangular screen. The size of a screen is often described by giving its diagonal. The length of this screen should be 3 inches greater than its width, w.

    a. Sketch the monitor screen and its diagonal. Label the dimensions of the screen.

    b. Use the Pythagorean theorem to write an equation relating the width of the screen and its diagonal, D.

    c. A particular model must have a diagonal of 15 inches. Write an equation to express this requirement.

    d. Solve your equation. Which of your two solutions makes sense for this problem situation?

59. Icon Industries produces all kinds of electronic equipment. The cost, C, in dollars, of producing a piece of equipment depends on the number of hours, t, it takes to build it, where

$$C = 8t^2 - 32t - 16$$

How many hours does it take to build a transformer that costs $80 to produce?

60. During a softball game, the cleanup batter for the Rockets slugs one deep into the outfield. The height, h, of the ball t seconds after it is hit is given, in feet, by

$$h = -16t^2 + 64t + 4$$

How long does it take the ball to reach a height of 52 feet?

4.9 More About Factoring

Reading

In Section 4.6, we graphed the equation

$$h = -16t^2 + 36t + 360$$

This equation gives the height of Delbert's algebra book above the ground t seconds after being thrown off a cliff. From the graph of the equation, you found that it takes the book 6 seconds to reach the ground. In other words, 6 is one of the solutions of the quadratic equation

$$0 = -16t^2 + 36t + 360$$

Can we solve this equation algebraically, without using a graph?

We have already solved quadratic equations of the form $x^2 + bx + c = 0$ by factoring. However, in this equation, the coefficient of the quadratic term is -16, not 1. In order to factor quadratic trinomials $ax^2 + bx + c$, where a is not 1, we need to look more closely at the products of binomials, represented as areas of rectangles.

A Property of Binomial Products

Consider the product

$$(3t + 2)(t + 3)$$

This product is represented by the following rectangle diagram:

	t	3
$3t$	$3t^2$	$9t$
2	$2t$	6

We see that the product of the two binomials is

$$3t^2 + 9t + 2t + 6 = 3t^2 + 11t + 6$$

However, products of binomials have another interesting property that will help us with factoring. Let's compute the product of the expressions along each diagonal of the rectangle diagram:

$$3t^2 \cdot 6 = 18t^2$$
$$9t \cdot 2t = 18t^2$$

The products of the diagonal entries are equal. This is not surprising when you think about it, because each diagonal product is actually the product of *all four* terms of the binomials—namely, $3t$, 2, t, and 3—just multiplied in a different order. In the figure in Activity 2 on page 491, you can see where the diagonal entries came from in our example:

$$18t^2 = 3t^2 \cdot 6 = 3t \cdot t \cdot 2 \cdot 3$$
$$18t^2 = 9t \cdot 2t = 3t \cdot 3 \cdot 2 \cdot t$$

We'll use this observation about the diagonals as part of our factoring strategy.

> When the product of two binomials is represented by a rectangle diagram, the products of the entries on the two diagonals are equal.

EXERCISE 1

Compute each product using the area of a rectangle, and verify that the products of the diagonal entries are equal.

a. $(2x - 5)(3x - 4)$

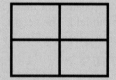

b. $(4t + 15)(t - 6)$

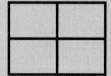

Skills Review

Use the given areas to find the length and width of each rectangle.

1.

$6x^2$	$9x$
$10x$	15

2.

$8t^2$	$-14t$
$-12t$	21

3.

$12m^2$	$-10m$
$30m$	-25

4.

$9a^2$	$21a$
$-21a$	-49

Answers: *1.* $3x + 5, 2x + 3$ *2.* $2t - 3, 4t - 7$ *3.* $2m + 5, 6m - 5$
4. $3a - 7, 3a + 7$

Lesson

We can use rectangle diagrams to help us factor quadratic trinomials. Recall that factoring is the opposite, or reverse, of multiplying, so we must first understand how multiplication works.

Look carefully at the rectangle diagram for the following product:

$$(3x + 4)(x + 2) = 3x^2 + 10x + 8$$

	x	2
$3x$	$3x^2$	$6x$
4	$4x$	8

Note that the quadratic term of the product, $3x^2$, appears in the upper left cell. The constant term of the product, 8, appears in the lower right cell. (The linear term, $10x$, is the sum of the entries in the other two cells.)

Factoring Quadratic Trinomials of the Form $ax^2 + bx + c$

Now, suppose we are asked to factor the quadratic trinomial

$$3x^2 + 10x + 8$$

We'll try to reverse the steps for multiplication. Instead of starting with the factors on the outside of the rectangle diagram, we begin by filling in the cells. We know that the quadratic term, $3x^2$, goes in the upper left, and the constant term, 8, goes in the lower right:

$3x^2$	
	8

What about the other two cells? We know that their sum must be $10x$, but we don't know what expression goes in each. There is nothing in the original trinomial, $3x^2 + 10x + 8$,   to tell us whether to use $x + 9x$,   or $2x + 8x$,   or $3x + 7x$,   or some other combination that gives a sum of $10x$. This is where we apply our observation about binomial products from the reading assignment:

When the product of two binomials is represented by a rectangle diagram, ***the products of the entries on the two diagonals are equal.***

We'll call this diagonal product D. We can easily compute the product of the entries on the diagonal from upper left to lower right:

$$D = 3x^2 \cdot 8 = 24x^2$$

Thus the product of the entries on the other diagonal must also be $24x^2$. We now know two things about those entries:

1. Their product is $24x^2$.

2. Their sum is $10x$.

The easiest way to find the two unknown entries is to list all the ways to factor $D = 24x^2$ into factors px and qx, and then choose the factors whose sum is $10x$. (You may be able to do this in your head.)

Factors of $D = 24x^2$		*Sum of Factors*
x	$24x$	$x + 24x = 25x$
$2x$	$12x$	$2x + 12x = 14x$
$3x$	$8x$	$3x + 8x = 11x$
$4x$	$6x$	$4x + 6x = 10x$

We see that the sum of the last pair of factors, $4x$ and $6x$, is $10x$. These are the expressions we enter in the remaining cells (it doesn't matter which one goes in which spot):

$3x^2$	$4x$
$6x$	8

Finally, we factor each row of the rectangle diagram and write the factors on the outside. Start with the top row, factoring out x and writing the result, $3x + 4,$ at the top. We get the same result when we factor 2 from the bottom row:

	$3x$	4
x	$3x^2$	$4x$
2	$6x$	8

The factors of $3x^2 + 10x + 8$ are thus

$$3x^2 + 10x + 8 = (x + 2)(3x + 4)$$

Here is a summary of this factoring method.

To Factor $ax^2 + bx + c$ Using a Rectangle Diagram

1. Write the quadratic term ax^2 in the upper left cell and the constant term c in the lower right cell.
2. Multiply these two terms to find the diagonal product, D.
3. List all possible factors px and qx of D, and choose the pair whose sum is the linear term, bx, of the quadratic trinomial.
4. Write the factors px and qx in the remaining cells.
5. Factor each row of the rectangle diagram, writing the factors on the outside. These are the factors of the quadratic trinomial.

EXAMPLE 1

Factor $2x^2 - 11x + 15$.

Solution

Step 1 We enter $2x^2$ and 15 on the diagonal of the rectangle diagram:

$2x^2$	
	15

Step 2 We compute the diagonal product:

$$D = 2x^2 \cdot 15 = 30x^2$$

Step 3 We list all possible factors of D, and compute the sum of each pair of factors. (Note that both factors must be negative.)

Factors of $D = 30x^2$		Sum of Factors
$-x$	$-30x$	$-x - 30x = -31x$
$-2x$	$-15x$	$-2x - 15x = -17x$
$-3x$	$-10x$	$-3x - 10x = -13x$
$-5x$	$-6x$	$-5x - 6x = -11x$

The correct factors are $-5x$ and $-6x$.

Step 4 We enter the factors $-5x$ and $-6x$ into the rectangle diagram:

$2x^2$	$-6x$
$-5x$	15

Step 5 We factor $2x$ from the top row of the rectangle diagram, and write the result, $x - 3$,   at the top:

	x	-3
$2x$	$2x^2$	$-6x$
	$-5x$	15

Finally, we factor $x - 3$ from the bottom row and write the result, -5, on the left:

	x	-3
$2x$	$2x^2$	$-6x$
-5	$-5x$	15

The correct factorization is

$$2x^2 - 11x + 15 = (2x - 5)(x - 3)$$

Now try Exercise 2.

EXERCISE 2

Factor $4x^2 + 4x - 3$.

Step 1 Enter the correct terms on the diagonal of a rectangle diagram.

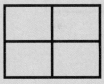

Step 2 Compute the diagonal product, D.

Step 3 List all possible factors of D, and compute the sum of each pair of factors. Note that the factors must have opposite signs. (Complete each pair of factors below.)

Factors of D		Sum of Factors
$-x$	____	____
$-2x$	____	____
$-3x$	____	____

The correct factors are ____ ____

Step 4 Enter the correct factors into the rectangle diagram.

Step 5 Factor the top row of the rectangle diagram, and write the result at the top. Finally, factor the bottom row and write the result on the left. The correct factorization is

$$4x^2 + 4x - 3 = \text{_____}$$

EXERCISE 3

Use a rectangle diagram to factor each quadratic trinomial.

a. $3z^2 + 10z + 3$

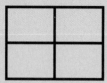

b. $4p^2 - 11p + 6$

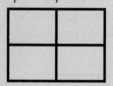

EXERCISE 4

Solve $2x^2 = 7x + 15$.

First, write the equation in standard form. Then factor the left side. (Use the rectangle diagram.) Finally, set each factor equal to zero and solve each equation. (You can verify that both solutions satisfy the original equation.)

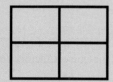

Now try Exercise 3.

Solving Equations

You may want to review the steps given in Section 4.7 for solving a quadratic equation by factoring. Then complete Exercise 4.

We can now use algebra to solve the problem posed in the reading assignment, instead of reading values from a graph.

EXAMPLE 2

The equation

$$h = -16t^2 + 36t + 360$$

gives the height, in feet above the ground, of a book t seconds after it is thrown off a cliff. How long will it take the book to reach the ground?

Solution

We set $h = 0$ to obtain the equation

$$-16t^2 + 36t + 360 = 0$$

It is more convenient if the coefficient of t^2 is positive, so we'll factor out -4:

$$-4(4t^2 - 9t - 90) = 0 \quad \text{Divide both sides by } -4.$$
$$4t^2 - 9t - 90 = 0$$

Now we are ready to factor the left side of the equation. You can verify that the factorization is

$$(4t + 15)(t - 6) = 0$$

Applying the zero factor principle yields

$$4t + 15 = 0 \qquad t - 6 = 0$$
$$t = \frac{-15}{4} \qquad t = 6$$

The solutions are $\frac{-15}{4}$ and 6. Because a negative time does not make sense for the problem, we discard that solution. The book takes 6 seconds to reach the ground.

ANSWERS TO 4.9 EXERCISES

1a. $6x^2 - 23x + 20$; diagonal products $= 120x^2$

1b. $4t^2 - 9t - 90$; diagonal products $= -360t^2$

2. $(2x - 1)(2x + 3)$

3a. $(3z + 1)(z + 3)$

3b. $(4p - 3)(p - 2)$

4. $x = \frac{-3}{2}, x = 5$

HOMEWORK 4.9

Factor each quadratic trinomial in Problems 1–6.

1. $2x^2 + 11x + 5$

2. $7w^2 + 15w + 2$

3. $5t^2 + 7t + 2$

4. $3y^2 + 5y + 2$

5. $3x^2 - 8x + 5$

6. $2x^2 - 7x + 3$

Use a rectangle to factor each quadratic trinomial in Problems 7–18.

7. $2x^2 - 13x + 18$

8. $3x^2 - 22x + 24$

9. $5x^2 + 16x - 16$

10. $4a^2 + 8a + 3$

11. $6h^2 + 7h + 2$

12. $9m^2 - 6m + 1$

13. $9n^2 - 8n - 1$

14. $4p^2 - 4p - 15$

15. $6t^2 - 5t - 25$

16. $6r^2 + 5r - 6$

17. $5x^2 - 14x - 24$

18. $5x^2 - 24x - 36$

Arrange each trinomial in Problems 19–24 in standard form; then factor it.

19. $9x^2 - 8 - 21x$

20. $-3 + 2y^2 + y$

21. $-5 - 2z + 16z^2$

22. $-11z - 5 + 16z^2$

23. $23a + 4a^2 - 6$

24. $15 + 24c^2 - 38c$

Solve each equation in Problems 25–36.

25. $3n^2 - n = 4$

26. $4m^2 + 4m = 3$

27. $11t = 6t^2 + 3$

28. $8t = 12t^2 - 15$

29. $1 = 4y - 4y^2$

30. $1 = 6x - 9x^2$

31. $12z^2 + 26z = 10$

32. $18z^2 + 18z = 8$

33. $y(3y + 4) = 4$

34. $y(2y - 3) = -1$

35. $(2x - 1)(x - 2) = -1$

36. $(2x + 5)(x - 4) = -18$

37. A competitive diver executes a double somersault with a twist from a platform. Her height, h, above the water t seconds after leaving the board is given, in feet, by

$$h = -16t^2 + 8t + 24$$

How long does she have to complete her somersault before entering the water?

38. The cost C, in dollars, of producing a wool rug depends on the time t, in hours, that it takes to weave it and is given by

$$C = 3t^2 - 4t + 100$$

How many hours did it take to weave a rug that costs $120?

In Problems 39–44:
a. Draw and label a sketch to illustrate the problem.
b. Write an equation about the problem situation.
c. Solve your equation and complete the problem.

39. The area of a computer circuit board must be 60 square centimeters. The length of the circuit board should be 2 centimeters shorter than twice its width. Find the dimensions of the circuit board.

40. An artist wants a rectangular canvas whose length is 4 inches less than twice its width and whose area is 240 square inches. What should the dimensions of the canvas be?

41. A paper airplane in the shape of a triangle has an area of 40 square inches. Its base is 11 inches longer than its altitude. Find the base and altitude of the triangle.

42. The base of a triangle is 4 feet shorter than twice its altitude. If the area of the triangle is 48 square feet, find the base and altitude.

43. Steve's boat locker is 2 feet longer than twice its width. Find the dimensions of the locker if the 13-foot mast of Steve's boat fits exactly lying diagonally across the floor of the locker.

44. Nigel's umbrella is 34 inches long. It fits exactly lying diagonally across the bottom of his suitcase. If the length of the suitcase is 2 inches shorter than twice its width, what are its dimensions?

In Problems 45–50:
a. Find the *x*- and *y*-intercepts of each parabola.
b. Find the vertex of each parabola.
c. Graph both parabolas on the same grid and compare them.

45. $y = x^2 - 2x - 15$
$y = x^2 + 2x - 15$

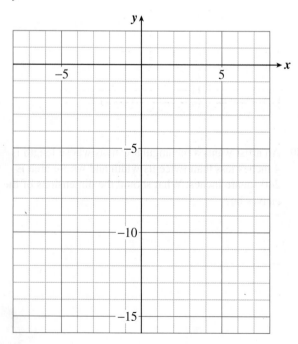

46. $y = x^2 - 8x + 15$
$y = x^2 + 8x + 15$

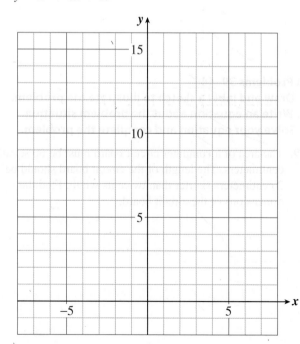

47. $y = x^2 - x - 2$
  $y = -x^2 + x + 2$

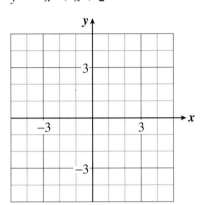

48. $y = 2x^2 - 8$
  $y = 8 - 2x^2$

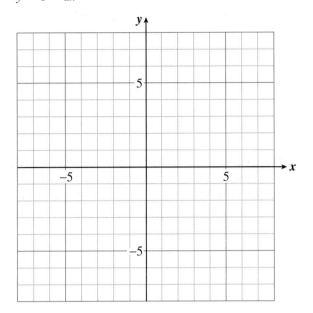

49. $y = 3x^2 - 6x + 3$
  $y = 3x^2 + 6x + 3$

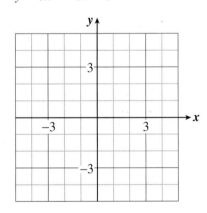

50. $y = x^2 + 1$
  $y = -x^2 - 1$

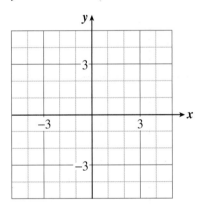

Solve each proportion in Problems 51–54.

51. $\dfrac{p-3}{2} = \dfrac{7}{p+2}$

52. $\dfrac{3}{p+5} = \dfrac{p+5}{3}$

53. $\dfrac{m-4}{2m+1} = \dfrac{m+1}{2m-1}$

54. $\dfrac{3m+2}{m+2} = \dfrac{3m-1}{m-3}$

In Problems 55–64, find the other factor of each quadratic trinomial mentally, without using pencil, paper, or calculator.

55. $x^2 + 15x + 56 = (x+7)\,(\underline{\hspace{2cm}})$

56. $v^2 + 16v + 63 = (v+9)(\underline{\hspace{2cm}})$

57. $b^2 + 8b - 240 = (b-12)(\underline{\hspace{2cm}})$

58. $h^2 + 14h - 480 = (h+30)(\underline{\hspace{2cm}})$

59. $n^2 - 97n - 300 = (n-100)(\underline{\hspace{2cm}})$

60. $a^2 - 42a - 400 = (a-50)(\underline{\hspace{2cm}})$

61. $3u^2 - 17u - 6 = (u-6)(\underline{\hspace{2cm}})$

62. $4t^2 - 13t - 12 = (t-4)(\underline{\hspace{2cm}})$

63. $2t^2 - 21t + 54 = (t-6)(\underline{\hspace{2cm}})$

64. $2z^2 - 27z + 91 = (z-7)(\underline{\hspace{2cm}})$

4.10 The Quadratic Formula

Reading

In this chapter, you have learned techniques for solving a quadratic equation by factoring. You have also seen that the x-intercepts of the graph of

$$y = ax^2 + bx + c$$

are given by the solutions of the equation

$$ax^2 + bx + c = 0$$

Now, let's consider the graph of the equation

$$y = x^2 + 2x - 5$$

By reading the graph, shown in Figure 4.13, we can estimate the x-intercepts at approximately 1.5 and -3.5. However, to find their exact values algebraically, we must solve the equation

$$x^2 + 2x - 5 = 0$$

But, as you will discover when you check, the trinomial $x^2 + 2x - 5$ cannot be factored.

So far, we have used two methods for solving quadratic equations: extracting roots and factoring. Neither of these methods applies to every quadratic equation. In this section, we consider a technique that works on *all* quadratic equations.

Try Exercise 1.

Skills Review

Solve each equation by the easiest method.

1. $x^2 = 36$ 2. $x^2 - 9x = 36$ 3. $x^2 - 9x = 0$
4. $9x^2 - 36 = 0$ 5. $3x^2 - 24x + 36 = 0$ 6. $3x^2 = 31x - 36$

Answers: *1.* $x = 6, x = -6$ *2.* $x = 12, x = -3$ *3.* $x = 0, x = 9$

4. $x = 2, x = -2$ *5.* $x = 2, x = 6$ *6.* $x = \dfrac{4}{3}, x = 9$

Lesson

The graph of the quadratic equation

$$y = ax^2 + bx + c$$

depends on its coefficients, a, b, and c. In particular, the x-intercepts of the graph are determined by the values of these coefficients. In fact, there is a formula that uses the coefficients to calculate the solutions of the equation $ax^2 + bx + c = 0$. It is called the **quadratic formula**.

> ### The Quadratic Formula
>
> The solutions of the equation
>
> $$ax^2 + bx + c = 0, \quad \text{where} \quad a \neq 0$$
>
> are given by the formula
>
> $$x = \frac{-b \pm \sqrt{b^2 - 4ac}}{2a}$$

Figure 4.13

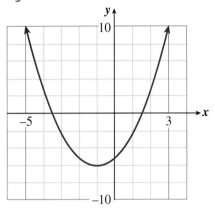

EXERCISE 1

a. Here are equations for three parabolas. Factor each trinomial if possible.

  i. $y = x^2 - 4x + 3$

  ii. $y = x^2 - 4x + 4$

  iii. $y = x^2 - 4x + 5$

b. Match each equation from part (a) with one of the graphs shown. Explain your reasoning.

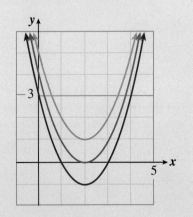

The symbol $\pm$ that appears in the formula is read "plus or minus." It is a shorthand notation used in this context to combine two similar formulas into one. It means that the quadratic equation has two solutions:

$$\frac{-b + \sqrt{b^2 - 4ac}}{2a} \quad \text{and} \quad \frac{-b - \sqrt{b^2 - 4ac}}{2a}$$

(If you would like to know how the quadratic formula is derived, see Appendix B.2, Completing the Square.)

EXAMPLE 1

Use the quadratic formula to solve the equation

$$x^2 + 2x - 5 = 0$$

Solution

We identify the coefficients a, b, and c: a is the coefficient of the quadratic term, b is the coefficient of the linear term, and c is the constant term. For this example,

$$a = 1, \quad b = 2, \quad c = -5$$

We substitute the values of a, b, and c into the quadratic formula and simplify according to the order of operations.

$$\begin{aligned}
x &= \frac{-b \pm \sqrt{b^2 - 4ac}}{2a} \\
&= \frac{-2 \pm \sqrt{2^2 - 4(1)(-5)}}{2(1)} \qquad \text{Simplify under the radical first.} \\
&= \frac{-2 \pm \sqrt{4 + 20}}{2} = \frac{-2 \pm \sqrt{24}}{2}
\end{aligned}$$

The solutions of the equation are thus $\dfrac{-2 + \sqrt{24}}{2}$ and $\dfrac{-2 - \sqrt{24}}{2}$. We can use a calculator to approximate each solution to two decimal places, as follows.

$$\frac{-2 + \sqrt{24}}{2} \approx \frac{-2 + 4.90}{2} = 1.45$$

$$\frac{-2 - \sqrt{24}}{2} \approx \frac{-2 - 4.90}{2} = -3.45$$

These values are very close to our estimates from the graph in Figure 4.13.

Using the Quadratic Formula

You should write a quadratic equation in standard form in order to identify the coefficients a, b, and c. Of course, if a quadratic equation can be solved by factoring, the quadratic formula will give the same solutions. You can use whichever technique seems faster for a particular equation.

EXAMPLE 2

Solve the equation $\quad 3x^2 - x = 4$.

Solution

First write the equation in standard form:

$$3x^2 - x - 4 = 0$$

Thus

$$a = 3, \qquad b = -1, \qquad c = -4$$

Use the quadratic formula to find:

$$x = \frac{-b \pm \sqrt{b^2 - 4ac}}{2a} = \frac{-(-1) \pm \sqrt{(-1)^2 - 4(3)(-4)}}{2(3)}$$

$$= \frac{1 \pm \sqrt{1 + 48}}{6} = \frac{1 \pm \sqrt{49}}{6}$$

$$= \frac{1 \pm 7}{6}$$

This last expression is equivalent to two values:

$$x = \frac{1 + 7}{6} = \frac{8}{6} = \frac{4}{3}$$

$$x = \frac{1 - 7}{6} = \frac{-6}{6} = -1$$

The solutions are $\frac{4}{3}$ and -1.

The equation can also be solved by factoring:

$$(3x - 4)(x + 1) = 0$$

$$3x - 4 = 0 \qquad x + 1 = 0$$

$$x = \frac{4}{3} \qquad\qquad x = -1$$

We obtain the same solutions as before, $\frac{4}{3}$ and -1.

Now try Exercise 2.

If some or all of the coefficients are fractions, it is usually easier to clear the entire equation of fractions before applying the quadratic formula. Recall that this can be done by multiplying each term of the equation by the lowest common denominator, or LCD, of all the fractions involved. Try Exercise 3.

Nonreal Solutions

The graph of

$$y = x^2 - 4x + 5$$

has no *x*-intercepts, as you can see in Figure 4.14. Nonetheless, we can use the quadratic formula to solve the equation

$$x^2 - 4x + 5 = 0$$

Applying the formula with $a = 1$, $b = -4$, and $c = 5$, we find

$$x = \frac{-b \pm \sqrt{b^2 - 4ac}}{2a} = \frac{-(-4) \pm \sqrt{(-4)^2 - 4(1)(5)}}{2(1)}$$

$$= \frac{4 \pm \sqrt{16 - 20}}{2} = \frac{4 \pm \sqrt{-4}}{2}$$

EXERCISE 2

a. Use the quadratic formula to solve $2x^2 = 7 - 4x$.

b. Find decimal approximations to two decimal places for the solutions.

EXERCISE 3

a. Solve $x = \frac{2}{3} - \frac{x^2}{6}$. (Clear the fractions first.)

b. Find approximations to two decimal places for the solutions.

Figure 4.14

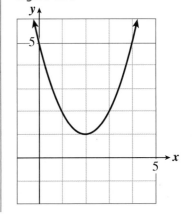

EXERCISE 4

Give as much information as you can about the x-intercepts of each parabola.

a. $y = x^2 + 4$

b. $y = 3x^2 + 4x - 2$

c. $y = -x^2 + 2x - 1$

d. $y = 2x^2 + 4x + 3$

We cannot continue with the calculation because $\sqrt{-4}$ is undefined; it is not equal to any real number. This particular quadratic equation does not have any real-valued solutions, and that is why the graph of $y = x^2 - 4x + 5$ does not have any x-intercepts.

The solutions to the equation above are called *complex numbers*. You can learn more about complex numbers in Appendix B.5. For now, it is enough to observe that if a quadratic equation does not have real-valued solutions, the corresponding parabola does not have x-intercepts.

Every quadratic equation has two solutions. The nature of those solutions determines the nature of the x-intercepts of the graph. There are three possibilities:

1. Both solutions are real numbers and unequal. The graph has two x-intercepts.

2. The solutions are real and equal. The graph has one x-intercept, which is also its vertex.

3. Both solutions are nonreal complex numbers. The graph has no x-intercepts.

Now try Exercise 4.

ANSWERS TO 4.10 EXERCISES

1a. i: $(x - 3)(x - 1)$; ii: $(x - 2)^2$; iii: Cannot be factored

1b. From top to bottom: iii, ii, i

2a. $x = \dfrac{-4 \pm \sqrt{72}}{4}$

2b. $x = 1.12, x = -3.12$

3a. $x = \dfrac{-6 \pm \sqrt{52}}{2}$

3b. $x = 0.61, x = -6.61$

4a. There are no x-intercepts.

4b. There are two x-intercepts: $x = \dfrac{-4 \pm \sqrt{40}}{6}$ or 0.39 and -1.72.

4c. There is one x-intercept: $x = 1$.

4d. There are no x-intercepts.

HOMEWORK 4.10 www

In Problems 1–8:
a. Solve the equation by using the quadratic formula.
b. Give approximate values for your solutions, rounded to the nearest hundredth.

1. $x^2 - 8x + 4 = 0$

2. $x^2 + 2x - 2 = 0$

3. $3s^2 + 2s = 2$

4. $9r^2 = 2 - 6r$

5. $n^2 = n + 1$

6. $p^2 + 1 = 3p$

7. $-4z^2 + 2z + 1 = 0$

8. $-2w^2 + 6w - 3 = 0$

Solve each equation in Problems 9–14 two ways:
a. Use the quadratic formula.
b. Use either factoring or extraction of roots. Round your answers to the nearest hundredth, if necessary.

9. $3t^2 - 5 = 0$

10. $2q^2 - 7 = 0$

11. $z = 3z^2$

12. $7v^2 = 5v$

13. $2w^2 + 6 = 7w$

14. $3y^2 + y = 2$

Solve Problems 15–22.

15. $\dfrac{x^2}{6} + x = \dfrac{2}{3}$

16. $2m^2 + \dfrac{7}{3}m = 1$

17. $v^2 + 3v - 2 = 9v^2 - 12v + 5$

18. $3s^2 + 2s = \dfrac{3}{4}s^2 + \dfrac{s}{2} + 2$

19. $m^2 - 3m = \dfrac{1}{3}(m^2 - 1)$

20. $\dfrac{1}{2}(t + 1)(t - 1) = \dfrac{1}{3}t(t + 1)$

21. $-0.2x^2 + 3.6x - 9 = 0$

22. $0.1x^2 - 2.5x - 6.5 = 0$

For Problems 23–28:
a. Find the *x*-intercepts of each parabola. (Round your answers to the nearest hundredth.)
b. Sketch the graph.

23. $y = x^2 - 2x - 2$

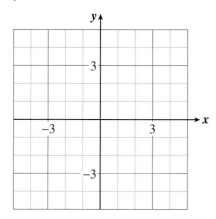

24. $y = 1 - 6x - 3x^2$

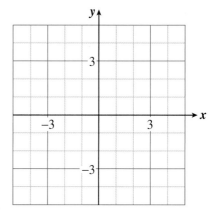

25. $y = -3x^2 + 2x - 1$

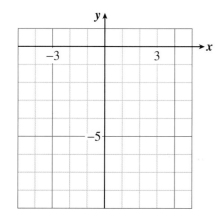

26. $y = 3 + 2x + x^2$

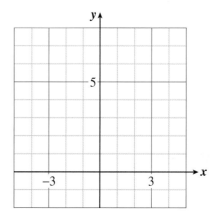

27. $y = x^2 - 4x - 1$

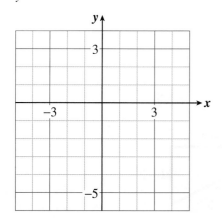

28. $y = -x^2 - 4x + 3$

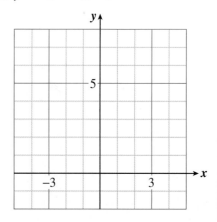

29. Delbert throws a penny from the top of the Texas Building in Fort Worth. After t seconds, the height of the penny, in feet, is given by

$$h = -16t^2 + 8t + 380$$

   a. When does the penny pass a window 300 feet above the ground?

   b. How long does the penny take to reach the ground?

30. Francine throws a diamond engagement ring from the top of the First National Bank in Louisville. After t seconds, the height of the ring, in feet, is given by

$$h = -16t^2 - 4t + 512$$

   a. When does the ring pass a balcony 400 feet above the ground?

   b. How long does the ring take to reach the ground?

31. The volume of a cedar chest is 12,000 cubic inches. The chest is 20 inches high, and its length is 5 inches less than three times its width. Find the dimensions of the chest.

32. The volume of a large aquarium at a zoo is 2160 cubic feet. The tank is 10 feet wide, and its length is 6 feet less than twice its height. Find the dimensions of the aquarium.

33. The perimeter of a rectangle is 42 inches, and its diagonal is 15 inches. Find the dimensions of the rectangle.

34. The perimeter of a rectangle is 46 centimeters, and its diagonal is 17 centimeters. Find the dimensions of the rectangle.

Find the unknown sides of each right triangle in Problems 35–38.

35.

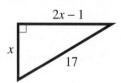

36.

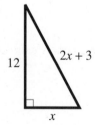

37.

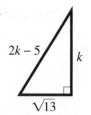

38.

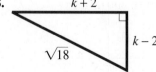

CHAPTER SUMMARY AND REVIEW

Glossary: Write a sentence explaining the meaning of each term.

exponent	principal square root	rational number	monomial	axis of symmetry
base	radical	irrational number	binomial	factoring
power	radicand	real numbers	trinomial	common factor
volume	quadratic equation	hypotenuse	parabola	quadratic formula
surface area	extraction of roots	cube root	vertex	

Properties and formulas

1. State the Pythagorean theorem, and explain what it means.

2. State the zero factor principle, and explain what it means.

3. State a property of binomial products that is useful for factoring.

4. State the quadratic formula. What is it used for?

Techniques and procedures

5. State the order of operations, including powers and roots.

6. What is the standard form of a quadratic equation?

7. Explain how to solve a quadratic equation by extracting roots.

8. How can you tell from its decimal form whether a number is rational or irrational?

9. Explain how to multiply two binomials.

10. What steps should you use to solve a quadratic equation by factoring?

11. If you know the *x*-intercepts of a parabola, how can you find the coordinates of its vertex?

12. Explain how to factor a quadratic trinomial.

Simplify each expression in Problems 13–16.

13. $4 - 2 \cdot 3^2$

14. $-2 - 3(-3)^3 - 2$

15. $\dfrac{-6 - \sqrt{6^2 - 4(2)(4)}}{2(2)}$

16. $18 - 2\sqrt[3]{\dfrac{4}{3}(48)}$

Evaluate the expressions in Problems 17 and 18 for the given values of the variables, and simplify.

17. For $x = -4$:
  a. $3 - x^2$

  b. $3(-x)^2$

18. For $a = 6, \quad b = -2$:
  a. $\dfrac{3b^3}{6} - \dfrac{4 - a^2}{8}$

  b. $\dfrac{(a - b)^2}{ab^2}$

19. Explain the difference between a *coefficient* and an *exponent*. Use the expressions $2x^3$ and $3x^2$ to illustrate your explanation.

20. Is the statement $\sqrt{81} = \sqrt{9} = 3$ correct? Explain why or why not.

Simplify each expression in Problems 21 and 22, if possible.

21. a. $6t^2 - 8t^2$ **b.** $6t - 8t^2$ **c.** $6t(-8t^2)$

22. a. $w^2 + w^2$ **b.** $-w - w$ **c.** $w^2 - w$

Give a decimal approximation, rounded to the nearest thousandth, for each expression in Problems 23 and 24.

23. $-8 - 5\sqrt{6}$

24. $\dfrac{3 + \sqrt[3]{3}}{3}$

Write expressions for the volume and surface area of each figure in Problems 25 and 26.

25.

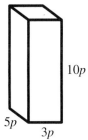

$10p$

$5p$ $3p$

26.

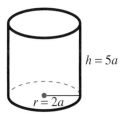

$h = 5a$

$r = 2a$

Solve the equations in Problems 27 and 28 by extraction of roots.

27. $9k^2 + 21 = 25$

28. $6a^2 + 3 = 4a^2 + 19$

29. a. The volume of a spherical communications satellite is 65.45 cubic feet. What is the radius of the satellite?

 b. What is the surface area of the satellite?

30. a. Francine's coffee cup is a cylinder 10 centimeters tall with a radius of 3.36 centimeters. What is its volume in cubic centimeters?

 b. One fluid ounce is about 29.56 cubic centimeters. Find the volume of Francine's coffee cup in ounces, rounded to the nearest ounce.

 c. Delbert's coffee cup is also 10 centimeters tall, but its volume is 24 fluid ounces. What is its radius?

In Problems 31–34, solve each formula for the indicated variable.

31. $S = \dfrac{n}{2}(a + f)$ for a

32. $s = vt + \dfrac{1}{2}at^2$ for a

33. $C = bh^2r$ for h

34. $G = \dfrac{np}{r^2}$ for r

In Problems 35–38:
a. Decide whether each number is rational or irrational.
b. Find a decimal equivalent for each number. If the number is irrational, round the decimal equivalent to the nearest hundredth.

35. $\sqrt{300}$

36. $\sqrt[3]{512}$

37. $5 + \sqrt[3]{15}$

38. $\dfrac{7}{\sqrt{81}}$

39. If you are standing at an altitude of h meters, the distance, d, in miles, that you can see to the horizon is given by

$$d = \sqrt{12h}$$

a. Complete the table and graph the equation.

h	d
0	
1000	
2000	
3000	
4000	
5000	

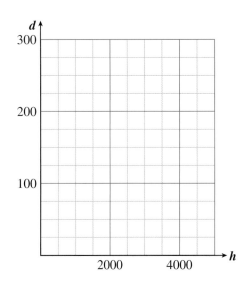

b. Use your graph to estimate the distance you can see from the top of Mt. Whitney, at an altitude of 4,149 meters.

c. Use the equation to find the distance in part (b) algebraically.

d. If you want to see for 100 miles, at what altitude do you need to be?

40. a. Complete the table of values for the equation $y = \sqrt{x-1}$.

x	−1	0	1	2	3	4	5
y							

b. Graph the equation.

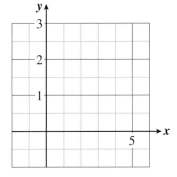

Simplify the expressions in Problems 41–44.

41. $(\sqrt{3x})(\sqrt{3x})$

42. $3\sqrt{x}(3\sqrt{x})$

43. $(5\sqrt{a})^2$

44. $(\sqrt[3]{B})^2(\sqrt[3]{B})$

45. A rectangle is 7 meters wide and 9 meters long. What is the length of its diagonal?

46. The size of a TV screen is the length of its diagonal. What is the size of a TV screen whose sides measure 12 inches by 16 inches?

47. A 26-meter guy wire is attached to a radio antenna for support. One end of the wire is attached to the antenna 24 meters above the ground, and the other is attached to an iron ring in a cement slab on the ground. How far is the ring from the base of the antenna?

48. Find the height of an equilateral triangle whose sides are each 8 centimeters long.

In Problems 49 and 50, multiply the given factors.

49. $3xy(2x - 4 - y)$

50. $-2a(-a^2 - 2a + 4)$

Simplify each expression in Problems 51 and 52.

51. $5a(2a - 3) - 4(3a^2 - 2) + 6a$

52. $2x(x - 3y) - 3y(x - 3y)$

In Problems 53–58, multiply the given factors.

53. $(u - 5)(u - 2)$

54. $(3r + 2)(r - 4)$

55. $(a + 6)^2$

56. $(6y + 5)(6y - 5)$

57. $(2a - 5c)(3a + 2c)$

58. $-3(x - 4)(2x + 5)$

59. If you drop a stone from a bridge 100 feet above the water, the height h of the stone t seconds after you drop it is given in feet by

$$h = 100 - 16t^2$$

a. Complete the table.

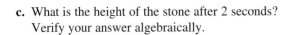

t	0	0.5	1	1.25	1.5	1.75	2	2.25	2.5
h									

b. Sketch the graph of the equation.

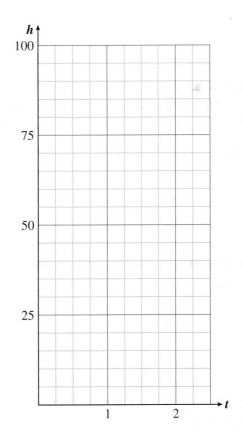

c. What is the height of the stone after 2 seconds? Verify your answer algebraically.

d. When is the stone at a height of 75 feet? Write an algebraic equation you can solve to verify your answer.

60. Sportsworld sells $180 - 3p$ pairs of its name-brand running shoes per week when the shoes are priced at p dollars per pair.

  a. Write an equation for Sportsworld's revenue, R, in terms of p.

  b. Fill in the table, and graph your equation.

p	0	10	20	30	40	50	60
R							

  c. At what price(s) will Sportsworld's revenue be zero?

  d. At what price will Sportsworld's revenue be maximum? What is the maximum revenue the company can expect?

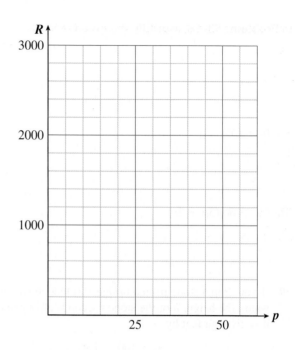

In Problems 61 and 62, graph each pair of parabolas on the same grid.

61. $y = x^2 - 1$
$y = 1 - x^2$

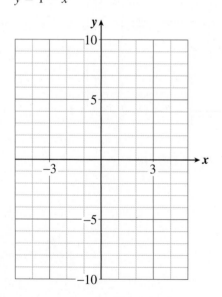

62. $y = (x - 2)^2$
$y = (x + 2)^2$

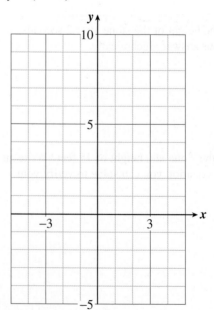

Factor each expression in Problems 63–74.

63. $24x^2 - 18x$

64. $-32y^2 + 24y$

65. $a^2 - 18a + 45$

66. $x^2 - 14x - 51$

67. $4y^2 + 16y + 15$

68. $8b^2 - 18b + 9$

69. $14w - 5 + 3w^2$

70. $-3 + 2p^2 - 5p$

71. $z^2 - 121$

72. $81 - 4t^2$

73. $6x^2 + 21x + 9$

74. $8y^2 - 6y - 2$

Solve each equation in Problems 75–80.

75. $0 = m^2 + 10m + 25$

76. $b^2 - 25 = 0$

77. $4p^2 = 16p$

78. $11t = 6t^2 + 3$

79. $(x - 5)(x + 1) = -8$

80. $2q(3q - 1) = 4$

In Problems 81 and 82:
a. Find the *x*- and *y*-intercepts of the graph of the given equations.
b. Find the vertex of the graph.
c. Sketch the graph.

81. $y = x^2 + 6x$

82. $y = x^2 + 3x - 4$

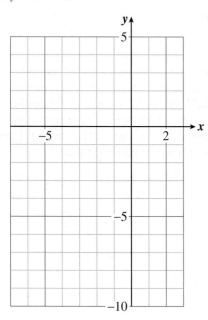

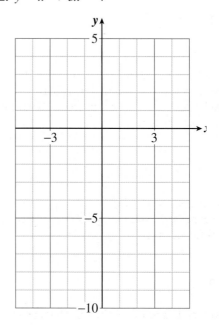

Solve each equation in Problems 83 and 84 by using the quadratic formula.

83. $2t^2 + 6t + 3 = 0$

84. $\dfrac{x^2}{4} + 1 = \dfrac{13}{12}x$

85. Audrey launches her experimental hydraulic rocket from the top of her apartment building. The height of the rocket after t seconds is given, in feet, by

$$h = -16t^2 + 40t + 80$$

How long does the test flight last before the rocket hits the ground?

86. The formula $S = \dfrac{1}{2}n^2 + \dfrac{1}{2}n$ gives the sum of the first n counting numbers. How many counting numbers must you add together to get a sum of 325?

87. The Corner Market sells $160 - 2p$ pounds of bananas per week at a price of p cents per pound.
  a. Write an expression for the Corner Market's weekly revenue, R, from bananas.

  b. It costs the market $80 + 24p$ cents to buy and display the bananas. Write an expression for the market's profit, M, from selling the bananas.

  c. If the market made a profit of $22 (or 2200 cents) on bananas last week, how much did it charge per pound?

88. The town of Amory lies due north of Chester, and Bristol lies due east of Chester. If you drive from Amory to Bristol by way of Chester, the distance is 17 miles, but if you take the back road directly from Amory to Bristol, you save 4 miles. How far is it from Bristol to Chester? (*Hint:* Make a sketch.)

Find the sides of each right triangle in Problems 89 and 90.

89.

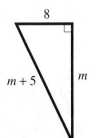

90.

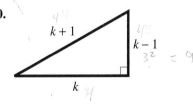

Polynomials and Rational Expressions

5.1 Polynomials

Reading

The linear and quadratic expressions we have studied are examples of *polynomials*. You'll recall that the words *monomial*, *binomial*, and *trinomial* refer to the number of terms in an expression; similarly, a **polynomial** is an expression with several terms. (There are no special names for polynomials with more than three terms.) Each term is a power of a variable with a constant coefficient. The exponent must be a whole number; that is, the expression has no radicals containing variables and no variables in the denominators of fractions. Here are some examples of polynomials:

$$4x^3 + 2x^2 - 7x - 5 \qquad \frac{1}{2}at^2 + vt$$

$$2a^2 - 6ab + 3b^2 \qquad \pi r^2 h$$

Try Exercise 1.

EXERCISE 1

Which of the following expressions are polynomials?

a. $t^2 - 7t + 3$

b. $2 + \dfrac{6}{x}$

c. $-a^3 b^2 c^4$

d. $3w^2 - w + 5\sqrt{w}$

Degree

In a term containing only one variable, the exponent of the variable is called the **degree** of the term. (The degree of a constant term is zero, for reasons that will become clear in Chapter 6.) For example,

$$3x^2 \quad \text{is of second degree}$$
$$8y^3 \quad \text{is of third degree}$$
$$4x \quad \text{is of first degree (the exponent on } x \text{ is 1)}$$
$$5 \quad \text{is of zero degree}$$

The degree of a polynomial in one variable is the largest exponent that appears in any term. For example,

$$2x + 1 \qquad \text{is of first degree}$$
$$3y^2 - 2y + 2 \qquad \text{is of second degree}$$
$$m - 2m^2 + m^5 \qquad \text{is of fifth degree}$$

Try Exercise 2.

Polynomials in one variable are usually written in **descending powers** of the variable. This means that the term with the largest exponent comes first, then the term with the next highest exponent, and so on down to the constant term, if there is one. For example, the following polynomial is written in descending powers of x:

Exponents decrease from left to right.

$$3x^4 - 2x^2 + x - 5$$

Constant term is written last.

Now try Exercise 3.

Skills Review

Replace the comma in each pair by the proper symbol: >, <, or =.

1. $(-2)^3, -2^3$ 2. $-5^4, (-5)^4$ 3. $7 - 3^2, 4^2$
4. $2 + 4^3, 6^3$ 5. $4 \cdot 2^3, 8^3$ 6. $2 \cdot 5^2, 10^2$
7. $\dfrac{1}{2}, \left(\dfrac{1}{2}\right)^2$ 8. $\dfrac{3}{4}, \left(\dfrac{3}{4}\right)^2$

Answers: 1. = 2. < 3. < 4. < 5. < 6. < 7. > 8. >

Lesson

Evaluating Polynomials

We evaluate polynomials in the same way we evaluate any other algebraic expression: by substituting the given values for the variables and then following the order of operations to simplify.

EXAMPLE 1

Evaluate $16t^3 - 6t + 20$ for $t = \dfrac{3}{2}$

Solution

Substitute $\dfrac{3}{2}$ for t, and simplify.

$$16\left(\frac{3}{2}\right)^3 - 6\left(\frac{3}{2}\right) + 20 \qquad \text{Compute the power.}$$

$$= 16\left(\frac{27}{8}\right) - 6\left(\frac{3}{2}\right) + 20 \quad \text{Perform all multiplications.}$$

$$= 54 - 9 + 20 \qquad\qquad \text{Add.}$$

$$= 65$$

Now try Exercise 4.

Applications of Polynomials

Polynomials occur in a variety of applications.

EXAMPLE 2

Most banks give compound interest on savings accounts. If you invest P dollars in an account that earns interest rate r compounded annually, then after 3 years your account balance, B, will be given by

$$B = P(r^3 + 3r^2 + 3r + 1)$$

Find the balance after 3 years if \$600 is invested in an account that earns 5% interest compounded annually.

Solution

We evaluate the given polynomial for $P = 600$ and $r = 0.05$.

$$B = 600[(0.05)^3 + 3(0.05)^2 + 3(0.05) + 1]$$
$$= 600[0.000125 + 0.0075 + 0.15 + 1]$$
$$= 600[1.157625] = 694.575$$

Rounding to the nearest penny, we see that the balance in the account after 3 years is \$694.58.

Like Terms

We can add two polynomials together or subtract one polynomial from another. When we do this, the result is a new polynomial. To add or subtract polynomials, we must be able to add and subtract like terms.

In Section 2.8, you learned that *like terms* are any terms that are exactly alike in their variable factors. The exponents on the variable factors must also match. For example, the following are not like terms because their variable factors are different:

$$x^3 \qquad \text{and} \qquad 5x^2$$
$$2x^2y \qquad \text{and} \qquad 2xy^2$$

However, the following *are* like terms, because x^2y and yx^2 are equivalent expressions:

$$\frac{1}{2}x^2y \qquad \text{and} \qquad -3yx^2$$

EXERCISE 4
Evaluate $2x^3y^2 - 3xy^3 - 3$ for $x = -2$, $y = -3$

Which (if either) of the following two expressions can be simplified?

$$3x^2 + 5x^3 \qquad \text{and} \qquad 3x^2 + 5x^2$$

In the first expression, the two terms have different exponents, even though the base, x, is the same. Powers of the same variable with different exponents are not like terms. The first expression, $3x^2 + 5x^3$, cannot be simplified. In the second expression, $3x^2$ and $5x^2$ are like terms. Recall that to add or subtract like terms, we add or subtract their numerical coefficients. The variable factors of the terms remain unchanged. Thus

$$3x^2 + 5x^2 = 8x^2$$

Note that in the sum, we do *not* change the exponent on x; it is still 2.

Try Exercise 5 now.

Adding Polynomials

To add two polynomials, we need only remove parentheses and combine like terms.

EXAMPLE 3

$$\begin{aligned}
&(4a^3 - 2a^2 - 3a + 1) + (2a^3 + 4a - 5) && \text{Remove parentheses.}\\
&= 4a^3 - 2a^2 - 3a + 1 + 2a^3 + 4a - 5 && \text{Combine like terms.}\\
&= 6a^3 - 2a^2 + a - 4
\end{aligned}$$

Try Exercise 6.

We can also use a vertical format to add polynomials.

EXAMPLE 4

Add $4x^2 + 2 - 7x$ and $5x - 5 + 2x^2$.

Solution

First, we write each polynomial in descending powers of x. Then we write the second polynomial beneath the first, aligning like terms.

$$\begin{array}{r}
4x^2 - 7x + 2 \quad \text{Combine like terms vertically.}\\
+\ 2x^2 + 5x - 5 \\
\hline
6x^2 - 2x - 3
\end{array}$$

Subtracting Polynomials

In Section 2.8, you learned that if an expression in parentheses is preceded by a minus sign, you must change the sign of each term within parentheses when you remove the parentheses. This rule applies to subtracting polynomials as well.

EXAMPLE 5

$$(4x^2 + 2x - 5) - (2x^2 - 3x - 2)$$

Each sign is changed.
Combine like terms.

$$\begin{aligned}
&= 4x^2 + 2x - 5 - 2x^2 + 3x + 2\\
&= 2x^2 + 5x - 3
\end{aligned}$$

Now try Exercise 7.

To use a vertical format for subtraction, we change the sign of *each term* in the polynomial on the bottom (the one being subtracted).

EXAMPLE 6

Subtract $2n^2 - 3n - 2$ from $4n^2 + 2n - 5$.

Solution

$$
\begin{array}{l}
4n^2 + 2n - 5 \\
\underline{- (2n^2 - 3n - 2)}
\end{array}
\quad \longrightarrow \quad
\begin{array}{l}
\text{Change sign} \\
\text{of each term.}
\end{array}
\quad \longrightarrow \quad
\text{(Add.)}
\begin{array}{l}
4n^2 + 2n - 5 \\
\underline{- 2n^2 + 3n + 2} \\
2n^2 + 5n - 3
\end{array}
$$

Try Exercise 8.

Applied problems may involve addition or subtraction of polynomials.

EXAMPLE 7

It costs The Cookie Company $200 + 2x$ dollars to produce x bags of cookies per week, and the company earns $8x - 0.01x^2$ dollars from the sale of x bags of cookies.

a. Write a polynomial for the profit earned by The Cookie Company on x bags of cookies.

b. Find the company's profit on 300 bags and on 600 bags of cookies.

Solution

a. Applying the formula   ***profit = revenue − cost***   to this situation, we subtract polynomials to find

$$\text{profit} = (8x - 0.01x^2) - (200 + 2x) \quad \text{Change signs of second polynomial.}$$
$$= 8x - 0.01x^2 - 200 - 2x \quad \text{Combine like terms.}$$
$$= -0.01x^2 + 6x - 200$$

b. Evaluate the profit polynomial for $x = 300$ and $x = 600$.   For 300 bags of cookies, the profit is

$$-0.01(300)^2 + 6(300) - 200 = -900 + 1800 - 200 = 700$$

or \$700. For 600 bags of cookies, the profit is

$$-0.01(600)^2 + 6(600) - 200 = -3600 + 3600 - 200 = -200$$

The company loses \$200 if it produces 600 bags of cookies.

EXERCISE 7
Subtract the polynomials.
a. $(y^2 - 3y + 5) - (y^2 + 4y - 3)$

b. $(2m^4 - 4m^3 + 3m - 1) - (m^3 - 4m^2 - 6m + 2)$

EXERCISE 8
Use a vertical format to add or subtract the polynomials.
a. Add $5y^2 - 3y + 1$ to $-3y^2 + 6y - 7$.

b. Subtract $2x^2 + 5 - 2x$ from $7 - 3x - 4x^2$.

ANSWERS TO 5.1 EXERCISES

1. a and c **2a.**   Degree 9   **2b.**   Degree 4

3. $-x^6 - 6x^3 + 3x - 8$ **4.** -309

5a. $-x^3 + 6x^2 + 4x$ **5b.** $6a^2b - 2ab$

6a. $9x^2 - 2x - 2$ **6b.** $2m^4 - 3m^3 + 4m^2 - 3m + 1$

7a. $-7y + 8$ **7b.** $2m^4 - 5m^3 + 4m^2 + 9m - 3$

8a. $2y^2 + 3y - 6$ **8b.** $-6x^2 - x + 2$

HOMEWORK 5.1

Which of the features listed in Problems 1–4 are not allowed in a polynomial?

1. More than three terms

2. Coefficients that are fractions

3. Division by a variable

4. A term without variables

5. Explain the difference between a polynomial of degree 3 and a trinomial. Give examples.

6. What does it mean to write a polynomial in descending powers of the variable?

Give an example of each type of polynomial listed in Problems 7–10.

7. A monomial of fourth degree

8. A binomial of first degree

9. A trinomial of degree 2

10. A monomial of degree 0

Which of the expressions in Problems 11 and 12 are polynomials? If an expression is not a polynomial, explain why not.

11. a. $5x^4 - 3x^2$

b. $3x + 1 + \dfrac{2}{x^2}$

c. $\dfrac{1}{2a^2 + 5a - 6}$

d. $\dfrac{2}{3}t^2 + \dfrac{1}{4}t^3 + \dfrac{5}{8}$

12. a. $27y^8$

b. $2a^2 - 6ab + 3b^2$

c. $\dfrac{3}{x^4} - \dfrac{7}{x^3} + \dfrac{5}{3}$

d. $9z^9 - \dfrac{1}{2}z^2 + 8z^6$

Give the degree of each polynomial in Problems 13–18.

13. $x^2 + 4x - \dfrac{1}{4}$

14. $-\dfrac{3}{4}y^5 - y + 2y^2$

15. $y - 2.8y^7$

16. $2.2 - 2x - 2x^2 - 2x^3$

17. $\dfrac{z}{4} - 3z^4 + 4z^3$

18. $\dfrac{5}{2}z^4$

Write each polynomial in Problems 19–22 in descending powers of x.

19. $x - 1.9x^3 + 6.4$

20. $3x^2 - 5x^4 + 2 - x^3$

21. $6xy - 2x^2 + 2y^3$

22. $4xy^2 + 4x^2y - 1$

Evaluate each polynomial in Problems 23–28 for the given values of the variables.

23. $2 - z^2 - 2z^3$, for $z = -2$

24. $t^3 + 6t^2 - 4t + 1$, for $t = \dfrac{3}{2}$

25. $2a^4 + 3a^2 - 3a$, for $a = 1.6$

26. $-x^2 - \dfrac{4}{3}xy + \dfrac{3}{2}y^2$, for $x = 3$, $y = -2$

27. $-abc^2$,   for $a = -3$, $b = 2$, $c = 2$

28. $0.3a^2b - 0.2b^2 + d^3$,   for $a = -6$, $b = -2$, $d = -4$

29. Elizabeth invests $1000 in an account that pays 7% interest compounded annually. How much money will be in the account 3 years from now? (See Example 2.)

30. Suppose you want to choose three items from a list of n possible items. The number of different ways you can make your choice is given by the polynomial

$$\frac{1}{6}n^3 - \frac{1}{2}n^2 + \frac{1}{3}n$$

a. How many ways can you pick three elective courses from a list of eight approved courses?

b. How many ways can you pick 3 cards from a deck of 52?

c. How many different 3-person committees can be picked from the 20 members of a club?

31. After its brakes are applied, a small car traveling at s miles per hour can stop in approximately $0.04s^2 + 0.6s$ feet. When traveling at 50 miles per hour on the freeway, will the car be able to avoid hitting a stalled car in the same lane 100 feet ahead?

32. On a wet road surface, a small car traveling at s miles per hour needs approximately

$$0.08s^2 + 0.6s$$

feet to stop after the brakes are applied. How much following distance should the driver allow when traveling at 35 miles per hour on a rainy day?

Evaluate each polynomial in Problems 33–36 for *n* = 10. Try to do the calculations mentally. (What do you notice?)

33. $5n^2 + 6n + 7$

34. $5n^3 + n^2 + 3n + 3$

35. $n^3 + 1$

36. $8n^4 + 8n$

Expand each expression in Problems 37–40 by removing parentheses and brackets. (What do you notice?)

37. $x[x(x + 3) + 4] + 1$

38. $[(x - 2)x + 6]x - 5$

39. $x(x[x(x - 7) - 5] + 8) - 3$

40. $([(x - 2)x]x + 2)x - 9$

41. a. Evaluate the expression in Problem 37 for $x = 2$. Can you do this mentally? Is it easier to evaluate the expression before or after expanding it?

b. Use a calculator to evaluate the expression for $x = 0.8$.

42. a. Evaluate the expression in Problem 38 for $x = 3$. Can you do this mentally? Is it easier to evaluate the expression before or after expanding it?

b. Use a calculator to evaluate the expression for $x = 0.6$.

Simplify the expressions in Problems 43–46 by combining like terms.

43. $6b^3 - 2b^3 - (-8b^3)$

44. $-7z^5 - (-z^5) - 3z^5$

45. $6x - 3y + 5xy - 6y + xy$

46. $4ab^2 + 3a^2b - 2a^2b - 5ab^2$

Explain why each calculation in Problems 47–52 is *incorrect*, and give the correct answer.

47. $6w^3 + 8w^3 \rightarrow 14w^6$

48. $8c^5 - 5c^2 \rightarrow 3c^3$

49. $6 + 3x^2 \rightarrow 9x^2$

50. $8 - 2z^3 \rightarrow 6z^3$

51. $4t^2 + 7 - (3t^2 - 5) \rightarrow t^2 + 2$

52. $5b^3 + 2b - (b^3 + b) \rightarrow 4b^3 + 3b$

Add or subtract the polynomials in Problems 53–56.

53. $(2y^3 - 4y^2 - y) + (6y^2 + 2y + 1)$

54. $(z^2 - 4z + 1) - (2z^2 + z - 1)$

55. $(5x^3 + 3x^2 - 4x + 8) - (2x^3 - 4x - 3)$

56. $(2ab^2 + 6ab - a - 3b^2) - (2ab + a + 3b^2 + 2)$

Use a vertical format to add or subtract the polynomials in Problems 57–60.

57. Add $8x^2 - 3x + 4$ to $-2x^2 + 5x - 7$.

58. Add $y^3 - 4y^2 - y + 6$ to $2y + y^3 - 3 - 3y^2$.

59. Subtract $4x^2 - 3x - 1$ from $-3x^2 + 4x - 2$.

60. Subtract $1 - 2y - 2y^2 - y^3$ from $-2y^3 + 3y^2 + 2y - 1$.

61. Myra sells mugs at the sidewalk fair every week. Her revenue from selling x mugs is $12x - 0.3x^2$ dollars, and the cost of producing x mugs is $50 + 3x$ dollars.
   a. Write a polynomial for the profit Myra earns from selling x mugs.

   b. Find Myra's profit from the sale of 10 mugs, 15 mugs, and 20 mugs.

62. Colonial Lamps manufactures and sells table lamps. It costs the company $200 + 8x$ dollars to produce x units of the most popular model, and the revenue from selling those x lamps is $80x - 0.1x^2$ dollars.
   a. Write a polynomial for the profit Colonial Lamps earns from selling x lamps.

   b. Find the profit from the sale of 300 lamps, 400 lamps, and 500 lamps.

63. The Flying Linguine Brothers are working on a new act for the circus. Mario swings from a trapeze and catches Alfredo, who has somersaulted off a trampoline, in midair. Mario's height, in feet, at time t is given approximately by $12t^2 - 24t + 34$, and Alfredo's height, also in feet, at time t is $-16t^2 + 32t + 6$.
   a. Write a polynomial for the difference in the brothers' heights above the ground at any time t.

   b. Find the difference in the brothers' heights above the ground when they start (at time $t = 0$), and after $\dfrac{1}{2}$ second.

   c. When will Mario and Alfredo be at the same height?

64. In skeet shooting, the clay pigeons are launched at time $t = 0$, and their height t seconds after launch is given by $-16t^2 + 64t + 4$. The skeet shooter fires 1 second later, and the height of the bullet t seconds after the pigeon is launched is given by $-16t^2 + 88t - 32$.
   a. Write a polynomial for the difference in the height of the clay pigeon and the bullet at any time t.

   b. What is the difference in height at $t = 1$ (when the gun is fired) and at $t = 2$?

   c. When are the bullet and the clay pigeon at the same height?

65. The owner of the Koffee Shop pours the remainder of her old house blend, which is 30% Colombian beans, into a 50-pound bin and fills the bin up with her new house blend, which is 25% Colombian beans. Let *h* stand for the number of pounds of the old house blend. Write algebraic expressions to answer the following questions.

   a. How many pounds of Colombian beans are in the old house blend?

   b. How many pounds of new blend does the owner pour into the bin?

   c. How many pounds of Colombian beans are in this amount of new house blend?

   d. How many pounds of Colombian beans are in the 50-pound bin?

66. Harding College has a student population of 2400, enrolled in two schools. Women make up 28% of the School of Science and Engineering and 56% of the School of Liberal Arts. Let *s* stand for the number of students enrolled in the School of Science and Engineering. Write algebraic expressions to answer the following questions.

   a. How many women are enrolled in the School of Science and Engineering?

   b. How many students are enrolled in the School of Liberal Arts?

   c. How many women are enrolled in the School of Liberal Arts?

   d. How many of the students at Harding College are women?

67. Ralph and Waldo start in towns that are 20 miles apart and travel in opposite directions for 2 hours. Ralph travels 30 miles per hour faster than Waldo. Let *w* stand for Waldo's speed. Write algebraic expressions to answer the following questions.

   a. How far did Waldo travel?

   b. What was Ralph's speed?

   c. How far did Ralph travel?

   d. How far apart are Ralph and Waldo after 2 hours?

68. Henry and David are taking a walking tour of New England. Starting from the Bayberry Bed and Breakfast one morning, David walks for 2 hours at speed *r*, has a snack, and then walks 10 more miles. Henry walks for 4 hours on the same route but at a rate 2 miles per hour faster than David. Write algebraic expressions to answer the following questions.

   a. How far did David walk?

   b. What was Henry's speed?

   c. How far did Henry walk?

   d. If Henry walked farther, what is the distance between them now?

69. Let S_n stand for the sum of the first n integers. For example,

$$S_1 = 1$$
$$S_2 = 1 + 2 = 3$$
$$S_3 = 1 + 2 + 3 = 6$$

and so on.

a. Fill in the first 10 values of S_n in the table.

n	S_n	$\frac{1}{2}n^2 + \frac{1}{2}n$
1		
2		
3		
4		
5		
6		
7		
8		
9		
10		

b. Evaluate the polynomial $\frac{1}{2}n^2 + \frac{1}{2}n$ for integer values of n from 1 to 10, and fill in the rest of the table.

c. Compare your answers to parts (a) and (b).

70. Let T_m stand for the sum of the squares of the first m integers:

$$T_1 = 1^2 = 1$$
$$T_2 = 1^2 + 2^2 = 5$$
$$T_3 = 1^2 + 2^2 + 3^2 = 14$$

and so on.

a. Fill in the first 10 values of T_m in the table.

m	T_m	$\frac{1}{3}m^3 + \frac{1}{2}m^2 + \frac{1}{6}m$
1		
2		
3		
4		
5		
6		
7		
8		
9		
10		

b. Evaluate the polynomial

$$\frac{1}{3}m^3 + \frac{1}{2}m^2 + \frac{1}{6}m$$

for integer values of m from 1 to 10, and fill in the rest of the table.

c. Compare your answers to parts (a) and (b).

Evaluate each expression in Problems 71 and 72. (*Hint:* There is a hard way and an easy way.)

71. $(2x + y - z) + (3x - 4y + 6z) - (5x - 9y - 3z)$ for $x = 2.8,$ $y = -3.6,$ $z = 1.8$

72. $(2a^2 - 3a + 6) - (a^2 + 5a - 6) - (3a^2 - 4a + 2)$ for $a = -5$

5.2 Products of Polynomials

Reading

In Section 4.5, you learned how to compute the product of two first-degree binomials to get a quadratic trinomial. For example,

$$(2x + 3)(x - 4) = 2x^2 - 5x - 12$$

In this section, we consider products of polynomials of higher degree. We first investigate products of two powers with the same base.

Products of Powers

Suppose we would like to multiply together two powers with the same base. For instance, consider the product $(x^3)(x^4)$, which can be written as

$$(x^3)(x^4) = xxx \cdot xxxx = x^7$$

because x occurs as a factor 7 times. We see that the number of x's in the product is the *sum* of the number of x's in each factor.

On the other hand, if we want to multiply x^3 by y^4, we cannot simplify the product because the two powers do not have the same base:

$$(x^3)(y^4) = xxx \cdot yyyy = x^3y^4$$

These observations illustrate the following rule.

> ### First Law of Exponents
>
> To multiply two powers with the same base, add the exponents and leave the base unchanged. In symbols,
>
> $$a^m \cdot a^n = a^{m+n}$$

Try Exercise 1 and 2.

Products of Monomials

We can use the first law of exponents to multiply two monomials together.

EXAMPLE 1

Multiply $(2x^2y)(5x^4y^3)$.

Solution

We rearrange the factors to group together the numerical coefficients and the powers of each base:

$$(2x^2y)(5x^4y^3) = (2)(5)(x^2x^4)(yy^3)$$

We multiply the two coefficients and use the first law of exponents to find the products of the variable factors:

$$(2)(5)(x^2x^4)(yy^3) = 10x^6y^4$$

(Which law of algebra allows us to rearrange the factors in a product?)

Try Exercise 3 on page 567.

EXERCISE 1
Use the first law of exponents to find each product.
a. $k^2 \cdot k^8$

b. $y^3(y^3)$

EXERCISE 2
Explain why each product is *incorrect*. Then give the correct product.
a. $3^4 \cdot 3^3 \rightarrow 9^7$ ← Incorrect!

b. $t^3 \cdot t^5 \rightarrow t^{15}$ ← Incorrect!

Skills Review

Simplify. (See Lesson 2.9.)

1. $3(b - 4) - 2b(3 - 2b)$
2. $5x - 2x(1 - 2x) - 3(x - 2)$
3. $6 + 3[x - 2x(x - 4)]$
4. $a - 2[a - 2(a - 2)]$
5. $4\left[-2\left(t - \dfrac{1}{2}\right)(t - 1)\right]$
6. $5\left[-6\left(w + \dfrac{2}{3}\right)\left(w - \dfrac{3}{2}\right)\right]$

Answers: *1.* $4b^2 - 3b - 12$ *2.* $4x^2 + 6$ *3.* $-6x^2 + 27x + 6$
4. $3a - 8$ *5.* $-8t^2 + 12t - 4$ *6.* $-30w^2 + 25w + 30$

EXERCISE 3
Multiply $-3a^4b(-4a^3b)$.

Lesson

Multiplying by a Monomial

To multiply a polynomial by a monomial, we use the distributive law. To simplify products of powers with the same base, we apply the first law of exponents.

EXAMPLE 2

Multiply $-3xy^2(4x^2 - 2xy + 2)$.

Solution

We apply the distributive law and multiply each term of the polynomial by the monomial $-3xy^2$:

$$-3xy^2(4x^2 - 2xy + 2) = -3xy^2(4x^2) - 3xy^2(-2xy) - 3xy^2(2)$$
$$= -3 \cdot 4 \cdot x \cdot x^2 \cdot y^2 - 3(-2) \cdot x \cdot x \cdot y^2 \cdot y - 3 \cdot 2 \cdot xy^2$$
$$= -12x^3y^2 + 6x^2y^3 - 6xy^2$$

To simplify each term, we grouped together powers with the same base, just as we did in Example 1, and applied the first law of exponents.

EXERCISE 4
Simplify $2a^2(3 - a + 4a^2) - 3a(5a - a^2)$ by following the steps below:

Step 1 Apply the distributive law to remove parentheses.

Step 2 Use the first law of exponents to simplify each term.

Step 3 Combine like terms.

Products of Polynomials

We can also use the distributive law to help us compute products of two or more polynomials.

EXAMPLE 3

Multiply $(2x - 1)(3x^2 - x + 2)$.

Solution

We must multiply each term of the first polynomial by each term of the second polynomial. This involves six multiplications: First we multiply each term of the trinomial by $2x$; then we multiply each term by -1.

$$(2x - 1)(3x^2 - x + 2) = 2x(3x^2) + 2x(-x) + 2x(2) - 1(3x^2) - 1(-x) - 1(2)$$

$$= 6x^3 - 2x^2 + 4x - 3x^2 + x - 2 \quad \text{Combine like terms.}$$
$$= 6x^3 - 5x^2 + 5x - 2$$

Now try Exercise 5.

EXERCISE 5
Multiply $(3x + 2)(3x^2 + 4x - 2)$.

If a product contains both polynomial and monomial factors, it is a good idea to multiply the polynomial factors together first, and save the monomial factor for last.

EXAMPLE 4

Multiply $2x(x + 2)(3x - 5)$.

Solution

We begin by multiplying the binomial factors, $(x + 2)(3x - 5)$:

$$2x[(x + 2)(3x - 5)] = 2x[3x^2 - 5x + 6x - 10]$$
$$= 2x(3x^2 + x - 10)$$

Then we use the distributive law to multiply the result by $2x$:

$$2x(3x^2 + x - 10) = 6x^3 + 2x^2 - 20x$$

Try Exercise 6 now.

EXERCISE 6
Multiply $s^2t^2(2s + 1)(3s - 1)$.

In a product of three or more polynomials, we start by multiplying any two of the three factors.

EXAMPLE 5

Multiply $(3a - 1)(a + 2)(2a - 3)$.

Solution

We begin by multiplying the last two binomials, $(a + 2)(2a - 3)$.

$$(3a - 1)[(a + 2)(2a - 3)] = (3a - 1)[2a^2 - 3a + 4a - 6]$$
$$= (3a - 1)(2a^2 + a - 6)$$

Then we use the distributive law to multiply each term of the trinomial by each term of the binomial, as shown in Example 3:

$$(3a - 1)(2a^2 + a - 6) = 6a^3 + 3a^2 - 18a - 2a^2 - a + 6 \quad \text{Combine like terms.}$$
$$= 6a^3 + a^2 - 19a + 6$$

Try Exercise 7.

EXERCISE 7
Multiply $(x + 2)(3x - 2)(2x - 1)$.

ANSWERS TO 5.2 EXERCISES

1a. k^{10} **1b.** y^6

2a. The base should not be changed; 3^7

2b. The exponents should be added, not multiplied; t^8

3. $12a^7b^2$ **4.** $8a^4 + a^3 - 9a^2$ **5.** $9x^3 + 18x^2 + 2x - 4$

6. $6s^4t^2 + s^3t^2 - s^2t^2$ **7.** $6x^3 + 5x^2 - 12x + 4$

HOMEWORK 5.2

In Problems 1–6, apply the first law of exponents to find the product.

1. $x^3 \cdot x^6$

2. $t^2 \cdot t^5$

3. $5^6 \cdot 5^8$

4. $2^4 \cdot 2^5$

5. $b^3(b)(b^5)$

6. $c^2(c^4)(c^4)$

Find a value of n that makes the expressions in Problems 7–12 equivalent.

7. $y^3 \cdot y^n = y^8$

8. $x^n \cdot x^2 = x^{12}$

9. $a^n \cdot a^4 = a^8$

10. $b^5 \cdot b^n = b^{10}$

11. $3 \cdot 3^n = 3^3$

12. $4^n \cdot 4^2 = 4^3$

13. Simplify each of the following products.
    a. $2x^4(-3x^4)$ **b.** $-x^4(-2x^2)$

    c. $-x^4 \cdot y^3$ **d.** $-3x^5(3y^5)$

14. Explain in your own words when a product can be simplified and how to do it.

15. Simplify each of the following sums.
    a. $2x^4 - 3x^4$ **b.** $-x^4 - 2x^2$

    c. $-x^4 + y^3$ **d.** $-3x^5 + 3y^5$

16. Explain in your own words when a sum can be simplified and how to do it.

Find (a) the product and (b) the sum of each pair of expressions in Problems 17–22.

17. $3a, 8a$

18. $4x, -6x$

19. $-2b^3, 5b^3$

20. $-3z^4, -7z^4$

21. $4p^2q, -7p^2q$

22. $-9cd^3, -cd^3$

For each rectangle in Problems 23–26:
a. Find its perimeter.
b. Find its area.

23.
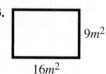
$9m^2$
$16m^2$

24.
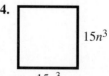
$15n^3$
$15n^3$

25.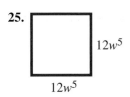

$12w^5$

$12w^5$

26.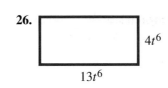

$4t^6$

$13t^6$

Simplify each expression in Problems 27–30, if possible. If an expression cannot be simplified, say so.

27. a. $x^2 + x^2$ **b.** $x^2(x^2)$ **28. a.** $-x - x$ **b.** $-x(-x)$

   c. $x^2 - x^2$ **d.** $x^2(-x^2)$ **c.** $-x^2 - x^2$ **d.** $-x^2(-x^2)$

29. a. $x + x^2$ **b.** $x(x^2)$ **30. a.** $-x^3 \cdot x^3$ **b.** $x^3(-x^2)$

   c. $x^2 - x$ **d.** $x^2(-x)$ **c.** $x^3 - x^2$ **d.** $(-x)^3(-x)^2$

Multiply the monomials in Problems 31–36.

31. $w^3(-8w^4)$ **32.** $4y^9(-9y^4)$

33. $-5s^2(2s^5t)$ **34.** $-4x^4(3x^3y^3)$

35. $-6xy^2(-3xy^3)(-2xy)$ **36.** $-3a^2b(-5a^3b^2)(ab^2)$

Simplify each expression in Problems 37–40, if possible. If an expression cannot be simplified, say so.

37. a. $2x + 2x$ **b.** $2x(-2x)$ **38. a.** $3x^2 + 2x^3$ **b.** $3x^2(2x^3)$

c. $2(x - 2x)$ **d.** $-2x - 2 - x$ **c.** $x^3(3x^3)$ **d.** $x^3 + 3x^3$

39. a. $3b + 4b$ **b.** $3b(4b)$ **40. a.** $4z^2 - 6z^2$ **b.** $4z^2(-6z^2)$

c. $b^3 + b^4$ **d.** $b^3(b^4)$ **c.** $6z^4 - 4z^2$ **d.** $-6(-z^2) - 4(-z)^2$

Find the products in Problems 41–44.

41. $-xy(x^2 + xy + y^2)$ **42.** $(2a^2 - 6ab + b^2)(3b^3)$

43. $(6 - st + 3s^2t^2)(-3s^2t^2)$ **44.** $-4p^2q(2p^3q - 3p^2q^2 + pq^3)$

Simplify the expressions in Problems 45–48.

45. $ax(x^2 + 2x - 3) - a(x^3 + 2x^2)$ **46.** $3x(ax^2 + ax - a) - 2a(x^3 - 2x)$

47. $3ab^2(2 + 3a) - 2ab(3ab + 2b)$ **48.** $8x^2y(y - 3) + 5y^2(x - 3x^2)$

Multiply the polynomials in Problems 49–58.

49. $4a(a - 1)(a + 5)$

50. $2r(3r - 5)(3r + 5)$

51. $-x(2x - 1)^2$

52. $-2w^3(3w + 2)^2$

53. $s^2t^2(2s + t)(3s - t)$

54. $3x^2y^3(6x - 2y)(6x + 2y)$

55. $(x - 2)(x^2 - 3x + 2)$

56. $(x - 1)(x^2 + x - 1)$

57. $(3x - 1)(9x^2 - 3x + 1)$

58. $(3x + 2)(3x^2 + 4x - 2)$

59. a. Multiply $(x + 1)(x + 2)(x + 3)$.

b. Evaluate your product for $x = -1$, $x = -2$, and $x = -3$.

60. a. Multiply $(x - 2)(x - 2)(x - 2)$.

b. Evaluate your product for $x = 2$, $x = 0$, and $x = 3$.

61. a. Multiply $(2x - 1)(2x - 1)(2x - 1)$.

    b. What value of x makes the product equal to zero?

62. a. Multiply $(x + 2)(3x - 2)(2x - 1)$.

    b. Find three values of x that make the product equal to zero.

63. The sum of two numbers is 16.
    a. If one of the numbers is n, write an expression for the other number.

    b. Write a polynomial for the product of the two numbers.

64. The difference of two numbers is 5.
    a. If the larger number is p, write an expression for the smaller number.

    b. Write a polynomial for the product of the two numbers.

65. A large wooden box is 3 feet longer than it is wide, and its height is 2 feet shorter than its width.
    a. If the width of the box is w, write expressions for its length and its height.

    b. Write a polynomial for the volume of the box.

    c. Write a polynomial for the surface area of the box.

66. A cardboard box is 4 inches wider than it is tall, and its length is 6 inches greater than its width.
    a. If the height of the box is h, write expressions for its width and its length.

    b. Write a polynomial for the volume of the box.

    c. Write a polynomial for the surface area of the box.

Simplify each expression in Problems 67–72 mentally, without using paper, pencil, or calculator.

67. $10^3(8 \cdot 10^4)$

68. $10^2(5 \cdot 10^3)$

69. $(3 \cdot 10^2)(2 \cdot 10^2)$

70. $(2 \cdot 10^3)(4 \cdot 10^2)$

71. $(3.3 \cdot 10^2)(2 \cdot 10^2)$

72. $(1.6 \cdot 10^3)(2 \cdot 10^3)$

5.3 | Factoring Polynomials

Reading

In Chapter 4, you learned to factor quadratic expressions such as $-16t^2 + 64t$ and $2x^2 + 9x - 5$. In this section, we factor some polynomials of higher degree. To begin, we consider quotients of powers with the same base.

Quotients of Powers

Recall that we can reduce a fraction by dividing numerator and denominator by any common factors. For example,

$$\frac{10}{15} = \frac{2 \cdot \cancel{5}}{3 \cdot \cancel{5}} = \frac{2}{3}$$

We first factored the numerator and denominator of the fraction and then canceled the 5's by dividing. We can apply the same technique to quotients of powers. For example, to simplify the quotient $\dfrac{a^5}{a^3}$, we first write the numerator and denominator in factored form:

$$\frac{a^5}{a^3} = \frac{\cancel{a} \cdot \cancel{a} \cdot \cancel{a} \cdot a \cdot a}{\cancel{a} \cdot \cancel{a} \cdot \cancel{a}} = \frac{a^2}{1} = a^2$$

You may have observed that the exponent of the quotient can be obtained by subtracting the exponent of the denominator from the exponent of the numerator. In other words,

$$\frac{a^5}{a^3} = a^{5-3} = a^2$$

What if the larger power occurs in the denominator of the fraction? Consider this quotient:

$$\frac{a^4}{a^8} = \frac{\cancel{a} \cdot \cancel{a} \cdot \cancel{a} \cdot \cancel{a}}{\cancel{a} \cdot \cancel{a} \cdot \cancel{a} \cdot \cancel{a} \cdot a \cdot a \cdot a \cdot a} = \frac{1}{a^4}$$

We see that we can subtract the exponent of the numerator from the exponent of the denominator. That is,

$$\frac{a^4}{a^8} = \frac{1}{a^{8-4}} = \frac{1}{a^4}$$

These examples suggest the following property.

Second Law of Exponents

To divide two powers with the same base, subtract the smaller exponent from the larger one, and keep the base.

1. If the larger exponent occurs in the numerator, put the power in the numerator.

2. If the larger exponent occurs in the denominator, put the power in the denominator.

In symbols,

1. $\dfrac{a^m}{a^n} = a^{m-n}$ if $n < m$

2. $\dfrac{a^m}{a^n} = \dfrac{1}{a^{n-m}}$ if $n > m$

EXERCISE 1

Use the second law of exponents to find each quotient.

a. $\dfrac{x^6}{x^2}$

b. $\dfrac{b^2}{b^3}$

EXERCISE 2

Divide $\dfrac{8x^2y}{12x^5y^3}$.

Try Exercise 1.

Quotients of Monomials

To divide one monomial by another, we apply the second law of exponents to the powers of each variable.

EXAMPLE 1

Divide $\dfrac{3x^2y^4}{6x^3y}$.

Solution

We consider the numerical coefficients and the powers of each base separately. We use the second law of exponents to simplify each quotient of powers:

$$\frac{3x^2y^4}{6x^3y} = \frac{3}{6} \cdot \frac{x^2}{x^3} \cdot \frac{y^4}{y} \qquad \text{Subtract exponents on each base.}$$

$$= \frac{1}{2} \cdot \frac{1}{x^{3-2}} \cdot y^{4-1}$$

$$= \frac{1}{2} \cdot \frac{1}{x} \cdot \frac{y^3}{1} = \frac{y^3}{2x} \qquad \text{Multiply factors.}$$

Try Exercise 2.

Skills Review

Factor each polynomial.

1. $3x^2 - 6x$ 2. $-9ay^2 + 6y$ 3. $p^2 - 12p - 45$
4. $b^2 + b - 72$ 5. $5t^2 + 7t + 2$ 6. $5x^2 - 7x - 24$

Answers: **1.** $3x(x - 2)$ **2.** $3y(-3ay + 2)$ **3.** $(p - 15)(p + 3)$
4. $(b + 9)(b - 8)$ **5.** $(5t + 2)(t + 1)$ **6.** $(5x + 8)(x - 3)$

Lesson

Greatest Common Factors

Now, we consider several techniques for factoring polynomials. The first of these involves removing the **greatest common factor (GCF)** from each term. For example, the greatest common factor for

$$4a^3b^2 + 6ab^3 - 18a^2b^4$$

is $2ab^2$. This is the largest factor that divides evenly into each term of the polynomial. Note that the exponent on each variable of $2ab^2$ is the *smallest* exponent that appears on that variable among the terms of the polynomial.

Try Exercise 3.

EXERCISE 3
Find the greatest common factor for
$15x^2y^2 - 12xy + 6xy^3$.

Factoring Out the Greatest Common Factor

Once we have found the greatest common factor for a polynomial, we can write each term as a product of the GCF and another factor. For example, the GCF of $8x^2 - 6x$ is $2x$, and we can write

$$8x^2 - 6x = 2x \cdot 4x - 2x \cdot 3$$

We can then use the distributive law to write the expression on the right side as a product:

$$2x \cdot 4x - 2x \cdot 3 = 2x(4x - 3)$$

We have *factored out* the greatest common factor from the polynomial.

For more complicated polynomials, we can divide the GCF into each term to find the remaining factors.

EXAMPLE 2

Factor $4a^3b^2 + 6ab^3 - 18a^2b^4$.

Solution

We first identify the greatest common factor for the polynomial; in this case, as noted above, it is $2ab^2$. We factor out the GCF from each term and write the polynomial as a product:

$$2ab^2(\qquad\qquad)$$

To find the factor inside parentheses, we divide each term of the polynomial by the GCF:

$$\frac{4a^3b^2}{2ab^2} = 2a^2, \qquad \frac{6ab^3}{2ab^2} = 3b, \qquad \frac{-18a^2b^4}{2ab^2} = -9ab^2$$

We apply the distributive law to factor $2ab^2$ from each term:

$$4a^3b^2 + 6ab^3 - 18a^2b^4 = 2ab^2 \cdot 2a^2 + 2ab^2 \cdot 3b - 2ab^2 \cdot 9ab^2$$
$$= 2ab^2(2a^2 + 3b - 9ab^2)$$

Try Exercise 4.

EXERCISE 4
Follow the suggested steps to factor
$15x^2y^2 - 12xy + 6xy^3$.
Step 1 Find the GCF.

Step 2 Write the GCF outside a set
of parentheses.

Step 3 Divide each term by the GCF.

Step 4 Write the quotients inside the
parentheses.

Sometimes the greatest common factor is not a monomial but instead has two or more terms.

EXAMPLE 3

Factor $x^2(2x + 1) - 3(2x + 1)$.

Solution

The given expression has two terms, and $(2x + 1)$ is a factor of each. We factor out the entire binomial, $(2x + 1)$:

$$x^2(2x + 1) - 3(2x + 1) = (2x + 1)(\qquad\qquad)$$

To find the factor inside the parentheses, we divide $(2x + 1)$ into each term of the given expression:

$$\frac{x^2(2x + 1)}{(2x + 1)} = x^2, \qquad \frac{-3(2x + 1)}{(2x + 1)} = -3$$

Thus

$$x^2(2x + 1) - 3(2x + 1) = (2x + 1)(x^2 - 3).$$

Try Exercise 5.

EXERCISE 5
Factor $9(x^2 + 5) - x(x^2 + 5)$.

Combining Factoring Techniques

Sometimes a polynomial can be factored further after we have removed the greatest common factor.

EXAMPLE 4

Factor $2b^3 - 8b^2 - 10b$ completely.

Solution

We begin by factoring out the greatest common factor, $2b$:

$$2b^3 - 8b^2 - 10b = 2b(b^2 - 4b - 5)$$

The remaining factor, $b^2 - 4b - 5$, is a quadratic trinomial. To factor it, we must find two numbers p and q such that $pq = -5$ and $p + q = -4$. You can check that $p = -5$ and $q = 1$ will work. Thus

$$b^2 - 4b - 5 = (b - 5)(b + 1),$$

and

$$2b^3 - 8b^2 - 10b = 2b(b - 5)(b + 1)$$

As a general rule, you should always begin factoring by checking to see whether there is a common factor that can be factored out.

Try Exercise 6.

EXERCISE 6
Factor $4a^6 - 10a^5 + 6a^4$ completely.

Quadratic Trinomials in Two Variables

So far, we have factored quadratic trinomials in one variable—that is, polynomials of the form

$$ax^2 + bx + c$$

The method we have applied can also be used to factor trinomials in two variables of the form

$$ax^2 + bxy + cy^2$$

Note that the first and last terms are quadratic terms, whereas the middle term is a *cross-term* consisting of the product of the two variables.

EXAMPLE 5

Factor $x^2 + 5xy + 6y^2$.

Solution

As usual, we begin by factoring the first term into x times x:

$$x^2 + 5xy + 6y^2 = (x + \underline{\quad\quad})(x + \underline{\quad\quad})$$

Next, we look for factors of the last term, $6y^2$. In order to obtain the xy-term in the middle, we need a y in each factor. Thus the possibilities are

$$y \text{ and } 6y \qquad \text{or} \qquad 2y \text{ and } 3y$$

We check the sum of **O**utside and **I**nside products for each possibility:

$$(x + y)(x + 6y) \quad \text{O} + \text{I} = 6xy + xy = 7xy$$
$$(x + 2y)(x + 3y) \quad \text{O} + \text{I} = 3xy + 2xy = 5xy$$

The second possibility gives the correct middle term, so the factorization is

$$x^2 + 5xy + 6y^2 = (x + 2y)(x + 3y)$$

In Exercise 7, you'll apply your factoring skills to trinomials in two variables. For part (a), you should factor out the GCF first. For part (b), you may want to review the technique you learned in Section 4.9.

EXERCISE 7
Factor each polynomial completely.
a. $2a^3b - 24a^2b^2 - 90ab^3$

b. $3x^2 - 8xy + 4y^2$

ANSWERS TO 5.3 EXERCISES

1a. x^4

1b. $\dfrac{1}{b}$

2. $\dfrac{2}{3x^3y^2}$

3. $3xy$

4. $3xy(5xy - 4 + 2y^2)$

5. $(x^2 + 5)(9 - x)$

6. $2a^4(2a - 3)(a - 1)$

7a. $2ab(a - 15b)(a + 3b)$

7b. $(3x - 2y)(x - 2y)$

HOMEWORK 5.3 ⟨www⟩

In Problems 1–6, find each quotient by using the second law of exponents.

1. $\dfrac{a^6}{a^3}$

2. $\dfrac{c^{12}}{c^4}$

3. $\dfrac{3^9}{3^4}$

4. $\dfrac{8^6}{8^2}$

5. $\dfrac{z^6}{z^9}$

6. $\dfrac{w^4}{w^8}$

In Problems 7–12, choose a value for the variable and evaluate each pair of expressions to show that they are *not* equivalent.

7. $t^2 \cdot t^3, \ t^6$

8. $w^3 \cdot w^3, \ w^9$

9. $\dfrac{v^8}{v^2}, \ v^4$

10. $\dfrac{u^6}{u^3}, \ u^2$

11. $\dfrac{n^3}{n^5}, \ n^2$

12. $\dfrac{p^3}{p^7}, \ p^4$

In Problems 13–18, perform the division.

13. $\dfrac{2x^3y}{8x^4y^5}$

14. $\dfrac{8x^2y}{12x^5y^3}$

15. $\dfrac{-12bx^4}{8bx^2}$

16. $\dfrac{-12ax^6}{20ax}$

17. $\dfrac{-15x^3y^2}{-3x^3y^4}$

18. $\dfrac{-25w^3z^2}{-5w^8z^2}$

Find the greatest common factor for each polynomial in Problems 19–22.

19. $2x^4 - 4x^2 + 8$

20. $12x^2y + 18xy^2$

21. $16a^3b^3 - 12a^2b + 8ab^2$

22. $18a^2b^3 + 6a^3b^2 - 9a^2b$

Factor out the greatest common factor in Problems 23–26.

23. $9x^2 - 12x^5 + 3x^3$

24. $48t^7 + 27t^5 - 45t^{11}$

25. $14x^3y - 35x^2y^2 + 21xy^3$

26. $25h^2k^3 - 5hk - 30h^3k^2$

Factor out a negative monomial in Problems 27–30.

27. $-b^2 - bc - ab$

28. $-6h^3 - 3h^2 - 3h$

29. $-4k^4 + 4k^2 - 2k$

30. $-3x^2 + 3xy - 3xy^2$

Factor out the common factor in Problems 31–34.

31. $2x(x + 6) - 3(x + 6)$

32. $4(3x - 2) - 3x(3x - 2)$

33. $3x^2(2x + 3) - (2x + 3)$

34. $4x^2(6x - 5) + (6x - 5)$

For Problems 35–38, arrange the trinomial in descending powers of the variable and then factor. Begin by factoring -1 from each term, if necessary.

35. $21 - 4x - x^2$

36. $24 + 10z - z^2$

37. $24a + 81 - a^2$

38. $7u - 120 + u^2$

In Problems 39–46, factor each expression completely.

39. $2x^2 + 10x + 12$

40. $5c^2 - 25c + 30$

41. $4a^2b + 12ab - 7b$

42. $3p^4 - 30p^3 + 63p^2$

43. $4z^3 + 10z^2 + 6z$

44. $12x^5 + 14x^4 - 10x^3$

45. $18a^2b - 9ab - 27b$

46. $36b^7 - 54b^6 + 20b^5$

In Problems 47–60, factor each expression completely.

47. $x^2 - 5xy + 6y^2$

48. $p^2 - 5pq - 14q^2$

49. $x^2 + 4ax - 77a^2$

50. $s^2 - 8st + 16t^2$

51. $4x^3 + 12x^2y + 8xy^2$

52. $6x^4 - 12x^3y - 18x^2y^2$

53. $9a^3b + 9a^2b^2 - 18ab^3$

54. $12p^3w - 36p^2w^2 + 24pw^3$

55. $2t^2 - 5st - 3s^2$

56. $3x^2 - 7ax + 2a^2$

57. $4b^2y^2 + 5by + 1$

58. $9a^2b^2 + 9ab - 4$

59. $12ab^2 + 15a^2b + 3a^3$

60. $12h^4k^2 - 30h^3k^3 + 12h^2k^4$

5.4 Special Products and Factors

Reading

A handful of special binomial products occur so frequently that it is useful to recognize their forms. This will enable you to write the factored forms directly, without trial and error. To prepare for these special products, we first consider the squares of monomials.

Squares of Monomials

Study the following squares of monomials. Do you see a quick way to find each product?

$$(a^2)^2 = a^2 \cdot a^2 = a^4$$
$$(w^5)^2 = w^5 \cdot w^5 = w^{10}$$
$$(4x^3)^2 = 4x^3 \cdot 4x^3 = 4 \cdot 4 \cdot x^3 \cdot x^3 = 16x^6$$
$$(-6h^6)^2 = -6h^6(-6h^6) = -6(-6) \cdot h^6 \cdot h^6 = 36h^{12}$$

In the first two examples, we doubled the exponent and kept the same base. (This is a special case of a more general law of exponents, which we'll study in Chapter 6.) In the last two examples, we squared the numerical coefficient and doubled the exponent.

Try Exercises 1 and 2.

EXERCISE 1
Simplify each square.
a. $(6t^4)^2$

b. $(12st^8)^2$

EXERCISE 2
Find a monomial whose square is $64b^6$.

Skills Review

Express each product as a polynomial.

1. $(z - 3)^2$ 2. $(x + 4)^2$ 3. $(3a + 5)^2$

4. $(2b - 7)^2$ 5. $(2n - 5)(2n + 5)$ 6. $(4m + 9)(4m - 9)$

Answers: **1.** $z^2 - 6z + 9$ **2.** $x^2 + 8x + 16$ **3.** $9a^2 + 30a + 25$
4. $4b^2 - 28b + 49$ **5.** $4n^2 - 25$ **6.** $16m^2 - 81$

Lesson

Squares of Binomials

We can use the distributive law to verify each of the following special products.

$$(a + b)^2 = (a + b)(a + b) \qquad (a - b)^2 = (a - b)(a - b)$$
$$= a^2 + ab + ab + b^2 \qquad\qquad = a^2 - ab - ab + b^2$$
$$= a^2 + 2ab + b^2 \qquad\qquad = a^2 - 2ab + b^2$$

We have the following results.

Squares of Binomials

1. $(a + b)^2 = a^2 + 2ab + b^2$
2. $(a - b)^2 = a^2 - 2ab + b^2$

We can use these two products as formulas to compute squares of binomials. For example, the formula for the square of a sum says to square the first term, add twice the product of the two terms, and then add the square of the second term.

EXAMPLE 1

Expand $(2x^3 + 3y)^2$ as a polynomial.

Solution

We use formula 1, replacing a by $2x^3$ and b by $3y$:

$$(a + b)^2 = a^2 + 2\,ab + b^2$$
$$(2x^3 + 3y)^2 = (2x^3)^2 + 2(2x^3)(3y) + (3y)^2$$

square of first term   twice their product   square of second term

$$= 4x^6 + 12x^3y + 9y^2$$

You can verify that you get the same answer for Example 1 if you compute the square as a product of binomials, $(2x^3 + 3y)(2x^3 + 3y)$.

Try Exercise 3.

EXERCISE 3
Use formula 2 to expand $(4 - 3t)^2$.

Difference of Two Squares

Now let's consider the product

$$(a + b)(a - b) = a^2 - ab + ab - b^2$$
$$= a^2 - b^2$$

In this product, the two middle terms cancel each other, and we are left with a *difference of two squares*:

> ### Difference of Two Squares
>
> $(a + b)(a - b) = a^2 - b^2$

EXAMPLE 2

Multiply $(2y + 9w)(2y - 9w)$.

Solution

The product has the form $(a + b)(a - b)$,   with a replaced by $2y$ and b replaced by $9w$. We use formula 3 to write the product as a polynomial:

$$(a + b)(a - b) =   a^2 - b^2$$
$$(2y + 9w)(2y - 9w) = (2y)^2 - (9w)^2$$

square of   square of
first term   second term

$$= 4y^2 - 81w^2$$

Try Exercise 4.

EXERCISE 4
Multiply $(5x^4 + 4)(5x^4 - 4)$.

Factoring Special Products

The three special products we have just studied are useful as patterns for factoring certain polynomials. For this purpose, it is helpful to switch the right and left sides of the formulas.

> ### Special Factorizations
>
> 1. $a^2 + 2ab + b^2 = (a + b)^2$
> 2. $a^2 - 2ab + b^2 = (a - b)^2$
> 3. $a^2 - b^2 = (a + b)(a - b)$

To use these formulas, we must first recognize a given trinomial as being one of the three forms on the left side. We then use the right side of the appropriate formula to write the factored form. All three special products involve two squared terms, a^2 and b^2, so we first look for two squared terms in a trinomial.

EXAMPLE 3

Factor $x^2 + 24x + 144$.

Solution

This trinomial has two squared terms, x^2 and 144. If these terms are a^2 and b^2, then $a = x$ and $b = 12$.   We must check whether the middle term is equal to $2ab$.

$$2ab = 2(x)(12) = 24x$$

This is the correct middle term, so the given trinomial has the form of formula 1, with $a = x$ and $b = 12$. Thus

$$a^2 + 2ab + b^2 = (a + b)^2 \quad \text{\small Replace } a \text{ by } x \text{ and } b \text{ by 12.}$$
$$x^2 + 24x + 144 = (x + 12)^2$$

Try Exercise 5.

Note that the sum of two squares, $a^2 + b^2$, cannot be factored. For example, none of these expressions can be factored:

$$x^2 + 16 \qquad 9x^2 + 4y^2 \qquad 25y^4 + w^4$$

You should check, for instance, that $x^2 + 16 \neq (x + 4)(x + 4)$.

> The sum of two squares, $a^2 + b^2$, cannot be factored.

As always when factoring, we should first check for common factors.

EXAMPLE 4

Factor $98 - 28x^4 + 2x^8$ completely.

Solution

Each term has a factor of 2, so we begin by factoring out 2:

$$98 - 28x^4 + 2x^8 = 2(49 - 14x^4 + x^8)$$

The polynomial in parentheses has the form $(a - b)^2$, with $a = 7$ and $b = x^4$. [Note that the middle term is $-2ab = -2(7)(x^4)$.] We use formula 2 to write

$$a^2 - 2ab + b^2 = (a - b)^2 \quad \text{\small Replace } a \text{ by 7 and } b \text{ by } x^4$$
$$49 - 14x^4 + x^8 = (7 - x^4)^2$$

Thus $98 - 28x^4 + 2x^8 = 2(7 - x^4)^2$.

Try Exercise 6.

EXERCISE 5

Use one of the three formulas and follow the suggested steps to factor each polynomial.

a. $25y^2 - w^2$

b. $m^6 - 18m^3 + 81$

Step 1 Choose the appropriate formula.

Step 2 Identify a and b.

Step 3 Verify that the middle term (if any) equals $2ab$.

Step 4 Replace a and b in the formula with your answers to Step 2.

EXERCISE 6

Factor $x^6 - 16x^2$ completely.

ANSWERS TO 5.4 EXERCISES

1a. $36t^8$

1b. $144s^2t^{16}$

2. $8b^3$

3. $16 - 24t + 9t^2$

4. $25x^8 - 16$

5a. $(5y - w)(5y + w)$

5b. $(m^3 - 9)^2$

6. $x^2(x^2 + 4)(x + 2)(x - 2)$

HOMEWORK 5.4   www

Square each monomial in Problems 1–6.

1. $(8t^4)^2$

2. $(7x^3)^2$

3. $(-12a^2)^2$

4. $(-4z^4)^2$

5. $(10h^2k)^2$

6. $(11p^3n^2)^2$

Use the formulas for squares of binomials to expand each square in Problems 7–12.

7. $(3a + b)^2$

8. $(2t - v)^2$

9. $(7b^3 - 6)^2$

10. $(6m^5 + 2)^2$

11. $(2h + 5k^4)^2$

12. $(11s^2 - 2t)^2$

Use the formula for the difference of two squares to multiply the binomials in Problems 13–18.

13. $(3p - 4)(3p + 4)$

14. $(5y + 6)(5y - 6)$

15. $(2x^2 - 1)(2x^2 + 1)$

16. $(r - 6w^3)(r + 6w^3)$

17. $(h^2 + 7t)(h^2 - 7t)$

18. $(8pq - m^4)(8pq + m^4)$

Factor the squares of binomials in Problems 19–24.

19. $y^2 + 6y + 9$

20. $a^2 - 12a + 36$

21. $m^2 - 30m + 225$

22. $1 - 16v + 64v^2$

23. $x^2 + 4xy + 4y^2$

24. $4p^2 - 12pq + 9q^2$

Factor each polynomial in Problems 25–30.

25. $x^2 - 9$

26. $x^2 - 16y^2$

27. $36 - a^2b^2$

28. $4b^2 - 9$

29. $64y^2 - 49x^2$

30. $1 - 100a^2b^2$

Factor each polynomial in Problems 31–36.

31. $a^4 + 10a^2 + 25$

32. $x^6 - 64$

33. $36y^8 - 49$

34. $b^{10} - 12b^5 + 36$

35. $16x^6 - 9y^4$

36. $49a^{10}b^{12} - 144$

Factor each polynomial in Problems 37–48 completely.

37. $3a^2 - 75$

38. $5b^2 - 80$

39. $2a^3 - 12a^2 + 18a$

40. $3b^4 + 12b^3 + 12b^2$

41. $9x^7 - 81x^3$

42. $8a^2b^4 - 18a^4b^2$

43. $12h^2 + 3k^6$

44. $-4m^4 - 9g^8$

45. $81x^8 - y^4$

46. $48x^2y^2z^2 - 3$

47. $162a^4b^8 - 2a^8$

48. $4p^2q^4 + 32p^3q^3 + 64p^4q^2$

49. Use areas of rectangles to explain why the figure illustrates the product

$$(a + b)^2 = a^2 + 2ab + b^2$$

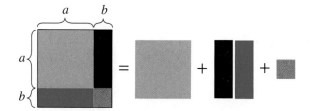

50. Use areas of rectangles to explain why the figure illustrates the factorization

$$a^2 - b^2 = (a + b)(a - b)$$

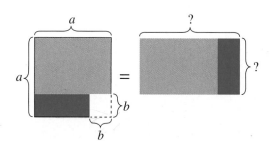

51. a. Expand $(a - b)^3$ by multiplying.

b. Use part (a) as a formula to expand $(2x - 3)^3$.

c. Substitute $a = 5$ and $b = 2$ to show that $(a - b)^3$ is not equivalent to $a^3 - b^3$.

53. a. Multiply $(a + b)(a^2 - ab + b^2)$.

b. Factor $a^3 + b^3$.

c. Factor $x^3 + 8$.

52. a. Expand $(a + b)^3$ by multiplying.

b. Use part (a) as a formula to expand $(3x + 4)^3$.

c. Substitute $a = 2$ and $b = 4$ to show that $(a + b)^3$ is not equivalent to $a^3 + b^3$.

54. a. Multiply $(a - b)(a^2 + ab + b^2)$

b. Factor $a^3 - b^3$.

c. Factor $x^3 - 27$.

5.5 Inverse Variation

Reading

In Section 3.3, we saw that the equation relating two proportional variables has the form $y = kx$ and that the graph of this equation is a straight line through the origin. For example, if gasoline costs $1.50 per gallon, the price p of g gallons of gas is given by

$$p = 1.5g$$

The graph of this equation is shown in Figure 5.1.

This type of relationship between variables is called *direct variation*, because when one variable increases, the other variable also increases. Actually, we can describe the relationship more precisely: If we double one variable, then the other variable doubles as well. You can check this by completing the table in Exercise 1.

In fact, if we multiply the value of g by any constant, then the value of p will also be multiplied by that same constant. This property characterizes direct variation.

In this section, we consider variables that vary *inversely*. The product of such variables is constant: When one variable increases, the other decreases. More specifically, if we double one variable, then the other variable will be halved. Consider the variables in Exercise 2.

Figure 5.1

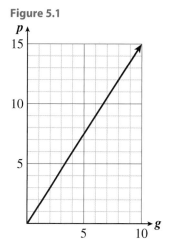

EXERCISE 1
Use the equation $p = 1.5g$ to complete the table. For each row, check that the new value of p is double the old value.

g	p	Double g	Compute new p
2			
3			
5			

EXERCISE 2
Arlen would like to build a rectangular rabbit pen with an area of 60 square feet. If the length and width of the pen are l and w, you know that

$$lw = 60$$

a. Complete the table, and use it to answer the following questions.

w	2	3	4	5	6	10	12	15
l								

b. What happens to the value of l if you double the value of w? What happens to the value of l if you triple the value of w?

Skills Review

Decide whether each equation, table, or graph describes direct variation. Explain why or why not in each case.

1. $y = 5.2 + 2.4x$

2. $y = 25 - x$

3.
x	2	4	5	7
y	1.5	3	3.75	5.25

4.
x	15	25	30	50
y	10	20	25	45

5.

6.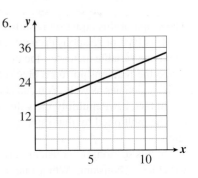

Answers: **1.** No   **2.** No   **3.** Yes   **4.** No   **5.** No   **6.** No

Lesson

In the following Activity, we consider the graphs of variables that vary inversely.

Activity

1. In Exercise 2, you made a table of possible lengths and widths for a rectangle of area 60 square feet.

   a. Complete the following copy of that table.

   b. Write an equation for the length, l, of the rectangle in terms of its width, w.

   c. Use the values in the table to graph your equation.

   d. What will happen to the value of l if you continue to increase the value of w?

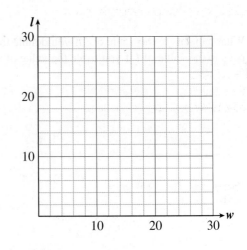

w	l
2	
3	
4	
5	
6	
10	
12	
15	
20	

2. Matt wants to travel 360 miles to visit a friend over spring break. He is deciding whether to ride his bike or drive. If he travels at an average speed of r miles per hour, the trip will take t hours.

    a. Fill in the table.

    b. Write an equation for t in terms of r.

    c. Use the values in the table to graph your equation.

    d. If Matt doubles his speed, what happens to his travel time? Show this on your graph, starting with $r = 20$.

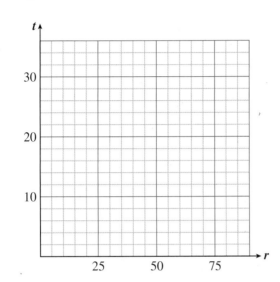

r	t
10	
20	
30	
40	
60	
80	
90	

The equations you graphed in the Activity represent **inverse variation**. In general, we have the following definition.

Inverse Variation

We say that **y varies inversely with x** if

$$y = \frac{k}{x}$$

where k is a constant, called the **constant of variation**.

EXERCISE 3

Suppose that y varies inversely with x and that $y = 12$ when $x = 4$.

a. What is the constant of variation?

b. What is the value of y when $x = 3$?

EXERCISE 4

Look at the tables of values you completed in the Activities. How can you recognize inverse variation from a table of values?

EXERCISE 5

Which of the following tables represents inverse variation?

a.

h	1.6	2	3.2	8
w	6	4.8	3	1.2

b.

V	320	270	230	160
R	80	130	170	240

ANSWERS TO 5.5 EXERCISES

1.

g	p	*Double g*	*Compute new p*
2	3	4	6
3	4.5	6	9
5	7.5	10	15

2a.

w	2	3	4	5	6	10	12	15
l	30	20	15	12	10	6	5	4

2b. If you double w, then l is divided by 2. If you triple w, then l is divided by 3.

3a. 48   **3b.** 16

4. The product of the variables is constant.

5. The table in part (a)

HOMEWORK 5.5

1. To celebrate their graduation, Janel and her friends want to charter a boat for a dinner cruise on the river. It costs $1200 to rent the boat for the evening (not including dinner), and they will split the cost equally.
   a. Fill in the table showing each person's share of the rental fee, s, if n people go on the cruise.

n	s
5	
6	
10	
12	
20	
25	
30	
40	

   b. Write an equation for s in terms of n.

   c. Graph your equation.

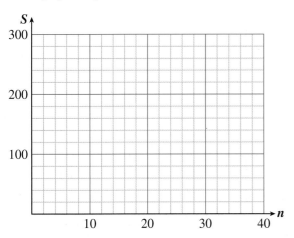

   d. At the last minute, one-third of the people who signed up for the cruise back out. By what factor does the price per person increase?

2. Every Friday, Takuya takes his company's weekly employee bulletin to the copy center and asks for 1000 copies. The time it takes to make the copies depends on the speed of the copier used.
   a. Fill in the table, showing the time, t, in minutes, it takes to make the copies on a copier that produces c copies per minute.

c	t
5	
10	
20	
25	
50	
100	
125	

   b. Write an equation for t in terms of c.

   c. Graph your equation.

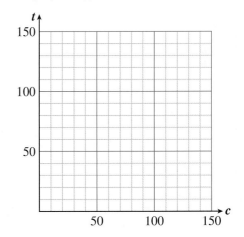

   d. If the speed of the copier is increased by 25%, by what factor will the time required to make the copies decrease?

3. The manager at Cut 'n' Style finds that the number of customers per week varies inversely with the price he charges for a haircut. During a week when he charged $10, he had 36 customers.
    a. Find the constant of variation, and write an equation for the number of customers, c, expected when the price of a haircut is p dollars.

    b. Fill in the table.

p	c
6	
9	
10	
12	
15	
18	
20	

    c. Graph your equation.

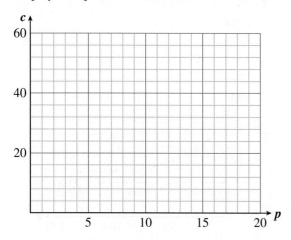

    d. If the manager increases the price of a haircut by 50%, by what factor will the number of customers decrease?

4. When two people balance on a seesaw, the distance each sits from the fulcrum varies inversely with his or her weight. Shannon weighs 120 pounds and is sitting 3 feet from the fulcrum. She wants to give each child in her preschool class a turn with her on the seesaw.
    a. If a child weighs w pounds, write an equation for the distance, d, the child should sit from the fulcrum.

    b. Fill in the table.

w	d
30	
36	
40	
45	
60	

    c. Graph your equation.

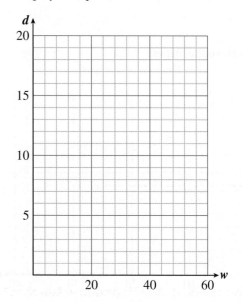

    d. Shawn's weight is $\frac{4}{3}$ of Shannon's weight. Where should he sit in order to balance each child on the seesaw?

Each table in Problems 5–8 represents inverse variation. Find the constant of variation, write an equation relating the variables, and complete the table.

5.

w	t
4.5	8
7.2	
	4.8

6.

z	h
32	21
42	
	14

7.

R	C
20	200
	250
125	

8.

m	T
0.2	50
0.8	
	6.25

Each graph in Problems 9–12 represents inverse variation. Find an equation relating the variables.

9.

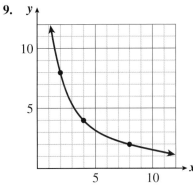

10.

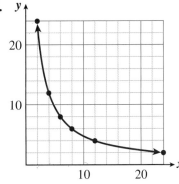

11.

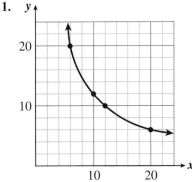

12.

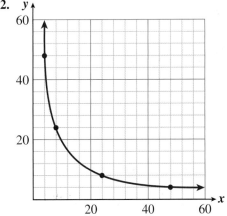

In Problems 13–16, each table presents data about the problem situation. Could the variables be related by inverse variation? Explain why or why not.

13. The frequency, F, of the note produced by a guitar string of length L

L	F
50	286
55	260
71.5	200

14. The concentration, c, of pollutant in a lake (in parts per million), after m months

m	c
2	76.8
5	39.32
10	4.3

15. The amount, g, of a radioactive substance remaining after t years

t	g
20	74.6
50	67.3
200	40

16. The pressure, P, exerted by a gas compressed to a volume V

V	P
40	1.2
32	1.5
19.2	2.5

In Problems 17–20, graph the equation of the form $y = \dfrac{k}{x}$, using both positive and negative values for x.

Note that the expression $\dfrac{k}{x}$ is undefined when $x = 0$.

a. First consider large positive values of x. What happens to y as x becomes large and positive?

b. Next evaluate $\dfrac{1}{x}$ for small positive values of x. What happens to y as x becomes small and positive?

c. Now look at negative values of x. What happens to y as x becomes "large" and negative?

d. Finally, consider negative values of x close to zero. What happens to y as x approaches zero from the negative side?

17. $y = \dfrac{1}{x}$

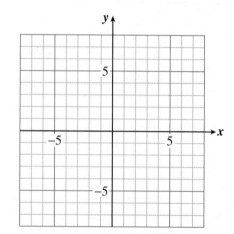

18. $y = \dfrac{-1}{x}$

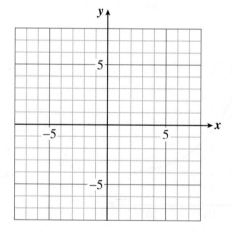

19. $y = \dfrac{-3}{x}$

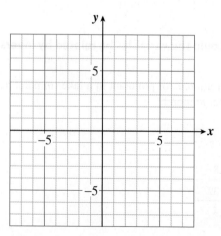

20. $y = \dfrac{2}{x}$

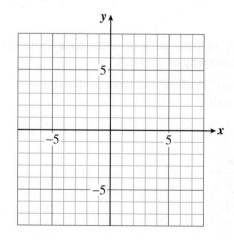

Match each equation in Problems 21–26 with its graph.

21. $y = 2x$

22. $y = x^2$

23. $y = x + 2$

24. $y = \dfrac{x}{2}$

25. $y = \dfrac{2}{x}$

26. $y = 2 - x$

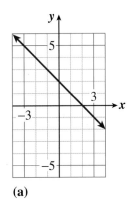

(a)

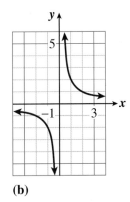

(b)

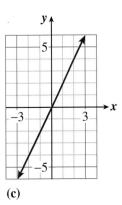

(c)

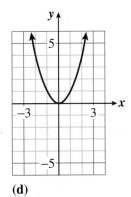

(d)

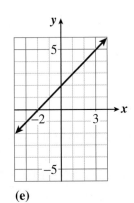

(e)

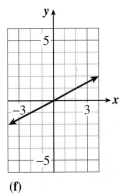

(f)

MIDCHAPTER REVIEW

In Problems 1–10, use complete sentences to answer the questions.

1. Explain why $\dfrac{x}{2} + 3$ is a polynomial but $\dfrac{2}{x} + 3$ is not.

2. A classmate says that $\sqrt{2}x^2 + 3\sqrt{2}x + 1$ is a polynomial, but another classmate disagrees. Who is correct? Explain.

3. If a polynomial is written in descending powers of the variable and the first term has degree 5, what is the degree of the polynomial? Give an example.

4. If a polynomial of degree 3 is added to a polynomial of degree 2, what is the degree of the sum? Give an example.

5. A classmate tells you that, depending on the binomials used, the sum of two binomials can have one, two, three, or four terms. Give an example of each.

6. If a polynomial of degree 3 is multiplied by a polynomial of degree 2, what is the degree of the product? Give an example.

7. How many terms are there in the square of a monomial? Of a binomial? Give examples.

8. What is a difference of two squares? Give examples.

9. Explain the difference between the sum of two squares and the square of a binomial. Give examples of each.

10. A classmate claims that if the variables x and y satisfy the equation $xy = k$ for some constant k, then y varies inversely with x. Explain why she is correct.

In Problems 11 and 12, write each polynomial in descending powers of the variable, and state the degree of the polynomial.

11. $1 - \dfrac{x^2}{2} + \dfrac{x^4}{24} - \dfrac{x^6}{720}$

12. $10n^4 + 10^6 n + 10^8 n^2$

Evaluate each polynomial in Problems 13–18 for the given value(s) of the variable(s).

13. $-16t^2 + 50t + 5$,　for　$t = \dfrac{1}{2}$

14. $\dfrac{1}{2}n^2 + \dfrac{1}{2}n$,　for　$n = 100$

15. $\dfrac{1}{6}z^3 + \dfrac{1}{2}z^2 + z + 1$,　for　$z = -1$

16. $p^4 + 4p^3 + 6p^2 + 4p + 1$,　for　$p = -2$

17. $4x^2 - 12xy + 9y^2$,　for　$x = -3$, $y = -2$

18. $2R^4 S$,　for　$R = 150$, $S = 0.01$

19. Suppose you want to choose four items from a list of n possible items. The number of different ways you can make your choice is given by the polynomial

$$\frac{1}{24}n^4 - \frac{1}{4}n^3 + \frac{11}{24}n^2 - \frac{1}{4}n$$

a. How many different sets of four compact disks can be chosen from a collection of 20 compact disks?

b. Of course, you cannot choose four different items from a list of only three possible items. What do you get when you evaluate the polynomial for　$n = 3$? $n = 2$?　$n = 1$?

c. Evaluate the polynomial for　$n = 4$.　Explain why your answer makes sense in terms of what the polynomial represents.

20. Brenda was flying at a constant altitude of 4000 feet before starting her descent for landing. Her altitude in feet is given by the polynomial

$$125x^3 - 750x^2 + 4000$$

where x is how far she traveled horizontally, in miles, after beginning her descent.

a. What was Brenda's altitude after she had traveled 2 miles horizontally?

b. What do you get when you evaluate the polynomial for　$x = 0$?　How does this compare with the information previously stated?

c. Evaluate the polynomial for　$x = 4$.　Explain what your answer means in the context of this problem.

In Problems 21–32, add, subtract, multiply, or divide the polynomials, as indicated.

21. $(3a^2 - 4a - 7) - (a^2 - 5a + 2)$

22. $(5ab^2 + 6a^2b) - (3ab^2 + 5ab)$

23. $\dfrac{13c^2d}{26c^3d}$

24. $\dfrac{12m^4 + 4m}{4m}$

25. $7q^2(8 - 7q^2 - q^4)$

26. $2k^2(-3km)(m^3k)$

27. $(9v + 5w)(9v - 5w)$

28. $(x^2 + 1)^2$

29. $-3p^2(p + 2)(p - 5)$

30. $12rs^2(3r - s)(r + 4s)$

31. $(2x - 3)(4x^2 + 6x + 9)$

32. $(3x + 2)(3x + 2)(3x + 2)$

For each rectangle in Problems 33–36:
a. Find its perimeter.
b. Find its area.

33.

$3a^4$

$2a^4$

34.

$7xy$

$5xy$

35.

$$m + 2n$$

$3m - n$ ▭

36.

$$4w + 9$$

$4w - 9$ ▭

37. Evaluate $(4a - 3b - 2c) - (a + 6b - 5c) + (3a + 9b - 2c)$ for $a = -6.3, \quad b = -4.8, c = 5.2.$

38. *Newsday* magazine surveyed 400 people and asked each the question "Do you think the government is spending too much on defense?" The magazine reported the following results: 72% of the college-educated respondents answered yes, and 48% of those without a college education answered yes. Suppose you would like to know how many of the 400 people surveyed answered yes. You will need to know how many of the 400 have a college education. Let x represent this unknown value. Write and simplify expressions in terms of x to answer each of the following questions.

a. How many of the people surveyed do not have a college education?

b. How many of the college-educated respondents answered yes?

c. How many of those without a college education answered yes?

d. How many people answered yes?

In Problems 39–50, decide whether each of the following is an equation or a polynomial. If it is an equation, solve it. If it is a polynomial, factor it.

39. $2x^2 + x - 3 = 0$

40. $a^2 - 9$

41. $2x^2 + x - 3$

42. $a^2 = 9$

43. $2x^2 + x = 0$

44. $a - 9 = 0$

45. $2x + 3 = 0$

46. $a^2 = 9a$

47. $2x^3 - 2x$

48. $a^4 - 16$

49. $p^3 - p = 0$

50. $n(n - 3)(n + 3) = 0$

In Problems 51 and 52, factor out the greatest common factor.

51. $12x^5 - 8x^4 + 20x^3$

52. $9a^4b^2 + 6a^3b^3 - 3a^2b^4$

In Problems 53 and 54, factor out a negative monomial.

53. $-10d^4 + 20d^3 - 5d^2$

54. $-6m^3n - 18m^2n + 6mn$

In Problems 55 and 56, factor out the common binomial factor.

55. $7q(q - 3) - (q - 3)$

56. $-r^3(3r + 2) + 4(3r + 2)$

Factor each expression in Problems 57–64 completely.

57. $3z^3 - 12z$

58. $-4x^3y + 8x^2y - 4xy$

59. $a^4 + 10a^2 + 25$

60. $4x^8 - 64$

61. $2a^2b^6 + 32a^6b^2$

62. $4p^2q^4 + 32p^3q^3 + 64p^4q^2$

63. $-2a^4b - 4a^3b^2 + 30a^2b^3$

64. $15r^3s^2 + 39r^2s^3 - 18rs^4$

Each table in Problems 65 and 66 represents inverse variation. Find the constant of variation, write an equation relating the variables, and complete the table.

65.

v	t
2.4	7.5
3.6	
	1.2

66.

P	V
48	75
60	
	15

Each graph in Problems 67 and 68 represents inverse variation. Find an equation relating the variables.

67.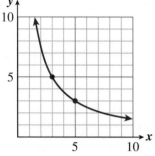

68.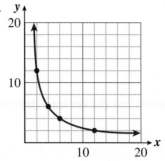

Each table in Problems 69 and 70 represents data about the problem situation. Could the variables be related by inverse variation? Explain why or why not.

69. The number of pages Emily has left to read in her book depends on the number of pages she has already read.

Pages already read	Pages left to read
25	263
56	232
72	216
97	191
112	176
149	139
178	110
201	87
222	66

70. The time required for a trip upstream depends on the speed of the current.

Speed (miles/hour)	Time (hours)
1	2
2	2.5
2.5	2.4
2.8	2.5
4	3
5	3.6
6	4.5
7	6
8	9
9	18

5.6 Algebraic Fractions

Reading

An **algebraic fraction** (or *rational expression*, as they are sometimes called) is a fraction in which both the numerator and the denominator are polynomials. Here are some examples of algebraic fractions:

$$\frac{3}{x} \qquad \frac{a^2 + 1}{a - 2} \qquad \frac{z - 1}{2z + 3}$$

Algebraic fractions can be evaluated just like any other algebraic expression.

EXAMPLE 1

Evaluate $\dfrac{a^2 + 1}{a - 2}$ for $a = 4$.

Solution

We substitute 4 for a in the fraction.

$$\frac{a^2 + 1}{a - 2} = \frac{4^2 + 1}{4 - 2} \qquad \text{Simplify numerator and denominator.}$$

$$= \frac{17}{2}$$

If we try to evaluate the fraction in Example 1 for $a = 2$, we get $\dfrac{2^2 + 1}{2 - 2}$, or $\dfrac{5}{0}$, which is undefined. When working with an algebraic fraction, we must exclude any values of the variable that make the denominator equal to zero.

Try Exercise 1.

Applications of Algebraic Fractions

Envirogreen Technology, Inc. decides to produce a new type of water filter for home use. The company spends $5000 for start-up costs, and each filter costs $50 to produce. So that she can decide on a selling price for the filters, the marketing manager would like to know the average cost per filter if the company produces x filters. She first computes the total cost of producing x filters:

$$\text{total cost} = \text{start-up cost} + \text{cost of } x \text{ filters}$$
$$= 5000 + 50x$$

Then, to find the average cost per filter, she divides the total cost by the number of filters produced:

$$\text{average cost} = \frac{\text{total cost}}{\text{number of filters}} = \frac{5000 + 50x}{x}$$

This expression is an algebraic fraction. We can evaluate the average cost for various production levels, x, as shown in the table on page 608.

EXERCISE 1

a. Evaluate $\dfrac{z - 1}{2z + 3}$ for $z = -3$.

b. For what value of z is the fraction undefined?

EXERCISE 2

Use the graph of average cost in Figure 5.2 and the table of values of x and A to help you answer the following questions.

a. Approximately how many filters should Envirogreen produce so that the average cost is $75?

b. What happens to the average cost per filter as x increases?

c. As x increases, the average cost appears to be approaching a limiting value. What is that value?

d. A market analysis concludes that Envirogreen can sell 1250 filters this year. How much should the company charge for one filter if management would like to make a total profit of $100,000?

x	50	100	200	400	500	1000	1250	2000
A	150	100	75	62.50	60	55	54	52.50

Figure 5.2 shows the graph of the equation

$$A = \frac{5000 + 50x}{x}$$

Figure 5.2

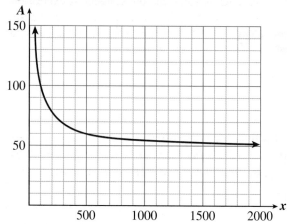

Try Exercise 2.

Skills Review

Factor completely.

1. $x^2 - 4$ 2. $x^2 - 4x$ 3. $x^2 - 4x + 4$
4. $4x^2 - 1$ 5. $4x^2 - 4x$ 6. $4x^2 + 4x + 1$

Answers: **1.** $(x - 2)(x + 2)$ **2.** $x(x - 4)$ **3.** $(x - 2)^2$
4. $(2x - 1)(2x + 1)$ **5.** $4x(x - 1)$ **6.** $(2x + 1)^2$

Lesson

Reducing Fractions

You'll recall from your study of arithmetic that we can **reduce** a fraction if we can divide both numerator and denominator by a common factor. For example,

$$\frac{6}{8} = \frac{\cancel{2} \cdot 3}{\cancel{2} \cdot 4} = \frac{3}{4}$$

In this case, both numerator and denominator have been divided by 2. The new fraction has the same value as the old one, namely 0.75, but it is simpler (the numbers are smaller). Reducing is an application of the fundamental principle of fractions.

Fundamental Principle of Fractions

We can multiply or divide the numerator and denominator of a fraction by the same nonzero factor, and the resulting fraction will be equivalent to the original fraction.

$$\frac{a \cdot c}{b \cdot c} = \frac{a}{b}, \quad \text{if} \quad b, c \neq 0$$

When we apply the fundamental principle, we often say that we are *canceling* common factors. Note carefully that "canceling" in this context means *dividing*: We can cancel common *factors* (expressions that are multiplied), but not common *terms* (expressions that are added or subtracted).

Try Exercise 3.

CAUTION!

You must be very careful to cancel only common factors, never common terms. For this reason, you should think of reducing fractions in two steps: first, *factor* the numerator and denominator completely; then *divide* by any common factors.

EXAMPLE 2

Reduce $\dfrac{3x + 12}{6x + 24}$.

Solution

We first factor the numerator and denominator completely:

$$\frac{3x + 12}{6x + 24} = \frac{3(x + 4)}{2 \cdot 3(x + 4)}$$

Then we divide numerator and denominator by any common factors:

$$\frac{3(x + 4)}{2 \cdot 3(x + 4)} = \frac{\cancel{3}\cancel{(x + 4)}}{2 \cdot \cancel{3}\cancel{(x + 4)}} = \frac{1}{2}$$

Note that if all the factors are canceled from the numerator, we replace them by 1, because any expression divided by itself is 1.

Try Exercise 4.

We summarize the procedure for reducing an algebraic fraction as follows:

To Reduce an Algebraic Fraction:

1. Factor the numerator and the denominator completely.

2. Divide both numerator and denominator by any common factors.

Try Exercises 5 and 6.

EXERCISE 3
Use your calculator to determine which calculation in each pair is correct. (In part b, choose a value for the variable and evaluate.)

a. $\dfrac{12}{8} = \dfrac{4 \cdot 3}{4 \cdot 2} \to \dfrac{3}{2},$

$\dfrac{7}{6} = \dfrac{4 + 3}{4 + 2} \to \dfrac{3}{2}$

b. $\dfrac{5x}{8x} \to \dfrac{5}{8},$

$\dfrac{x + 5}{x + 8} \to \dfrac{5}{8}$

EXERCISE 4
If you evaluate $\dfrac{3x + 12}{6x + 24}$ for $x = 2$ and for $x = -5$, what answer do you expect to get? (Refer to Example 2.)

EXERCISE 5
Reduce $\dfrac{x^2 - x - 6}{x^2 - 9}$.

Negative of a Binomial

Recall that any number (except zero) divided by itself is 1 and that any number divided by its opposite is -1. For example,

$$\frac{5}{5} = 1 \qquad \text{and} \qquad \frac{-5}{5} = -1$$

The same is true for binomials and other algebraic expressions. The opposite of an expression can be found by multiplying it by -1. Thus the opposite of $a - b$ is

$$-(a - b) = -a + b = b - a$$

and, consequently,

$$\frac{b-a}{a-b} = \frac{-(a-b)}{a-b} = -1$$

Opposites sometimes arise when we reduce fractions.

EXAMPLE 3

Reduce $\dfrac{2x - 4y}{6y - 3x}$.

Solution

First, we factor the numerator and the denominator:

$$\frac{2x - 4y}{6y - 3x} = \frac{2(x - 2y)}{3(2y - x)}$$

We see that $x - 2y$ is the opposite of $2y - x$; that is, $x - 2y = -(2y - x)$. Thus

$$\frac{2(x - 2y)}{3(2y - x)} = \frac{-2(2y - x)}{3(2y - x)} = \frac{-2}{3}$$

Try Exercises 7 and 8.

HOMEWORK 5.6

In Problems 1–4:
a. Evaluate each fraction for the given values of the variable.
b. For what values of the variable is the fraction undefined?

1. $\dfrac{x + 1}{x - 3}$,   for $x = 4, -4$

2. $\dfrac{x - 5}{x + 2}$,   for $x = 2, -3$

3. $\dfrac{2a - a^2}{a^2 + 1}$,   for $a = 3, -1$

4. $\dfrac{a^2 + 4}{3a - a^2}$,   for $a = 1, -5$

5. Which of the following fractions are undefined for $x = 1$?

  a. $\dfrac{1 - x}{x + 1}$ **b.** $\dfrac{1 + x}{1 - x}$

  c. $\dfrac{2x}{x^2 - 1}$ **d.** $\dfrac{x - 2}{x^2 - 2x + 1}$

6. For each fraction in Problem 5, find a value of x for which the fraction is equal to zero.

In Problems 7–12, write an algebraic fraction and then evaluate it.

7. Sharelle's car still had 4 gallons in the gas tank when she paid $18 for enough gas to fill it.
  a. If the gas tank holds x gallons, what was the price per gallon of the gasoline?

  b. Evaluate your fraction for $x = 14$. What does your answer mean in the context of the problem?

8. An office clerk ordered three boxes of envelopes plus 20 extra envelopes for a total cost of $40.
  a. If each box holds n envelopes, how much did each envelope cost?

  b. Evaluate your fraction for $n = 100$. What does your answer mean in the context of the problem?

9. Morgan drove across country in h hours, but he estimates that he spent 10 hours stopped for rest and meals.
  a. If he drove a total of 2800 miles, what was Morgan's average speed?

  b. Evaluate your fraction for $h = 50$. What does your answer mean in the context of the problem?

10. The Lake Michigan Ferry Company bought a new ferry that can travel 12 miles per hour faster than the old one, which had a top speed of s miles per hour.
  a. How long does the 90-mile trip from Ludington to Kewaunee take on the new ferry at its top speed?

  b. Evaluate your fraction for $s = 18$. What does your answer mean in the context of the problem?

11. The volume of a test tube is given by its height times the area of its cross-section. A test tube that holds 200 cubic centimeters is $2x - 1$ centimeters long.
  a. What is the area of its cross-section?

  b. Evaluate your fraction for $x = 13$. What does your answer mean in the context of the problem?

12. It took Rashid $2x + 3$ hours to type his autobiography.
  a. What fraction of the autobiography did Rashid type in 1 hour?

  b. What fraction of the autobiography did he type in 6 hours?

In Problems 13–18, divide the monomial into each term of the polynomial. Write your answer as the sum of a polynomial and one or more algebraic fractions.

13. $\dfrac{9x^5 - 6x^2 - 2}{3x^2}$

14. $\dfrac{6x^4 - 6x^2 - 4}{12x^2}$

15. $\dfrac{3n^3 - 3n^2 + 2n - 3}{3n^2}$

16. $\dfrac{2m^3 + 8m^2 + 2m - 1}{2m^2}$

17. $\dfrac{2x^2y^2 - 4xy^2 + 6xy}{2xy^2}$

18. $\dfrac{8x^3y + 4x^2y - 4xy}{x^2y}$

Reduce each fraction in Problems 19–30 to lowest terms.

19. $\dfrac{6x}{8x}$

20. $\dfrac{15y}{-12y}$

21. $\dfrac{-3a^3b^3}{15a^4b}$

22. $\dfrac{-5a^2b^2}{-25a^5b}$

23. $\dfrac{3x + 12}{6}$

24. $\dfrac{8 - 4y}{2y}$

25. $\dfrac{2a^2}{2a^2 - 6a}$

26. $\dfrac{-3b}{2b + 4b^2}$

27. $\dfrac{3(a + b)}{4(a + b)}$

28. $\dfrac{2x - 6y}{3x - 9y}$

29. $\dfrac{x - 4}{x^2 - 3x - 4}$

30. $\dfrac{u^2 + u - 6}{u^2 - 9}$

In Problems 31–36, reduce each fraction if possible, and decide whether answer a or b is the correct answer.

31. $\dfrac{x+2}{y+2}$ **a.** $\dfrac{x}{y}$ **b.** Cannot be reduced

32. $\dfrac{2x+3}{2y}$ **a.** $\dfrac{x+3}{y}$ **b.** Cannot be reduced

33. $\dfrac{2x+4}{4}$ **a.** $\dfrac{x+2}{2}$ **b.** $2x$

34. $\dfrac{3a+a^2}{3a}$ **a.** a^2 **b.** $\dfrac{3+a}{3}$

35. $\dfrac{y^2-1}{y-1}$ **a.** $y+1$ **b.** y

36. $\dfrac{a^3}{a^4-a^3}$ **a.** $\dfrac{1}{a-1}$ **b.** $\dfrac{1}{a^4}$

In Problems 37–40, decide whether each fraction is equivalent to 1, is equivalent to −1 , or cannot be reduced.

37. $\dfrac{x+4}{x-4}$ **38.** $\dfrac{t-5w}{5w-t}$

39. $\dfrac{x+3z}{z+3x}$ **40.** $\dfrac{-(m-1)}{1-m}$

41. The crew team can row at a steady pace of 10 miles per hour in still water. Every afternoon, the team's training includes a 5-mile row upstream on the river. If the current in the river on a given day flows at v miles per hour, then the time required for this workout, in minutes, is given by

$$t = \frac{300}{10 - v}$$

Use the graph of this equation shown below to answer the questions in parts (a) and (b).

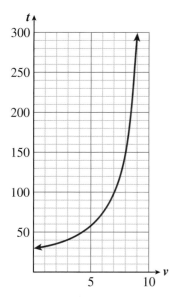

a. How long does the workout take if there is no current in the river?

b. How long does the workout take if the current is flowing at 4 miles per hour?

c. Find an exact answer for part (b) by using the equation.

d. If the workout took $2\frac{1}{2}$ hours, how fast was the current in the river?

e. As the speed of the current increases, what happens to the team's workout time? What happens when the current is flowing at 10 miles per hour?

42. A small lake in a state park has become polluted by runoff from a factory upstream. The cost of removing p percent of the pollutant from the lake, in thousands of dollars, is given by

$$C = \frac{25p}{100 - p}$$

Use the graph of this equation shown below to answer the questions in parts (a) and (b).

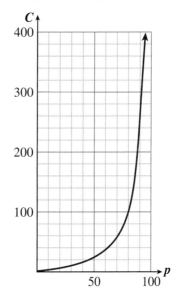

a. How much will it cost to remove 40% of the pollutant?

b. How much will it cost to remove 90% of the pollutant?

c. Find an exact answer for part (b) by using the equation. (Note that $p = 90$, not 0.90.)

d. How much of the pollutant can be removed for $100,000?

e. What happens to the cost as the amount of pollutant removed increases? How much will it cost to remove all of the pollutant?

Reduce each fraction in Problems 43–50, if possible.

43. $\dfrac{b - 2}{4 - 2b}$

44. $\dfrac{3 - y}{3y - 9}$

45. $\dfrac{a - b}{a^2 - b^2}$

46. $\dfrac{b + a}{b^2 - a^2}$

47. $\dfrac{(3x + 2y)^2}{4y^2 - 9x^2}$

48. $\dfrac{16y^2 - x^2}{(x - 4y)^2}$

49. $\dfrac{3a - a^2}{a^2 - 2a - 3}$

50. $\dfrac{a^2 - 3a + 2}{2a - a^2}$

Determine mentally which expressions in Problems 51–56 are equivalent to the given fraction. Do not use pencil, paper, or calculator.

51. $\dfrac{2}{3}$

a. $\dfrac{2 + 4}{3 + 4}$

b. $\dfrac{2 \cdot 4}{3 \cdot 4}$

c. $\dfrac{22}{33}$

d. $\dfrac{2 \div 4}{3 \div 4}$

52. $\dfrac{24}{32}$

a. $\dfrac{24 - 12}{32 - 12}$

b. $\dfrac{24 \div 6}{32 \div 8}$

c. $\dfrac{2.4}{3.2}$

d. $\dfrac{24^2}{32^2}$

53. $\dfrac{3}{5}$

    a. $\dfrac{3x}{5x}$

    b. $\dfrac{3-x}{5-x}$

    c. $\dfrac{30+x}{x+50}$

    d. $\dfrac{30x}{50x}$

54. $\dfrac{v-1}{v+1}$

    a. -1

    b. $\dfrac{1-v}{1+v}$

    c. $-\dfrac{1-v}{v+1}$

    d. $\dfrac{v+1}{v-1}$

55. $\dfrac{2w-6}{w}$

    a. -4

    b. $\dfrac{6-2w}{-w}$

    c. $2-\dfrac{6}{w}$

    d. $-\dfrac{2w+6}{w}$

56. $\dfrac{3a+2}{3a-2}$

    a. $\dfrac{3a-2}{3a+2}$

    b. $\dfrac{3a-2}{2-3a}$

    c. -1

    d. $\dfrac{2a+3}{2a-3}$

In Problems 57–60, explain what the error is in each calculation.

57. $\dfrac{3x+4}{3} \rightarrow x+4$

58. $\dfrac{y+6}{y-5} \rightarrow \dfrac{6}{-5}$

59. $\dfrac{2(5z-4)}{5z} \rightarrow \dfrac{2(-4)}{1} = -8$

60. $\dfrac{54}{15} \rightarrow \dfrac{4}{1} = 4$

5.7 Operations on Algebraic Fractions

Reading

Before considering operations on algebraic fractions, let's review how to add, subtract, multiply, and divide arithmetic fractions. Study each example; then try the corresponding exercise. One part of each exercise involves only arithmetic fractions, and the other part applies the same technique to simple algebraic fractions. If you have forgotten any of these skills, please refer to Appendices A.2 and A.3.

Multiplying Fractions

EXERCISE 1

a. $\dfrac{2}{5} \cdot \dfrac{2}{3} =$

b. $\dfrac{2}{w} \cdot \dfrac{z}{3} =$

EXAMPLE 1

$$\frac{5}{8} \cdot \frac{3}{4} = \frac{5 \cdot 3}{8 \cdot 4} = \frac{15}{32}$$ Multiply the numerators together; multiply the denominators together.

EXERCISE 2

a. $\dfrac{5}{4} \cdot \dfrac{8}{9} =$

b. $\dfrac{5}{2a} \cdot \dfrac{4a}{9} =$

EXAMPLE 2

$$\frac{3}{4} \cdot \frac{5}{6} = \frac{\cancel{3}}{4} \cdot \frac{5}{2 \cdot \cancel{3}} = \frac{5}{8}$$ Divide out common factors first; then multiply.

EXERCISE 3

a. $\dfrac{7}{3} \cdot 12 =$

b. $\dfrac{7}{x} \cdot 4x =$

EXAMPLE 3

$$\frac{3}{4} \cdot 6 = \frac{3}{4} \cdot \frac{6}{1} = \frac{3}{2 \cdot \cancel{2}} \cdot \frac{\cancel{2} \cdot 3}{1} = \frac{9}{2}$$

Dividing Fractions

EXERCISE 4

a. $\dfrac{8}{3} \div \dfrac{2}{9} =$

b. $\dfrac{4a}{3b} \div \dfrac{2a}{3} =$

EXAMPLE 4

$$\frac{12}{5} \div \frac{8}{5} = \frac{12}{5} \cdot \frac{5}{8}$$ Take the reciprocal of the second fraction; then multiply.

$$= \frac{\cancel{4} \cdot 3}{\cancel{5}} \cdot \frac{\cancel{5}}{\cancel{4} \cdot 2} = \frac{3}{2}$$

EXERCISE 5

a. $\dfrac{2}{3} \div 4 =$

b. $\dfrac{2}{3y} \div 4y =$

EXAMPLE 5

$$\frac{3}{5} \div 6 = \frac{3}{5} \cdot \frac{1}{6}$$

$$= \frac{\cancel{3}}{5} \cdot \frac{1}{\cancel{3} \cdot 2} = \frac{1}{10}$$

Adding or Subtracting Like Fractions

EXAMPLE 6

$$\frac{5}{4} + \frac{3}{4} = \frac{5+3}{4} = \frac{8}{4} = 2$$ Combine the numerators; keep the same denominator.

Skills Review

Simplify each expression.

1. $2x + 5 - (3x + 2)$
2. $4(x - 3) - 3(2x - 5)$
3. $2(3x - x^2) + x(2x - 4)$
4. $(x - 3)2x + (x + 2)4$
5. $x^2(x - 1) - 2x(x + 2)$
6. $3x(x + 4) - 2(x^2 - 16) - 4x$

Answers: **1.** $-x + 3$ **2.** $-2x + 3$ **3.** $2x$ **4.** $2x^2 - 2x + 8$
5. $x^3 - 3x^2 - 4x$ **6.** $x^2 + 8x + 32$

Lesson

Products of Fractions

To multiply two fractions, we multiply their numerators together and then multiply their denominators together.

Product of Fractions

If $b \neq 0$ and $d \neq 0$, then

$$\frac{a}{b} \cdot \frac{c}{d} = \frac{ac}{bd}$$

If a common factor occurs in a numerator and a denominator of either fraction, we can divide it out either before or after multiplying. For example,

$$\frac{3a}{4} \cdot \frac{5}{6a^2} = \frac{\cancel{3} \cdot \cancel{a}}{4} \cdot \frac{5}{\cancel{3} \cdot 2 \cdot \cancel{a} \cdot a} = \frac{5}{8a}$$

or

$$\frac{3a}{4} \cdot \frac{5}{6a^2} = \frac{15a}{24a^2} = \frac{\cancel{3} \cdot 5 \cdot \cancel{a}}{\cancel{3} \cdot 8 \cdot \cancel{a} \cdot a} = \frac{5}{8a}$$

When we are multiplying algebraic fractions, it is usually easier to cancel any common factors before multiplying.

EXAMPLE 7

Multiply $\dfrac{2x-4}{3x+6} \cdot \dfrac{6x+9}{x-2}$.

Solution

First, we factor each numerator and denominator. Then we divide numerator and denominator by any common factors.

$$\frac{2x-4}{3x+6} \cdot \frac{6x+9}{x-2} = \frac{2(x-2)}{3(x+2)} \cdot \frac{3(2x+3)}{x-2} \qquad \text{Multiply remaining factors of numerators and of denominators.}$$

$$= \frac{2(2x+3)}{x+2} \qquad \text{or} \qquad \frac{4x+6}{x+2}$$

To multiply a fraction by a whole number, we can write the whole number with a denominator of 1.

$$\frac{2x}{3} \cdot 4 = \frac{2x}{3} \cdot \frac{4}{1} = \frac{8x}{3}$$

The same applies to the product of an algebraic fraction and any nonfractional expression.

Try Exercise 7 now.

We summarize the procedure for multiplying algebraic fractions as follows:

To Multiply Algebraic Fractions:

1. Factor each numerator and denominator completely.
2. If any factor appears in both a numerator and a denominator, divide out that factor.
3. Multiply the remaining factors of the numerator and the remaining factors of the denominator.
4. Reduce the product if necessary.

Quotients of Fractions

To divide one fraction by another, we multiply the first fraction by the **reciprocal** of the second fraction. For example,

$$\frac{m}{2} \div \frac{2p}{3} = \frac{m}{2} \cdot \frac{3}{2p} = \frac{3m}{4p}$$

(To review the meaning of *reciprocal*, please see Appendix A.2.)

Quotient of Fractions

If $b, c, d \neq 0$, then

$$\frac{a}{b} \div \frac{c}{d} = \frac{a}{b} \cdot \frac{d}{c}$$

Thus, to divide two algebraic fractions, we take the reciprocal of the divisor and then follow the rules for multiplying algebraic fractions.

EXAMPLE 8

Divide $\dfrac{a-2}{6a^2} \div \dfrac{4-a^2}{4a^2-2a}$.

Solution

First, we change the operation to multiplication by taking the reciprocal of the divisor:

$$\frac{a-2}{6a^2} \cdot \frac{4a^2-2a}{4-a^2}$$

Now, we follow the rules for multiplication. We factor each numerator and denominator:

$$\frac{a-2}{6a^2} \cdot \frac{4a^2-2a}{4-a^2} = \frac{a-2}{3 \cdot 2 \cdot a \cdot a} \cdot \frac{2a(2a-1)}{(2-a)(2+a)}$$

Divide out common factors; note that $a - 2 = -1(2 - a)$.

$$= \frac{-1(2-a)}{3 \cdot 2 \cdot a \cdot a} \cdot \frac{2a(2a-1)}{(2-a)(2+a)}$$

Multiply.

$$= \frac{-1(2a-1)}{3a(2+a)} = \frac{1-2a}{3a(2+a)}$$

The denominators of algebraic fractions are often left in factored form.

If a nonfractional expression appears in a quotient, we can write it with a denominator of 1, just as we did for products. For example,

$$\frac{3}{2} \div 6a = \frac{3}{2} \div \frac{6a}{1}$$

$$= \frac{3}{2} \cdot \frac{1}{2 \cdot 3a} = \frac{1}{4a}$$

Try Exercise 8 now.

We summarize the procedure for dividing algebraic fractions as follows:

EXERCISE 8

Divide $\dfrac{6ab^2}{2a+3b} \div 4a^2b$.

> ## To Divide One Algebraic Fraction by Another:
>
> 1. Take the reciprocal of the second fraction, and change the operation to multiplication.
> 2. Follow the rules for multiplication of fractions.

Adding Like Fractions

Fractions with the same denominator are called **like fractions**. The following are pairs of like fractions:

$$\frac{5}{8} \text{ and } \frac{9}{8}, \qquad \frac{4}{5x} \text{ and } \frac{3}{5x}, \qquad \frac{1}{a-2} \text{ and } \frac{a}{a-2}$$

These pairs are unlike fractions:

$$\frac{2}{3} \text{ and } \frac{2}{5}, \qquad \frac{5}{x+1} \text{ and } \frac{2x}{x-1}$$

We can add or subtract two fractions only if they are like fractions, for the same reason that we can add only like terms. Just as

$$3x + 4x = 7x$$

we can think of the sum

$$\frac{3}{5} + \frac{4}{5}$$

as

$$3\left(\frac{1}{5}\right) + 4\left(\frac{1}{5}\right) = 7\left(\frac{1}{5}\right)$$

or $\frac{7}{5}$. The denominators of the terms must be the same, because they tell us what kind of quantity we are adding. We can add only quantities of the same kind.

When we add like fractions, we add their numerators and keep their denominators the same. For example,

$$\frac{10}{3x} + \frac{4}{3x} = \frac{10 + 4}{3x} = \frac{14}{3x} \qquad \text{Add the numerators; keep the same denominator.}$$

The same holds true for all algebraic fractions.

Sum or Difference of Fractions

If $c \neq 0$, then

$$\frac{a}{c} + \frac{b}{c} = \frac{a + b}{c}$$

and

$$\frac{a}{c} - \frac{b}{c} = \frac{a - b}{c}$$

EXAMPLE 9

Add $\dfrac{2x - 5}{x + 2} + \dfrac{x + 4}{x + 2}$.

Solution
Combine the numerators over a single denominator.

$$\frac{2x - 5}{x + 2} + \frac{x + 4}{x + 2} = \frac{(2x - 5) + (x + 4)}{x + 2} \qquad \text{Add like terms in the numerator.}$$

$$= \frac{3x - 1}{x + 2}$$

We summarize the procedure for adding or subtracting like fractions as follows:

To Add or Subtract Like Fractions:

1. Add or subtract the numerators.
2. Keep the same denominator.
3. Reduce the sum or difference if necessary.

Try Exercise 9 now.

Subtracting Like Fractions

We must be careful when subtracting algebraic fractions: A subtraction sign in front of a fraction applies to the *entire* numerator.

EXAMPLE 10

Subtract $\dfrac{x - 3}{x - 1} - \dfrac{3x - 5}{x - 1}$.

Solution

We combine the numerators over a single denominator. We use parentheses around $3x - 5$ to show that the subtraction applies to the entire numerator.

$$\frac{x - 3}{x - 1} - \frac{3x - 5}{x - 1} = \frac{(x - 3) - (3x - 5)}{x - 1} \qquad \text{Remove parentheses; distribute negative sign.}$$

$$= \frac{x - 3 - 3x + 5}{x - 1} \qquad \text{Combine like terms in the numerator.}$$

$$= \frac{-2x + 2}{x - 1}$$

We always check to see whether the fraction can be reduced. In this case, we can factor the numerator:

$$\frac{-2x + 2}{x - 1} = \frac{-2(x - 1)}{x - 1} = -2$$

Try Exercises 10 and 11.

EXERCISE 9

Add $\dfrac{2n}{n - 3} + \dfrac{n + 2}{n - 3}$.

EXERCISE 10

Evaluate the sum in Example 10 for $x = -9$. How do you know that the answer will be -2 without carrying out any calculations?

EXERCISE 11

Subtract

$$\frac{3}{x^2 + 2x + 1} - \frac{2 - x}{x^2 + 2x + 1}.$$

ANSWERS TO 5.7 EXERCISES

1a. $\dfrac{4}{15}$ **1b.** $\dfrac{2z}{3w}$ **2a.** $\dfrac{10}{9}$ **2b.** $\dfrac{10}{9}$ **3a.** 28 **3b.** 28

4a. 12 **4b.** $\dfrac{2}{b}$ **5a.** $\dfrac{1}{6}$ **5b.** $\dfrac{1}{6y^2}$ **6a.** $\dfrac{4}{3}$ **6b.** $\dfrac{8k}{n}$

7a. $\dfrac{-5a}{b}$ **7b.** $\dfrac{12}{x - 1}$ **8.** $\dfrac{3b}{4a^2 + 6ab}$ **9.** $\dfrac{3n + 2}{n - 3}$

10. The sum simplifies to -2. **11.** $\dfrac{1}{x + 1}$

HOMEWORK 5.7

Multiply the fractions in Problems 1–8.

1. $\dfrac{2}{3x^3} \cdot \dfrac{9x^2}{4}$

2. $\dfrac{-24a}{5b} \cdot \dfrac{15ab}{14}$

3. $\dfrac{2b}{3} \cdot \dfrac{4}{b+1}$

4. $\dfrac{-v}{v+1} \cdot \dfrac{v}{v-1}$

5. $\dfrac{3x-9}{5x-15} \cdot \dfrac{10x-5}{8x-4}$

6. $\dfrac{2x+4}{3x-12} \cdot \dfrac{2x-8}{5x+10}$

7. $\dfrac{5a+25}{5a} \cdot \dfrac{10a}{2a+10}$

8. $\dfrac{3b-18}{3b} \cdot \dfrac{6b}{4b-24}$

Write each product in Problems 9–12 as a fraction.

9. $\dfrac{2}{3}x$

10. $-2 \cdot \dfrac{y}{y+1}$

11. $\dfrac{3}{4}(a-b)$

12. $-5\left(\dfrac{x+2}{x-3}\right)$

In Problems 13–16, multiply.

13. $\dfrac{-2}{t^2}\left(4t^3 - \dfrac{t^2}{8} + \dfrac{3t}{2}\right)$

14. $\dfrac{3}{z}\left(z^3 - \dfrac{1}{3} - \dfrac{2}{z}\right)$

15. $\dfrac{4}{3}v\left(\dfrac{2}{3}v - \dfrac{6}{v^2} - \dfrac{3}{4v}\right)$

16. $\dfrac{n^3}{8}\left(\dfrac{2}{n^4} + \dfrac{4}{n^2} - \dfrac{1}{2}\right)$

Multiply the fractions in Problems 17–22.

17. $\dfrac{4V}{D} \cdot \dfrac{LR}{DV}$

18. $\dfrac{1}{2}MR^2 \cdot \dfrac{a}{R}$

19. $\dfrac{2L}{c}\left(1 + \dfrac{V^2}{c^2}\right)$

20. $\dfrac{4\pi}{c^2}\left(\dfrac{c}{4\pi}H + cM\right)$

21. $\dfrac{q}{8\pi}\left(\dfrac{3}{R} - \dfrac{a^2}{R^3}\right)$

22. $\dfrac{a^2}{d}\left(1 - \dfrac{at}{2d}\right)$

Divide the fractions in Problems 23–28.

23. $\dfrac{12c}{21d} \div \dfrac{24c}{27d}$

24. $\dfrac{a^4}{b^4} \div \dfrac{ab^2}{b^3}$

25. $\dfrac{2ab^3}{3} \div 4a^2b$

26. $\dfrac{3xy}{4x^4} \div (-12y^2)$

27. $1 \div \dfrac{x}{2y}$

28. $y \div \dfrac{3x}{-y^3}$

Divide the fractions in Problems 29–36.

29. $\dfrac{a^2 - ab}{ab} \div \dfrac{2a - 2b}{3ab}$

30. $\dfrac{2x - 2y}{5xy} \div \dfrac{4x - 4y}{xy}$

31. $\dfrac{3xy + x}{y^2 - y} \div \dfrac{3y + 1}{xy}$

32. $\dfrac{a^2 + ab}{2a^2 - ab} \div \dfrac{ab^2 + b^3}{4a - 2b}$

33. $\dfrac{6a^2 - 12a}{3a + 9} \div \dfrac{8a^2 - 4a^3}{15 + 5a}$

34. $\dfrac{4s^2 - t^2}{s^2 - 4t^2} \div \dfrac{6s^2 - 3st}{6t - 3s}$

35. $\dfrac{c^2 - 6c + 5}{c^2 + 2c - 15} \div \dfrac{c^2 - 3c - 10}{c^2 + 3c + 2}$

36. $\dfrac{2z^2 + 3z - 2}{2z^2 - 3z - 2} \div \dfrac{2z^3 - z^2}{z^2 - 4}$

Add or subtract the fractions in Problems 37–48.

37. $\dfrac{5}{2a} + \dfrac{3}{2a}$

38. $\dfrac{7}{2b} - \dfrac{5}{2b}$

39. $\dfrac{3}{x - 1} + \dfrac{5}{x - 1}$

40. $\dfrac{3w}{w + 5} - \dfrac{2}{w + 5}$

41. $\dfrac{x - 2y}{3x} + \dfrac{x + 3y}{3x}$

42. $\dfrac{p - 1}{p + 2} + \dfrac{p + 1}{2 + p}$

43. $\dfrac{m^2 + 1}{m - 1} - \dfrac{2m}{m - 1}$

44. $\dfrac{2a + b}{a - b} - \dfrac{a - 2b}{a - b}$

45. $\dfrac{z^2 - 2}{z + 2} - \dfrac{z + 4}{2 + z}$

46. $\dfrac{z^2 - 2z}{z - 3} - \dfrac{3z - 6}{z - 3}$

47. $\dfrac{b + 1}{b^2 - 2b + 1} - \dfrac{5 - 3b}{b^2 - 2b + 1}$

48. $\dfrac{2a + 1}{a^2 - a - 6} - \dfrac{a - 1}{a^2 - a - 6}$

49. a. Multiply $\dfrac{2m^2 - 8}{3m^2 - 3} \cdot \dfrac{6 - 6m}{2m^2 + 4m}$.

50. a. Multiply $\dfrac{n^2 - n - 20}{4n^2 - 9} \cdot \dfrac{2n^2 + 11n + 12}{25 - n^2}$.

b. Evaluate the expression in part (a) for $m = 3$.

b. Evaluate the expression in part (a) for $n = -4$.

In Problems 51 and 52, multiply.

51. $2x(x + 2)\left(\dfrac{1}{2x} + \dfrac{x}{x + 2} - 1\right)$

52. $3(x^2 - 1)\left(\dfrac{2}{3} - \dfrac{2}{x - 1} - \dfrac{4}{x + 1}\right)$

In Problems 53 and 54, add.

53. $\dfrac{y(y + 1)}{(y - 1)(y + 1)} + \dfrac{3(y - 1)}{(y + 1)(y - 1)}$

54. $\dfrac{2a(a + 1)}{2a(a - 3)} - \dfrac{4(a - 3)}{2a(a - 3)}$

The figure illustrates the following product:

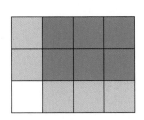

$$\frac{2}{3} \cdot \frac{3}{4} = \frac{6}{12} = \frac{1}{2}$$

Sketch similar figures to illustrate the products in Problems 55 and 56.

55. $\dfrac{5}{6} \cdot \dfrac{3}{4}$

56. $\dfrac{5}{8} \cdot \dfrac{2}{3}$

Write an algebraic expression for each phrase in Problems 57–59, and simplify the expression.

57. a. One-half of x

b. x divided by one-half

c. One-half divided by x

59. a. The reciprocal of $a + b$

b. Three-fourths of the reciprocal of $a + b$

c. The reciprocal of $a + b$ divided by three-fourths

58. a. Two-thirds of y

b. y divided by two-thirds

c. Two-thirds divided by y

60. Simplify each expression:

a. $\left(\dfrac{1}{c} \cdot \dfrac{1}{5}\right) \div \dfrac{2}{5}$

b. $\dfrac{1}{c} \cdot \left(\dfrac{1}{5} \div \dfrac{2}{5}\right)$

c. $\dfrac{1}{c} \div \left(\dfrac{1}{5} \div \dfrac{2}{5}\right)$

d. $\left(\dfrac{1}{c} \div \dfrac{1}{5}\right) \div \dfrac{2}{5}$

Simplify the expressions in Problems 61–66 mentally, without using pencil and paper.

61. $\dfrac{x + 3}{4} \cdot 4$

62. $\dfrac{2z + 5}{z} \cdot z$

63. $6n \cdot \dfrac{2n + 3}{2n}$

64. $-7s \cdot \dfrac{9s - 2}{7s}$

65. $\dfrac{6w - 1}{-3w} \cdot 9w$

66. $\dfrac{5c^2 + 3c - 2}{c - 8}(8 - c)$

5.8 Lowest Common Denominators

Reading

Sometimes mathematics can help us solve problems when our intuition fails us. Consider the following problem:

On a weekday afternoon, when traffic is always extremely heavy, Kathy left her home in the San Fernando Valley north of Los Angeles and drove south 120 miles along the San Diego Freeway to San Juan Capistrano. Her average speed was 40 miles per hour. She returned home on Saturday, at an average speed of 60 miles per hour. What was her average speed for the round trip?

If you said that Kathy's average speed was 50 miles per hour, you were wrong! Let's do some calculations. Kathy's average speed for the round trip is given by

$$\text{average speed} = \frac{\text{total distance}}{\text{total time}}$$

The total distance she drove is 240 miles, but to find the total time, we must compute the time she drove on each part of the trip. We use the formula $d = rt$ and solve for t.

	d	r	t
Driving South	120	40	3
Driving North	120	60	2

The total time for the round trip was $3 + 2 = 5$ hours, so Kathy's average speed was

$$\frac{240}{5} = 48 \text{ miles per hour}$$

Why does the average speed turn out to be less than 50 miles per hour? Because Kathy spent more time driving at 40 miles per hour (3 hours) than she did driving at 60 miles per hour (2 hours).

By generalizing the problem above, we can find an algebraic formula for the average speed on a two-part trip. Suppose that the distances traveled on the two parts of the trip are d_1 and d_2 and that the corresponding speeds on the two parts are r_1 and r_2. We fill in a table to find the time required for each part:

	Distance	*Rate*	*Time*
First part	d_1	r_1	$\dfrac{d_1}{r_1}$
Second part	d_2	r_2	$\dfrac{d_2}{r_2}$

The total distance traveled on the trip is $d_1 + d_2$, and the total time required is $\dfrac{d_1}{r_1} + \dfrac{d_2}{r_2}$. Thus the average speed for the entire trip is

$$\text{average speed} = \frac{\text{total distance}}{\text{total time}} = \frac{d_1 + d_2}{\dfrac{d_1}{r_1} + \dfrac{d_2}{r_2}}$$

EXERCISE 1
Bruce drove 30 miles in the city at an average speed of 40 miles per hour. Then he drove 105 miles on the highway at an average speed of 70 miles an hour. Use the formula we discovered in the reading to find his average speed for the entire trip.

The last expression is called a *complex fraction*, one of the topics we'll consider in the lesson.

Try Exercise 1.

Skills Review

Review Appendix A.3 on finding the lowest common denominator. Then find the LCD for each pair of fractions.

1. $\dfrac{1}{6}, \dfrac{3}{8}$ 2. $\dfrac{3}{5}, \dfrac{2}{25}$ 3. $\dfrac{7}{20}, \dfrac{7}{12}$ 4. $\dfrac{2}{7}, \dfrac{4}{9}$

Review Appendix A.1 on building fractions. Then, for each of the following fractions, write an equivalent fraction with the given denominator.

5. $\dfrac{2}{3} = \dfrac{?}{15}$ 6. $\dfrac{5}{8} = \dfrac{?}{24}$ 7. $\dfrac{1}{6} = \dfrac{?}{36}$ 8. $\dfrac{5}{16} = \dfrac{?}{144}$

Answers: **1.** 24 **2.** 25 **3.** 60 **4.** 63 **5.** $\dfrac{10}{15}$ **6.** $\dfrac{15}{24}$ **7.** $\dfrac{6}{36}$

8. $\dfrac{45}{144}$

Lesson

Adding and Subtracting Unlike Fractions

To add or subtract fractions with unlike denominators, we must first convert the fractions into equivalent forms with the same denominator. For example, to compute the sum

$$\frac{2}{3} + \frac{3}{4}$$

we first find the lowest common denominator for the two fractions: The smallest number that is a multiple of both 3 and 4 is 12. Next, we "build" each fraction to form an equivalent fraction with a denominator of 12:

$$\frac{2}{3} = \frac{2 \cdot 4}{3 \cdot 4} = \frac{8}{12}$$

$$\frac{3}{4} = \frac{3 \cdot 3}{4 \cdot 3} = \frac{9}{12}$$

Building is an application of the fundamental principle of fractions,

$$\frac{a}{b} = \frac{a \cdot c}{b \cdot c}, \quad \text{if} \quad c \neq 0$$

The new fractions are like fractions, and we can add them by combining their numerators:

$$\frac{2}{3} + \frac{3}{4} = \frac{8}{12} + \frac{9}{12} = \frac{17}{12}$$

We add or subtract algebraic fractions in the same way.

To Add or Subtract Algebraic Fractions:

1. Find the lowest common denominator (LCD) for the fractions.
2. Build each fraction to form an equivalent one with the same denominator.
3. Add or subtract the numerators of the resulting like fractions, and keep the same denominator.
4. Reduce the sum or difference if necessary.

EXAMPLE 1

Subtract $\dfrac{x + 2}{6} - \dfrac{x - 1}{15}$.

Solution

Step 1 We find the LCD: The smallest multiple of 6 and 15 is 30.

Step 2 We must build each fraction to form an equivalent one with a denominator of 30. The building factor is 5 for the first fraction and 2 for the second fraction.

$$\frac{x + 2}{6} = \frac{(x + 2) \cdot 5}{6 \cdot 5} = \frac{5x + 10}{30}$$

$$\frac{x - 1}{15} = \frac{(x - 1) \cdot 2}{15 \cdot 2} = \frac{2x - 2}{30}$$

Step 3 We subtract the resulting like fractions:

$$\frac{x + 2}{6} - \frac{x - 1}{15} = \frac{5x + 10}{30} - \frac{2x - 2}{30} \qquad \text{Combine the numerators;}$$
$$\text{keep the same denominator.}$$

$$= \frac{(5x + 10) - (2x - 2)}{30} \qquad \text{Simplify the numerator.}$$

$$= \frac{5x + 10 - 2x + 2}{30} = \frac{3x + 12}{30}$$

Step 4 We reduce the fraction:

$$\frac{3x + 12}{30} = \frac{\cancel{3}(x + 4)}{\cancel{3} \cdot 10} = \frac{x + 4}{10}$$

Try Exercise 2.

Finding the Lowest Common Denominator

The *lowest common denominator* for two or more algebraic fractions is the simplest algebraic expression that is a multiple of each denominator. If neither denominator factors, the LCD is the product of the two denominators. For example, the LCD for the fractions

$$\frac{6}{x} \qquad \text{and} \qquad \frac{x}{x - 2}$$

is $x(x - 2)$.

EXERCISE 2

Add $\dfrac{5x}{6} + \dfrac{3 - 2x}{4}$.

EXAMPLE 2

Add $\dfrac{6}{x} + \dfrac{x}{x-2}$.

Solution

Step 1 The LCD is $x(x-2)$.

Step 2 We build each fraction to form an equivalent one with denominator $x(x-2)$. To find the building factors, we compare the given denominator with the desired LCD. The building factor for each fraction is the "missing" factor. Thus the building factor for $\dfrac{6}{x}$ is $x-2$, and the building factor for $\dfrac{x}{x-2}$ is x.

$$\frac{6}{x} = \frac{6(x-2)}{x(x-2)} = \frac{6x-12}{x(x-2)}$$

$$\frac{x}{x-2} = \frac{x(x)}{(x-2)(x)} = \frac{x^2}{x(x-2)}$$

Step 3 We combine the resulting like fractions:

$$\frac{6}{x} + \frac{x}{x-2} = \frac{6x-12}{x(x-2)} + \frac{x^2}{x(x-2)} \qquad \text{Combine the numerators; keep the same denominator.}$$

$$= \frac{x^2 + 6x - 12}{x(x-2)}$$

Step 4 The numerator of this fraction cannot be factored, so the sum cannot be reduced.

Try Exercise 3 now.

If the denominators contain any common factors, their product is not the simplest common denominator. For example, the LCD for

$$\frac{5}{12} + \frac{7}{18}$$

is not $12(18) = 216$. It is true that 216 is a multiple of both 12 and 18, but it is not the smallest one! We can find a smaller common denominator by factoring each denominator.

$$12 = 2 \cdot 2 \cdot 3$$
$$18 = 2 \cdot 3 \cdot 3$$

To find a number that both 12 and 18 divide into evenly, we need only enough factors to "cover" each of them. In this case, two 2's and two 3's are sufficient, so

$$\text{LCD} = 2 \cdot 2 \cdot 3 \cdot 3 = 36$$

(Note that 36 is a multiple of both 12 and 18.)

In general, we can find the LCD in the following way.

EXERCISE 3

Follow the suggested steps to subtract $\dfrac{2}{x+2} - \dfrac{3}{x-2}$.

Step 1 Find the LCD.

Step 2 Build each fraction.

Step 3 Subtract like fractions.

Step 4 Reduce if possible.

To Find the LCD:

1. Factor each denominator completely.

2. Include in the LCD each factor the greatest number of times that it occurs in any single denominator.

Now try Exercise 4.

E X A M P L E   3

Add $\dfrac{x-4}{x^2-2x}+\dfrac{4}{x^2-4}$.

EXERCISE 4

Find the LCD for $\dfrac{3}{4x^2}+\dfrac{5}{6xy}$.

Solution

Step 1 To find the LCD, we first factor each denominator completely.

$$x^2-2x=x(x-2)$$
$$x^2-4=(x-2)(x+2)$$

The LCD is $x(x-2)(x+2)$.

Step 2 We build each fraction to form an equivalent one with the LCD as its denominator.

$$\frac{x-4}{x(x-2)}=\frac{(x-4)(x+2)}{x(x-2)(x+2)}=\frac{x^2-2x-8}{x(x-2)(x+2)}$$
$$\frac{4}{(x-2)(x+2)}=\frac{4x}{(x-2)(x+2)x}$$

Step 3 Because the fractions are now "like" fractions, we add them by combining their numerators:

$$\frac{x-4}{x^2-2x}+\frac{4}{x^2-4}=\frac{x^2-2x-8}{x(x-2)(x+2)}+\frac{4x}{x(x-2)(x+2)}$$
$$=\frac{x^2+2x-8}{x(x-2)(x+2)}$$

Step 4 We reduce the fraction by first factoring numerator and denominator:

$$\frac{x^2+2x-8}{x(x-2)(x+2)}=\frac{(x+4)\cancel{(x-2)}}{x\cancel{(x-2)}(x+2)}=\frac{x+4}{x(x+2)}$$

Complex Fractions

A fraction that contains one or more fractions in its numerator or denominator or both is called a **complex fraction**. For example, these are complex fractions:

$$\frac{\dfrac{1}{2}}{\dfrac{3}{4}}\qquad\text{and}\qquad\frac{\dfrac{1}{a}+\dfrac{a}{a-2}}{a-\dfrac{1}{a}}$$

All complex fractions can be simplified into standard fractions to make them easier to work with. For example, the first complex fraction above is really equal to $\frac{2}{3}$. You can see this by recalling that all fractions represent quotients, so

$$\frac{\dfrac{1}{2}}{\dfrac{3}{4}}=\frac{1}{2}\div\frac{3}{4}=\frac{1}{2}\cdot\frac{4}{3}=\frac{2}{3}$$

Any complex fraction can be simplified by dividing the denominator into the numerator. However, this can be a long process for algebraic fractions. Consider the second complex fraction above. If we rewrite this fraction as a division, we have

$$\frac{\dfrac{1}{a} + \dfrac{a}{a-2}}{a - \dfrac{1}{a}} = \left(\frac{1}{a} + \frac{a}{a-2}\right) \div \left(a - \frac{1}{a}\right)$$

To simplify this last expression, we must first perform the additions and subtractions inside parentheses and then divide the results. Luckily, there is a shorter way to simplify complex fractions.

Using an LCD to Simplify a Complex Fraction

Consider again the first complex fraction given above:

$$\frac{\dfrac{1}{2}}{\dfrac{3}{4}}$$

The complex fraction is made up of the simple fractions $\dfrac{1}{2}$ and $\dfrac{3}{4}$. The lowest common denominator for these fractions is 4. We multiply the numerator and the denominator of the complex fraction by 4. We are applying the fundamental principle of fractions, so the result will be equivalent to the original fraction.

$$\frac{\dfrac{1}{2} \cdot 4}{\dfrac{3}{4} \cdot 4} = \frac{\dfrac{1}{2} \cdot \dfrac{4}{1}}{\dfrac{3}{4} \cdot \dfrac{4}{1}} = \frac{2}{3}$$

The result is a simple fraction, as desired. Multiplying by the LCD clears all the denominators of the simple fractions within the complex fraction. This method works on algebraic fractions as well.

To Simplify a Complex Fraction:

1. Find the LCD for all the simple fractions within the complex fraction.
2. Multiply the numerator and denominator of the complex fraction by the LCD.
3. Reduce the result if necessary.

Try Exercise 5.

EXERCISE 5

Simplify $\dfrac{\dfrac{2x}{y^3}}{\dfrac{x}{3y}}$.

EXAMPLE 4

Simplify $\dfrac{\dfrac{1}{2x} - \dfrac{1}{x^2}}{\dfrac{1}{4} - \dfrac{1}{2x}}$.

Solution

This complex fraction contains the simple fractions $\dfrac{1}{2x}, \dfrac{1}{x^2}, \dfrac{1}{4}$, and $\dfrac{1}{2x}$. The LCD of these fractions is $4x^2$. We multiply each term of the numerator and the denominator of the complex fraction by $4x^2$, applying the distributive law:

$$\frac{4x^2\left(\dfrac{1}{2x} - \dfrac{1}{x^2}\right)}{4x^2\left(\dfrac{1}{4} - \dfrac{1}{2x}\right)} = \frac{\dfrac{4x^2}{1} \cdot \dfrac{1}{2x} - \dfrac{4x^2}{1} \cdot \dfrac{1}{x^2}}{\dfrac{4x^2}{1} \cdot \dfrac{1}{4} - \dfrac{4x^2}{1} \cdot \dfrac{1}{2x}} = \frac{2x - 4}{x^2 - 2x}$$

Finally, we reduce the result to obtain

$$\frac{2x - 4}{x^2 - 2x} = \frac{2(x - 2)}{x(x - 2)} = \frac{2}{x}$$

Try Exercises 6 and 7.

Try Exercises 6 and 7.

EXERCISE 6

Evaluate the complex fraction in Example 4 for $x = 8$. Explain how you know that the answer is $\dfrac{1}{4}$ without doing a lot of calculations.

EXERCISE 7

Follow the suggested steps to simplify $\dfrac{1 + \dfrac{b}{a}}{1 - \dfrac{bc}{ad}}$.

Step 1 Find the LCD of $\dfrac{b}{a}$ and $\dfrac{bc}{ad}$.

Step 2 Multiply *each term* of the numerator and the denominator of the complex fraction by the LCD from Step 1.

Step 3 Reduce your result if necessary.

ANSWERS TO 5.8 EXERCISES

1. 60 miles per hour

2. $\dfrac{4x + 9}{12}$

3. $\dfrac{-x - 10}{(x + 2)(x - 2)}$

4. $12x^2 y$

5. $\dfrac{6}{y^2}$

6. The complex fraction is equivalent to $\dfrac{2}{x}$, and evaluating $\dfrac{2}{x}$ for $x = 8$ gives $\dfrac{1}{4}$.

7. $\dfrac{ad + bd}{ad - bc}$

HOMEWORK 5.8

Add or subtract the fractions in Problems 1–12.

1. $\dfrac{3}{x} - \dfrac{4}{y}$

2. $\dfrac{5}{a} + \dfrac{2}{b}$

3. $1 - \dfrac{3}{a}$

4. $\dfrac{1}{b} + 2$

5. $\dfrac{1}{x} + \dfrac{x}{y} + \dfrac{x}{z}$

6. $\dfrac{u}{v} - \dfrac{v}{w} - \dfrac{w}{u}$

7. $\dfrac{u - 4}{3} + \dfrac{6}{v}$

8. $\dfrac{t + 2}{t} - \dfrac{t + 3}{2}$

9. $\dfrac{3}{x} - \dfrac{2}{x + 1}$

10. $\dfrac{1}{2} + \dfrac{m}{m - 2}$

11. $\dfrac{3}{n + 3} + \dfrac{4n}{n - 3}$

12. $\dfrac{x}{x + 1} - \dfrac{2}{x - 2}$

Write algebraic fractions in simplest form for Problems 13–18.

13. The dimensions of a rectangular rug are $\dfrac{12}{x}$ feet and $\dfrac{12}{x-2}$ feet.

    a. Write an expression for the area of the rug.

    b. Write an expression for the perimeter of the rug.

14. The dimensions of a rectangular wading pool are $\dfrac{20}{x}$ feet and $\dfrac{20}{x+3}$ feet.

    a. Write an expression for the area of the pool.

    b. Write an expression for the perimeter of the pool.

15. Colonial Airline has a commuter flight between Richmond, Virginia and Washington, DC, a distance of 100 miles. The plane flies at x miles per hour in still air. Today, there is a steady wind from the north at 10 miles per hour.

    a. How long will the flight north from Richmond to Washington take today?

    b. How long will the flight from Washington to Richmond take?

    c. How long will a round trip take?

    d. Evaluate your fractions in parts (a), (b), and (c) for $x = 150$.

16. Vacation Cruise Company is planning a cruise on the Savannah River from Augusta downstream to Savannah, a distance of 150 miles, and back. The boat travels at x miles per hour in still water, and the current in the river is 6 miles per hour.

    a. How long will the cruise downstream take?

    b. How long will the cruise upstream take?

    c. How long will the round trip take?

    d. Evaluate your fractions in parts (a), (b), and (c) for $x = 40$.

17. Francine's cocker spaniel eats a large bag of dog food in d days. Delbert's sheepdog takes 5 fewer days to eat the same size bag.
  a. What fraction of a bag of dog food does Francine's cocker spaniel eat in 1 day?

  b. What fraction of a bag of dog food does Delbert's sheepdog eat in 1 day?

  c. If Delbert and Francine get married, what fraction of a bag of dog food will their dogs eat in 1 day?

  d. If $d = 25$, how often will Delbert and Francine have to buy more dog food?

18. The *Daily Bugle* bought a new printing press last week. The new press can print the morning edition 2 hours faster than the old press can. The old press requires h hours for the printing.
  a. What fraction of the morning edition can the old press print in 1 hour?

  b. What fraction of the morning edition can the new press print in 1 hour?

  c. If the *Bugle* runs both presses, what fraction of the morning edition can be printed in 1 hour?

  d. If $h = 4$, how long will it take to print the morning edition with both presses running?

In Problems 19–24, add or subtract, as indicated.

19. $h - \dfrac{3}{h + 2}$

20. $d - \dfrac{4}{d + 3}$

21. $\dfrac{v + 1}{v} + \dfrac{1}{v - 1}$

22. $\dfrac{3k}{3k - 4} - \dfrac{5}{k + 6}$

23. $\dfrac{2}{x} + \dfrac{x}{x - 2} - 2$

24. $\dfrac{3y}{y + 2} - \dfrac{2}{y} + 2$

In Problems 25–30:
a. Find the lowest common denominator for the fractions.
b. Add or subtract the fractions.

25. $\dfrac{5}{2x} + \dfrac{3}{4x^2}$

26. $\dfrac{1}{6a} - \dfrac{3}{8b}$

27. $\dfrac{2z - 3}{8z} + \dfrac{z - 2}{6z}$

28. $\dfrac{5r - 2}{2} - \dfrac{3r + 1}{6r}$

29. $\dfrac{3}{2a - b} + \dfrac{1}{8a - 4b}$

30. $\dfrac{7}{5w - 10} - \dfrac{5}{3w - 6}$

In Problems 31–42:
a. Find the lowest common denominator for the fractions.
b. Add or subtract the fractions.

31. $\dfrac{1}{x} + \dfrac{1}{d - x}$

32. $\dfrac{1}{n} - \dfrac{1}{n - 1}$

33. $a + \dfrac{N - a^2}{2a}$

34. $\dfrac{s + 1}{t + 1} - \dfrac{s}{t}$

35. $\dfrac{1}{2} \cdot \dfrac{a}{t} - \dfrac{m}{a}$

36. $\dfrac{-H}{RT} + \dfrac{S}{R}$

37. $\dfrac{q}{4\pi r} + \dfrac{qa}{2\pi r^2}$

38. $\dfrac{1}{LC} - \left(\dfrac{R}{2L}\right)^2$

39. $\dfrac{L}{c - V} + \dfrac{L}{c + V}$

40. $\left(\dfrac{a}{a + b}\right)x + \left(\dfrac{b}{a + b}\right)y$

41. $\dfrac{q}{r - a} - \dfrac{2q}{r} + \dfrac{q}{r + a}$

42. $\dfrac{2r^2}{a^2} + \dfrac{2r}{a} + 1$

Simplify each complex fraction in Problems 43–54.

43. $\dfrac{\dfrac{3x}{y}}{\dfrac{x}{2y^2}}$

44. $\dfrac{\dfrac{4a}{5b^2}}{\dfrac{8a^3}{15b}}$

45. $\dfrac{1 - \dfrac{1}{6}}{2 + \dfrac{2}{3}}$

46. $\dfrac{\dfrac{1}{2} + \dfrac{1}{3}}{\dfrac{1}{3} - \dfrac{1}{6}}$

47. $\dfrac{n}{\dfrac{p}{q} + 1}$

48. $\dfrac{1}{1 - \dfrac{2m}{r}}$

49. $\dfrac{4 - \dfrac{1}{x^2}}{2 - \dfrac{1}{x}}$

50. $\dfrac{a + b}{\dfrac{1}{a} + \dfrac{1}{b}}$

51. $\dfrac{\dfrac{b^2}{d} + d}{2}$

52. $\dfrac{\dfrac{u}{x} - \dfrac{v}{x}}{v}$

53. $\dfrac{1 + \dfrac{x^2}{y^2}}{\dfrac{x}{y}}$

54. $\dfrac{\dfrac{x}{t} - V}{1 - V\left(\dfrac{x}{t}\right)}$

55. Find the slope of the line through $\left(\dfrac{-4}{3}, 2\right)$ and $\left(\dfrac{3}{2}, \dfrac{-5}{3}\right)$.

56. Find the slope of the line through $\left(4, \dfrac{5}{6}\right)$ and $\left(-1, \dfrac{8}{3}\right)$.

57. a. On the figure below, locate the points $P\left(\dfrac{a}{2}, 0\right)$, $Q\left(a, \dfrac{b}{2}\right)$, $R\left(\dfrac{a}{2}, b\right)$, and $S\left(0, \dfrac{b}{2}\right)$. Use line segments to connect the points in the order *PQRS* to form a four-sided figure.

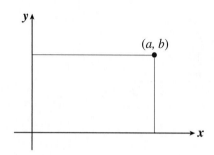

b. Compute the slope of each side of the figure.

58. a. Suppose x and y are two positive numbers. Write an expression for their average and another for the reciprocal of their average.

b. Write an expression for the average of the reciprocals of the same numbers x and y.

c. Are your expressions in parts (a) and (b) the same? Choose values for x and y, and evaluate both expressions.

Write each expression in Problems 59–64 as a single fraction in simplest form.

59. $\left(1 - \dfrac{k}{n}\right)\left(1 + \dfrac{k}{n}\right)$

60. $m\left(\dfrac{v - V}{t}\right)\left(\dfrac{v + V}{2}\right)$

61. $\dfrac{2d}{c} \cdot \dfrac{1}{1 - \left(\dfrac{u}{c}\right)^2}$

62. $\dfrac{1}{\dfrac{1}{n}\left(\dfrac{1}{n} - 1\right)}$

63. $\dfrac{\dfrac{L}{F}}{\dfrac{L}{F} - 1} \cdot \dfrac{K}{N}$

64. $b\left(\dfrac{a - b}{a + b}\right) + b$

Add or subtract the fractions in Problems 65–74.

65. $\dfrac{2x+3}{x-1}+\dfrac{2x-5}{1-x}$

66. $\dfrac{w-3}{2w-3}-\dfrac{2w}{3-2w}$

67. $\dfrac{5}{2p-4}-\dfrac{2}{6-3p}$

68. $\dfrac{3}{4h-12}-\dfrac{4}{9-3h}$

69. $\dfrac{-2}{m^2+3m}+\dfrac{1}{m^2-9}$

70. $\dfrac{-5}{r^2-2r}+\dfrac{2}{r^2-4}$

71. $\dfrac{4}{k^2-3k}+\dfrac{1}{k^2+k}$

72. $\dfrac{6}{q^2+2q}+\dfrac{3}{q^2-q}$

73. $\dfrac{x-1}{x^2+3x}+\dfrac{x}{x^2+6x+9}$

74. $\dfrac{3t}{t^2+3t-10}-\dfrac{2t}{t^2+t-6}$

Simplify each complex fraction in Problems 75–78.

75. $\dfrac{A(T_2 - T_1)}{\dfrac{L_1}{K_1} + \dfrac{L_2}{K_2}}$

76. $\dfrac{1}{\dfrac{1}{C_1} + \dfrac{1}{C_2} + \dfrac{1}{C_3}}$

77. $\dfrac{1 - \dfrac{2h}{m}}{m - \dfrac{4h^2}{m}}$

78. $\dfrac{\dfrac{6}{h+2} - \dfrac{3}{h}}{\dfrac{4}{h} + \dfrac{3}{h+2}}$

5.9   Equations with Algebraic Fractions

Reading

In Section 5.6, we considered the cost of producing x water filters. The start-up cost for making the filters was $5000, and each filter cost $50 to produce. To find the average cost per water filter, we divided the total cost, $5000 + 50x$,   by the number of filters produced, x, to get

$$A = \frac{5000 + 50x}{x}$$

We used the graph of this equation, shown in Figure 5.2, to find the number of water filters that Envirogreen should produce if it wants the cost per filter to be $75. In other words, we used the graph to solve the equation

$$\frac{5000 + 50x}{x} = 75$$

Can we solve the same equation algebraically?

In Chapter 1, you learned that you can multiply both sides of an equation by the same (nonzero) quantity without changing the solution of the equation. For example, to solve the equation

$$\frac{3x}{4} = 9$$

we first multiply both sides by 4 (the denominator of the fraction) to get

$$4\left(\frac{3x}{4}\right) = (9)4$$
$$3x = 36$$

At this stage the equation no longer involves any fractions. We finish solving by dividing both sides by 3, to get $x = 12$.

We can use the same principle to solve the equation

$$\frac{5000 + 50x}{x} = 75$$

We multiply both sides of the equation by x, the denominator of the fraction, to get

$$\cancel{x}\left(\frac{5000 + 50x}{\cancel{x}}\right) = (75)x$$

$$5000 + 50x = 75x$$

Then we can proceed as usual to finish the solution. We subtract $50x$ from both sides to find

$$5000 = 25x$$

$$200 = x$$

Envirogreen should produce 200 water filters if the average price is to be \$75 per filter.

EXERCISE 1

The manager of a new health club kept track of the number of active members over the club's first few months of operation. The following equation gives the number, N, of active members, in hundreds, t months after the club opened.

$$N = \frac{10t}{4 + t^2}$$

The graph of this equation is shown below.

a. Use the equation to find out in which months the club had 200 active members.

b. Verify your answers on the graph.

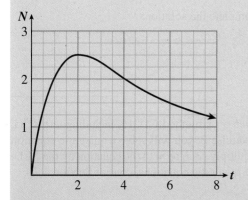

Skills Review

Write each expression as a single fraction in simplest form.

1. $\dfrac{1}{x-2} + \dfrac{2}{x}$ 2. $\dfrac{1}{x-2} - \dfrac{2}{x}$ 3. $\dfrac{1}{x-2} \cdot \dfrac{2}{x}$

4. $\dfrac{1}{x-2} \div \dfrac{2}{x}$ 5. $\dfrac{1}{\dfrac{1}{x-2} - \dfrac{2}{x}}$ 6. $\dfrac{\dfrac{2}{x} + \dfrac{1}{2}}{\dfrac{1}{x-2} + 1}$

Answers: **1.** $\dfrac{3x-4}{x^2-2x}$ **2.** $\dfrac{4-x}{x^2-2x}$ **3.** $\dfrac{2}{x^2-2x}$ **4.** $\dfrac{x}{2x-4}$

5. $\dfrac{x^2-2x}{4-x}$ **6.** $\dfrac{x^2+2x-8}{2x^2-2x}$

Lesson

Using an LCD to Clear Fractions

If an equation contains more than one fraction, we can multiply both sides of the equation by the LCD of all the fractions. This clears all the denominators at once.

E X A M P L E   1

Solve $\dfrac{x}{3} - 2 = \dfrac{4}{5} + \dfrac{x}{5}$.

Solution

The LCD of $\dfrac{x}{3}, \dfrac{4}{5}$, and $\dfrac{x}{5}$ is 15. We multiply both sides of the equation by 15:

$$15\left(\frac{x}{3} - 2\right) = \left(\frac{4}{5} + \frac{x}{5}\right)15 \qquad \text{Apply the distributive law.}$$

$$15\left(\frac{x}{3}\right) - 15(2) = 15\left(\frac{4}{5}\right) + 15\left(\frac{x}{5}\right)$$

$$5x - 30 = 12 + 3x$$

Now we can proceed as usual to complete the solution:

$$5x - 30 = 12 + 3x \qquad \text{Subtract } 3x \text{ from both sides.}$$
$$ \qquad \text{Add 30 to both sides.}$$
$$2x = 42 \qquad \text{Divide both sides by 2.}$$
$$x = 21$$

EXERCISE 2

Solve $\dfrac{x^2}{2} + \dfrac{5x}{4} = 3$ by first clearing the fractions.

In Example 1, we multiplied *each term* by the LCD, 15, including terms that are not fractions, namely -2. In Exercise 2, you will solve a quadratic equation that involves fractions. Be sure to multiply *each term* of the equation by the LCD. (If you have forgotten how to solve quadratic equations, consult Section 4.7.)

Variables in the Denominator

Equations that involve algebraic fractions can also be solved using an LCD.

EXAMPLE 2

Solve $\dfrac{3}{4} = 8 - \dfrac{2x + 11}{x - 5}$.

Solution

The LCD for the two fractions in the equation is $4(x - 5)$. We multiply both sides of the equation by $4(x - 5)$:

$$4(x - 5)\left(\frac{3}{4}\right) = \left(8 - \frac{2x + 11}{x - 5}\right) \cdot 4(x - 5) \quad \text{Apply the distributive law.}$$

$$\cancel{4}(x - 5)\left(\frac{3}{\cancel{4}}\right) = 4(x - 5)(8) - 4\cancel{(x - 5)}\left(\frac{2x + 11}{\cancel{x - 5}}\right)$$

$$3(x - 5) = 32(x - 5) - 4(2x + 11)$$

We proceed as usual to complete the solution. First, we use the distributive law to remove parentheses:

$$3x - 15 = 32x - 160 - 8x - 44 \quad \text{Combine like terms.}$$
$$3x - 15 = 24x - 204$$
$$-21x = -189$$
$$x = 9$$

Now try Exercise 3.

Remember that you can multiply both sides of an equation by a *nonzero* number in order to obtain an equivalent equation. In Example 2, we multiplied by $4(x - 5)$. After solving, we found that $x = 9$, so $4(x - 5) = 4(9 - 5) = 16$; because this is not zero, the multiplication step was valid. The next example illustrates what can go wrong if we multiply by zero.

EXAMPLE 3

Solve $6 + \dfrac{4}{x - 3} = \dfrac{x + 1}{x - 3}$.

Solution

We multiply both sides of the equation by the LCD, $x - 3$, to clear the fractions:

$$(x - 3)\left(6 + \frac{4}{x - 3}\right) = \left(\frac{x + 1}{x - 3}\right)(x - 3) \quad \text{Apply the distributive law.}$$

$$(x - 3)(6) + \cancel{(x - 3)}\left(\frac{4}{\cancel{x - 3}}\right) = \left(\frac{x + 1}{\cancel{x - 3}}\right)\cancel{(x - 3)}$$

$$6(x - 3) + 4 = x + 1$$

We complete the solution as usual.

$$6x - 18 + 4 = x + 1$$
$$6x - 14 = x + 1$$
$$5x = 15$$
$$x = 3$$

EXERCISE 3
Follow the suggested steps to solve
$$\frac{1}{x - 2} + \frac{2}{x} = 1.$$
Step 1 Find the LCD for all the fractions in the equation.

Step 2 Multiply *each term* of the equation by the LCD, and simplify.

Step 3 Solve the resulting equation with no fractions as usual.

The solution appears to be $x = 3$. But here we have a problem, because the LCD, $x - 3$, equals zero when $x = 3$. We have multiplied both sides of the equation by zero. When we try to check the solution, we find

$$6 + \frac{4}{3 - 3} = \frac{3 + 1}{3 - 3}$$

or

$$6 + \frac{4}{0} = \frac{4}{0}$$

Because division by zero is undefined, 3 is *not* a solution after all. The original equation does not have a solution.

In Example 3, when we multiplied both sides of the equation by zero, we found a false "solution" for the equation. Such solutions are called **extraneous solutions**, and there is always a danger that an extraneous solution may be introduced when we multiply by an expression that contains the variable. Consequently, *always check the solutions to equations that involve algebraic fractions.* You can do this by verifying that the proposed solution does not cause any of the denominators in the equation to equal zero.

Try Exercise 4.

Formulas

We can also solve formulas that involve algebraic fractions.

EXAMPLE 4

Solve $\dfrac{1}{T} = \dfrac{PR}{A - P}$ for P.

Solution

The LCD for the two fractions in the equation is $T(A - P)$. We multiply both sides of the equation by the LCD to obtain

$$\cancel{T}(A - P)\frac{1}{\cancel{T}} = \frac{PR}{\cancel{A - P}}\,T\cancel{(A - P)}$$

or

$$A - P = PRT$$

Next, we get all the terms containing the desired variable, P, on one side of the equation. We add P to both sides to get

$$A = P + PRT$$

We now have two unlike terms that contain the desired variable. To proceed, we *factor out* this variable and then divide both sides by the remaining factor:

$$A = P(1 + RT) \qquad \text{Divide both sides by } 1 + RT.$$

$$\frac{A}{1 + RT} = \frac{P(1 + RT)}{1 + RT}$$

Thus

$$P = \frac{A}{1 + RT}$$

Try Exercise 5.

Applications

Recall the formula $d = rt$, which is useful in solving problems about motion. Note that we can write this equation in other forms that may be more appropriate for a particular problem:

$$r = \frac{d}{t} \quad \text{and} \quad t = \frac{d}{r}$$

EXAMPLE 5

A cruise boat travels 18 miles downstream and back in $4\frac{1}{2}$ hours. If the speed of the current is 3 miles per hour, what is the speed of the boat in still water?

Solution

Let x stand for the speed of the boat in still water. We make a table showing information about distance, rate, and time for each part of the trip. We begin by filling in the given information: the distance traveled and the speed of the boat on each part of the journey.

	Distance	Rate	Time
Downstream trip	18	$x + 3$	
Upstream trip	18	$x - 3$	

We use the formula $t = \dfrac{d}{r}$ to fill in the last column of the table:

	Distance	Rate	Time
Downstream trip	18	$x + 3$	$\dfrac{18}{x + 3}$
Upstream trip	18	$x - 3$	$\dfrac{18}{x - 3}$

We did not use the $4\frac{1}{2}$ hours in the table, because it was not the trip upstream or the trip downstream that took $4\frac{1}{2}$ hours, but the total trip. In other words, the *sum* of the times for the upstream and downstream trips was $4\frac{1}{2}$ hours:

$$\frac{18}{x + 3} + \frac{18}{x - 3} = \frac{9}{2}$$

To solve the equation, we multiply both sides by the LCD, $2(x - 3)(x + 3)$. We apply the distributive law to multiply *each term* of the equation by the LCD:

$$2(x - 3)(x + 3)\left(\frac{18}{x + 3} + \frac{18}{x - 3}\right) = \left(\frac{9}{2}\right)2(x - 3)(x + 3)$$

$$2(x - 3)(x + 3)\left(\frac{18}{x + 3}\right) + 2(x - 3)(x + 3)\left(\frac{18}{x - 3}\right) = \left(\frac{9}{2}\right)2(x - 3)(x + 3)$$

$$36(x - 3) + 36(x + 3) = 9(x - 3)(x + 3)$$

We simplify each side of the equation and write it in standard form:

$$36x - 108 + 36x + 108 = 9x^2 - 81$$

$$9x^2 - 72x - 81 = 0$$

This is a quadratic equation, and we can solve it by factoring.

$$9(x^2 - 8x - 9) = 0$$

$$9(x - 9)(x + 1) = 0 \quad \text{Set each factor equal to zero.}$$

$$x - 9 = 0 \qquad x + 1 = 0$$

$$x = 9 \qquad x = -1$$

Neither solution is extraneous. However, because the speed of the boat cannot be a negative number, we discard the solution $x = -1$. The boat travels 9 miles per hour in still water.

Rate Problems

Problems involving other types of rates can be solved with similar techniques. Suppose it takes you 8 hours to type a term paper for your history class. If you worked at a constant rate, in 1 hour you would complete $\frac{1}{8}$ of your task. The rate at which you work, or your *work rate*, is one-eighth job per hour. In 3 hours, you would complete

$$3 \cdot \frac{1}{8} = \frac{3}{8}$$

of the job. In t hours, you would complete

$$t \cdot \frac{1}{8} = \frac{t}{8}$$

of the job. In general, the amount of work done, expressed as a fraction of the whole job, is given by the work formula

$$\text{work rate} \times \text{time} = \text{work completed}$$

or

$$rt = w$$

Follow the suggestions to solve the problem in Exercise 6. In this application, the "work" being done is the water flowing into (or out of) the reservoir, so the work rate is the flow rate of water through the pipes.

EXERCISE 6

The city reservoir was completely emptied for repairs and is being refilled. Water flows in through the intake pipe at a steady rate that can fill the reservoir in 120 days. However, water is also being drained from the reservoir as people use it, so the reservoir actually takes 150 days to fill. After the reservoir is filled, the intake pipe is turned off, but people continue to use water at the same rate. How long will it take to drain the reservoir dry?

To solve the problem, compute the flow rate for each pipe, intake and outflow, separately. Use the following steps.

a. Let d stand for the number of days it takes for the outflow pipe to drain the full reservoir. Write an expression for the fraction of the reservoir that is drained in one day. (This is the rate at which water leaves the reservoir.)

b. Now, imagine that the intake pipe is open, but the outflow pipe is closed. Write an expression for the fraction of the reservoir that is filled in one day. (This is the rate at which water enters the reservoir.)

	Flow rate	Time	Fraction of reservoir
Water entering			
Water leaving			

c. Consider the time period described in the problem, when both pipes are open. Fill in the table.

d. Write an equation.

$$\left(\begin{array}{c} \text{fraction of} \\ \text{reservoir filled} \end{array} \right) - \left(\begin{array}{c} \text{fraction of} \\ \text{reservoir drained} \end{array} \right) = \text{one whole reservoir}$$

e. Solve your equation.

ANSWERS TO 5.9 EXERCISES

1a. In months 1 and 4   **2.** $x = \dfrac{3}{2}, x = -4$ **3.** $x = 1, x = 4$

4. There is no solution.   **5.** $q = \dfrac{wT_h}{T_h - T_c}$

6a. $\dfrac{1}{d}$ **6b.** $\dfrac{1}{120}$

6c.

	Flow rate	Time	Fraction of reservoir
Water entering	$\dfrac{1}{120}$	150	$\dfrac{150}{120}$
Water leaving	$\dfrac{1}{d}$	150	$\dfrac{150}{d}$

6d. $\dfrac{150}{120} - \dfrac{150}{d} = 1$ **6e.** The reservoir will be emptied in 600 days.

HOMEWORK 5.9   www

In Problems 1–8, solve each equation.

1. $\dfrac{5x}{2} - 1 = x + \dfrac{1}{2}$

2. $\dfrac{2t}{3} - \dfrac{1}{4} = \dfrac{25}{12} + \dfrac{t}{3}$

3. $\dfrac{t}{6} - \dfrac{7}{3} = \dfrac{2t}{9} - \dfrac{t}{4}$

4. $\dfrac{2y}{3} - \dfrac{2y + 5}{6} = \dfrac{1}{2}$

5. $\dfrac{2}{3}(x - 1) + x = 6$

6. $\dfrac{2}{5}(3x + 2) - \dfrac{2}{3}(x - 1) = 4$

7. $\dfrac{3x^2}{2} - \dfrac{x}{4} = \dfrac{1}{2}$

8. $\dfrac{3}{4}x^2 + \dfrac{5}{2}x - 2 = 0$

Solve the equations in Problems 9–16.

9. $2 + \dfrac{5}{2x} = \dfrac{3}{x} + \dfrac{3}{2}$

10. $\dfrac{y - 2}{y} = \dfrac{14}{3y} - \dfrac{1}{3}$

11. $1 + \dfrac{1}{x(x - 1)} = \dfrac{3}{x}$

12. $\dfrac{4}{x} - 3 = \dfrac{5}{2x + 3}$

13. $\dfrac{4}{x-1} - \dfrac{4}{x+2} = \dfrac{3}{7}$

14. $\dfrac{4}{x-2} - \dfrac{7}{x-3} = \dfrac{1}{3}$

15. $\dfrac{1}{x-2} + \dfrac{1}{x+2} = \dfrac{4}{x^2-4}$

16. $\dfrac{2}{n^2-2n} + \dfrac{1}{2n} = \dfrac{-1}{n^2+2n}$

Solve the equations in Problems 17 and 18.

17. $\dfrac{15x}{1+x^2} = 6$

18. $\dfrac{x+4}{x^2-2x} = \dfrac{1}{4}$

19. A small lake in a state park has become polluted by runoff from a factory upstream. The cost of removing p percent of the pollutant from the lake, in thousands of dollars, is given by

$$C = \frac{25p}{100-p}$$

How much of the pollutant can be removed for $25,000?

20. The crew team can row at a steady pace of 10 miles per hour in still water. Every afternoon, the team's training includes a 5-mile row upstream on the river. If the current in the river on a given day flows at v miles per hour, then the time required for this workout, in minutes, is given by

$$t = \frac{300}{10-v}$$

If the workout took 50 minutes, how fast was the current in the river?

21. During the baseball season so far, Pete got hits 34 times out of 164 times at bat.

   a. What is Pete's batting average so far? (Batting average is the fraction of at-bats that result in a hit.)

   b. Write an expression for the new batting average that Pete will have if he gets a hit on every one of his next x at-bats.

   c. How many consecutive hits does Pete need to raise his batting average to .350?

22. Penny has earned 132 out of 240 possible homework points so far this semester.

   a. What is Penny's homework average so far?

   b. Write an expression for the new homework average that Penny will have if she earns all of the next x possible homework points.

   c. How many homework points must Penny earn to raise her average to 80%?

23. The rectangle *ABCD* is divided into a square and a smaller rectangle, *CDEF*, as shown in the figure. The two rectangles *ABCD* and *CDEF* are similar. A rectangle like *ABCD* that can be divided in this way is called a *golden rectangle*, and the ratio of its length to its width is called the *golden ratio*. The golden ratio appears frequently in art and nature and is considered to give the most pleasing proportions to many figures. Compute the golden ratio as indicated in the following steps.

   a. Let *AB* = 1 and *AD* = x. What are the lengths of *AE*, *ED*, and *CD*?

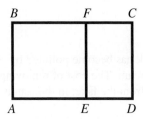

   c. Solve your proportion for x. Find the golden ratio, $\dfrac{AD}{AB} = \dfrac{x}{1}$.

   b. Write a proportion in terms of x for the similarity of rectangles *ABCD* and *CDEF*. Be careful to match up the corresponding sides.

24. The figure shows the graphs of two equations, $y = x$ and $y = \dfrac{1}{x} + 1.$

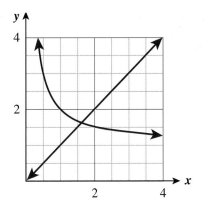

a. Find the x-coordinate of the intersection point of the two graphs.

b. Compare your answer to the golden ratio you computed in Problem 23.

In Problems 25–38, solve each formula for the specified variable.

25. $V = \dfrac{hT}{P},$ for T

26. $A = \dfrac{h}{2}(b + c),$ for b

27. $a = \dfrac{F}{m + M},$ for m

28. $S = \dfrac{a(1 - r^3)}{1 - r},$ for a

29. $m = \dfrac{y - k}{x - h},$ for x

30. $y = \dfrac{3x - 1}{2 - x},$ for x

31. $\dfrac{1}{R} = \dfrac{1}{A} + \dfrac{1}{B},$ for A

32. $\dfrac{x}{a} + \dfrac{y}{b} = 1,$ for x

33. $w = 0.622 \dfrac{e}{P - e},$ for e

34. $C = \dfrac{rR}{r + R},$ for R

35. $I = \dfrac{E}{R + \dfrac{r}{n}}$, for R

36. $E = \dfrac{Ff}{(P - x)p}$, for x

37. $r = \dfrac{dc}{1 - ec}$, for e

38. $\dfrac{1}{Q} + \dfrac{1}{I} = \dfrac{2}{r}$, for r

Solve Problems 39–46.

39. An express train travels 180 miles in the same time as a freight train travels 120 miles. If the express train travels 20 miles per hour faster than the freight train, find the speed of each.

40. Rani went hiking in the Santa Monica Mountains last weekend. She drove 10 miles in her car and then walked 4 miles, arriving at a small lake $2\frac{1}{4}$ hours after she left home. If Rani drives 20 times faster than she walks, how fast does she walk?

41. Sam Scholarship and Reginald Privilege each travel the 360 miles to Fort Lauderdale on spring break, but Reginald drives his Porsche while Sam hitches a ride on a vegetable truck. Reginald travels 20 miles per hour faster than Sam does and arrives in 3 hours less time. How fast does each travel?

42. The crew team rowed 6 miles downstream and then returned upstream to the dock. The round trip took 4 hours, and the speed of the current was 2 miles per hour. How fast would the crew team travel in still water?

43. Periwinkle Printing can print the first run of a volume of poems in 10 hours on its new press; the same job takes 18 hours on the old machine. The new press is finishing another job, so the press manager starts the old machine 4 hours ahead of the new one. How many hours are needed with both presses running to finish the printing?

44. Walt and Irma use a tank of heating oil every 25 days in their furnace during the winter. If they use their space heater also, the tank lasts 40 days. This winter they use the furnace for the first 10 days after filling the oil tank, but it becomes so cold that they decide to light their space heater as well. How long will the remaining oil in the tank last if they run both heaters?

45. It takes 30 minutes to fill a large water tank. However, the tank has a small leak that would completely drain it in 4 hours. How long will it take to fill the tank if the leak is not plugged?

46. An underground spring fills a small pond in 12 days. Evaporation from the surface of the pond can empty the pond in 28 days. If the pond is completely dry, how long will it take to fill again?

Complete Problems 47 and 48, which compare the procedures for adding fractions and for solving fractional equations. Explain how the LCD is used differently for adding fractions and for solving equations.

47. a. Add $\dfrac{x-1}{4} + \dfrac{3x}{5}$.

48. a. Add $\dfrac{2}{x} + \dfrac{2}{x+3}$.

b. Solve $\dfrac{x-1}{4} + \dfrac{3x}{5} = 1$.

b. Solve $\dfrac{2}{x} + \dfrac{2}{x+3} = 1$.

49. Find the error in the following "proof" that $1 = 0$. Start by letting $x = 1$.

$$x = 1 \qquad \text{Multiply both sides by } x.$$
$$x^2 = x \qquad \text{Subtract 1 from both sides.}$$
$$x^2 - 1 = x - 1 \qquad \text{Factor the left side.}$$
$$(x - 1)(x + 1) = x - 1 \qquad \text{Divide both sides by } x - 1.$$
$$\frac{(x - 1)(x + 1)}{x - 1} = \frac{x - 1}{x - 1} \qquad \text{Simplify both sides.}$$
$$x + 1 = 1 \qquad \text{Subtract 1 from both sides.}$$
$$x = 0$$

Because $x = 1$ and $x = 0$, $1 = 0$.

50. What is wrong with the following solution to an addition problem?

Add $\dfrac{3}{8} + \dfrac{1}{2}$.

$$8 \cdot \frac{3}{8} + 8 \cdot \frac{1}{2} = 3 + 4 \qquad \text{Multiply by the LCD, 8.}$$
$$= 7 \qquad \text{The answer is 7.}$$

In Problems 51–54, find the error in each "solution" and correct it.

51.
$$\frac{a}{2} - \frac{a}{3} = 1$$
$$6\left(\frac{a}{2} - \frac{a}{3}\right) = 1$$
$$3a - 2a = 1$$
$$a = 1$$

52.
$$\frac{z}{4} - 3 = \frac{2}{5}$$
$$20 \cdot \frac{z}{4} - 3 = \frac{2}{5} \cdot 20$$
$$5z - 3 = 8$$
$$z = \frac{11}{5}$$

53.
$$m^2 - 6m = 0$$
$$m^2 = 6m$$
$$\frac{m^2}{m} = \frac{6m}{m}$$
$$m = 6$$

54.
$$\frac{3}{p} - \frac{2}{p + 2} = 1$$
$$(p + 2)\left(\frac{3}{p} - \frac{2}{p + 2}\right) = 1(p + 2)$$
$$6 - 2 = p + 2$$
$$2 = p$$

55. If two hens lay two eggs in 2 days, how long will it take six hens to lay six eggs?

56. If 20 musicians play a symphony in 80 minutes, how long will it take 30 musicians to play the same symphony?

CHAPTER SUMMARY AND REVIEW

Glossary: Write a sentence explaining the meaning of each term.

polynomial	degree	greatest common factor
inverse variation	constant of variation	like fractions
lowest common denominator	complex fraction	extraneous solution

Properties and formulas

1. State the first law of exponents, and give an example.

2. State the two cases of the second law of exponents, and give examples.

3. State two formulas for the squares of binomials.

4. State a formula for factoring the difference of two squares.

5. State a formula for inverse variation.

6. State the fundamental principle of fractions, and explain what it is used for.

Techniques and procedures

7. Explain how to add or subtract two polynomials.

8. Explain how to multiply two monomials.

9. Explain how to divide one monomial by another.

10. When you are trying to factor a polynomial, what is the best way to start?

11. Give the factored form of each expression: $a^2 - b^2$, $a^2 + b^2$.

12. How can you tell from a table of values whether two variables vary inversely?

13. Explain how to reduce an algebraic fraction.

14. Explain how to multiply two algebraic fractions.

15. Explain how to divide one algebraic fraction by another.

16. Explain how to add or subtract like fractions.

17. Give four steps for adding or subtracting unlike fractions.

18. Explain how to simplify a complex fraction.

19. What is the first step in solving an equation that contains fractions?

20. State a formula that is useful for solving problems that involve work.

21. Evaluate the polynomial $\frac{5}{4}x^2 + \frac{1}{2}xy - 2y^2$ for $x = -4$, $y = 2$.

22. The sum of the cubes of the first n counting numbers is given by $\frac{n^4}{4} + \frac{n^3}{2} + \frac{n^2}{4}$.   Find the sum of the cubes of the first five counting numbers.

23. Which of the following expressions are polynomials?
  a. $-25y^{10} - 1$
  b. $\sqrt{3x} + 2$

  c. $\dfrac{4}{x^3} - \dfrac{3}{x^2} + \dfrac{1}{x}$
  d. $t^4 + 5t^3 - 2\sqrt{t} + 3$

24. Write $x(3x[2x(x-2)+1]-4)$ as a polynomial.

Simplify the expressions in Problems 25 and 26 by combining like terms.

25. $8p^2q - 3pq^2 - 2pq^2 + p^2q$

26. $1.7m^3 + 2.6 - 0.3m - 1.4m^2 - 1.2m^3 + 4.5m^2 + 1.1$

Add or subtract the polynomials in Problems 27 and 28.

27. $(4b^3 + 2b^2 - 3b + 7) - (-2b + 3 - b^3 - 7b)$

28. $(8w^6 - 5w^4 - 3w^2) + (2w^4 - 8w^2 + 4)$

29. On spring break, Johann and Sebastian both walk from the university to the next town. Johann leaves at noon and Sebastian leaves 1 hour later, but Sebastian walks 1 mile per hour faster. Let r stand for Johann's walking speed. Write polynomials to answer the following questions.
  a. What is Sebastian's walking speed?

  b. How far has Sebastian walked at 3 P.M.?

  c. How far has Johann walked at 3 P.M.?

  d. How far apart are Johann and Sebastian at 3 P.M.?

30. Evaluate the following expression for $x = -3$:
$$(2x^3 + x^2 + 5x - 7) + (2x^2 + 3x - 2) - (x^3 + 5x^2 - 2x + 9)$$

Simplify the expressions in Problems 31–36, if possible.

31. a. $3x^2 + x^2$

    b. $3x^2 \cdot x^2$

32. a. $5b^3 - b^3$

    b. $5b^3(-b^3)$

33. a. $7a^4 + a^6$

    b. $7a^4 \cdot a^6$

34. a. $2r^4 - r^3$

    b. $2r^4(-r^3)$

35. a. $\dfrac{3b^9}{9b^3}$

    b. $\dfrac{9b^3}{3b^9}$

36. a. $\dfrac{4m^8}{m^8}$

    b. $\dfrac{m^4}{8m^4}$

In Problems 37 and 38, multiply.

37. $(5m^2n)(-6m^3n^2)$

38. $-7qr^2(2q^4 - 1)$

In Problems 39 and 40, divide.

39. $\dfrac{-21x^4y^3}{3xy^3}$

40. $\dfrac{3a^2b}{6a^4b^3}$

In Problems 41–46, multiply.

41. $-2y(y - 4)(y + 3)$

42. $6p^4(2p + 1)(p - 4)$

43. $(d - 2)(d^2 - 4d + 4)$

44. $(2k + 1)(k - 2)(k + 3)$

45. $9x^3y(4x - 3y)^2$

46. $(a - 3)^3$

47. Write a polynomial for the volume of a box whose width is 10 centimeters less than its length and 2 centimeters more than its height. Let w represent the width of the box.

48. Nova Cosmetics sells $140 - 2p$ cans of styling mousse each month at p dollars per can.
   a. Write a polynomial for the company's monthly revenue from mousse.

   b. Find the revenue if each can costs $4.

In Problems 49–52, factor out the greatest common factor.

49. $30w^9 - 42w^4 + 54w^8$

50. $45x^2y^2 + 18x^2y^3 - 27x^3y^3$

51. $-vw^5 - vw^4 + vw^2$

52. $5(x - 2) - x^2(x - 2)$

Factor Problems 53–60 completely.

53. $2q^4 + 6q^3 - 80q^2$

54. $32 - b^4 - 14b^2$

55. $x^2 - 3xy + 2y^2$

56. $y^2 - 3by - 28b^2$

57. $3a^2b + 12ab^2 + 9b^3$

58. $80t^4 - 28t^3 - 24t^2$

59. $9x^2y^2 + 3xy - 2$

60. $4a^3x - 2a^2x^2 - 12ax^3$

Factor Problems 61–68, if possible.

61. $h^2 - 24h + 144$

62. $9t^2 - 30tv + 25v^2$

63. $x^2 + 144$

64. $98n^4 - 8n^2$

65. $w^8 + 12w^4 + 36$

66. $q^6 - 14q^3 + 49$

67. $3s^4 - 48$

68. $2y^4 - 2$

69. Which graph describes inverse variation? Find an equation for that graph.

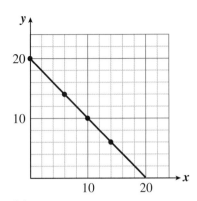

(a)

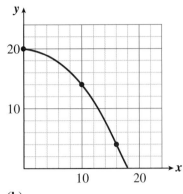

(b)

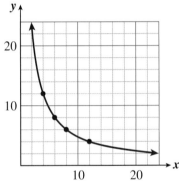

(c)

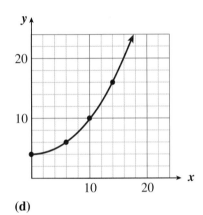

(d)

70. Which table represents inverse variation?

a.

x	2	8	30
y	0.4	1.6	6

b.

x	0.5	3	8
y	12	2	0.75

c.

x	1.8	1.3	0.4
y	0.2	0.7	1.6

71. The electrical current in a wire varies inversely with the resistance in the circuit. An appliance with 6 ohms of resistance draws 20 amperes of current.

   a. Write an equation that gives the current, I, in terms of the resistance, R.

   b. How much current flows through a circuit that has 4 ohms of resistance?

72. Explain the difference between inverse variation and direct variation.

73. a. Evaluate the fraction $\dfrac{s^2 - s}{s^2 + 3s - 10}$ for $s = -2$

    b. For what value(s) of s is the fraction undefined?

74. At his diner, Ed uses a package of coffee filters every $x + 5$ days.
    a. What fraction of a package does Ed use every day?

    b. What fraction of a package does Ed use in one week?

In Problems 75–82, reduce each fraction, if possible.

75. $\dfrac{a + 3}{b + 3}$

76. $\dfrac{5x + 7}{5x}$

77. $\dfrac{10 + 2y}{2y}$

78. $\dfrac{3x^2 - 1}{1 - 3x^2}$

79. $\dfrac{v - 2}{v^2 - 4}$

80. $\dfrac{q^5 - q^4}{q^4}$

81. $\dfrac{-3x}{6x^2 + 9x}$

82. $\dfrac{x^2 + 5x + 6}{x^2 - 4}$

In Problems 83–88:
a. Add the fractions.
b. Multiply the fractions.

83. $\dfrac{3}{8}, \dfrac{5}{12}$

84. $\dfrac{2}{x}, \dfrac{1}{x+2}$

85. $\dfrac{3x}{2x+2}, \dfrac{x+1}{6x}$

86. $\dfrac{x+1}{x-1}, \dfrac{1}{x^2-1}$

87. $2, \dfrac{1}{x}$

88. $x, \dfrac{1}{x+2}$

In Problems 89–94, write each expression as a single fraction in lowest terms.

89. $\dfrac{4c^2 d}{3} \div 6cd^2$

90. $\dfrac{u^2 - 2uv}{uv} \div \dfrac{3u - 6v}{2uv}$

91. $\dfrac{2m^2 - m - 1}{m + 1} - \dfrac{m^2 - m}{m + 1}$

92. $\dfrac{3}{2p} + \dfrac{7}{6p^2}$

93. $\dfrac{5q}{q-3} - \dfrac{7}{q} + 3$

94. $\dfrac{2w}{w^2 - 4} + \dfrac{4}{w^2 + 4w + 4}$

95. On Saturday mornings, Olive takes her motorboat 5 miles upstream to the general store for supplies and then returns home. The current in the river flows at a rate of 2 miles per hour. Let x represent Olive's speed in still water, and write algebraic fractions to answer each question.

    a. How long does it take Olive to get to the store?

    b. How long does the return trip take?

    c. How long does the round trip take?

96. Choose from the following list of operations to answer the questions:

    adding fractions
    subtracting fractions
    multiplying fractions
    dividing fractions
    simplifying complex fractions
    solving equations with fractions

    a. Which operations use an LCD?

    b. Which operations use building factors?

    c. Which operations involve multiplying by the LCD?

Simplify each complex fraction in Problems 97 and 98.

97. $\dfrac{2 - \dfrac{a}{b}}{2 - \dfrac{b}{a}}$

98. $\dfrac{3p - \dfrac{q^2}{3p}}{1 - \dfrac{q}{3p}}$

Solve each equation in Problems 99–102.

99. $q - \dfrac{16}{q} = 6$

100. $\dfrac{2 - x}{5x} = \dfrac{4}{15x} - \dfrac{1}{6}$

101. $\dfrac{9}{m + 2} + \dfrac{2}{m} = 2$

102. $\dfrac{15}{x^2 - 3x} + \dfrac{4}{x} = \dfrac{5}{x - 3}$

In Problems 103 and 104, solve for the indicated variable.

103. $\dfrac{1}{x} + \dfrac{1}{y} = \dfrac{1}{z}$, for x

104. $y = \dfrac{2x + 3}{1 - x}$, for x

105. On a walking tour, Nora walks uphill 5 miles to an inn where she has lunch. After lunch, she increases her speed by 2 miles per hour and walks for 8 more miles. If she walked for 1 hour longer before lunch than after lunch, what was her speed before lunch?

106. Brenda can fill her pool in 30 hours using the normal intake pipe. She can instead fill the pool in 45 hours using the garden hose. How long will it take to fill the empty pool if both the pipe and the garden hose are running?

More About Exponents and Radicals

6.1 Laws of Exponents

Reading

In Sections 5.2 and 5.3, you learned the first and second laws of exponents, which enabled you to compute products and quotients of powers. Recall that to multiply two powers with the same base, we add the exponents and leave the base unchanged. For example,

$$5^2 \cdot 5^6 = 5^8$$

To divide two powers with the same base, we subtract the smaller exponent from the larger. If the larger exponent occurs in the numerator, we put the power in the numerator. If the larger exponent occurs in the denominator, we put the power in the denominator.

$$\frac{3^7}{3^2} = 3^5 \qquad \text{and} \qquad \frac{2^3}{2^5} = \frac{1}{2^2}$$

larger exponent in numerator   larger exponent in denominator

These laws are expressed in symbols as follows:

First Law of Exponents

$$a^m \cdot a^n = a^{m+n}$$

Second Law of Exponents

$$\frac{a^m}{a^n} = a^{m-n} \qquad (n < m)$$

$$\frac{a^m}{a^n} = \frac{1}{a^{n-m}} \qquad (n > m)$$

EXERCISE 1
Use the laws of exponents to simplify each of the following.

a. $(4x^4)(5x^5)$

b. $\dfrac{8x^8}{4x^4}$

c. $\dfrac{3^2 a^3 (3a^2)}{2^3 a(2^3 a^4)}$

EXERCISE 2
Write $(y^2)^5$ as a repeated product, and apply the first law of exponents to simplify it.

EXERCISE 3
a. $(5^3)^6 =$

b. $(y^4)^2 =$

Try Exercise 1.

In the lesson, we'll study three additional laws of exponents.

Skills Review

Simplify each expression according to the order of operations.

1. $10 - 6^2$	2. $10(-6)^2$	3. $(10 - 6)^2$
4. $2 \cdot 5^2 - 3^2$	5. $-2 \cdot 5^2(-3^2)$	6. $-2^2 - (5 - 3)^2$

Answers: *1.* -26 *2.* 360 *3.* 16 *4.* 41 *5.* 450 *6.* -8

Lesson

Power of a Power

Consider the expression $(x^4)^3$, which means the cube of x^4. We can simplify this expression as follows:

$$(x^4)^3 = (x^4)(x^4)(x^4) = x^{\overbrace{4 + 4 + 4}^{\text{add exponents}}} = x^{12}$$

Because cubing is really just repeated multiplication, we wrote $(x^4)^3$ as a repeated product and applied the first law of exponents by adding the three exponents.

Try Exercise 2.

Of course, because repeated addition is actually multiplication, we can obtain the same result by multiplying the exponents:

$$(x^4)^3 = x^{4 \cdot 3} = x^{12}$$

This gives us another law of exponents.

> ### Third Law of Exponents
>
> To raise a power to a power, keep the same base and multiply the exponents. In symbols,
>
> $$(a^m)^n = a^{mn}$$

EXAMPLE 1

a. $(x^4)^3 = x^{4 \cdot 3} = x^{12}$
b. $(y^2)^5 = y^{2 \cdot 5} = y^{10}$

Try Exercise 3.

CAUTION!

Note carefully the difference between the following two expressions:

$$(x^3)(x^4) = x^{3+4} = x^7$$

but

$$(x^3)^4 = x^{3 \cdot 4} = x^{12}$$

The first expression is a product, so we add the exponents. The second expression raises a power to a power, so we multiply the exponents.

Try Exercise 4.

Power of a Product

To simplify the expression $(2x)^3$, we can use the commutative and associative properties of multiplication to write

$$(2x)^3 = (2x)(2x)(2x)$$
$$= 2 \cdot 2 \cdot 2 \cdot x \cdot x \cdot x$$
$$= 2^3 x^3$$

Thus, to raise a product to a power, we can raise each factor to the power. This example illustrates the following law.

> **Fourth Law of Exponents**
>
> To raise a product to a power, raise each factor to the power. In symbols,
>
> $$(ab)^n = a^n b^n$$

EXAMPLE 2

a. $(5ab)^2 = 5^2 a^2 b^2 = 25a^2 b^2$ Square each factor.

b. $(-xy^3)^4 = (-x)^4(y^3)^4$ Raise each factor to the fourth power; apply the third law of exponents.
$$= x^4 y^{12}$$

Now try Exercise 5.

CAUTION!

Note carefully the difference between the two expressions $3a^2$ and $(3a)^2$:

$$3a^2$$

cannot be simplified, but

$$(3a)^2 = 3^2 a^2 = 9a^2$$

In the expression $3a^2$, only the factor a is squared, but in $(3a)^2$, both 3 and a are squared.

Try Exercise 6.

In general, a power of a negative number with an odd exponent is negative, and a power of a negative number with an even exponent is positive. For example,

$$(-2)^4 = 2^4 = 16$$
$$(-2)^5 = -2^5 = -32$$

Power of a Quotient

To simplify the expression $\left(\dfrac{x}{2}\right)^4$, we multiply together four copies of the fraction $\dfrac{x}{2}$.

$$\left(\frac{x}{2}\right)^4 = \frac{x}{2} \cdot \frac{x}{2} \cdot \frac{x}{2} \cdot \frac{x}{2} = \frac{x \cdot x \cdot x \cdot x}{2 \cdot 2 \cdot 2 \cdot 2}$$
$$= \frac{x^4}{2^4} = \frac{x^4}{16}$$

EXERCISE 4
Simplify each expression using the laws of exponents. Then use your calculator to verify that your answers are correct.

a. $(5^4)^2 =$

b. $(5^4)(5^2) =$

EXERCISE 5
Simplify $(6q^6)^2$.

EXERCISE 6
Recall that $-x$ can be interpreted as $-1 \cdot x$. Use this idea to simplify each expression, if possible.

a. $-x^4$

b. $(-x)^4$

c. $-x^5$

d. $(-x)^5$

This example suggests the following law of exponents.

Fifth Law of Exponents

To raise a quotient to a power, raise both the numerator and the denominator to the power. In symbols,

$$\left(\frac{a}{b}\right)^n = \frac{a^n}{b^n}$$

EXAMPLE 3

a. $\left(\dfrac{x}{y}\right)^5 = \dfrac{x^5}{y^5}$ Raise numerator and denominator to the fifth power.

b. $\left(\dfrac{2}{y^2}\right)^3 = \dfrac{2^3}{(y^2)^3}$ Raise numerator and denominator to the third power.

$\qquad\qquad = \dfrac{2^3}{y^{2\cdot 3}} = \dfrac{8}{y^6}$ Apply the third law of exponents to the denominator.

Try Exercise 7.

EXERCISE 7

Simplify $\left(\dfrac{n^3}{k^4}\right)^8$.

Using the Laws of Exponents

For convenience, the five laws of exponents are presented together below. All of the laws are valid when the base a is not equal to zero and when the exponents m and n are positive integers.

Laws of Exponents

1. $a^m \cdot a^n = a^{m+n}$

2. $\dfrac{a^m}{a^n} = a^{m-n}$ $(n < m)$

$\qquad \dfrac{a^m}{a^n} = \dfrac{1}{a^{n-m}}$ $(n > m)$

3. $(a^m)^n = a^{mn}$

4. $(ab)^n = a^n b^n$

5. $\left(\dfrac{a}{b}\right)^n = \dfrac{a^n}{b^n}$

Here are some more examples of how these laws are used to simplify algebraic expressions.

EXAMPLE 4

Simplify $2x^2y(3xy^2)^4$.

Solution

According to the order of operations, we should compute powers before products, so we first simplify the factor $(3xy^2)^4$:

$$(3xy^2)^4 = 3^4x^4(y^2)^4 \quad \text{Fourth law: raise each factor to the power.}$$
$$= 81x^4y^8 \quad \text{Third law: } (y^2)^4 = y^{2 \cdot 4}$$

Now, we multiply the result by $2x^2y$ and apply the first law to obtain

$$2x^2y(81x^4y^8) = 2 \cdot 81 \cdot x^2 \cdot x^4 \cdot y \cdot y^8$$
$$= 162x^6y^9$$

EXAMPLE 5

Simplify $(-x)^4(-xy)^3$.

Solution

Each power should be simplified before we compute the product. Because 4 is an even exponent, $(-x)^4$ is positive; that is,

$$(-x)^4 = x^4$$

To simplify $(-xy)^3$, we apply the fourth law of exponents:

$$(-xy)^3 = (-x)^3y^3 \quad \text{3 is an odd exponent, so } (-x)^3 = -x^3.$$
$$= -x^3y^3$$

Thus

$$(-x)^4(-xy)^3 = x^4(-x^3y^3) \quad \text{Apply the first law.}$$
$$= -x^7y^3$$

Now try Exercises 8 and 9.

ANSWERS TO 6.1 EXERCISES

1a. $20x^9$ **1b.** $2x^4$ **1c.** $\dfrac{27}{64}$

2. $y^2 \cdot y^2 \cdot y^2 \cdot y^2 \cdot y^2 = y^{2+2+2+2+2} = y^{10}$

3a. 5^{18} **3b.** y^8

4a. 5^8 **4b.** 5^6 **5.** $36q^{12}$

6a. $-x^4$ **6b.** x^4 **6c.** $-x^5$ **6d.** $-x^5$

7. $\dfrac{n^{24}}{k^{32}}$ **8.** $\dfrac{-8a^3b^{12}}{27c^{15}}$ **9.** $7x^3y^6 - xy^3$

EXERCISE 8

Follow the suggested steps to simplify $\left(\dfrac{-2ab^4}{3c^5}\right)^3$.

Step 1 Apply the fifth law: raise numerator and denominator to the third power.

Step 2 Apply the fourth law: raise each factor to the third power.

Step 3 Apply the third law: simplify each power of a power.

EXERCISE 9

Follow the suggested steps to simplify $3x(xy^3)^2 - xy^3 + 4x^3(y^2)^3$.

Step 1 Identify the three terms of the expression by underlining each separately.

Step 2 Simplify the first term: apply the fourth law and then the third law.

Step 3 Simplify the last term: apply the third law.

Step 4 Add or subtract any like terms.

HOMEWORK 6.1

Simplify each power in Problems 1 and 2 using the third law of exponents.

1. a. $(t^3)^5$ 　　　　　　　　　**b.** $(b^4)^2$ 　　　　　　　　　**c.** $(w^{12})^{12}$

2. a. $(3^4)^3$ 　　　　　　　　　**b.** $(4^3)^3$ 　　　　　　　　　**c.** $(6^6)^6$

Simplify each power in Problems 3 and 4 using the fourth law of exponents.

3. a. $(5x)^3$ 　　　　　　　　　**b.** $(-3wz)^4$ 　　　　　　　　**c.** $(-ab)^5$

4. a. $(4 \cdot 100)^2$ 　　　　　　　**b.** $(-2 \cdot 0.1)^3$ 　　　　　　**c.** $(1.93 \times 10^3)^4$

Simplify each power in Problems 5 and 6 using the fifth law of exponents.

5. a. $\left(\dfrac{w}{2}\right)^6$ 　　　　　　　**b.** $\left(\dfrac{5}{v}\right)^4$ 　　　　　　　**c.** $\left(\dfrac{-m}{p}\right)^3$

6. a. $\left(\dfrac{2}{5}\right)^4$ 　　　　　　　**b.** $\left(\dfrac{3}{100}\right)^2$ 　　　　　　**c.** $\left(\dfrac{-1}{3}\right)^5$

7. Simplify each expression.

　　a. $x^3 \cdot x^6$ 　　　　**b.** $(x^3)^6$ 　　　　**c.** $\left(\dfrac{x^3}{x^6}\right)$ 　　　　**d.** $\dfrac{x^6}{x^3}$

8. Simplify each expression.

    a. $5 \cdot 5^{15}$ **b.** $(5^{15})^{15}$ **c.** $\dfrac{5}{5^{15}}$ **d.** $\dfrac{5^{15}}{5}$.

9. Explain the difference between the first and third laws of exponents. Use examples.

10. Explain the difference between the second and fifth laws of exponents. Use examples.

In Problems 11–16, find and correct the error in each calculation.

11. $2 \cdot 3^2 \rightarrow 36$

12. $(4z^3)^2 \rightarrow 8z^6$

13. $-10^2 \rightarrow 100$

14. $(9w)^3 \rightarrow 9w^3$

15. $a^4 \cdot a^3 \rightarrow a^{12}$

16. $a^3(-b)^3 \rightarrow a^3 - b^3$

Use the laws of exponents to simplify each expression in Problems 17–22.

17. $(2p^3)^5$

18. $(-4x^2y^4)^4$

19. $\left(\dfrac{-3}{q^4}\right)^5$

20. $\left(\dfrac{k^3}{3}\right)^3$

21. $\left(\dfrac{-2h^2}{m^3}\right)^4$

22. $\left(\dfrac{-a^3b}{5z^4}\right)^2$

Simplify each expression in Problems 23–34.

23. $x^3(x^2)^5$

24. $(x^2y)^3(xy^3)$

25. $(2x^3y)^2(xy^3)^4$

26. $(ab^2)^3(3bc^3)$

27. $[ab^2(a^2b)^3]^3$

28. $[-2a(-2a)^2]^2$

29. $-a^2(-a)^2$

30. $(-x)^3(-x^3)^3$

31. $-(-xy)^2(xy^2)$

32. $-2^2(-x^3y)^2(-y^2)$

33. $-4p(-p^2q^2)^2(-q^3)^2$

34. $-a(-ab^2)^3(-ab^3)^4$

Simplify each expression in Problems 35–40.

35. $2y(y^3)^2 - 2y^4(3y)^3$

36. $6t^2(2t)^3 + 4t(2t^2)^2$

37. $2a(a^2)^4 + 3a^2(a^6) - a^2(a^2)^3$

38. $b^2(b^3)^2 + 3b^4(b^2)^2 - (b^2 \cdot 2b)^2b^2$

39. $-3v^2(2v^3 - v^2) + v(-4v)^2$

40. $8w^3 - 2w^3(w^4 - w^2) - (w^5 - 5w^3)$

In Problems 41–50, simplify each expression as much as possible.

41. a. $4x^2 + 2x^4$ **b.** $4x^2(2x^4)$ **42. a.** $(-x)^3 - x^4$ **b.** $(-x)^3(-x^4)$

43. a. $(-x)^3 x^4$ **b.** $[(-x^3)(-x)]^4$ **44. a.** $x^2(-3x^2)^3$ **b.** $x^2 - (-3x^2)^3$

45. a. $(3x^2)^4(2x^4)^2$ **b.** $(3x^2)^4 - (2x^4)^2$ **46. a.** $3x^2 y - (3x)^2 y$ **b.** $3xy^2 - (3xy)^2$

47. a. $6x^3 - 3x^6$ **b.** $6x^3(-3x^6)$ **48. a.** $(x^3 y^3)^6$ **b.** $(x^3 + y^3)^6$

49. a. $6x^3 - 3x^3(x^3)$ **b.** $(6x^3 - 3x^3)x^3$ **50. a.** $x^3(x^3 - x^6)$ **b.** $x^3[x^3(-x^6)]$

In Problems 51–56, find the value of *n*.

51. $b^3 \cdot b^n = b^9$ **52.** $(b^3)^n = b^6$

53. $\dfrac{c^8}{c^n} = c^2$ **54.** $2^4 n^4 = 10{,}000$

55. $\dfrac{n^3}{3^3} = 8$ **56.** $\dfrac{z^n}{z^8} = z$

In Problems 57–60, factor out the indicated monomial.

57. $x^4 + x^6 = x^2(\underline{\hspace{2cm}})$

58. $a^3 - a^9 = a^3(\underline{\hspace{2cm}})$

59. $4m^4 - 4m^8 + 8m^{16} = 4m^4(\underline{\hspace{2cm}})$

60. $6h^3 + 12h^4 - 3h^7 = 3h^3(\underline{\hspace{2cm}})$

Mental exercise: In Problems 61–72, replace the comma with the appropriate symbol, <, >, or =. Do not use pencil, paper, or calculator.

61. $-8^2, \quad 64$

62. $3 \cdot 4^2, \quad 144$

63. $(-3)^5, \quad -3^5$

64. $(-27)^8, \quad 0$

65. $\left(\dfrac{-7}{4}\right)^{11}, \quad 0$

66. $\dfrac{36^7}{9^7}, \quad 4^7$

67. $6^{10} \cdot 4^{10}, \quad 24^{10}$

68. $5^4 \cdot 2^4, \quad 10{,}000$

69. $(8 - 2)^3, \quad 8 - 2^3$

70. $(-2^3)^2, \quad (-2^2)^3$

71. $(17^4)^5, \quad (17^5)^4$

72. $3^3 - 4^3, \quad (3 - 4)^3$

6.2 Negative Exponents

Reading

In this section, we extend our use of exponential notation to include negative numbers as exponents. This will simplify a number of calculations involving exponents and enable us to study a useful form for writing very large or very small numbers.

First, what do negative exponents mean? In order for negative exponents to be useful, their properties must fit with what we already know about exponents. Consider the two lists below, and try to fill in the unknown values by following the pattern. As you move down the list, you can find each new entry by dividing the previous entry by the base.

$2^4 = 16$ 　　　　　　　 $5^4 = 625$
　　　　　 Divide by 2. 　　　　　　　　　 Divide by 5.
$2^3 = 8$ 　　　　　　　 $5^3 = 125$
　　　　　 Divide by 2. 　　　　　　　　　 Divide by 5.
$2^2 = 4$ 　　　　　　　 $5^2 = 25$
$2^1 =$ 　　　　　　　　 $5^1 =$
$2^0 =$ 　　　　　　　　 $5^0 =$
$2^{-1} =$ 　　　　　　　 $5^{-1} =$
$2^{-2} =$ 　　　　　　　 $5^{-2} =$
$2^{-3} =$ 　　　　　　　 $5^{-3} =$

Try Exercise 1.

Skills Review

Decide whether each statement is true or false.

1. The reciprocal of $x + 2$ is $\dfrac{1}{x} + \dfrac{1}{2}$.

2. The reciprocal of $\dfrac{1}{3} + \dfrac{1}{4}$ is 7.

3. $\dfrac{3}{\frac{1}{5}} = 15$ 　　　　　　 4. $\dfrac{3}{\frac{1}{5} + \frac{1}{2}} = 15 + 6$

5. The reciprocal of $\dfrac{1}{x}$ is x.

6. The reciprocal of $\dfrac{1}{x + y}$ is $x + y$.

Answers: **1.** False **2.** False **3.** True **4.** False **5.** True **6.** True

Lesson

Zero as an Exponent

Recall that a^n means the product of n factors of a. For example, a^3 means $a \cdot a \cdot a$. Is there a reasonable meaning for a^0? The reading presented evidence that $2^0 = 1$ and $5^0 = 1$. Here is another way to understand the meaning of zero as an exponent: Consider the quotient $\dfrac{a^4}{a^4}$. If a does not equal zero, this quotient equals 1, because any nonzero number divided by itself is 1. On the other hand, we can also think of the expression as a quotient of powers and subtract the exponents. If we extend the second law of exponents to include the case $m = n$, we have

$$\frac{a^4}{a^4} = a^{4-4} = a^0$$

EXERCISE 1

Answer the following questions about the lists:

a. What did you find for the values of 2^0 and 5^0?

b. If you make a list with another base (say, 3, for example), what will you find for the value of 3^0?

c. Can you explain why this is true?

d. Compare the values of 2^3 and 2^{-3}. Do you see a relationship between them?

e. What about the values of 5^2 and 5^{-2}? Try to state a general rule about powers with negative exponents.

f. Use your rule to guess the value of 3^{-4}.

Thus it seems reasonable to make the following definition.

$$a^0 = 1, \qquad \text{if } a \neq 0$$

For example,

$$3^0 = 1 \qquad (-427)^0 = 1 \qquad \text{and} \qquad (5xy)^0 = 1 \quad (\text{if } x, y \neq 0)$$

Negative Exponents

We can also use the second law of exponents to give meaning to negative exponents. Consider the quotient $\dfrac{a^4}{a^7}$. According to the second law of exponents,

$$\frac{a^4}{a^7} = \frac{1}{a^{7-4}} = \frac{1}{a^3}$$

However, if we allow negative numbers as exponents, we can apply the first half of the second law to obtain

$$\frac{a^4}{a^7} = a^{4-7} = a^{-3}$$

We therefore define a^{-3} to mean $\dfrac{1}{a^3}$. In general, for $a \neq 0$, we have the following definition.

$$a^{-n} = \frac{1}{a^n}, \quad a \neq 0$$

For example,

$$2^{-4} = \frac{1}{2^4} \qquad \text{and} \qquad x^{-5} = \frac{1}{x^5}$$

We see that *a power with a negative exponent is the reciprocal of the corresponding power with a positive exponent.*

EXAMPLE 1

Write each expression without exponents.

    a. 10^{-4} **b.** $\left(\dfrac{1}{4}\right)^{-3}$ **c.** $\left(\dfrac{3}{5}\right)^{-2}$

Solution

    a. $10^{-4} = \dfrac{1}{10^4} = \dfrac{1}{10,000},$ or 0.0001

    b. To compute a power of a fraction with a negative exponent, we can compute the corresponding power of its reciprocal with a positive exponent. (We can do this because of the fifth law of exponents.) Thus

$$\left(\frac{1}{4}\right)^{-3} = 4^3 = 64$$

c. As in part (b), we compute the corresponding power of the reciprocal of $\frac{3}{5}$:

$$\left(\frac{3}{5}\right)^{-2} = \left(\frac{5}{3}\right)^{2} = \frac{25}{9}$$

CAUTION!

A negative exponent does *not* mean that the power is negative. For example,

$$2^{-4} \neq -2^4$$

Try Exercise 2.

We can also rewrite fractional expressions by using negative exponents.

EXAMPLE 2

Write each expression using negative exponents.

a. $\dfrac{1}{3^4}$ **b.** $\dfrac{7}{10^2}$ **c.** $\dfrac{8}{x}$

Solution

a. $\dfrac{1}{3^4} = 3^{-4}$ **b.** $\dfrac{7}{10^2} = 7 \cdot \dfrac{1}{10^2} = 7 \cdot 10^{-2}$ **c.** $\dfrac{8}{x} = 8 \cdot \dfrac{1}{x} = 8x^{-1}$

CAUTION!

Recall that an exponent applies only to its base. The same is true of negative exponents. For example, the exponent -2 in the expression $3x^{-2}$ applies only to x, but in $(3x)^{-2}$, the exponent applies to $3x$. Thus

$$3x^{-2} = 3 \cdot \frac{1}{x^2} = \frac{3}{x^2}$$

but

$$(3x)^{-2} = \frac{1}{(3x)^2} = \frac{1}{9x^2}$$

Try Exercise 3.

Laws of Exponents

The laws of exponents discussed in Section 6.1 also apply to negative exponents. In particular, if we allow negative exponents, we can write the second law as a single rule.

Laws of Exponents

1. $a^m \cdot a^n = a^{m+n}$

2. $\dfrac{a^m}{a^n} = a^{m-n} \quad a \neq 0$

3. $(a^m)^n = a^{mn}$

4. $(ab)^n = a^n b^n$

5. $\left(\dfrac{a}{b}\right)^n = \dfrac{a^n}{b^n}, \quad b \neq 0$

EXERCISE 2
Write each expression without exponents.

a. -6^2

b. 6^{-2}

c. $(-6)^{-2}$

EXERCISE 3
Write each expression without negative exponents.

a. $4t^{-2}$

b. $(4t)^{-2}$

c. $\left(\dfrac{x}{3}\right)^{-4}$

EXAMPLE 3

Simplify each expression using the first or second law of exponents.

a. $x^5 \cdot x^{-8}$ 　　　　　　　　**b.** $\dfrac{5^2}{5^{-6}}$

Solution

a. We apply the first law and add the exponents.
$$x^5 \cdot x^{-8} = x^{5-8} = x^{-3}$$

b. We apply the second law and subtract the exponents.
$$\frac{5^2}{5^{-6}} = 5^{2-(-6)} = 5^8$$

Try Exercise 4.

As a special case of the second law, note that $1 = a^0$. Thus, we can write
$$\frac{1}{a^{-n}} = \frac{a^0}{a^{-n}} = a^{0-(-n)} = a^n$$

And we have the following general properties.

$$\frac{1}{a^{-n}} = a^n \quad \text{and} \quad \frac{b}{a^{-n}} = b \cdot a^n, \quad a \neq 0$$

For example,
$$\frac{1}{2^{-3}} = 2^3 \quad \text{and} \quad \frac{8}{x^{-5}} = 8x^5$$

Here are some examples of using the third and fourth laws of exponents.

Try Exercise 5.

EXAMPLE 4

Simplify each expression using the third or fourth law of exponents.

a. $(2^{-3})^{-3}$ 　　　　　　　**b.** $(ab)^{-3}$

Solution

a. We apply the third law and multiply the exponents.
$$(2^{-3})^{-3} = 2^{-3(-3)} = 2^9$$

b. We apply the fourth law and raise each factor to the power.
$$(ab)^{-3} = a^{-3}b^{-3}$$

Try Exercise 6.

EXERCISE 4
Simplify each expression using the first or second law of exponents.

a. $3^{-3} \cdot 3^{-6}$

b. $\dfrac{b^{-7}}{b^{-3}}$

EXERCISE 5
Write each expression without negative exponents, and simplify.

a. $\dfrac{1}{15^{-2}}$

b. $\dfrac{3k^2}{m^{-4}}$

EXERCISE 6
Simplify each expression using the third or fourth law of exponents.

a. $(3y)^{-2}$

b. $(a^{-3})^{-2}$

It is often necessary to apply more than one law to simplify a given expression.

EXERCISE 7

Explain the following simplification of $\dfrac{(3z^{-4})^{-2}}{2z^{-3}}$. State the law of exponents or other property used in each step.

$$\frac{(3z^{-4})^{-2}}{2z^{-3}} = \frac{3^{-2}(z^{-4})^{-2}}{2z^{-3}}$$

$$= \frac{3^{-2}z^8}{2z^{-3}}$$

$$= \frac{3^{-2}z^{8-(-3)}}{2}$$

$$= \frac{z^{11}}{3^2 \cdot 2} = \frac{z^{11}}{18}$$

HOMEWORK 6.2

In Problems 1–16, write each expression without using zero or negative exponents, and simplify.

1. 5^{-2}

2. 4^{-3}

3. x^{-6}

4. y^{-2}

5. $(8x)^0$

6. $(3y)^0$

7. $\left(\dfrac{3}{4}\right)^{-3}$

8. $\left(\dfrac{5}{2}\right)^{-4}$

9. $\left(\dfrac{b}{3}\right)^{-4}$

10. $\left(\dfrac{w}{9}\right)^{-2}$

11. $(2q)^{-5}$

12. $(4k)^{-3}$

13. $3 \cdot 4^{-3}$

14. $4 \cdot 7^{-2}$

15. $4x^{-2}$

16. $7x^{-4}$

In Problems 17–28, write each expression using negative exponents.

17. $\dfrac{1}{2^3}$

18. $\dfrac{1}{4^3}$

19. $\dfrac{3}{5^2}$

20. $\dfrac{7}{3^4}$

21. $\dfrac{1}{27}$

22. $\dfrac{1}{32}$

23. $\dfrac{x}{625}$

24. $\dfrac{y}{216}$

25. $\dfrac{2}{z^2}$

26. $\dfrac{3}{r^4}$

27. $\left(\dfrac{z}{10}\right)^5$

28. $\left(\dfrac{w}{10}\right)^3$

In Problems 29–34, find and correct the error in each calculation.

29. $x^0 \to 0$

30. $-29^0 \to 1$

31. $w^{-3} \to -w^3$

32. $\dfrac{5}{p^{-2}} \to -5p^2$

33. $2x^{-4} \to \dfrac{1}{2x^4}$

34. $\left(\dfrac{3}{4}\right)^{-2} \to \dfrac{-9}{16}$

Simplify each product in Problems 35–40 using the first law of exponents.

35. $x^{-3} \cdot x^8$

36. $y^2 \cdot y^{-6}$

37. $5^{-4} \cdot 5^{-3}$

38. $4^{-2} \cdot 4^{-6}$

39. $(3b^{-5})(5b^2)$

40. $(2a^{-8})(3a^4)$

Simplify each quotient in Problems 41–46 using the second law of exponents.

41. $\dfrac{c^{-7}}{c^{-4}}$

42. $\dfrac{m^{-9}}{m^{-2}}$

43. $\dfrac{8b^{-4}}{4b^{-8}}$

44. $\dfrac{9n^{-3}}{3n^{-9}}$

45. $\dfrac{6^6}{6^{-2}}$

46. $\dfrac{3^{10}}{3^{-5}}$

Simplify each quotient in Problems 47–52.

47. $\dfrac{1}{6^{-3}}$

48. $\dfrac{1}{5^{-4}}$

49. $\dfrac{3}{2^{-6}}$

50. $\dfrac{5}{4^{-3}}$

51. $\dfrac{8x^3}{y^{-5}}$

52. $\dfrac{7a^2}{b^{-8}}$

Simplify each power in Problems 53–58 using the third law of exponents.

53. $(8^{-2})^5$

54. $(5^{-4})^3$

55. $(w^{-6})^{-3}$

56. $(t^{-1})^{-4}$

57. $(d^6)^{-4}$

58. $(z^{12})^{-3}$

Simplify the expressions in Problems 59–64 using the fourth law of exponents.

59. $(pq)^{-5}$

60. $(uv)^{-7}$

61. $(3x)^{-2}$

62. $(2y)^{-3}$

63. $5(2r)^{-3}$

64. $2(3b)^{-4}$

Simplify the expressions in Problems 65–72.

65. $(a^{-4}c^2)^{-3}$

66. $(y^2z^{-5})^{-2}$

67. $(2u^2)^{-3}(u^{-4})^2$

68. $(3h^{-4})^{-2}(h^2)^{-3}$

69. $\dfrac{5k^{-3}(k^4)^{-3}}{6k^{-5}}$

70. $\dfrac{4v^{-5}(v^{-2})^{-4}}{3v^{-8}}$

71. $\left(\dfrac{2p^{-3}}{p^2}\right)^{-2}$

72. $\left(\dfrac{q^4}{3q^{-2}}\right)^{-3}$

In Problems 73–80, write each expression as a complex fraction, and simplify.

73. $\dfrac{a^{-1}+2}{a^{-1}-2}$

74. $\dfrac{3+b^{-1}}{2+b^{-1}}$

75. $\dfrac{c^{-2}-1}{c^{-1}+1}$

76. $\dfrac{4-d^{-2}}{2+d^{-1}}$

77. $\dfrac{x^{-2}-y^{-2}}{x^{-1}-y^{-1}}$

78. $\dfrac{x+y^{-1}}{y+x^{-1}}$

79. $\dfrac{v^{-2}+v^{-1}}{w^{-1}v^{-1}+w^{-1}}$

80. $\dfrac{3+pn^{-1}}{9p^{-2}-n^{-2}}$

Mental exercise: Simplify each quotient in Problems 81–90 without using pencil, paper, or calculator. Write your answer as a power of 10.

81. $\dfrac{10^3}{10^{-2}}$

82. $\dfrac{10^{-4}}{10^3}$

83. $\dfrac{10^{-5}}{10^{-3}}$

84. $\dfrac{10^{-2}}{10^{-4}}$

85. $\dfrac{10}{10^{-1}}$

86. $\dfrac{10^{-1}}{10^{0}}$

87. $\dfrac{10^{-3} \times 10^{-2}}{10^{-6}}$

88. $\dfrac{10^{-8} \times 10^{4}}{10^{-2}}$

89. $\dfrac{10^{-5}}{10^{-4} \times 10^{7}}$

90. $\dfrac{10^{-4}}{10^{-3} \times 10^{10}}$

6.3 ■ Scientific Notation

Reading

In their work, scientists and engineers often encounter very large numbers such as

$$5,980,000,000,000,000,000,000,000$$

(the mass of the earth in kilograms) and very small numbers such as

$$0.000\ 000\ 000\ 000\ 000\ 000\ 000\ 001\ 67$$

(the mass of a hydrogen atom in grams). These numbers can be written in a more compact and useful form using powers of 10.

Multiplying by a Power of 10

Recall that multiplying a number by a power of 10 with a positive exponent has the effect of moving the decimal point k places to the *right*, where k is the exponent of 10. For example,

$$2.358 \times 10^{2} = 235.8 \qquad \text{and} \qquad 17 \times 10^{4} = 170,000.$$

Multiplying a number by a power of 10 with a negative exponent has the effect of moving the decimal point to the *left*. For example,

$$5452 \times 10^{-3} = 5.452 \qquad \text{and} \qquad 2.3 \times 10^{-5} = 0.000023$$

Try Exercise 1.

We can also reverse the process just discussed to write a number in factored form, where one factor is a power of 10.

EXERCISE 1
Compute each product.

a. 1.47×10^{5}

b. 5.2×10^{-2}

EXAMPLE 1

Fill in the correct power of 10 for each factored form.

a. $38{,}400 = 3.84 \times$ _____   b. $0.005\,7 = 5.7 \times$ _____

Solution

a. To obtain 38,400 from 3.84, we must move the decimal point four places to the right, so we multiply by 10^4.

$$38{,}400 = 3.84 \times 10^4$$

b. To obtain 0.0057 from 5.7, we must move the decimal point three places to the left, so we multiply by 10^{-3}.

$$0.005\,7 = 5.7 \times 10^{-3}$$

There is an easy way to decide whether the exponent on 10 should be positive or negative. Note that the number in part (a) of Example 1—namely, 38,400—is a large number, so the exponent on 10 in its factored form is positive. The number in part (b), 0.0057, is a decimal fraction less than 1. The exponent on 10 in its factored form is negative.

Try Exercise 2.

Skills Review

Simplify each expression and write each without any negative exponents.

1. a. $(m^4)^6$ b. $m^4 m^6$ 2. a. $s^{-2} + t^{-2}$ b. $(s + t)^{-2}$

3. a. an^{-3} b. $(an)^{-3}$ 4. a. $\dfrac{3}{v^{-2}}$ b. $\left(\dfrac{3}{v}\right)^{-2}$

5. a. $2x^{-4} + 2x^{-4}$ b. $2x^{-4}(2x^{-4})$ 6. a. $(2z)^3(2z)^{-3}$ b. $(2z)^3 - (2z)^{-3}$

Answers: *1. a.* m^{24} *b.* m^{10} *2. a.* $\dfrac{1}{s^2} + \dfrac{1}{t^2}$ *b.* $\dfrac{1}{(s + t)^2}$

3. a. $\dfrac{a}{n^3}$ *b.* $\dfrac{1}{a^3 n^3}$ *4. a.* $3v^2$ *b.* $\dfrac{v^2}{9}$ *5. a.* $\dfrac{4}{x^4}$ *b.* $\dfrac{4}{x^8}$

6. a. 1 *b.* $8z^3 - \dfrac{1}{8z^3}$

Lesson

In each factored form in Example 1, there is just one digit to the left of the decimal point. This form for writing a number is called **scientific notation**. In other words, a number is written in scientific notation if it is expressed as the product of a number between 1 and 10 and a power of 10. For example, the following numbers are written in scientific notation:

$$4.18 \times 10^{12}, \qquad 2.9 \times 10^{-8}, \qquad \text{and} \qquad 4 \times 10^1$$

To write a number in scientific notation, we first position the decimal point and then determine the correct power of 10.

EXERCISE 2

Fill in the correct power of 10 for each factored form.

a. $0.004\,27 = 4.27 \times$ _____

b. $4800 = 4.8 \times$ _____

EXAMPLE 2

Write each number in scientific notation.

    a. 62,000,000   **b.** 0.000 431

Solution

    a. First, we position the decimal point so that there is just one nonzero digit to the left of the decimal.

$$62,000,000 = 6.2 \times \underline{\hspace{1cm}}$$

    To obtain 62,000,000 from 6.2, we must move the decimal point seven places to the right. Therefore, we multiply 6.2 by 10^7.

$$62,000,000 = 6.2 \times 10^7$$

    b. First, we position the decimal point so that there is just one nonzero digit to the left of the decimal.

$$0.000\ 431 = 4.31 \times \underline{\hspace{1cm}}$$

    To obtain 0.000 431 from 4.31, we must move the decimal point four places to the left. Therefore, we multiply 4.31 by 10^{-4}.

$$0.000\ 431 = 4.31 \times 10^{-4}$$

Try Exercise 3.

Calculators and Scientific Notation

Your calculator displays numbers in scientific notation if they have too many digits to fit in the display screen. For example, try computing the square of 12,345,678 on your calculator. You should enter

$$12345678 \quad \boxed{x^2}$$

(and press $\boxed{\text{ENTER}}$ if you have a graphing calculator). A scientific calculator will display the result as

$$\boxed{\text{1.524157653} \qquad \text{14}}$$

A graphing calculator will display

$$\boxed{\text{1.524157653 E 14}}$$

(Some calculators may round the result to fewer decimal places.) Both of these displays represent the number $1.524157653 \times 10^{14}$. Because scientific notation always involves a power of 10, most calculators do not display the 10, but only its exponent.

    If you now press the $\boxed{1/x}$ key, your calculator will display the reciprocal of $1.524157653 \times 10^{14}$ as

$$\boxed{\text{6.56100108} \qquad -15} \quad \text{or} \quad \boxed{\text{6.56100108 E} -15}$$

This is how a calculator displays the number $6.56100108 \times 10^{-15}$.

    Try Exercise 4.

EXERCISE 3
Write each number in scientific notation.
a. The largest living animal is the blue whale, with an average weight of 120,000,000 grams.

b. The smallest animal is the fairy fly beetle, which weighs about 0.000 005 gram.

EXERCISE 4
Perform the following calculations on your calculator. Write the results in scientific notation, and round to two decimal places.
a. $6,565,656 \times 34,567$

b. $0.000\ 000\ 123 \div 98,765$

To enter a number in scientific notation into a calculator, we use the key marked either $\boxed{\text{EXP}}$ or $\boxed{\text{EE}}$. For instance, to enter 6.02×10^{23} ,   we key in the sequence

$$6.02 \boxed{\text{EXP}} \; 23$$

To enter a number with a negative exponent, such as 1.66×10^{-27} ,   on a scientific calculator, we key in

$$1.66 \boxed{\text{EXP}} \; 27 \boxed{+/-}$$

On a graphing calculator, we key in

$$1.66 \boxed{\text{EE}} \; \boxed{(-)} \; 27$$

EXAMPLE 3

The People's Republic of China encompasses about 2,317,400,000 acres of land and has a population of 1,190,500,000 people. How many acres of land per person are there in China?

Solution

We must divide the number of acres of land by the number of people. To do this, we first write each number in scientific notation:

$$2{,}317{,}400{,}000 = 2.3174 \times 10^9$$
$$1{,}190{,}500{,}000 = 1.1905 \times 10^9$$

We enter the division into the calculator as follows:

$$2.3174 \boxed{\text{EXP}} \; 9 \boxed{\div} \; 1.1905 \boxed{\text{EXP}} \; 9 \boxed{=}$$

The calculator displays the quotient, 1.946577068. Thus there are just under 1.95 acres of land per person in China.

Mental Calculation with Scientific Notation

We can often obtain a quick estimate for a calculation by converting each figure to scientific notation and rounding to just one or two digits.

EXAMPLE 4

Use scientific notation to find this product:

$$(6{,}200{,}000{,}000)(0.000\ 000\ 3)$$

Solution

We convert each number to scientific notation. Then we combine the decimal numbers and the powers of 10 separately.

$$(6{,}200{,}000{,}000)(0.000\ 000\ 3) = (6.2 \times 10^9)(3 \times 10^{-7})$$
$$= (6.2 \times 3) \times (10^9 \times 10^{-7})$$
$$= 18.6 \times 10^2 = 1860$$

Try Exercise 5.

ANSWERS TO 6.2 EXERCISES

1a. 147,000 **1b.** 0.052

2a. 10^{-3} **2b.** 10^3

3a. 1.2×10^8 **3b.** 5×10^{-6}

4a. 2.27×10^{11}

4b. 1.25×10^{-12}

5. 0.002 1

HOMEWORK 6.3

Compute each product in Problems 1–6.

1. 4.3×10^{4}

2. 5.7×10^{-4}

3. 8×10^{-6}

4. 0.03×10^{3}

5. 0.002×10^{-2}

6. $670,000 \times 10^{-4}$

Complete each factored form in Problems 7–12.

7. $234 = 2.34 \times$ _____

8. $0.074 = 7.4 \times$ _____

9. $0.92 = 9.2 \times$ _____

10. $0.000\ 369 = 3.69 \times$ _____

11. $1,720,000 = 1.72 \times$ _____

12. $983,000,000 = 9.83 \times$ _____

Write each number in Problems 13–24 in scientific notation.

13. 4834

14. 55

15. 0.072

16. 0.25

17. $0.000\ 007$

18. $0.000\ 323\ 2$

19. $685,000,000$

20. $1,920,000$

21. 56.74×10^{4}

22. 0.368×10^{6}

23. 385×10^{-3}

24. 0.038×10^{-2}

Imagine that each of the numbers in Problems 25–32 is written in scientific notation, and decide whether the exponent on 10 is positive or negative.

25. The population of Los Angeles County

26. The weight of a blood cell in grams

27. The length of a football field in miles

28. The length of a football field in centimeters

29. The number of seconds in a year

30. The number of years in a second

31. The speed at which your hair grows in miles per hour

32. The rate at which water flows over Niagara Falls in gallons per minute

Write each number in Problems 33–36 in scientific notation.

33. The height of Mount Everest is 29,141 feet.

34. The galaxy closest to our own is the Andromeda Galaxy, which is 2,200,000 light-years away.

35. The wavelength of red light is 0.000 076 centimeters.

36. The slope of the riverbed of the lower portion of the Mississippi River is about 0.000 09.

Write each number in Problems 37–40 in standard notation.

37. An amoeba weighs about 5×10^{-6} gram.

38. The radius of a carbon atom is 2.6 angstroms, or 2.6×10^{-8} centimeter.

39. Our solar system has existed for about 5×10^{9} years.

40. In 1990, the total energy consumption in the United States was 8.13×10^{16} BTUs.

Use scientific notation to compute Problems 41–46.

41. $(2,000,000)(0.000\ 07)$

42. $(150,000,000)(0.000\ 003)$

43. $0.000\ 036 \div 0.000\ 9$

44. $0.000\ 000\ 06 \div 0.004$

45. $\dfrac{(80,000,000,000)(0.000\ 6)}{20,000}$

46. $\dfrac{(0.000\ 03)(7,000,000)}{600,000}$

Give answers to Problems 47–54 in scientific notation, rounded to two decimal places.

47. A light-year is the distance that light can travel in one year. Light travels at approximately 1.86×10^5 miles per second. How many miles are there in one light-year? (One year is approximately 3.16×10^7 seconds.)

48. How long does it take light to travel from the sun to the earth, a distance of approximately 9.3×10^7 miles? (See Problem 47 for the speed of light.)

49. A 1-foot high stack of dollar bills contains 3.6×10^3 bills.
   a. In 1994, the U.S. federal debt amounted to 4.676×10^{12} dollars. How tall a stack of one-dollar bills, in feet, would be needed to pay off the federal debt?

   b. Express the height of the stack of bills in part (a) in miles. (One mile equals 5280 feet.)

50. A Stealth bomber costs 5×10^8 dollars.
   a. At a pay rate of $20 per hour, how many hours would it take you to earn the price of a Stealth bomber?

   b. Express the answer to part (a) in years.

51. There are about 200 million insects for every person on earth. The world's population is about 5.6 billion people.
   a. How many insects are there on earth?

   b. If the average insect weighs 3×10^{-4} gram, how much do all the insects on earth weigh?

52. A sales representative would like to find the most efficient route for visiting all the cities in her territory. By testing every possible itinerary, a computer algorithm can identify the optimal route. The algorithm requires 2^n operations, where n is the number of cities on the route. The computer can perform each operation in 0.0006 second.
   a. If there are 20 cities on the route, how many operations does the algorithm require? How long will it take the computer to identify the optimal route?

   b. If there are 50 cities on the route, how many operations does the algorithm require? How long will it take the computer to identify the optimal route? How long is that in years?

53. There are 5,800,000 cubic miles of fresh water on earth. Each cubic mile is equal to 110,000,000,000 gallons of water. If the population of earth is about 5.6 billion people, how many gallons of fresh water are there for each person?

54. The total volume of the ocean is about 300 million cubic miles. A cubic mile of water weighs about 4.6 billion tons.
   a. Compute the weight of the ocean in tons.

   b. Each ton of sea water contains approximately 1.8×10^{-7} ounce of dissolved gold. How much gold does the ocean contain? Convert your answer to tons. (There are 16 ounces in a pound and 2000 pounds in a ton.)

MIDCHAPTER REVIEW

1. Explain how the expressions $b^2 \cdot b^4$ and $(b^2)^4$ are different.

2. Explain how the expressions $2b^3$ and $(2b)^3$ are different.

3. What does a negative exponent mean? Give an example.

4. Simplify the expression $\dfrac{n^3}{n^6}$ in two ways:

    a. using a negative exponent

    b. without using a negative exponent

5. Where should the decimal point be located when a number is written in scientific notation?

6. When you write a number in scientific notation, how can you tell whether the exponent on 10 should be positive or negative?

7. Find and explain the error in each incorrect calculation:

    a. $(x + 5)^4 \rightarrow x^4 + 5^4$ **b.** $\dfrac{m^{-8}}{m^{-4}} \rightarrow m^4$

8. Simplify each expression:

    a. $(a^{-1})^{-1}$ **b.** $\dfrac{1}{a^{-1}}$

    c. $\dfrac{1}{\frac{1}{a}}$ **d.** $\left(\dfrac{1}{a}\right)^{-1}$

Simplify each expression in Problems 9–14.

9. $(-2x^2)^3$

10. $\left(\dfrac{-v^3}{3w}\right)^4$

11. $(3b^4)^2(4b)^3$

12. $-x^4(-x^2)^4$

13. $[-2a(-3a)^2]^2$

14. $-2a^3 - 3a(-a^2)$

In Problems 15 and 16, factor out the greatest common factor.

15. $2t^6 - 12t^{12}$

16. $6g^3 + 12g^9 - 9g^6$

Simplify each expression in Problems 17–22, and write each without using negative exponents.

17. $(2x)^{-4}$

18. $3x^{-3}(-3x^4)$

19. $\dfrac{8x^{-6}}{6x^{-2}}$

20. $5^{-2} \cdot 5^4$

21. $(3b^{-4})^{-2}(b^3 \cdot 3b^5)$

22. $\dfrac{(-2c^{-2})^3}{4c^{-3}(c^9)}$

Write each expression in Problems 23 and 24 as a complex fraction, and simplify.

23. $\dfrac{2t^{-2} + t^{-1}}{2t^{-1} + t^0}$

24. $\dfrac{4km^{-1} - (km)^{-1}}{k^{-1} - 2}$

Compute each product in Problems 25 and 26.

25. 23.4×10^{-5}

26. 0.086×10^4

Complete each factored form in Problems 27 and 28.

27. $4800 = 4.8 \times$ _____

28. $0.00427 = 4.27 \times$ _____

Write each number in Problems 29 and 30 in scientific notation.

29. $0.006\,3$

30. $520{,}000{,}000$

In Problems 31 and 32, use scientific notation to compute the answers.

31. $\dfrac{(0.000\ 000\ 8)(4.8 \times 10^{24})}{(32,000,000)(1.2 \times 10^{-15})}$

32. $\dfrac{3.6 \times 10^{20}}{(0.000\ 000\ 000\ 25)(1.2 \times 10^{32})}$

33. Ten quarters weigh about 1 ounce. About how much would 16 million dollars in quarters weigh?

34. A 14-ounce box of Rice Toasties contains about 2100 kernels of rice. How much would 1 million kernels of Rice Toasties weigh?

6.4 Properties of Radicals

Reading

In earlier chapters, we have seen that a given algebraic expression may be written in different forms. We studied ways to simplify algebraic expressions and encountered various examples of equivalent expressions. For example, these expressions are equivalent:

$$(x - 2)(x + 2) \qquad \text{and} \qquad x^2 - 4$$

They have the same value for any value of x. Radicals with the same value may also be written in different forms. Consider the following calculation:

$$(2\sqrt{2})^2 = 2^2(\sqrt{2})^2 \quad \text{By the fourth law of exponents}$$
$$= 4(2) = 8$$

This calculation shows that $2\sqrt{2}$ must be equal to $\sqrt{8}$, because the square of $2\sqrt{2}$ is equal to 8.

Here is another example: Use your calculator to square $3\sqrt{5}$. You should find that

$$(3\sqrt{5})^2 = 45$$

Thus $3\sqrt{5}$ is equal to $\sqrt{45}$.

When working with expressions that involve radicals, you will find it helpful to write each radical as simply as possible. The expression $2\sqrt{2}$ is considered to be simpler than $\sqrt{8}$, because the radicand is a smaller number. Similarly, $3\sqrt{5}$ is simpler than $\sqrt{45}$. In this section, we investigate some properties of radicals that help in simplifying radical expressions. Use the following specific examples to decide whether the general statement is true for all nonnegative values of a and b.

Activity

1. Is it true that $\sqrt{a + b} = \sqrt{a} + \sqrt{b}$?

    a. Does $\sqrt{9 + 16} = \sqrt{9} + \sqrt{16}$?

    b. Does $\sqrt{2 + 7} = \sqrt{2} + \sqrt{7}$?

2. Is it true that $\sqrt{a - b} = \sqrt{a} - \sqrt{b}$?

    a. Does $\sqrt{100 - 36} = \sqrt{100} - \sqrt{36}$?

    b. Does $\sqrt{12 - 9} = \sqrt{12} - \sqrt{9}$?

3. Is it true that $\sqrt{ab} = \sqrt{a}\sqrt{b}$?

    a. Does $\sqrt{4 \cdot 9} = \sqrt{4} \cdot \sqrt{9}$?

    b. Does $\sqrt{3 \cdot 5} = \sqrt{3} \cdot \sqrt{5}$?

4. Is it true that $\sqrt{\dfrac{a}{b}} = \dfrac{\sqrt{a}}{\sqrt{b}}$?

    a. Does $\sqrt{\dfrac{100}{4}} = \dfrac{\sqrt{100}}{\sqrt{4}}$?

    b. Does $\sqrt{\dfrac{20}{3}} = \dfrac{\sqrt{20}}{\sqrt{3}}$?

You should have found that the examples for statements 1 and 2 are false. Thus it is *not* true that $\sqrt{a + b} = \sqrt{a} + \sqrt{b}$ or that $\sqrt{a - b} = \sqrt{a} - \sqrt{b}$ for all nonnegative values of a and b. On the other hand, the examples for statements 3 and 4 are true. These examples do not prove that the statements in 3 and 4 are *always* true, but in fact they are. (In the homework problems, you'll consider proofs of those two statements.)

We have established the following properties of radicals.

Product Rule for Radicals

$$\text{If} \quad a, b \geq 0, \quad \text{then} \quad \sqrt{ab} = \sqrt{a}\,\sqrt{b}.$$

Quotient Rule for Radicals

$$\text{If} \quad a \geq 0, b > 0, \quad \text{then} \quad \sqrt{\dfrac{a}{b}} = \dfrac{\sqrt{a}}{\sqrt{b}}.$$

CAUTION!

It is important for you to remember that there is no sum or difference rule for radicals. That is, in general,

$$\sqrt{a + b} \neq \sqrt{a} + \sqrt{b}$$

and

$$\sqrt{a - b} \neq \sqrt{a} - \sqrt{b}$$

Try Exercise 1.

Skills Review

Find each missing factor.

1. $60x^9 = 3x^3 \cdot \,?$ 2. $16z^{16} = 4z^4 \cdot \,?$ 3. $108a^5b^2 = 36a^4b^2 \cdot \,?$

4. $\dfrac{20}{7}m^7 = 4m^6 \cdot \,?$ 5. $\dfrac{5k^5}{9n} = \dfrac{k^4}{9} \cdot \,?$ 6. $\dfrac{a^2 + 4a^4}{8} = \dfrac{a^2}{4} \cdot \,?$

Answers: *1.* $20x^6$ *2.* $4z^{12}$ *3.* $3a$ *4.* $\dfrac{5m}{7}$ *5.* $\dfrac{5k}{n}$ *6.* $\dfrac{1 + 4a^2}{2}$

EXERCISE 1
Decide whether each statement is true or false, and then verify using your calculator.

a. $\sqrt{12} = \sqrt{4}\,\sqrt{3}$

b. $\sqrt{12} = \sqrt{8} + \sqrt{4}$

Lesson

Simplifying Square Roots

Exercise 1a used the product rule for radicals to write $\sqrt{12} = \sqrt{4}\sqrt{3}$. Note that $\sqrt{4} = 2$, so we can simplify $\sqrt{12}$ as

$$\sqrt{12} = \sqrt{4}\sqrt{3} = 2\sqrt{3}$$

The expression $2\sqrt{3}$ is considered a simplified form for $\sqrt{12}$ because the factor of 4, which is a perfect square, has been removed from the radical. Our strategy for simplifying radicals can be summarized as follows:

> **To Simplify a Square Root:**
>
> 1. Factor any perfect squares from the radicand.
> 2. Use the product rule to write the radical as a product of two square roots.
> 3. Simplify the square root of the perfect square.

EXAMPLE 1

Simplify $\sqrt{45}$.

Solution

We look for a perfect square that divides evenly into 45. The largest perfect square that divides into 45 is 9, so we factor 45 as $9 \cdot 5$. We use the product rule to write

$$\sqrt{45} = \sqrt{9 \cdot 5} = \sqrt{9}\sqrt{5}$$

Finally, we simplify $\sqrt{9}$ to get

$$\sqrt{45} = \sqrt{9}\sqrt{5} = 3\sqrt{5}$$

Try Exercise 2.

CAUTION!

Finding a decimal approximation for a radical is not the same as "simplifying" the radical. In Example 1, we could have used a calculator to find

$$\sqrt{45} \approx 6.708$$

However, 6.708 is not the *exact* value for $\sqrt{45}$. For long calculations, too much inaccuracy may be introduced by approximating each radical. But $3\sqrt{5}$ is equivalent to $\sqrt{45}$, which means that their values are exactly the same. We can replace the second expression by the first without losing accuracy.

EXERCISE 2
Simplify $\sqrt{75}$.

Square Root of a Variable Expression

In Section 5.4, we found that to square a power of a variable, we double the exponent. For example, the square of x^5 is

$$(x^5)^2 = x^{5 \cdot 2} = x^{10}$$

Because taking the square root of a number is the opposite of squaring a number, we can see that, as long as x is a nonnegative number,

$$\sqrt{x^{10}} = x^5 \qquad \text{because} \qquad (x^5)^2 = x^{10}$$

Similarly, if a is not negative,

$$\sqrt{a^6} = a^3 \qquad \text{because} \qquad (a^3)^2 = a^6$$

For the rest of this section, we assume that all variables are nonnegative. In that case, to simplify the square root of a power with an even exponent, we divide the exponent by 2.

Try Exercise 3.

CAUTION!

Note carefully that $\sqrt{a^{16}}$ is *not* equal to a^4. Compare the two radicals:

$$\sqrt{16} = 4 \qquad \text{but} \qquad \sqrt{a^{16}} = a^8$$

To simplify the square root of a power with an odd exponent, we write the power with two factors, one with an even exponent and one with the exponent 1.

EXAMPLE 2

Simplify $\sqrt{x^7}$.

Solution

We factor x^7 as $x^6 \cdot x$. Then we use the product rule to write

$$\sqrt{x^7} = \sqrt{x^6 \cdot x} = \sqrt{x^6} \cdot \sqrt{x}$$

Finally, we simplify the square root of x^6 to get

$$\sqrt{x^7} = \sqrt{x^6} \sqrt{x} = x^3 \sqrt{x}$$

If the radicand contains more than one power or a coefficient, we consider the constants and each power separately.

EXAMPLE 3

Simplify $\sqrt{20x^2 y^3}$.

Solution

We look for the largest perfect square that divides into 20; it is 4. We write the radicand as the product of two factors, one containing the perfect square and powers of the variables with the largest possible even exponents:

$$20x^2 y^3 = 4x^2 y^2 \cdot 5y$$

Now we write the radical as a product.

$$\sqrt{20x^2 y^3} = \sqrt{4x^2 y^2 \cdot 5y} = \sqrt{4x^2 y^2} \sqrt{5y}$$

Finally, we simplify the first of the two factors to find

$$\sqrt{20x^2 y^3} = \sqrt{4x^2 y^2} \sqrt{5y}$$
$$= 2xy \sqrt{5y}$$

Try Exercise 4.

Sums and Differences

How can we add or subtract radicals? In the reading, you saw that, in general,

$$\sqrt{a} + \sqrt{b} \neq \sqrt{a + b}$$

EXERCISE 3

Find the square root of each power.

a. $\sqrt{y^4}$

b. $\sqrt{a^{16}}$

EXERCISE 4

Simplify $\sqrt{72u^6 v^9}$.

This means, for example, that

$$\sqrt{3} + \sqrt{5} \neq \sqrt{8} \qquad \text{and} \qquad \sqrt{4} + \sqrt{16} \neq \sqrt{20}$$

What if the radicands are the same? You should verify on your calculator that

$$\sqrt{7} + \sqrt{7} \neq \sqrt{14}$$

However, we *can* write

$$\sqrt{7} + \sqrt{7} = 2\sqrt{7}$$

just as we write

$$x + x = 2x$$

If two square roots have identical radicands, they are said to be **like radicals**. We can add or subtract like radicals in the same way as we add or subtract like terms—namely, by adding or subtracting their coefficients. For example,

$$2r + 3r = 5r$$

where r is a variable that can stand for any real number. In particular, if $r = \sqrt{2}$, we have

$$2\sqrt{2} + 3\sqrt{2} = 5\sqrt{2}$$

Thus we add like radicals by adding their coefficients. The same idea applies to subtraction. For instance,

$$7\sqrt{3} - 2\sqrt{3} = 5\sqrt{3}$$

CAUTION!

We *cannot* simplify sums or differences of unlike radicals. Thus, for example, this sum *cannot* be combined into a single term:

$$3\sqrt{2} + 4\sqrt{3} \qquad \text{Cannot be simplified}$$

_____

Try Exercise 5.

CAUTION!

Note that the answer to Exercise 5b is *not* $6\sqrt{6x}$. When adding or subtracting like radicals, we do *not* add or subtract the radicands. For example,

$$3\sqrt{5} + 4\sqrt{5} = 7\sqrt{5} \qquad not \quad 7\sqrt{10}$$

_____

Sometimes we must simplify the square roots in a sum or difference before we can recognize like radicals.

EXAMPLE 4

Simplify $\sqrt{20} - 3\sqrt{50} + 2\sqrt{45}$.

Solution

We simplify each square root by removing perfect squares from the radicals:

$$\sqrt{20} - 3\sqrt{50} + 2\sqrt{45} = \sqrt{4 \cdot 5} - 3\sqrt{25 \cdot 2} + 2\sqrt{9 \cdot 5}$$
$$= 2\sqrt{5} - 3 \cdot 5\sqrt{2} + 2 \cdot 3\sqrt{5}$$
$$= 2\sqrt{5} - 15\sqrt{2} + 6\sqrt{5}$$

Then we combine the like radicals $2\sqrt{5}$ and $6\sqrt{5}$ to get

$$\sqrt{20} - 3\sqrt{50} + 2\sqrt{45} = 8\sqrt{5} - 15\sqrt{2}$$

_____

EXERCISE 5
Write each expression as a single term.
a. $13\sqrt{6} - 8\sqrt{6}$

b. $5\sqrt{3x} + \sqrt{3x}$

ANSWERS TO 6.4 EXERCISES
1a. True

1b. False

2. $5\sqrt{3}$

3a. y^2

3b. a^8

4. $6u^3v^4\sqrt{2v}$

5a. $5\sqrt{6}$

5b. $6\sqrt{3x}$

HOMEWORK 6.4

Decide whether each statement in Problems 1–6 is true or false. Then use a calculator to verify your answer.

1. $\sqrt{6} = \sqrt{2}\sqrt{3}$

2. $\sqrt{6} = \sqrt{2} + \sqrt{4}$

3. $\sqrt{16} = \sqrt{18} - \sqrt{2}$

4. $\sqrt{8} = \dfrac{\sqrt{72}}{\sqrt{9}}$

5. $\sqrt{5} + \sqrt{5} = \sqrt{10}$

6. $\sqrt{2}\,\sqrt{9} = \sqrt{18}$

Find each square root in Problems 7–18.

7. $\sqrt{y^8}$

8. $\sqrt{a^{64}}$

9. $\sqrt{n^{36}}$

10. $\sqrt{a^{12}}$

11. $\pm\sqrt{16x^4}$

12. $-\sqrt{a^2c^4}$

13. $-\sqrt{121a^2b^6}$

14. $\pm\sqrt{100p^{100}q^{50}}$

15. $\sqrt{9(x+y)^2}$

16. $-\sqrt{25(b+c)^2}$

17. $-\sqrt{\dfrac{64}{b^6}}$

18. $\sqrt{\dfrac{z^8}{144}}$

Simplify each square root in Problems 19–30.

19. $\sqrt{8}$

20. $\sqrt{18}$

21. $-\sqrt{20}$

22. $-\sqrt{24}$

23. $\sqrt{125}$

24. $\sqrt{80}$

25. $\sqrt{x^3}$

26. $\sqrt{z^5}$

27. $-\sqrt{b^{11}}$

28. $-\sqrt{a^{13}}$

29. $\sqrt{p^{25}}$

30. $\sqrt{q^{81}}$

Simplify each square root in Problems 31–48.

31. $\sqrt{8a^3}$

32. $-\sqrt{24y^5}$

33. $\pm\sqrt{72m^9}$

34. $\sqrt{80p^{12}}$

35. $\sqrt{\dfrac{x^8}{27}}$

36. $\sqrt{\dfrac{32}{n^7}}$

37. $\sqrt{48c^6d}$

38. $-\sqrt{20hk^4}$

39. $-\sqrt{\dfrac{45}{4}b^2d^3}$

40. $\sqrt{\dfrac{54}{121}m^4p^5}$

41. $\sqrt{\dfrac{9w^3}{28z}}$

42. $\sqrt{\dfrac{125x}{36y^4}}$

43. $3\sqrt{4x^3}$

44. $5\sqrt{18x}$

45. $-2a\sqrt{50a^3b^2}$

46. $-6b\sqrt{36ab^4}$

47. $-\dfrac{2}{3k}\sqrt{9b^3k^5}$

48. $\dfrac{3y}{5z}\sqrt{25y^3z^3}$

In Problems 49–62, simplify each expression, if possible.

49. $\sqrt{3} + 2\sqrt{3}$

50. $\sqrt{7} - 3\sqrt{7}$

51. $3\sqrt{5} - 3\sqrt{7}$

52. $2\sqrt{5} + 5\sqrt{2}$

53. $2\sqrt{6} - 9\sqrt{6}$

54. $3\sqrt{2} - 4\sqrt{2}$

55. $\sqrt{20} + \sqrt{45} - 2\sqrt{80}$

56. $\sqrt{12} + 2\sqrt{27} - 3\sqrt{48}$

57. $\sqrt{3} - 2\sqrt{12} - \sqrt{18}$

58. $\sqrt{18} - 5\sqrt{2} + 3\sqrt{12}$

59. $\sqrt{8a} + \sqrt{18a} - 7\sqrt{2a}$

60. $\sqrt{12b} - \sqrt{48b} + 5\sqrt{3b}$

61. $2\sqrt{5x^3} - x\sqrt{125x} - 3\sqrt{20x^2}$

62. $\sqrt{72y^3} + 4\sqrt{2y} - 2y\sqrt{32y}$

In Problems 63–68, find and correct the error in each calculation.

63. $\sqrt{36 + 64} \rightarrow 6 + 8$

64. $\sqrt{25 + 5} \rightarrow 5 + \sqrt{5}$

65. $\sqrt{3} + \sqrt{3} \rightarrow \sqrt{6}$

66. $\sqrt{x^9} \rightarrow x^3$

67. $\sqrt{9 + x^2} \rightarrow 3 + x$

68. $\sqrt{6x^6} \rightarrow 6x^3$

Mental exercise: In Problems 69–74, choose the best approximation for each square root. Do not use pencil, paper, or calculator.

69. $\sqrt{13}$ **a.** 6   **b.** 6.5   **c.** 3.5   **d.** 4

70. $\sqrt{18}$ **a.** 4   **b.** 4.3   **c.** 4.8   **d.** 5

71. $\sqrt{72}$ **a.** 64   **b.** 9   **c.** 36   **d.** 81

72. $\sqrt{156}$ **a.** 10   **b.** 12   **c.** 15   **d.** 16

73. $\sqrt{125.6}$ **a.** 11   **b.** 12   **c.** 15   **d.** 25

74. $\sqrt{50.13}$ **a.** 5.2 **b.** 7 **c.** 7.5 **d.** 25

In Problems 75–80, write each expression as the square root of an integer. (*Hint:* Square the given expression.)

75. $2\sqrt{5}$ **76.** $3\sqrt{2}$ **77.** $2\sqrt{3}$

78. $4\sqrt{3}$ **79.** $3\sqrt{6}$ **80.** $5\sqrt{5}$

Each calculation in Problems 81–84 is presented as a "proof" of one of the four statements in the reading assignment. In each problem, $a = r^2$ and $b = s^2$, where $a, b, r,$ and s are positive integers. Two of the calculations are valid arguments, and two are invalid. Identify the invalid arguments, and explain where the flaw lies.

81. $\sqrt{ab} = \sqrt{r^2 s^2} = \sqrt{(rs)^2} = rs = \sqrt{a}\sqrt{b}$

82. $\sqrt{a + b} = \sqrt{r^2 + s^2} = \sqrt{(r + s)^2} = r + s = \sqrt{a} + \sqrt{b}$

83. $\sqrt{a - b} = \sqrt{r^2 - s^2} = \sqrt{(r - s)^2} = r - s = \sqrt{a} - \sqrt{b}$

84. $\sqrt{\dfrac{a}{b}} = \sqrt{\dfrac{r^2}{s^2}} = \sqrt{\left(\dfrac{r}{s}\right)^2} = \dfrac{r}{s} = \dfrac{\sqrt{a}}{\sqrt{b}}$

Operations with Radicals

Figure 6.1

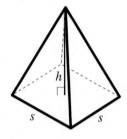

Figure 6.2

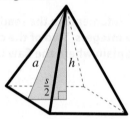

Reading

Radicals appear in a variety of applications, including geometric formulas. In Section 4.1, we studied formulas for the volume and surface area of several solid objects, such as cylinders and spheres. We can now develop a formula for the surface area of a pyramid with a square base, as shown in Figure 6.1.

The pyramid has five faces—namely, the square base and four triangular sides. To find the area of each triangular face, we need to know its base and altitude. We draw in a right triangle, as shown in Figure 6.2, with the height, h, of the pyramid as one of the legs and the unknown altitude, a, as the hypotenuse. The other leg is half the length of the base, so its dimension is $\dfrac{s}{2}$. Now we apply the Pythagorean theorem to find the length of a in terms of s and h.

$$a^2 = h^2 + \left(\frac{s}{2}\right)^2 \qquad \text{Take square roots.}$$

$$a = \sqrt{h^2 + \frac{s^2}{4}} \qquad \text{Add the fractions.}$$

$$= \sqrt{\frac{4h^2 + s^2}{4}} \qquad \text{Simplify the radical.}$$

$$= \frac{1}{2}\sqrt{4h^2 + s^2}$$

Thus, each lateral face of the pyramid is a triangle with base s and altitude $a = \dfrac{1}{2}\sqrt{4h^2 + s^2}$. The area of each triangle is given by

$$A = \frac{1}{2}(\text{base}) \times \text{altitude}$$

$$= \frac{1}{2}s \cdot \frac{1}{2}\sqrt{4h^2 + s^2}$$

$$= \frac{1}{4}s\sqrt{4h^2 + s^2}$$

EXERCISE 1
Find the surface area of the Sun Pyramid in Mexico, including the base. The pyramid is 210 feet high and has a square base that measures 689 feet on each side.

Finally, we find the surface area of the pyramid by adding the area of its base and the areas of the four triangular faces to get

$$S = (\text{area of base}) + 4(\text{area of triangular face})$$

$$= s^2 + 4 \cdot \frac{1}{4}s\sqrt{4h^2 + s^2}$$

$$= s^2 + s\sqrt{4h^2 + s^2}$$

We cannot simplify the radical any further because $4h^2$ and s^2 are terms, not factors. We know that, in general, $\sqrt{a + b} \neq \sqrt{a} + \sqrt{b}$.

Try Exercise 1.

Skills Review

Decide whether each statement is true or false.

1. $\sqrt{(2a+b)^2} = 2a + b$ 2. $\sqrt{4a^2 + b^2} = 2a + b$
3. $\sqrt{1 + 25r^2} = 1 + 5r$ 4. $\sqrt{4s^2 - 1} = 2s - 1$
5. $\sqrt{9q^2 + 36} = 3\sqrt{q^2 + 4}$ 6. $\sqrt{m^2 + \dfrac{1}{4}} = \dfrac{1}{2}\sqrt{4m^2 + 1}$

Answers: **1.** True **2.** False **3.** False **4.** False **5.** True **6.** True

Lesson

Products

In Section 6.4, we simplified square roots by using the product rule for radicals:

$$\sqrt{ab} = \sqrt{a}\,\sqrt{b}$$

We can also use the product rule to multiply two radicals.

> If $a, b \geq 0$,   then $\sqrt{a}\,\sqrt{b} = \sqrt{ab}$

For example,

$$\sqrt{2} \cdot \sqrt{3} = \sqrt{6} \quad \text{and} \quad \sqrt{5} \cdot \sqrt{x} = \sqrt{5x}$$

EXAMPLE 1

Find the product and simplify $\sqrt{2x}\,\sqrt{10xy}$.

Solution

$$\sqrt{2x}\,\sqrt{10xy} = \sqrt{20x^2y} \qquad \text{Factor out perfect squares.}$$
$$= \sqrt{4x^2}\sqrt{5y} = 2x\sqrt{5y}$$

Try Exercise 2.

We can use the distributive law to remove parentheses from products involving radicals. Compare the product involving radicals on the left with the more familiar use of the distributive law on the right:

$$6(\sqrt{3} + \sqrt{2}) = 6\sqrt{3} + 6\sqrt{2} \quad \text{and} \quad 6(x + y) = 6x + 6y$$

EXAMPLE 2

Multiply $5(3x + 4\sqrt{2})$.

Solution

We apply the distributive law to find

$$5(3x + 4\sqrt{2}) = 5 \cdot 3x + 5 \cdot 4\sqrt{2} = 15x + 20\sqrt{2}$$

Try Exercise 3.

EXERCISE 2
Find the product and simplify $(3x\sqrt{6x})(y\sqrt{15xy})$.

EXERCISE 3
Multiply $2a(2\sqrt{3} - \sqrt{a})$.

In the next example, we use the distributive law along with the product rule to simplify the product.

EXAMPLE 3

Multiply $2\sqrt{5}(3 + \sqrt{3})$.

Solution

We apply the distributive law to obtain

$$2\sqrt{5}(3 + \sqrt{3}) = 2\sqrt{5} \cdot 3 + 2\sqrt{5} \cdot \sqrt{3} \quad \text{\small Apply the product rule to the second term.}$$
$$= 6\sqrt{5} + 2\sqrt{15}$$

CAUTION!

Note the difference between the products

$$\sqrt{5} \cdot 3 = 3\sqrt{5} \quad \text{and} \quad \sqrt{5} \cdot \sqrt{3} = \sqrt{15}$$

The rule $\sqrt{a}\sqrt{b} = \sqrt{ab}$ applies to the second product, but not to the first.

Try Exercise 4.

To multiply binomials, we use the FOIL method discussed in Section 4.5.

EXERCISE 4
Multiply $3\sqrt{x}(\sqrt{2x} - 3x)$.

EXAMPLE 4

Multiply $(2 + \sqrt{3})(1 - 2\sqrt{3})$, and then simplify.

Solution

$$(2 + \sqrt{3})(1 - 2\sqrt{3}) = \underbrace{2 \cdot 1}_{F} - \underbrace{2 \cdot 2\sqrt{3}}_{O} + \underbrace{\sqrt{3} \cdot 1}_{I} - \underbrace{\sqrt{3} \cdot 2\sqrt{3}}_{L}$$
$$= 2 - 4\sqrt{3} + \sqrt{3} - 2 \cdot 3 \quad \text{\small Note that } \sqrt{3} \cdot \sqrt{3} = 3.$$
$$= 2 - 4\sqrt{3} + \sqrt{3} - 6 \quad \text{\small Combine like terms.}$$
$$= -4 - 3\sqrt{3}$$

Try Exercise 5.

EXERCISE 5
Expand and simplify $(\sqrt{2} - 2\sqrt{b})^2$.

Fractions

When we solved quadratic equations in Chapter 4, we encountered fractions that included square roots. Many such fractions can be simplified using techniques you have learned in this chapter.

EXAMPLE 5

One of the solutions of the equation $4x^2 - 8x = 1$ is $\dfrac{8 + \sqrt{80}}{8}$. Simplify this radical expression.

Solution

First, we simplify the square root.

$$\sqrt{80} = \sqrt{16 \cdot 5} = 4\sqrt{5}$$

Thus

$$\frac{8 + \sqrt{80}}{8} = \frac{8 + 4\sqrt{5}}{8} \qquad \text{Factor numerator and denominator.}$$

$$= \frac{\cancel{4}(2 + \sqrt{5})}{\cancel{4} \cdot 2} \qquad \text{Divide out common factors.}$$

$$= \frac{2 + \sqrt{5}}{2}$$

CAUTION!

In Example 5, note that

$$8 + 4\sqrt{5} \neq 12\sqrt{5}$$

because 8 and $4\sqrt{5}$ are not like terms. Also note that

$$\frac{8 + \sqrt{80}}{8} \neq \sqrt{80}$$

Because 8 is a *term* of the numerator, not a factor, we cannot cancel the 8's.

We can add or subtract like fractions involving radicals. If the fractions have unlike denominators, we must find the least common denominator before combining them.

EXAMPLE 6

Subtract $\dfrac{1}{2} - \dfrac{\sqrt{3}}{3}$.

Solution

The LCD for the two fractions is 6. We build each fraction to form an equivalent one with a denominator of 6 and then combine the numerators.

$$\frac{1}{2} - \frac{\sqrt{3}}{3} = \frac{1 \cdot 3}{2 \cdot 3} - \frac{\sqrt{3} \cdot 2}{3 \cdot 2}$$

$$= \frac{3}{6} - \frac{2\sqrt{3}}{6} = \frac{3 - 2\sqrt{3}}{6}$$

Try Exercise 6.

Extraction of Roots

We may encounter fractions when we solve quadratic equations by extraction of roots.

EXAMPLE 7

Solve by extraction of roots: $2(3x - 1)^2 = 36$.

Solution

First, we isolate the squared expression by dividing both sides by 2.

$$(3x - 1)^2 = 18 \qquad \text{Extract square roots.}$$

$$3x - 1 = \pm\sqrt{18} \qquad \text{Simplify the radical.}$$

$$3x - 1 = \pm 3\sqrt{2} \qquad \text{Write as two equations.}$$

EXERCISE 6

Add $\dfrac{\sqrt{x}}{3} + \dfrac{\sqrt{2}}{x}$.

Now we solve each equation to find two solutions.

$$3x - 1 = 3\sqrt{2} \qquad\qquad 3x - 1 = -3\sqrt{2}$$
$$3x = 1 + 3\sqrt{2} \qquad\qquad 3x = 1 - 3\sqrt{2}$$
$$x = \frac{1 + 3\sqrt{2}}{3} \qquad\qquad x = \frac{1 - 3\sqrt{2}}{3}$$

We can write the solutions as $\dfrac{1 \pm 3\sqrt{2}}{3}$.

Try Exercise 7.

Quotients

In Section 6.4, we used the quotient rule to simplify square roots of fractions by writing $\sqrt{\dfrac{a}{b}} = \dfrac{\sqrt{a}}{\sqrt{b}}$. We can also use the quotient rule to simplify quotients of square roots.

> If $a \geq 0$ and $b > 0$, then $\dfrac{\sqrt{a}}{\sqrt{b}} = \sqrt{\dfrac{a}{b}}$

EXAMPLE 8

Simplify $\dfrac{2\sqrt{30a}}{\sqrt{6a}}$.

Solution

We use the quotient rule to write the expression as a single radical. Then we simplify the fraction inside the radical.

$$\frac{2\sqrt{30a}}{\sqrt{6a}} = 2\sqrt{\frac{30a}{6a}} \qquad \text{Reduce the fraction.}$$
$$= 2\sqrt{5}$$

Try Exercise 8.

For some applications, it is easier to work with expressions that do not have radicals in the denominators of fractions. For example, we can express the fraction $\dfrac{\sqrt{2}}{\sqrt{3}}$ equivalently as follows:

$$\frac{\sqrt{2} \cdot \sqrt{3}}{\sqrt{3} \cdot \sqrt{3}} = \frac{\sqrt{6}}{3}$$

We multiply numerator and denominator by $\sqrt{3}$, the same root that appears in the denominator originally. This eliminates the radical from the denominator. This process is called **rationalizing the denominator**.

Try Exercise 9.

EXERCISE 7
Solve $(2x + 1)^2 = 8$.

EXERCISE 8
Simplify $\dfrac{3ab\sqrt{75a^3b}}{\sqrt{6ab^5}}$.

EXERCISE 9
Rationalize the denominator of $\dfrac{8}{\sqrt{x}}$.

HOMEWORK 6.5

Simplify the products in Problems 1–6.

1. $\sqrt{8}\,\sqrt{2}$

2. $\sqrt{3}\,\sqrt{27}$

3. $\sqrt{2x}\,\sqrt{3x}$

4. $\sqrt{5a}\,\sqrt{3a}$

5. $(3\sqrt{8a})(a\sqrt{18a})$

6. $(3b\sqrt{12b})(\sqrt{6b})$

Multiply the expressions in Problems 7–12.

7. $\sqrt{2}(3 + \sqrt{3})$

8. $\sqrt{3}(\sqrt{6} - 2)$

9. $\sqrt{5}(4 + 2\sqrt{15})$

10. $\sqrt{6}(\sqrt{3} - \sqrt{10})$

11. $2\sqrt{p}(\sqrt{2p} - p\sqrt{2})$

12. $\sqrt{3k}(3\sqrt{k} - k\sqrt{6k})$

Multiply the expressions in Problems 13–20.

13. $(3 + \sqrt{2})(1 - \sqrt{2})$

14. $(5 + \sqrt{g})(3 + \sqrt{g})$

15. $(4 - \sqrt{a})(4 + \sqrt{a})$

16. $(4 - 3\sqrt{6})(4 + 3\sqrt{6})$

17. $(2 + \sqrt{3})^2$ **18.** $(2\sqrt{n} - 1)^2$

19. $(2\sqrt{w} + \sqrt{5})(\sqrt{w} - 2\sqrt{5})$ **20.** $(3v\sqrt{5} + 2\sqrt{v})^2$

Reduce the fractions in Problems 21–26, if possible.

21. $\dfrac{9 - 3\sqrt{5}}{3}$ **22.** $\dfrac{6 - 9\sqrt{2}}{6}$ **23.** $\dfrac{-8 + \sqrt{8}}{4}$

24. $\dfrac{2 + \sqrt{72}}{4}$ **25.** $\dfrac{6a - \sqrt{18}}{6a}$ **26.** $\dfrac{2x - \sqrt{8}}{2x}$

Solve the equations in Problems 27–32 by extraction of roots.

27. $(2x + 1)^2 = 8$ **28.** $(3x - 2)^2 = 12$ **29.** $3(2x - 8)^2 = 60$

30. $2(3x - 6)^2 = 96$ **31.** $\dfrac{4}{3}(x + 3)^2 = 24$ **32.** $\dfrac{3}{4}(x - 6)^2 = 18$

Solve each formula in Problems 33–36 for *x*.

33. $x^2 + a^2 = b^2$

34. $x^2 - b^2 = c^2$

35. $\dfrac{x^2}{4} - y^2 = 1$

36. $y^2 + \dfrac{x^2}{9} = 1$

Write each expression in Problems 37–44 as a single fraction in simplest form.

37. $\dfrac{5}{4} + \dfrac{3\sqrt{2}}{2}$

38. $\dfrac{3\sqrt{5}}{4} - \dfrac{\sqrt{3}}{5}$

39. $\dfrac{3}{2a} + \dfrac{\sqrt{3}}{6a}$

40. $\dfrac{\sqrt{3}}{x} - \dfrac{1}{4x}$

41. $\dfrac{3\sqrt{3}}{2} + 3$

42. $4 + \dfrac{3\sqrt{2}}{2}$

43. $\dfrac{3}{4} - 2\sqrt{y}$

44. $\dfrac{2\sqrt{y}}{9} + \dfrac{\sqrt{x}}{12}$

In Problems 45–50, find and correct the error in each incorrect calculation.

45. $8\sqrt{7} - 2\sqrt{5} \rightarrow 6\sqrt{2}$

46. $4 + 3\sqrt{5} \rightarrow 7\sqrt{5}$

47. $6\sqrt{8} \rightarrow \sqrt{48}$

48. $2 + \sqrt{6} \rightarrow 2(1 + \sqrt{3})$

49. $\dfrac{5 - 10\sqrt{3}}{5} \rightarrow -10\sqrt{3}$

50. $3\sqrt{10} + 4\sqrt{10} \rightarrow 7\sqrt{20}$

In Problems 51–56, verify by substitution that the given value is a solution of the equation.

51. $t^2 - 2\sqrt{3}t + 3 = 0, \quad t = \sqrt{3}$

52. $3q^2 - 50 = \sqrt{5}q, \quad q = 2\sqrt{5}$

53. $s^2 + 1 = 4s, \quad s = \sqrt{3} + 2$

54. $v^2 + 10 = 8v, \quad v = 4 - \sqrt{6}$

55. $x^2 + 3x + 1 = 0, \quad x = \dfrac{-3 + \sqrt{5}}{2}$

56. $2a^2 - 2a - 3 = 0, \quad a = \dfrac{1 - \sqrt{7}}{2}$

Simplify the quotient in Problems 57–62.

57. $\dfrac{\sqrt{18}}{\sqrt{2}}$

58. $\dfrac{\sqrt{48}}{\sqrt{3}}$

59. $\dfrac{\sqrt{75x^3}}{\sqrt{3x}}$

60. $\dfrac{\sqrt{80a^3}}{\sqrt{5a}}$

61. $\dfrac{\sqrt{48b}}{\sqrt{27b}}$

62. $\dfrac{\sqrt{32a}}{\sqrt{18a}}$

Rationalize the denominator in Problems 63–70.

63. $\dfrac{5}{\sqrt{2}}$

64. $\dfrac{2}{\sqrt{3}}$

65. $\dfrac{a\sqrt{2}}{\sqrt{a}}$

66. $\dfrac{4\sqrt{3x}}{\sqrt{8}}$

67. $\dfrac{b\sqrt{21}}{\sqrt{3b}}$

68. $\dfrac{\sqrt{80}}{\sqrt{5x}}$

69. $\sqrt{\dfrac{7x}{12}}$

70. $\sqrt{\dfrac{50}{2x}}$

Simplify the complex fraction in Problems 71–76.

71. $\dfrac{\dfrac{\sqrt{5}}{3}}{\dfrac{1}{\sqrt{3}}}$

72. $\dfrac{\dfrac{2}{\sqrt{3}}}{\dfrac{3}{\sqrt{6}}}$

73. $\dfrac{1 - \dfrac{2}{\sqrt{6}}}{\dfrac{\sqrt{2}}{\sqrt{6}}}$

74. $\dfrac{2\left(\dfrac{1}{\sqrt{3}}\right)}{1 - \left(\dfrac{1}{\sqrt{3}}\right)^2}$

75. $\dfrac{1 + \dfrac{1}{\sqrt{2}}}{2\sqrt{3} - \dfrac{\sqrt{3}}{2}}$

76. $\dfrac{\dfrac{1}{2} - \dfrac{1}{\sqrt{3}}}{\dfrac{\sqrt{2}}{2} - \dfrac{\sqrt{2}}{3}}$

77. a. Find a formula for the height of an equilateral triangle of side w.

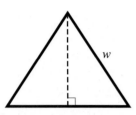

b. Find a formula for the area of the triangle.

78. a. Find an expression for the perimeter of the trapezoid.

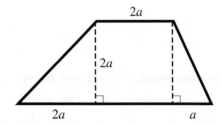

b. Find an expression for the area of the trapezoid.

79. a. Find the height of the pyramid in terms of k.

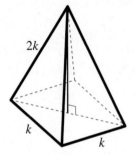

b. Find the volume of the pyramid in terms of k.

80. a. Find the diagonal, d, of the base of the box in terms of x.

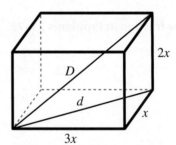

b. Find the diagonal, D, of the box in terms of x.

6.6 Equations with Radicals

Reading

Suppose you have been commissioned to design a new roller coaster for an amusement park. The roller coaster should be more spectacular than all existing roller coasters; in particular, it should include a vertical loop. In order to stay on the track through the loop, the cars must reach a speed, in miles per hour, given by

$$v = \sqrt{89.3r}$$

where r is the radius of the loop in feet. Table 6.1 lists the heights of three of the world's tallest vertical loops.

Table 6.1

Height of loop (feet)	Roller coaster	Location
188	Viper	Six Flags Magic Mountain, Valencia, California
173	Gash	Six Flags Great Adventure, Jackson, New Jersey
170	Shockwave	Six Flags Great Adventure, Gurnee, Illinois

Try Exercises 1 and 2.

Now suppose you know the maximum speed possible for the cars on a particular roller coaster. Can you calculate the height of the tallest vertical loop the cars can negotiate? Table 6.2 lists the speeds of the cars on three of the world's fastest roller coasters.

Table 6.2

Speed of cars (miles/hour)	Roller coaster	Location
86	Fujiyama	Fujikyu Highland Park, Japan
80	Steel Phantom	Kennywood, West Mifflin, Pennsylvania
79	Desperado	Buffalo Bill's, Jean, Nevada

To calculate the maximum safe loop height for the Desperado, we can substitute the speed of its cars, 79 miles per hour, into the formula. This gives us the equation

$$79 = \sqrt{89.3r}$$

This equation is called a **radical equation**, because the unknown value, r, appears under a radical sign. To solve the equation, we want to "undo" the operations performed on r. To undo the operation of taking a square root, we can square both sides of the equation. Remember that we undo operations in the opposite order. Thus we have

Operations Performed on r	*Steps for Solution*
1. Multiply by 89.3	**1.** Square both sides
2. Take square root	**2.** Divide by 89.3

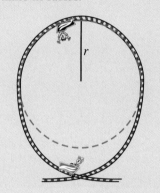

Here is the solution:

$$79^2 = (\sqrt{89.3r})^2 \quad \text{Square both sides.}$$

$$6241 = 89.3r \qquad \text{Divide both sides by 89.3.}$$

$$\frac{6241}{89.3} = \frac{89.3r}{89.3}$$

$$70 \approx r$$

The radius of the tallest negotiable loop would be about 70 feet, so the maximum safe height of the loop would be 2.5 times that distance, or 175 feet.

Try Exercise 3.

EXERCISE 3
Calculate the maximum safe loop height for the roller coaster called Fujiyama.

Skills Review

Square each expression.

1. a. $3x$
   b. $3 + x$

2. a. $\sqrt{3x}$
   b. $3\sqrt{x}$

3. a. $\sqrt{x + 3}$
   b. $\sqrt{x} + \sqrt{3}$

4. a. $x - \sqrt{3}$
   b. $3 - \sqrt{x}$

5. a. $2 - 3\sqrt{x}$
   b. $2\sqrt{2x - 3}$

6. a. $2 + \sqrt{x - 3}$
   b. $3 + 2\sqrt{x - 3}$

Answers: **1. a.** $9x^2$ **b.** $9 + 6x + x^2$ **2. a.** $3x$ **b.** $9x$ **3. a.** $x + 3$ **b.** $x + 2\sqrt{3x} + 3$ **4. a.** $x^2 - 2\sqrt{3}x + 3$ **b.** $9 - 6\sqrt{x} + x$ **5. a.** $4 - 12\sqrt{x} + 9x$ **b.** $8x - 12$ **6. a.** $x + 1 + 4\sqrt{x - 3}$ **b.** $4x - 3 + 12\sqrt{x - 3}$

Lesson

As you learned in the reading, a *radical equation* is one in which the variable appears under a radical, for example:

$$\sqrt{x - 3} = 4$$

We can solve this equation by squaring both sides to produce an equation without radicals.

$$(\sqrt{x - 3})^2 = 4^2$$

$$x - 3 = 16$$

$$x = 19$$

Try Exercise 4.

EXERCISE 4
Solve $\sqrt{x - 6} = 2$.

Extraneous Solutions

The technique of squaring both sides may introduce extraneous solutions. (Recall that an *extraneous solution* is a value that is not a solution to the original equation.) For example, consider the equation

$$\sqrt{x + 2} = -3$$

Squaring both sides gives

$$(\sqrt{x + 2})^2 = (-3)^2$$

$$x + 2 = 9$$

$$x = 7$$

However, if we substitute $x = 7$ into the original equation, we see that it is not a solution:

$$\sqrt{7 + 2} \stackrel{?}{=} -3$$

$$\sqrt{9} \stackrel{?}{=} -3$$

$$3 \neq -3$$

The value $x = 7$ is a solution to the squared equation, but not to the original equation. In this case, the original equation has no solution. Thus *whenever you square both sides of an equation, you must check the solutions* by substituting them in the original equation.

If a radical equation involves several terms, it is easiest to isolate the radical term on one side of the equation before squaring both sides.

EXAMPLE 1

Solve $4 + \sqrt{8 - 2x} = x$.

Solution

We first isolate the radical by subtracting 4 from both sides.

$$\sqrt{8 - 2x} = x - 4$$

Now we square both sides.

$$(\sqrt{8 - 2x})^2 = (x - 4)^2 \qquad \text{Recall that } (a - b)^2 = a^2 - 2ab + b^2.$$
$$8 - 2x = x^2 - 8x + 16 \qquad \text{Subtract } 8 - 2x \text{ from both sides.}$$
$$0 = x^2 - 6x + 8 \qquad \text{Factor the right side.}$$
$$0 = (x - 4)(x - 2)$$

Thus the possible solutions are $x = 4$ and $x = 2$. We must check both of these in the original equation. For $x = 4$, we obtain

$$4 + \sqrt{8 - 2(4)} \stackrel{?}{=} 4$$
$$4 + \sqrt{0} \stackrel{?}{=} 4$$
$$4 = 4$$

And for $x = 2$, we get

$$4 + \sqrt{8 - 2(2)} \stackrel{?}{=} 2$$
$$4 + 2 \stackrel{?}{=} 2$$
$$6 \neq 2$$

Thus $x = 2$ is an extraneous solution. The only solution to the original equation is $x = 4$.

CAUTION!

When squaring both sides of an equation, be careful to square the entire expression on either side of the equal sign. It is *incorrect* to square each term separately. Thus, in Example 1, it would *not* be correct to write

$$(\sqrt{8 - 2x})^2 = x^2 - 4^2 \qquad \leftarrow \text{Incorrect!}$$

Try Exercise 5.

Equations with Cube Roots

We can also solve equations in which the variable appears under a cube root.

EXERCISE 5
Solve $\sqrt{x - 3} + 5 = x$.

EXERCISE 6

Solve $3\sqrt[3]{4x - 1} = -15$.

ANSWERS TO 6.6 EXERCISES

1. Viper: 82 miles per hour; Gash: 79 miles per hour; Shockwave: 78 miles per hour

2. 85 miles per hour   **3.** 207 feet

4. $x = 10$ **5.** $x = 7$

6. $x = -31$

EXAMPLE 2

Solve the equation $15 - 2\sqrt[3]{x - 4} = 9$.

Solution

We first isolate the cube root.

$$15 - 2\sqrt[3]{x - 4} = 9 \qquad \text{Subtract 15 from both sides.}$$
$$-2\sqrt[3]{x - 4} = -6 \qquad \text{Divide both sides by } -2.$$
$$\sqrt[3]{x - 4} = 3$$

Next we "undo" the cube root by cubing both sides of the equation.

$$(\sqrt[3]{x - 4})^3 = 3^3$$
$$x - 4 = 27$$

Finally, we add 4 to both sides to find the solution, $x = 31$. We do not have to check for extraneous solutions when we cube both sides of an equation, but it is a good idea to check the solution for accuracy anyway.

$$15 - 2\sqrt[3]{31 - 4} = 15 - 2\sqrt[3]{27}$$
$$= 15 - 2(3)$$
$$= 15 - 6 = 9 \qquad \text{The solution checks.}$$

Try Exercise 6.

HOMEWORK 6.6

In Problems 1–10, solve each equation, and check the solutions.

1. $\sqrt{x + 4} = 5$

2. $\sqrt{x - 6} = 2$

3. $\sqrt{x} - 4 = 5$

4. $\sqrt{x} + 3 = 9$

5. $6 - \sqrt{x} = 8$

6. $5 - \sqrt{x} = 6$

7. $2 + 3\sqrt{x-1} = 8$

8. $1 + 2\sqrt{x+3} = 7$

9. $2\sqrt{3x+1} - 3 = 5$

10. $3\sqrt{5x-1} - 5 = 16$

11. Use the graph of $y = \sqrt{x}$ shown to solve the following equations. (You may have to estimate some of the solutions.) Check your answers algebraically.
   a. $\sqrt{x} = 4$ **b.** $\sqrt{x} = 2.5$

   c. $\sqrt{x} = -2$ **d.** $\sqrt{x} = 5.3$

12. Use the graph of $y = \sqrt[3]{x}$ shown to solve the following equations. (You may have to estimate some of the solutions.) Check your answers algebraically.
   a. $\sqrt[3]{x} = -2$ **b.** $\sqrt[3]{x} = 2.5$

   c. $\sqrt[3]{x} = 1.8$ **d.** $\sqrt[3]{x} = -3.5$

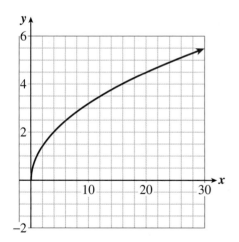

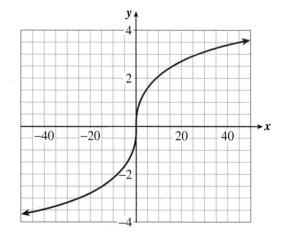

13. a. Complete the table of values, and graph
$y = \sqrt{x - 4}$.

x	4	5	6	10	16	19	24
y							

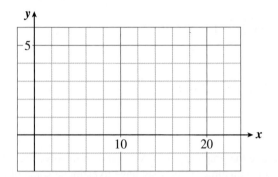

b. Solve $\sqrt{x - 4} = 3$ graphically and algebraically. Do your answers agree?

14. a. Complete the table of values, and graph
$y = 2 - \sqrt{x}$.

x	0	1	4	8	12	18	24
y							

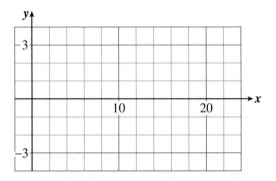

b. Solve $2 - \sqrt{x} = -1$ graphically and algebraically. Do your answers agree?

15. a. Complete the table of values, and graph
$y = 4 - \sqrt{x + 3}$.

x	−3	−2	0	1	4	8	16
y							

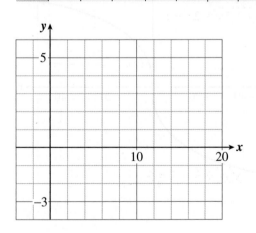

b. Solve $4 - \sqrt{x + 3} = 1$ graphically and algebraically. Do your answers agree?

16. a. Complete the table of values, and graph
$y = 2 + \sqrt{3x - 6}$.

x	3	4	5	8	10	16	20
y							

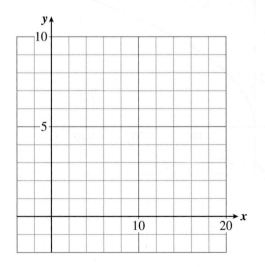

b. Solve $2 + \sqrt{3x - 6} = 8$ graphically and algebraically. Do your answers agree?

In Problems 17–26, solve each equation, and check for extraneous solutions.

17. $\sqrt{x} = 3 - 2x$

18. $\sqrt{x} = 2x - 6$

19. $\sqrt{x + 4} + 2 = x$

20. $\sqrt{x + 8} - 2 = x$

21. $x + \sqrt{2x + 7} = -2$

22. $\sqrt{3x + 10} - 2x = 6$

23. $\sqrt{x + 7} = 2x + 4$

24. $\sqrt{x + 10} = 2x - 1$

25. $6 + \sqrt{5x - 4} - x = 4$

26. $x + \sqrt{2x + 3} - 7 = 9$

27. The higher your altitude, the farther you can see to the horizon, if nothing blocks your line of sight. From a height of h meters, the distance d to the horizon in kilometers is given by

$$d = \sqrt{12h}$$

a. Mt. Wilson is part of the San Gabriel Mountains north of Los Angeles, and it has an elevation of 1740 meters. How far can you see from the top of Mt. Wilson?

b. The new Getty Center is built on the hills above Sunset Boulevard in Los Angeles, and from the patio on a clear day, you can see the city of Long Beach, 44 kilometers away. What does this tell you about the elevation of the patio at the Getty Center?

28. When an athlete or a dancer leaps into the air, the "hang time," t, in seconds, is related to the height of the leap, L, in feet, by the equation

$$t = \frac{\sqrt{L}}{2}$$

a. The American figure skater Michael Weiss has a leap height of 42 inches. How long is the hang time he has to complete a triple jump?

b. How high would a basketball player have to leap in order to have a hang time of 1.1 seconds?

29. The speed of a tsunami, in miles per hour, is given by

$$s = 3.9\sqrt{d}$$

where d is the depth of the ocean beneath the wave, in feet. A tsunami traveling along the Aleutian Trench off the coast of Alaska is clocked at a speed of over 615 miles per hour. What is the depth of the Aleutian Trench?

30. You can measure your reaction time by having a friend hold a ruler vertically, with 0 inches at the lower end and 12 at the top, just above your hand while you extend your thumb and forefinger. Have your friend let go of the ruler, and catch it between your thumb and forefinger as it falls. Note the distance, d, between the lower end of the ruler and your fingers. The time it takes you to react is given in seconds by

$$t = \frac{1}{8}\sqrt{\frac{d}{3}}$$

Kathy thinks her reaction time is better than 0.15 second. If it is, where will she catch the ruler?

31. The height of a cylindrical storage tank is four times its radius. If the tank holds V cubic inches of liquid, its radius in inches is

$$r = \sqrt[3]{\frac{V}{12.57}}$$

a. If the tank must be big enough to hold 340 cubic inches, what must its radius be?

b. If the radius of the tank is 5.5 inches, what is its capacity?

32. In order for a windmill to generate P watts of power, the velocity of the wind, in miles per hour, must be

$$v = \sqrt[3]{\frac{P}{0.015}}$$

a. What wind speed is necessary to generate 400 watts of power?

b. How much power would be generated by a wind blowing at 45 miles per hour?

33. The equation for the semicircle shown is

$$y = \sqrt{9 - x^2}$$

Find the x-coordinates of two points on the semicircle that have a y-coordinate of 2.

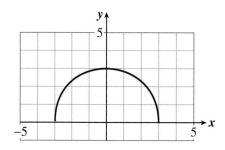

34. The top half of an ellipse is shown below. Its equation is

$$y = \frac{1}{2}\sqrt{16 - x^2}$$

Find the x-coordinates of two points on the graph that have a y-coordinate of $\sqrt{3}$.

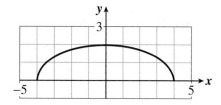

In Problems 35–40, solve each equation.

35. $2\sqrt[3]{x} + 15 = 5$

36. $20 - 3\sqrt[3]{x} = 14$

37. $\sqrt[3]{2x - 5} - 1 = 2$

38. $16 = 12 - \sqrt[3]{x + 24}$

39. $2 = 8 - 3\sqrt[3]{x^3 + 1}$

40. $4\sqrt[3]{x^3 - 8} + 7 = -5$

CHAPTER SUMMARY AND REVIEW

Properties and formulas

1. State the third law of exponents, and compare it with the first law.

2. State the fourth and fifth laws of exponents, and give examples.

3. Give the definition of a^{-n}, and give an example.

4. State the second law of exponents.

5. Explain how the quotient $\dfrac{2^5}{2^5} = 1$ illustrates the definition of a^0.

6. Describe the form of a number written in scientific notation.

7. State two properties of radicals that are useful in simplifying radical expressions.

8. State two similar "rules" for radicals that are false.

Techniques and procedures

9. Explain how to simplify a square root.

10. When can you simplify a sum or difference of square roots? How?

11. Explain how to rationalize the denominator of a fraction.

12. What is the procedure for solving a radical equation?

Simplify each expression in Problems 13–20.

13. a. $a^4 \cdot a^6$

b. $(a^4)^6$

14. a. $\dfrac{a^4}{a^6}$

b. $\dfrac{a^6}{a^4}$

15. a. $(2a^2)^3$

b. $2a^2(a^2)^3$

16. a. $\left(\dfrac{-3u}{v^2}\right)^4$

b. $\dfrac{-3u^4}{v^2(v^4)}$

17. $-4x(-2x^2)^3$

18. $-3w^2(-w^3)^2$

19. $4t^2(t^2)^3 - (6t^4)^2$

20. $(3v)^3(-v^3) - (2v)^2(-v^4)$

Simplify the expressions in Problems 21–32, and write the results without negative exponents.

21. a. $3x^{-2}$ **b.** $(3x)^{-2}$

22. a. $(4y)^0$ **b.** $4y^0$

23. a. $\left(\dfrac{5}{z}\right)^{-2}$ **b.** $\dfrac{5}{z^{-2}}$

24. a. $\dfrac{16c^{-4}}{8c^{-8}}$ **b.** $\dfrac{16c^{-4}}{-8c^8}$

25. $3p^{-4}(2p^{-3})$ **26.** $2q^{-4}(2q)^{-3}$

27. $\dfrac{(4k^{-3})^2}{2k^{-5}}$ **28.** $\dfrac{6h^{-4}(2h^{-2})}{3h^{-3}}$

29. $5g^{-6}(g^{-3})^{-2}$ **30.** $(8n)^{-2}(n^{-3})^{-4}$

31. $\dfrac{2}{x^{-1} + y^{-1}}$ **32.** $\dfrac{a^{-2} - b^{-2}}{a - b}$

Write each number in Problems 33–38 in scientific notation.

33. 586,000   **34.** 12,400,000

35. 0.000 7

36. 0.000 009

37. 483×10^3

38. $0.003\ 5 \times 10^2$

In Problems 39–42, use scientific notation to compute each answer.

39. $(48,000,000)(380,000,000)$

40. $(0.000\ 002\ 4)(1,900,000,000)$

41. $\dfrac{0.000\ 000\ 005}{0.000\ 2}$

42. $\dfrac{38,500,000}{(0.000\ 8)(0.001\ 7)}$

In Problems 43 and 44, give your answers in scientific notation, rounded to two decimal places.

43. One atomic unit is equal to 1.66×10^{-27} kilogram. What is the mass of 6.02×10^{23} atomic units?

44. The mass of an electron is 9.11×10^{-31} kilogram; the mass of a proton is 1.67×10^{-27} kilogram. How many electrons would you need to match the mass of one proton?

Simplify each radical in Problems 45–52, if possible.

45. a. $\sqrt{4x^6}$ **b.** $\sqrt{4 + x^6}$ **c.** $\sqrt{(4 + x)^6}$

46. a. $\sqrt{1 - w^9}$ **b.** $\sqrt{1 - w^8}$ **c.** $\sqrt{-w^8}$

47. $-\sqrt{27m^5}$

48. $\pm\sqrt{98q^{99}}$

49. $\sqrt{\dfrac{a^3c}{16}}$

50. $\sqrt{\dfrac{50b^7}{2g^4}}$

51. $\dfrac{2}{3}b\sqrt{12b^3}$

52. $\dfrac{4}{3a^2}\sqrt{45a^3}$

Simplify each expression in Problems 53–60.

53. $3\sqrt{24} + 2\sqrt{18} - 5\sqrt{6}$

54. $2x\sqrt{x} - 3\sqrt{x^3} - 6\sqrt{x}$

55. $\dfrac{\sqrt{54w^{12}}}{\sqrt{9w^6}}$

56. $\dfrac{\sqrt{24n^3}}{\sqrt{6n^5}}$

57. $\dfrac{6 - 3\sqrt{12}}{3}$

58. $\dfrac{\sqrt{8} - \sqrt{12}}{6}$

59. $\dfrac{2}{3} - \dfrac{\sqrt{3}}{2}$

60. $\dfrac{2\sqrt{3}}{5} - 1$

Solve Problems 61 and 62 by extraction of roots.

61. $(3a - 2)^2 = 24$

62. $5(2d + 1)^2 = 90$

Solve Problems 63 and 64 for the indicated variable.

63. $2a^2 + 4b^2 = c^2$, for b

64. $25w^2 - k = 16m$, for w

In Problems 65–70, multiply and simplify the answers.

65. $\sqrt{3}(\sqrt{2} - \sqrt{6})$

66. $3\sqrt{2}(8\sqrt{6} - \sqrt{12})$

67. $(2 - \sqrt{d})(2 + \sqrt{d})$

68. $(5 - 3\sqrt{2})(3 + \sqrt{2})$

69. $(\sqrt{7} + 3)^2$

70. $(3\sqrt{t} + 1)^2$

Simplify each expression in Problems 71–76, and rationalize the denominator, if necessary.

71. $\dfrac{2}{\sqrt{x}}$

72. $\sqrt{\dfrac{3a}{b}}$

73. $\dfrac{2\sqrt{5}}{\sqrt{8}}$

74. $\dfrac{a\sqrt{32}}{\sqrt{2a}}$

75. $\dfrac{2}{\sqrt{7}} + \dfrac{3\sqrt{7}}{7}$

76. $\dfrac{1}{2\sqrt{3}} - \dfrac{1}{3\sqrt{2}}$

In Problems 77 and 78, verify by substitution that the given value is a solution of the equation.

77. $2x^2 - 2x - 3 = 0, \quad x = \dfrac{1 + \sqrt{7}}{2}$

78. $x^2 + 4x - 1 = 0, \quad x = -2 - \sqrt{5}$

In Problems 79 and 80, find the length of the third side in each right triangle.

79.

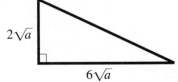

$2\sqrt{a}$

$6\sqrt{a}$

80.

$4b$

$2b$

Solve Problems 81–86.

81. $3\sqrt{x + 2} - 4 = 5$

82. $\sqrt{x - 3} + 4 = 2$

83. $\sqrt{2x + 1} = x - 7$

84. $4\sqrt{4x + 1} = 5x + 2$

85. $\sqrt[3]{3x + 2} - 4 = 1$

86. $9 - 4\sqrt[3]{1 - 2x} = 17$

87. The time, in seconds, that a pendulum takes to complete one full swing, from right to left and back again, is given by the formula

$$T = 2\pi\sqrt{\frac{L}{32}}$$

where L is the length of the pendulum in feet. The longest pendulum in the world is a reconstruction of Foucault's pendulum in the Convention Center in Portland, Oregon. The pendulum weighs 900 pounds and takes 10.54 seconds to complete one full swing. To the nearest foot, how long is the pendulum?

88. The velocity, v, in miles per hour, of a satellite orbiting the earth is given by

$$v = \sqrt{\frac{1.24 \times 10^{12}}{R + h}}$$

where h is the altitude of the satellite in miles, and R is the radius of the earth, about 3960 miles. The Russian space station Mir has an orbital velocity of 17,187 miles per hour. What is its altitude?

Review of Arithmetic Skills

A.1 **Reducing and Building Fractions**

A **fraction** is a quotient of two numbers. For example, $\frac{2}{3}$ and $\frac{7}{4}$ are fractions. The expression $\frac{2}{3}$ means "2 divided by 3," and $\frac{7}{4}$ means "7 divided by 4." The number above the fraction bar is called the **numerator,** and the number below the fraction bar is called the **denominator.** The denominator of a fraction can never be zero because we cannot divide by zero.

If the numerator is less than the denominator, then the fraction is called a **proper fraction.** A proper fraction is a number less than 1. These are examples of proper fractions:

$$\frac{2}{3} \quad \text{and} \quad \frac{5}{7}$$

If the numerator is greater than or equal to the denominator, then the fraction is called an **improper fraction.** An improper fraction represents a number greater than or equal to 1. For example, these are improper fractions:

$$\frac{7}{4} \quad \text{and} \quad \frac{5}{5}$$

Fractions are used to describe a portion or part of some whole amount. For example, the fraction $\frac{2}{3}$ indicates that some whole is divided into three equal parts and that two of them are being considered. Thus, to take $\frac{2}{3}$ of 12 items, we separate the 12 items into 3 groups of 4 items each and then take 2 of the groups, to get 8 items.

Figure A.1 illustrates that $\frac{2}{3}$ of 12 is equal to 8.

Figure A.1

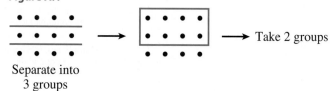

Separate into
3 groups

Take 2 groups

Equivalent Fractions

Any fraction can be rewritten in a different but equivalent form by multiplying both its numerator and its denominator by the same number (other than zero). For example,

$$\frac{2}{3} = \frac{2 \cdot 6}{3 \cdot 6} = \frac{12}{18}$$

The two fractions $\frac{2}{3}$ and $\frac{12}{18}$ are called **equivalent fractions.** Equivalent fractions are equal—that is, they represent the same portion of a whole.

The rule for forming equivalent fractions is called the **fundamental principle of fractions.**

Fundamental Principle of Fractions

If the numerator and denominator of a fraction are multiplied by the same (nonzero) number, the new fraction is equivalent to the old one.

Reducing Fractions

We can also form an equivalent fraction by dividing the numerator and the denominator by the same (nonzero) number. This is another version of the fundamental principle of fractions, and it enables us to reduce a fraction. A fraction is **reduced,** or in *lowest terms*, if no number greater than 1 divides into both the numerator and the denominator. For example, we can reduce $\frac{10}{15}$ to the equivalent fraction $\frac{2}{3}$.

To Reduce a Fraction:

1. Factor numerator and denominator completely.

2. Divide numerator and denominator by any common factors.

3. Simplify the result.

EXAMPLE 1

Reduce the fraction $\frac{21}{30}$ to lowest terms.

Solution

We follow the three steps given above:

$$\frac{21}{30} = \frac{3 \cdot 7}{2 \cdot 3 \cdot 5} \qquad \text{Factor numerator and denominator.}$$

$$= \frac{\cancel{3} \cdot 7}{2 \cdot \cancel{3} \cdot 5} \qquad \begin{array}{l} \text{Divide numerator and denominator} \\ \text{by any common factors.} \end{array}$$

$$= \frac{7}{2 \cdot 5} \qquad \begin{array}{l} \text{Multiply any remaining factors} \\ \text{in numerator and denominator.} \end{array}$$

$$= \frac{7}{10}$$

Because no number greater than 1 divides evenly into both 7 and 10, the fraction is reduced to lowest terms.

_____

Remember that any number divided by itself is 1.

EXAMPLE 2

Reduce each fraction to lowest terms.

    a. $\dfrac{2}{6}$ **b.** $\dfrac{18}{3}$

Solution

    a. $\dfrac{2}{6} = \dfrac{2}{2\cdot3} = \dfrac{\cancel{2}}{\cancel{2}\cdot3} = \dfrac{1}{3}$ **b.** $\dfrac{18}{3} = \dfrac{2\cdot3\cdot3}{3} = \dfrac{2\cdot3\cdot\cancel{3}}{\cancel{3}} = \dfrac{6}{1} = 6$

Building Fractions

You should always reduce a fraction if it is the final answer to a problem. However, when adding or subtracting fractions, you may need to **build** the fractions to equivalent forms with larger numerators and denominators before you can proceed with the computation. For example, we might want to build the fraction $\dfrac{3}{5}$ to an equivalent form with the denominator 100:

$$\frac{3}{5} = \frac{?}{100}$$

Instead of dividing numerator and denominator by common factors, as we do when reducing a fraction, when building a fraction we *multiply* numerator and denominator by the **building factor.** Here are the steps for building fractions.

To Build a Fraction:

1. Divide the new denominator by the original denominator. The quotient is called the *building factor.*

2. Multiply the numerator and denominator of the original fraction by the building factor.

EXAMPLE 3

Find the building factor (BF) to build the fraction as indicated:

$$\frac{3}{5} = \frac{3\cdot\text{BF}}{5\cdot\text{BF}} = \frac{?}{100}$$

Solution

We divide the new denominator by the original denominator:

$$100 \div 5 = 20$$

The building factor is 20. Multiply the numerator and denominator of the original fraction by 20.

$$\frac{3}{5} = \frac{3 \cdot 20}{5 \cdot 20} = \frac{60}{100}$$

The new fraction is $\frac{60}{100}$.

HOMEWORK A.1

Reduce the fractions in Problems 1–12 to lowest terms.

1. $\dfrac{6}{10}$

2. $\dfrac{14}{21}$

3. $\dfrac{12}{75}$

4. $\dfrac{45}{70}$

5. $\dfrac{24}{72}$

6. $\dfrac{30}{120}$

7. $\dfrac{36}{48}$

8. $\dfrac{100}{120}$

9. $\dfrac{63}{105}$

10. $\dfrac{75}{180}$

11. $\dfrac{216}{18}$

12. $\dfrac{175}{35}$

In Problems 13–24, find the building factor (BF), and build each fraction.

13. $\dfrac{2}{3} = \dfrac{2 \cdot BF}{3 \cdot BF} = \dfrac{?}{9}$

14. $\dfrac{5}{7} = \dfrac{5 \cdot BF}{7 \cdot BF} = \dfrac{?}{14}$

15. $\dfrac{3}{2} = \dfrac{3 \cdot BF}{2 \cdot BF} = \dfrac{?}{8}$

16. $\dfrac{4}{3} = \dfrac{4 \cdot BF}{3 \cdot BF} = \dfrac{?}{12}$

17. $\dfrac{2}{5} = \dfrac{2 \cdot BF}{5 \cdot BF} = \dfrac{?}{100}$

18. $\dfrac{1}{4} = \dfrac{1 \cdot BF}{4 \cdot BF} = \dfrac{?}{56}$

19. $\dfrac{1}{3} = \dfrac{1 \cdot BF}{3 \cdot BF} = \dfrac{?}{72}$

20. $\dfrac{4}{5} = \dfrac{4 \cdot BF}{5 \cdot BF} = \dfrac{?}{60}$

21. $\dfrac{5}{8} = \dfrac{5 \cdot BF}{8 \cdot BF} = \dfrac{?}{144}$

22. $\dfrac{5}{16} = \dfrac{5 \cdot BF}{16 \cdot BF} = \dfrac{?}{112}$

23. $\dfrac{0}{3} = \dfrac{0 \cdot BF}{3 \cdot BF} = \dfrac{?}{6}$

24. $\dfrac{0}{11} = \dfrac{0 \cdot BF}{11 \cdot BF} = \dfrac{?}{88}$

A.2 Multiplying and Dividing Fractions

Multiplying Fractions

To multiply two fractions, we multiply their numerators together and multiply their denominators together.

EXAMPLE 1

Multiply the fractions.

a. $\dfrac{2}{5} \cdot \dfrac{3}{7}$ **b.** $\dfrac{1}{3} \cdot \dfrac{3}{8}$

Solution

a. $\dfrac{2}{5} \cdot \dfrac{3}{7} = \dfrac{2 \cdot 3}{5 \cdot 7}$ Multiply numerators; multiply denominators.

$= \dfrac{6}{35}$

b. $\dfrac{1}{3} \cdot \dfrac{3}{8} = \dfrac{1 \cdot 3}{3 \cdot 8}$ Multiply numerators; multiply denominators.

$= \dfrac{3}{24} = \dfrac{\cancel{3}}{\cancel{3} \cdot 2 \cdot 2 \cdot 2}$ Reduce.

$= \dfrac{1}{8}$

We can save steps in Example 1b by dividing out, or canceling, any common factors from numerator and denominator of the product before we multiply, like this:

$$\frac{1}{3} \cdot \frac{3}{8} = \frac{1}{\cancel{3}} \cdot \frac{\cancel{3}}{8} = \frac{1}{8}$$

The common factors do not have to be in the numerator and denominator of the same fraction. In general, we have the following steps for multiplying fractions.

To Multiply Fractions:

1. Divide numerators and denominators by any common factors.
2. Multiply together the remaining factors in the numerators; multiply together the remaining factors in the denominators.
3. Reduce the product if necessary.

EXAMPLE 2

Multiply $\dfrac{9}{4} \cdot \dfrac{2}{5}$

Solution

Divide the denominator of the first fraction and the numerator of the second fraction by 2.

$$\frac{9}{4} \cdot \frac{2}{5} = \frac{9}{\underset{2}{\cancel{4}}} \cdot \frac{\cancel{2}}{5} \quad \text{\small Divide numerator and denominator by the common factor, 2.}$$

$$= \frac{9 \cdot 1}{2 \cdot 5} \quad \text{\small Multiply numerators; multiply denominators.}$$

$$= \frac{9}{10}$$

Dividing Fractions

The **reciprocal** of a fraction is found by interchanging the numerator and denominator.

For example,

the reciprocal of $\dfrac{5}{6}$ is $\dfrac{6}{5}$

the reciprocal of $\dfrac{1}{8}$ is $\dfrac{8}{1}$, or 8

We can change any division problem into a multiplication problem with the same answer by using reciprocals. This gives us a method for dividing fractions.

To Divide Fractions:

1. Replace the second fraction by its reciprocal, and change the division to multiplication.
2. Follow the rules for multiplication.

EXAMPLE 3

Divide the fractions.

    a. $\dfrac{5}{9} \div \dfrac{2}{3}$ **b.** $\dfrac{16}{3} \div 4$

Solution

    a. We replace $\dfrac{2}{3}$ by its reciprocal, $\dfrac{3}{2}$, and change to multiplication.

$$\frac{5}{9} \div \frac{2}{3} = \frac{5}{\cancel{9}} \cdot \frac{\cancel{3}}{2} \qquad \text{Divide numerator and denominator by the common factor, 3.}$$

$$= \frac{5 \cdot 1}{3 \cdot 2} \qquad \text{Multiply numerators; multiply denominators.}$$

$$= \frac{5}{6}$$

    b. We replace 4 by its reciprocal, $\dfrac{1}{4}$, and change to multiplication.

$$\frac{16}{3} \div 4 = \frac{\overset{4}{\cancel{16}}}{3} \cdot \frac{1}{\cancel{4}} \qquad \text{Divide numerator and denominator by the common factor, 4.}$$

$$= \frac{4 \cdot 1}{3 \cdot 1} \qquad \text{Multiply numerators; multiply denominators.}$$

$$= \frac{4}{3}$$

HOMEWORK A.2

Multiply the fractions in Problems 1–12.

1. $\dfrac{2}{3} \cdot \dfrac{5}{7}$ **2.** $\dfrac{4}{3} \cdot \dfrac{11}{5}$ **3.** $\dfrac{6}{7} \cdot \dfrac{14}{15}$

4. $\dfrac{15}{11} \cdot \dfrac{22}{35}$ **5.** $\dfrac{12}{16} \cdot \dfrac{18}{27}$ **6.** $\dfrac{8}{75} \cdot \dfrac{15}{20}$

7. $\dfrac{28}{56} \cdot \dfrac{10}{15}$ **8.** $\dfrac{16}{38} \cdot \dfrac{19}{12}$ **9.** $\dfrac{21}{48} \cdot \dfrac{88}{77}$

10. $\dfrac{18}{121} \cdot \dfrac{99}{90}$ **11.** $\dfrac{24}{20} \cdot \dfrac{24}{36} \cdot \dfrac{3}{4}$ **12.** $\dfrac{18}{30} \cdot \dfrac{6}{8} \cdot \dfrac{4}{20}$

Divide the fractions in Problems 13–24.

13. $\dfrac{3}{4} \div \dfrac{5}{8}$

14. $\dfrac{15}{32} \div \dfrac{25}{48}$

15. $\dfrac{7}{3} \div \dfrac{28}{5}$

16. $\dfrac{9}{2} \div \dfrac{18}{7}$

17. $\dfrac{4}{5} \div 6$

18. $\dfrac{5}{6} \div 10$

19. $4 \div \dfrac{2}{9}$

20. $5 \div \dfrac{15}{32}$

21. $\dfrac{11}{2} \div \dfrac{3}{4}$

22. $\dfrac{10}{3} \div \dfrac{3}{5}$

23. $\dfrac{30}{24} \div \dfrac{18}{72}$

24. $\dfrac{36}{42} \div \dfrac{48}{63}$

A.3   Adding and Subtracting Fractions

Like Fractions

Fractions that have the same denominator are called **like fractions**. For example,

$$\dfrac{7}{8} \quad \text{and} \quad \dfrac{3}{8} \quad \text{are like fractions}$$

$$\dfrac{2}{5} \quad \text{and} \quad \dfrac{4}{9} \quad \text{are } not \text{ like fractions}$$

It is easy to add or subtract like fractions.

To Add or Subtract Like Fractions:

1. Add or subtract the numerators.
2. Keep the same denominator.
3. Reduce if necessary.

EXAMPLE 1

Add the fractions.

a. $\dfrac{1}{5} + \dfrac{3}{5}$

b. $\dfrac{1}{12} + \dfrac{7}{12}$

Solution

a. $\dfrac{1}{5} + \dfrac{3}{5} = \dfrac{1+3}{5} = \dfrac{4}{5}$ Add the numerators; keep the same denominator.

b. $\dfrac{1}{12} + \dfrac{7}{12} = \dfrac{1+7}{12}$ Add the numerators; keep the same denominator.

$\quad\quad\quad = \dfrac{8}{12} = \dfrac{2}{3}$ Reduce.

EXAMPLE 2

Subtract the fractions.

a. $\dfrac{11}{5} - \dfrac{4}{5}$ **b.** $\dfrac{13}{4} - \dfrac{7}{4}$

Solution

a. $\dfrac{11}{5} - \dfrac{4}{5} = \dfrac{11-4}{5} = \dfrac{7}{5}$ Subtract the numerators; keep the same denominator.

b. $\dfrac{13}{4} - \dfrac{7}{4} = \dfrac{13-7}{4}$ Subtract the numerators; keep the same denominator.

$\quad\quad\quad = \dfrac{6}{4} = \dfrac{3}{2}$ Reduce.

Lowest Common Denominator

Before we can add two fractions with different denominators, we must rewrite both fractions as equivalent ones with the same denominator. This new denominator is called a *common denominator*. The first step in an addition problem involving unlike fractions is to discover a suitable common denominator, usually the smallest, or lowest, common denominator.

The **lowest common denominator (LCD)** for two or more fractions is the smallest number that each of the denominators will divide into evenly. For instance, the LCD for $\dfrac{1}{4}$ and $\dfrac{5}{6}$ is 12, because 12 is the smallest number that 4 and 6 both divide into evenly. Sometimes it is easy to find the LCD; if not, we can use the following steps.

To Find the Lowest Common Denominator (LCD):

1. Factor each denominator completely. List all the different factors that appear.
2. Choose one of the factors. In which denominator does that factor occur the most times? Write the factor that many times.
3. Repeat Step 2 for each of the different factors in your list.
4. Multiply all the factors from Steps 2 and 3 to get the LCD.

EXAMPLE 3

Find the LCD for the following fractions.

a. $\dfrac{7}{12}, \dfrac{5}{18}$ **b.** $\dfrac{2}{15}, \dfrac{13}{12}, \dfrac{23}{30}$

Solution

a. We factor each denominator completely:

$$12 = \boxed{2 \cdot 2} \cdot 3$$
$$18 = 2 \cdot \boxed{3 \cdot 3}$$

The factors involved are 2 and 3. The factor 2 occurs twice in the factorization of 12, so we use two 2's in the LCD. The factor 3 occurs twice in the factorization of 18, so we use two 3's in the LCD. Thus the LCD is

$$\text{LCD} = 2 \cdot 2 \cdot 3 \cdot 3 = 36$$

b. We factor each denominator completely:

$$15 = \boxed{3} \cdot \boxed{5}$$
$$12 = \boxed{2 \cdot 2} \cdot 3$$
$$30 = 2 \cdot 3 \cdot 5$$

The factors involved are 2, 3, and 5. The factor 2 occurs twice in the factorization of 12, so we use two 2's in the LCD. The factors 3 and 5 both occur at most once in any *one* denominator, so we use one 3 and one 5 in the LCD. Thus the LCD is

$$\text{LCD} = 2 \cdot 2 \cdot 3 \cdot 5 = 60$$

Unlike Fractions

Fractions with different denominators are called **unlike fractions**. To add unlike fractions, we must first build the fractions so that they have the same denominator.

To Add or Subtract Unlike Fractions:

1. Find the LCD for the fractions.
2. Build each fraction to form an equivalent one with the LCD as its denominator.
3. Add or subtract the numerators. Keep the same denominator.
4. Reduce if necessary.

EXAMPLE 4

Add $\dfrac{7}{10} + \dfrac{5}{6}$.

Solution

We find the LCD for the fractions by factoring each denominator:

$$10 = \boxed{2} \cdot \boxed{5}$$
$$6 = 2 \cdot \boxed{3}$$

The LCD is $2 \cdot 3 \cdot 5$, or 30. Build each fraction so that its denominator is 30. For the first fraction, the building factor is $30 \div 10$, or 3:

$$\frac{7}{10} \cdot \frac{3}{3} = \frac{21}{30}$$

For the second fraction, the building factor is $30 \div 6$, or 5:

$$\frac{5}{6} \cdot \frac{5}{5} = \frac{25}{30}$$

Finally, we add the two like fractions $\dfrac{21}{30}$ and $\dfrac{25}{30}$:

$$\frac{7}{10} + \frac{5}{6} = \frac{21}{30} + \frac{25}{30} = \frac{46}{30}$$

We reduce the sum to obtain

$$\frac{46}{30} = \frac{\cancel{2} \cdot 23}{\cancel{2} \cdot 15} = \frac{23}{15}$$

EXAMPLE 5

Subtract $\dfrac{7}{6} - \dfrac{3}{4}$.

Solution

We find the LCD by factoring each denominator:

$$6 = 2 \cdot \boxed{3} \qquad 4 = \boxed{2 \cdot 2}$$

The LCD is $2 \cdot 2 \cdot 3$, or 12. We build each fraction to an equivalent one with the denominator 12.

$$\frac{7}{6} \cdot \frac{2}{2} = \frac{14}{12} \quad \text{and} \quad \frac{3}{4} \cdot \frac{3}{3} = \frac{9}{12}$$

We subtract the new fractions to obtain

$$\frac{7}{6} - \frac{3}{4} = \frac{14}{12} - \frac{9}{12} = \frac{5}{12}$$

HOMEWORK A.3

Add or subtract the fractions in Problems 1–16.

1. $\dfrac{1}{11} + \dfrac{3}{11}$

2. $\dfrac{1}{9} + \dfrac{7}{9}$

3. $\dfrac{14}{15} - \dfrac{7}{15}$

4. $\dfrac{17}{20} - \dfrac{14}{20}$

5. $\dfrac{1}{6} + \dfrac{3}{6}$

6. $\dfrac{3}{8} + \dfrac{1}{8}$

7. $\dfrac{9}{10} - \dfrac{4}{10}$

8. $\dfrac{3}{4} - \dfrac{1}{4}$

9. $\dfrac{1}{6} + \dfrac{2}{6} + \dfrac{3}{6}$

10. $\dfrac{2}{10} + \dfrac{3}{10} + \dfrac{5}{10}$

11. $\dfrac{1}{5} + \dfrac{3}{5} - \dfrac{2}{5}$

12. $\dfrac{2}{3} + \dfrac{4}{3} - \dfrac{5}{3}$

13. $\dfrac{7}{8} - \dfrac{3}{8} - \dfrac{2}{8}$

14. $\dfrac{8}{9} - \dfrac{2}{9} - \dfrac{3}{9}$

15. $\dfrac{19}{25} - \dfrac{8}{25} + \dfrac{4}{25}$

16. $\dfrac{16}{21} - \dfrac{4}{21} + \dfrac{2}{21}$

Find the lowest common denominator (LCD) of the fractions in Problems 17–32.

17. $\dfrac{3}{6}, \dfrac{2}{3}$

18. $\dfrac{1}{2}, \dfrac{3}{4}$

19. $\dfrac{5}{4}, \dfrac{5}{6}$

20. $\dfrac{3}{10}, \dfrac{2}{15}$

21. $\dfrac{4}{3}, \dfrac{2}{7}$

22. $\dfrac{4}{5}, \dfrac{8}{11}$

23. $\dfrac{11}{12}, \dfrac{7}{30}$

24. $\dfrac{29}{30}, \dfrac{17}{45}$

25. $\dfrac{1}{2}, \dfrac{2}{3}, \dfrac{3}{5}$

26. $\dfrac{1}{4}, \dfrac{4}{5}, \dfrac{8}{7}$

27. $\dfrac{19}{6}, \dfrac{5}{9}, \dfrac{8}{15}$

28. $\dfrac{5}{8}, \dfrac{11}{12}, \dfrac{1}{20}$

29. $\dfrac{21}{9}, \dfrac{11}{4}, \dfrac{1}{12}$

30. $\dfrac{18}{25}, \dfrac{1}{50}, \dfrac{3}{4}$

31. $\dfrac{23}{24}, \dfrac{19}{36}, \dfrac{7}{12}$

32. $\dfrac{13}{60}, \dfrac{3}{40}, \dfrac{11}{30}$

Add or subtract the fractions in Problems 33–48.

33. $\dfrac{1}{2} + \dfrac{1}{3}$

34. $\dfrac{1}{3} + \dfrac{1}{4}$

35. $\dfrac{3}{4} - \dfrac{2}{3}$

36. $\dfrac{2}{3} - \dfrac{1}{2}$

37. $\dfrac{5}{8} + \dfrac{1}{12}$

38. $\dfrac{11}{18} + \dfrac{5}{24}$

39. $\dfrac{14}{15} - \dfrac{7}{10}$

40. $\dfrac{9}{14} - \dfrac{1}{3}$

41. $\dfrac{1}{2} + \dfrac{1}{3} + \dfrac{2}{5}$

42. $\dfrac{1}{7} + \dfrac{5}{12} + \dfrac{1}{4}$

43. $\dfrac{13}{20} + \dfrac{3}{8} - \dfrac{5}{12}$

44. $\dfrac{5}{6} + \dfrac{4}{9} - \dfrac{7}{15}$

45. $\dfrac{49}{50} - \dfrac{8}{25} - \dfrac{1}{4}$

46. $\dfrac{9}{4} - \dfrac{10}{9} - \dfrac{5}{12}$

47. $\dfrac{29}{30} - \dfrac{21}{40} + \dfrac{9}{70}$

48. $\dfrac{11}{36} - \dfrac{7}{72} + \dfrac{5}{24}$

A.4 Mixed Numbers and Improper Fractions

Writing a Mixed Number as an Improper Fraction

A **mixed number** is the sum of a whole number and a fraction, such as $5 + \dfrac{2}{3}$.

A mixed number is usually written without the addition sign, like this: $5\dfrac{2}{3}$. (We

have to remember that $5\dfrac{2}{3}$ means 5 *plus* $\dfrac{2}{3}$.) Because a mixed number is greater

than 1, we can also write it as an improper fraction. We do this simply by adding
the whole number to the fraction.

Recall that any whole number can be treated as a fraction by writing the whole number as a numerator over a denominator of 1. For our example, $5 = \dfrac{5}{1}$, and thus

$$5\frac{2}{3} = \frac{5}{1} + \frac{2}{3}$$

We can now add the fractions by building $\dfrac{5}{1}$ to an equivalent fraction with the denominator 3, as follows:

$$\frac{5}{1} + \frac{2}{3} = \frac{5 \cdot 3}{1 \cdot 3} + \frac{2}{3} \qquad \text{Building factor is 3.}$$

$$= \frac{15}{3} + \frac{2}{3} = \frac{17}{3}$$

Thus the mixed number $5\dfrac{2}{3}$ is equal to the improper fraction $\dfrac{17}{3}$.

We can state the following procedure for converting a mixed number to an improper fraction.

To Convert a Mixed Number to an Improper Fraction:

1. Write the whole number over a denominator of 1.
2. Multiply the numerator and the denominator of the whole number by the denominator of the fraction part.
3. Add the two fractions.

EXAMPLE 1

Convert the mixed number $2\dfrac{5}{7}$ to an improper fraction.

Solution

We write the mixed number as the sum of a whole number and a fraction; then add.

$$2\frac{5}{7} = \frac{2}{1} + \frac{5}{7} \qquad \text{Build the first fraction; the LCD is 7.}$$

$$= \frac{2 \cdot 7}{1 \cdot 7} + \frac{5}{7} \qquad \text{Simplify; add the like fractions.}$$

$$= \frac{14}{7} + \frac{5}{7} = \frac{19}{7}$$

Thus $2\dfrac{5}{7} = \dfrac{19}{7}$.

Operations with Mixed Numbers

It is usually easier to convert all mixed numbers to improper fractions before performing calculations.

EXAMPLE 2

Divide $\quad 3\dfrac{3}{4} \div 2\dfrac{1}{2}$

Solution

We convert each mixed number to an improper fraction.

$$3\frac{3}{4} = \frac{3}{1} + \frac{3}{4} = \frac{3 \cdot 4}{1 \cdot 4} + \frac{3}{4}$$

$$= \frac{12}{4} + \frac{3}{4} = \frac{15}{4}$$

$$2\frac{1}{2} = \frac{2}{1} + \frac{1}{2} = \frac{2 \cdot 2}{1 \cdot 2} + \frac{1}{2}$$

$$= \frac{4}{2} + \frac{1}{2} = \frac{5}{2}$$

We divide the improper fractions by taking the reciprocal of the divisor and multiplying.

$$3\frac{3}{4} \div 2\frac{1}{2} = \frac{15}{4} \div \frac{5}{2}$$

$$= \frac{15}{4} \cdot \frac{2}{5} = \frac{\overset{3}{\cancel{15}}}{\underset{2}{\cancel{4}}} \cdot \frac{\cancel{2}}{\cancel{5}}$$

$$= \frac{3}{2}$$

EXAMPLE 3

Add $\quad 1\dfrac{2}{5} + 2\dfrac{3}{8}$

Solution

We convert each mixed number to an improper fraction.

$$1\frac{2}{5} = \frac{1}{1} + \frac{2}{5} = \frac{1 \cdot 5}{1 \cdot 5} + \frac{2}{5}$$

$$= \frac{5}{5} + \frac{2}{5} = \frac{7}{5}$$

$$2\frac{3}{8} = \frac{2}{1} + \frac{3}{8} = \frac{2 \cdot 8}{1 \cdot 8} + \frac{3}{8}$$

$$= \frac{16}{8} + \frac{3}{8} = \frac{19}{8}$$

We add the improper fractions. The denominators are 5 and 8, so the LCD is $\;5 \cdot 8$, or 40. We build each fraction to an equivalent one with the denominator 40.

$$\frac{7}{5} = \frac{7 \cdot 8}{5 \cdot 8} = \frac{56}{40} \qquad \text{and} \qquad \frac{19}{8} = \frac{19 \cdot 5}{8 \cdot 5} = \frac{95}{40}$$

We add the new fractions.

$$\frac{56}{40} + \frac{95}{40} = \frac{56 + 95}{40} = \frac{151}{40}$$

Thus $\quad 1\dfrac{2}{5} + 2\dfrac{3}{8} = \dfrac{151}{40}$

Writing an Improper Fraction as a Mixed Number

Recall that the fraction bar is a division symbol; $\frac{3}{4}$ means "3 divided by 4." If the numerator and the denominator of a fraction are equal, the fraction is equal to 1. Thus, for example, $\frac{3}{3} = 1$ and $\frac{12}{12} = 1$. If the numerator is larger than the denominator, the fraction is a number greater than 1. By dividing the denominator into the numerator, we can write an improper fraction as a whole number or as a mixed number.

EXAMPLE 4

Write each improper fraction as a whole number or as a mixed number.

   a. $\dfrac{8}{4}$ **b.** $\dfrac{17}{5}$

Solution

   a. We divide the denominator into the numerator: $8 \div 4 = 2$.   Thus
      $\dfrac{8}{4} = 2$,   a whole number.

   b. When we divide the denominator into the numerator, there is a remainder.

$$
\begin{array}{r}
3 \leftarrow \text{quotient} \\
5\overline{)17} \\
-15 \\
\hline
2 \leftarrow \text{remainder}
\end{array}
$$

    The quotient, 3, is the whole number, and the remainder, 2, is the numerator of the proper fraction. Thus

$$
\frac{17}{5} = 3 + \frac{2}{5} = 3\frac{2}{5}
$$

The following procedure summarizes the method we used in Example 4.

To Convert an Improper Fraction to a Mixed Number:

1. Divide the denominator into the numerator.
2. The quotient is the whole-number part of the mixed number. The remainder is the numerator of the fraction part, with the original denominator.
3. If there is no remainder, then the improper fraction is equivalent to a whole number.

HOMEWORK A.4

In Problems 1–8, convert each mixed number to an improper fraction.

1. $3\dfrac{2}{3}$

2. $1\dfrac{2}{13}$

3. $12\dfrac{1}{2}$

4. $11\dfrac{2}{7}$

5. $20\dfrac{4}{5}$

6. $7\dfrac{11}{20}$

7. $4\dfrac{21}{50}$

8. $5\dfrac{5}{6}$

In Problems 9–24, convert each mixed number to an improper fraction; then perform the indicated operation.

9. $3\dfrac{1}{4} + 1\dfrac{5}{8}$

10. $5\dfrac{3}{4} + 2\dfrac{7}{8}$

11. $5\dfrac{2}{3} + 6\dfrac{3}{4}$

12. $4\dfrac{1}{3} + 2\dfrac{1}{4}$

13. $9\dfrac{3}{8} - 2\dfrac{1}{2}$

14. $5\dfrac{1}{5} - 4\dfrac{7}{10}$

15. $7\dfrac{3}{8} - 1\dfrac{7}{12}$

16. $6\dfrac{5}{8} - 2\dfrac{5}{6}$

17. $7\dfrac{1}{3} \cdot 2\dfrac{1}{4}$

18. $5\dfrac{2}{5} \cdot 3\dfrac{1}{3}$

19. $3\dfrac{3}{7} \cdot 2\dfrac{1}{12}$

20. $2\dfrac{2}{9} \cdot 3\dfrac{3}{5}$

21. $2\dfrac{1}{3} \div 5\dfrac{3}{5}$

22. $4\dfrac{1}{2} \div 2\dfrac{4}{7}$

23. $3\dfrac{3}{4} \div 1\dfrac{7}{8}$

24. $5\dfrac{5}{6} \div 2\dfrac{5}{8}$

In Problems 25–32, convert each improper fraction to a whole number or a mixed number.

25. $\dfrac{11}{3}$

26. $\dfrac{15}{4}$

27. $\dfrac{43}{8}$

28. $\dfrac{37}{8}$

29. $\dfrac{107}{16}$

30. $\dfrac{123}{16}$

31. $\dfrac{317}{32}$

32. $\dfrac{361}{32}$

A.5 Decimal Fractions

A **decimal fraction** is a fraction whose denominator is a power of 10. However, a decimal fraction is not written with a fraction bar. Instead, we show the denominator using place value notation. For example,

$$0.3 = \frac{3}{10}, \ 0.07 = \frac{7}{100}, \quad \text{and} \quad 0.241 = \frac{241}{1000}$$

Place Value Notation

In a whole number, the value of each digit is determined by its position. For example, the 6 in 68 stands for 6 tens; the 6 in 642 stands for 6 hundreds. In a whole number we usually do not write the decimal point.

Positions to the right of the decimal point are called **decimal places**, and digits in those positions represent decimal fractions. For example, the number 0.3 represents three tenths, or $\frac{3}{10}$, and the number 0.03 represents three hundredths, or $\frac{3}{100}$. The place values of the first few positions on either side of the decimal point are shown here:

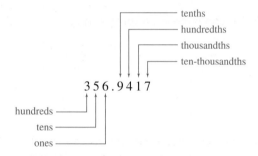

Converting a Decimal Fraction to a Common Fraction

The first place to the right of the decimal point is the tenths place, and the second is the hundredths place. For example, the digits of 0.57 stand for 5 tenths and 7 hundredths:

$$0.57 = \frac{5}{10} + \frac{7}{100}$$

If we add these fractions, we find that

$$\frac{5}{10} + \frac{7}{100} = \frac{5 \cdot 10}{10 \cdot 10} + \frac{7}{100} = \frac{57}{100}$$

Thus $0.57 = \frac{57}{100}$. We see that we can read a decimal fraction just as we do a whole number and then use the place value of the *last* digit for the denominator. This also gives us a rule for writing a decimal fraction as a common fraction.

> **To Convert a Decimal Fraction to a Common Fraction:**
>
> 1. Use the digits after the decimal point as the numerator of the fraction.
> 2. Use the place value of the last digit as the denominator of the fraction.
> 3. Reduce the fraction if possible.

EXAMPLE 1

Convert each decimal fraction to a common fraction.

    a. 0.007   **b.** 0.35   **c.** 4.9

Solution

    a. The numerator of the fraction is 007, or 7. The last digit, 7, is in the thousandths place, so the denominator of the fraction is 1000. Thus

$$0.007 = \frac{7}{1000}$$

    b. The numerator of the fraction is 35. The last digit, 5, is in the hundredths place, so the denominator of the fraction is 100. Thus

$$0.35 = \frac{35}{100}$$

    This fraction can be reduced as follows:

$$0.35 = \frac{35}{100} = \frac{\cancel{5} \cdot 7}{2 \cdot 2 \cdot \cancel{5} \cdot 5} = \frac{7}{20}$$

    c. The decimal number 4.9 represents a mixed number with whole-number part 4 and fraction part 0.9, or 9 tenths. Thus

$$4.9 = 4\frac{9}{10} = \frac{4 \cdot 10}{1 \cdot 10} + \frac{9}{10} = \frac{49}{10}$$

Converting a Common Fraction to a Decimal Fraction

To convert a common fraction to a decimal fraction, we divide the denominator into the numerator. (Recall that the fraction bar is a division symbol.) This is easily done with a calculator. However, if you must perform the division by hand, you must first write a decimal point and one or more zeros after the numerator.

EXAMPLE 2

Convert the fraction $\frac{4}{5}$ to decimal form.

Solution

First, we write a decimal point and a zero after the numerator, and then we divide the denominator into the numerator. To begin the division, we place another decimal point in the quotient directly above the decimal in the dividend.

$$\begin{array}{r} .8 \\ 5\overline{)4.0} \\ \underline{-4.0} \end{array}$$

After placing the decimal points, we may ignore them and proceed as though the numbers were whole numbers. We find that the decimal equivalent of $\frac{4}{5}$ is 0.8.

On a calculator, we enter the division in Example 2 as   4 [÷] 5 [=],   and the calculator displays 0.8.

If the division does not terminate after the first step, we must add more zeros after the numerator and continue dividing.

EXAMPLE 3

Convert each fraction to decimal form.

a. $\dfrac{3}{8}$ **b.** $\dfrac{6}{11}$

Solution

a. We divide the denominator into the numerator. We continue to add zeros after the numerator until the division terminates.

$$
\begin{array}{r}
.375 \\
8\overline{)3.000} \\
-24 \\
\hline
60 \\
-56 \\
\hline
40 \\
-40 \\
\hline
\end{array}
$$

Place decimal in quotient.

Thus $\dfrac{3}{8} = 0.375$.

b. We divide the denominator into the numerator.

$$
\begin{array}{r}
.5454\ldots \\
11\overline{)6.0000\ldots} \\
-55 \\
\hline
50 \\
-44 \\
\hline
60 \\
-55 \\
\hline
50\ldots
\end{array}
$$

Place decimal in quotient above decimal in numerator.

After a few steps, it becomes clear that the division is in an "infinite loop"—the same digits are repeated in the quotient over and over without ending. The quotient, 0.545454 . . . , is called a **repeating decimal**. We indicate a repeating decimal by writing a bar over the block of repeated digits: $0.\overline{54}$. Thus

$$
\frac{6}{11} = 0.545454\ldots \qquad \text{or} \qquad \frac{6}{11} = 0.\overline{54}
$$

You should be aware that your calculator may round off a repeating decimal to the last digit of its display. For example, the decimal equivalent of $\dfrac{8}{9}$ might be displayed as 0.8888889. This is an approximation to the actual decimal fraction:

$$
\frac{8}{9} = 0.\overline{8}
$$

HOMEWORK A.5 www

In Problems 1–16, convert each decimal fraction to a common fraction.

1. 0.17

2. 0.81

3. 0.07

4. 0.03

5. 0.023

6. 0.049

7. 0.6

8. 0.2

9. 0.26

10. 0.15

11. 0.375

12. 0.864

13. 2.25

14. 1.75

15. 3.60

16. 4.80

In Problems 17–32, convert each common fraction to a decimal fraction.

17. $\dfrac{21}{25}$

18. $\dfrac{7}{20}$

19. $\dfrac{23}{50}$

20. $\dfrac{19}{20}$

21. $\dfrac{31}{100}$

22. $\dfrac{73}{100}$

23. $\dfrac{31}{1000}$

24. $\dfrac{73}{1000}$

25. $\dfrac{5}{16}$

26. $\dfrac{13}{8}$

27. $\dfrac{3}{8}$

28. $\dfrac{19}{16}$

29. $\dfrac{5}{6}$

30. $\dfrac{2}{3}$

31. $\dfrac{3}{11}$

32. $\dfrac{9}{11}$

A.6 Rounding Decimal Numbers

Sometimes it is more practical to use an approximate value rather than the exact result of a calculation. A common method of approximating numbers is called **rounding**. For example, the sales tax on a purchase is usually rounded up to the nearest cent.

To round a number, we consider two of its digits, called the *target digit* and the *test digit*. Consider the number 12,469. If we want to round to thousands, we look for the multiple of 1000 that is closest to this number. The thousands digit, 2, is the target digit in this case. Is 12,469 closer to 12,000 or to 13,000? Because 469 is less than 500 (halfway between multiples of 1000), the number 12,469 is closer to 12,000. We used the number 4 to test which way to round, so 4 is the test digit.

target digit ─┐ ┌─ test digit

12,469 ⟶ 12,000

Here are the procedures for rounding decimal numbers. We consider two cases: rounding to a decimal place and rounding to whole number.

To Round a Number to a Decimal Place:

1. Circle the digit in the position you wish to round to; this is the target digit.
2. Underline the digit to the right of the target digit; this is the test digit.
3. If the test digit is 5 or greater (5, 6, 7, 8, or 9), increase the target digit by 1. If the test digit is less than 5 (0, 1, 2, 3, or 4), keep the original target digit.
4. Discard the test digit and any digits to its right.

To Round to a Whole Number:

1. Circle the digit in the position you wish to round to; this is the target digit.
2. Underline the digit to the right of the target digit; this is the test digit.
3. If the test digit is 5 or greater (5, 6, 7, 8, or 9), increase the target digit by 1. If the test digit is less than 5 (0, 1, 2, 3, or 4), keep the original target digit.
4. Replace the test digit and any digits between it and the decimal point with zeros. Discard any digits to the right of the decimal point.

EXAMPLE 1

Round 349.0258 to the indicated place.

   a. To the nearest tenth
   b. To the nearest thousandth

Solution

a. The target digit is the tenths digit, 0, and the test digit is 2. Because the test digit is less than 5, we keep the target digit. We then discard the test digit and the digits to its right. The result is 349.0.

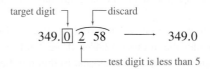

b. The target digit is the thousandths digit, 5, and the test digit is 8. Because the test digit is greater than 5, we increase the target digit by 1. We then discard the test digit. The result is 349.026.

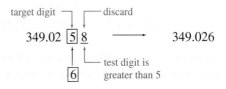

EXAMPLE 2

Round 2479.83 to the indicated place.

a. To the nearest thousand

b. To the nearest whole number

Solution

a. The target digit is the thousands digit, 2, and the test digit is 4. Because the test digit is less than 5, we keep the target digit. We then replace the test digit and the remaining digits up to the decimal point (4, 7, and 9) with zeros and discard the digits after the decimal point. The result is 2000.

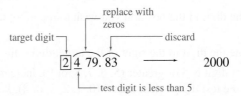

b. The target digit is the ones digit, 9, and the test digit is 8. Because the test digit is greater than 5, we add 1 to the target digit. (Note that this also changes the tens digit.) We then discard the digits after the decimal point. The result is 2480.

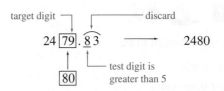

HOMEWORK A.6

In Problems 1–12, round each number four ways:
a. To the nearest ten
b. To the nearest tenth
c. To the nearest hundredth
d. To the nearest thousandth

1. 14.7742

2. 21.6344

3. 76.28256

4. 54.60791

5. 169.8991

6. 832.8196

7. 5545.9098

8. 9989.8982

9. 700.9597

10. 508.9595

11. 19.95059

12. 97.93965

In Problems 13–20, round each number three ways:
a. **To one decimal place**
b. **To two decimal places**
c. **To three decimal places**

13. 1.9069   **14.** 2.2591

15. 0.91994   **16.** 0.65232

17. 0.09857   **18.** 0.07579

19. 6.1695   **20.** 4.2945

A.7 Adding and Subtracting Decimal Fractions

Recall that when we add or subtract whole numbers, we align the numbers vertically so that digits with the same place value are in the same column. We also do this when we add decimal numbers. The easiest way to make sure that the digits are aligned correctly is to align the decimal points vertically.

EXAMPLE 1

Add   15.263 + 6.74.

Solution
We align the decimal points vertically and add just as we do with whole numbers. We place the decimal point for the sum directly below the decimals points in the two numbers.

$$
\begin{array}{r}
15.263 \\
+\ \ 6.740 \\
\hline
22.003
\end{array}
$$

We may need to add a zero at the end of one of the numbers, as we did to 6.74, so that both numbers have the same number of decimal places. The sum is 22.003.

EXAMPLE 2

Subtract $10 - 0.06$.

Solution

The decimal point in the whole number 10 is on the right, after the units place. We add a decimal point and two zeros to get 10.00, so that both numbers have the same number of decimal places. We align the decimal points and subtract.

$$
\begin{array}{r}
10.00 \\
-\ \ 0.06 \\
\hline
9.94
\end{array}
$$

The difference is 9.94.

Here is the procedure for adding and subtracting decimal numbers.

To Add or Subtract Decimal Numbers:

1. Write the numbers with the decimal points aligned vertically. If necessary, add zeros at the far right so that both numbers have the same number of decimal places.
2. Add or subtract just as you would for whole numbers. Place the decimal point in the sum or difference directly below the decimal points in the original numbers.

Of course, if you have a calculator, you can let it do the work.

EXAMPLE 3

Use a calculator to add or subtract.

    a. $15.263 + 6.74$ **b.** $10 - 0.06$

Solution

Key in the calculations as shown.

    a. 15.263 [+] 6.74 [=].   The calculator returns the sum, 22.003.

    b. 10 [−] 0.06 [=].   The calculator returns the difference, 9.94.

HOMEWORK A.7

Add the decimals in Problems 1–12.

  1. $1.46 + 3.27$ **2.** $8.01 + 1.90$ **3.** $3.64 + 0.36$

4. 11.22 + 1.08   **5.** 0.26 + 1.4   **6.** 0.91 + 7.3

7. 13 + 0.26   **8.** 5 + 0.55   **9.** 6.2 + 0.027

10. 7.1 + 0.109   **11.** 31.2 + 3.25   **12.** 42.3 + 2.48

Subtract the decimals in Problems 13–24.

13. 12.63 − 9.16   **14.** 6.31 − 2.26   **15.** 7 − 1.26

16. 10 − 7.11   **17.** 6.02 − 0.95   **18.** 5.03 − 0.87

19. 12.1 − 2.36   **20.** 14.4 − 8.43   **21.** 438.4 − 76.25

22. 587.3 − 32.91   **23.** 13 − 0.0006   **24.** 18 − 0.0002

Perform the additions and subtractions in Problems 25–30.

25. $2.49 + 6.3 - 5.08$

26. $7.92 - 1.48 + 3.6$

27. $573.1 - 28.96 - 2.094$

28. $348.8 - 59.29 - 6.123$

29. $465.5 - 127 + 238.58$

30. $651.3 - 249 + 367.42$

A.8 Multiplying and Dividing Decimal Fractions

Products

We multiply and divide decimal numbers in the same way we multiply and divide whole numbers, except that we must keep track of the decimal point. (Recall that a decimal place is a position of a digit *after*, or to the right of, the decimal point.) To find the number of decimal places in a product, we add the number of decimal places in the two numbers being multiplied.

EXAMPLE 1

Multiply the decimal numbers.

a. $(3.4)(0.68)$ **b.** $(0.4)(0.2)$

Solution

a. We multiply the numbers, ignoring their decimal points. (The decimal points do not have to be aligned, as they do in addition.)

$$
\begin{array}{r}
0.68 \\
\times \quad 3.4 \\
\hline
272 \\
204 \\
\hline
2312
\end{array}
$$

Next we locate the decimal point in the product. Because 3.4 has one decimal place and 0.68 has two decimal places, their product has three decimal places. We count three places *leftward* from the *right* side of 2312:

$$2.312$$

The product is 2.312.

b. We multiply the numbers, ignoring their decimal points:

$$
\begin{array}{r}
0.4 \\
\times 0.2 \\
\hline
8
\end{array}
$$

We locate the decimal point. Because 0.4 has one decimal place and 0.2 has one decimal place, their product has two decimal places. We count two places from the right side of 8, inserting zeros as necessary.

$$.08$$

The product is 0.08.

Here is the procedure for multiplying decimal numbers.

> ## To Multiply Two Decimal Numbers:
>
> 1. Multiply the numbers, ignoring their decimal points.
> 2. Add the numbers of decimal places in the two numbers being multiplied.
> 3. Count that many decimal places from the right side of the product, and place the decimal point there.

Quotients

We locate the decimal point in a product *after* we perform the computation. To find a quotient of decimal numbers, we locate the decimal point *before* we divide. If the divisor is not already a whole number, we must move the decimal point to the right side of the number, after the last digit. Then we *also* move the decimal in the dividend the *same number of places*. Once this is done, we write the decimal point for the quotient directly above the decimal of the dividend. Finally, we divide in the same way we divide whole numbers.

EXAMPLE 2

Divide the decimal numbers.

    a. $2.12 \div 0.25$ **b.** $3 \div 0.006$

Solution

    a. We move the decimal points in both divisor and dividend two places to the right:

$$25.\overline{)212.}$$

We then place a decimal point on the quotient bar directly above the decimal point following 212 and begin dividing. We add as many zeros to the right of the decimal point in the dividend as necessary.

$$
\begin{array}{r}
8.48 \\
25\overline{)212.00} \\
-200 \\
\hline
120 \\
-100 \\
\hline
200 \\
-200 \\
\hline
0
\end{array}
$$

The quotient is 8.48.

b. We move the decimal points in both divisor and dividend three places to the right, adding zeros to the dividend:

$$006.\overline{)3000.}$$

We place a decimal point for the quotient directly above the decimal point of the dividend and begin dividing.

$$
\begin{array}{r}
500. \\
6\overline{)3000.} \\
-30 \\
\hline
000
\end{array}
$$

The quotient is 500.

Here is the procedure for dividing decimal numbers.

To Divide One Decimal Number by Another:

1. If the divisor is not a whole number, move the decimal point to the right side of the number, after the last digit.

2. Move the decimal point of the dividend the same number of places to the right, adding zeros after the last digit as necessary.

3. Place the decimal point for the quotient directly above the decimal of the dividend.

4. Divide as usual.

We can find products and quotients of decimal numbers quite easily with a calculator. Example 3 shows the keying sequences for the products and quotients in Examples 1 and 2.

EXAMPLE 3

Use a calculator to find the following products and quotients.

    a. (3.4)(0.68)   **b.** (0.4)(0.2)

    c. $2.12 \div 0.25$ **d.** $3 \div 0.006$

Solution

We use the keying sequences shown.

a. 3.4 $\boxed{\times}$ 0.68 $\boxed{=}$. The calculator displays the product, 2.312.

b. 0.4 $\boxed{\times}$ 0.2 $\boxed{=}$. The calculator displays the product, 0.08.

c. 2.12 $\boxed{\div}$ 0.25 $\boxed{=}$. The calculator displays the quotient, 8.48.

d. 3 $\boxed{\div}$ 0.006 $\boxed{=}$. The calculator displays the quotient, 500.

Remember that a calculator rounds off any result to the number of digits it can display. For example, an eight-digit calculator displays the quotient

$$5 \boxed{\div} 3 \boxed{=}$$

as 1.6666667, instead of 1.66 . . . or $1.\overline{6}$.

HOMEWORK A.8

In Problems 1–16, multiply the decimal numbers.

1. (6.8)(0.6)

2. (4.7)(0.5)

3. (0.32)(0.4)

4. (0.47)(0.3)

5. (2.04)(0.02)

6. (3.07)(0.04)

7. (4.012)(0.03)

8. (5.007)(0.05)

9. (2.5)(1.3)

10. (4.7)(3.2)

11. (4.32)(2.4)

12. (5.07)(8.4)

13. (4.12)(0.42)

14. (6.28)(0.31)

15. (0.032)(0.12)

16. (0.041)(0.32)

In Problems 17–32, divide the decimal numbers.

17. 31.57 ÷ 7

18. 50.58 ÷ 9

19. 67.5 ÷ 16

20. 97.2 ÷ 17

21. 6.85 ÷ 0.5

22. 8.64 ÷ 0.4

23. 71.91 ÷ 4.7

24. 593.14 ÷ 9.4

25. 229.9 ÷ 0.38

26. 374.4 ÷ 0.78

27. 4.3776 ÷ 0.019

28. 10.9752 ÷ 0.024

29. 7 ÷ 1.54

30. 8 ÷ 2.45

31. 64 ÷ 0.08

32. 45 ÷ 0.09

A.9 Percents

A **percent** is just a fraction whose denominator is 100. *Percent* means "per one hundred," or "divided by 100." Thus 75% means $\dfrac{75}{100}$, or $\dfrac{3}{4}$.

Before using a percent in a calculation, we must rewrite it as a decimal fraction or as a common fraction.

Converting Percents to Decimal Fractions

To convert a percent to a decimal fraction, we simply divide by 100 and then discard the percent symbol. For example, 75% means $\dfrac{75}{100}$, so to find the decimal equivalent for 75%, we divide 75 by 100. We do not have to perform the division longhand, because dividing any number by 100 merely moves the decimal point two places to the left. That is,

$$75 \div 100 = 0.75$$

Thus 75% is equal to 0.75.

EXAMPLE 1

Convert each percent to a decimal fraction.

    a. 5%   **b.** 135%

    c. 0.8%   **d.** $37\dfrac{1}{2}\%$

Solution

For parts (a), (b), and (c), we move the decimal point two places to the left and remove the percent symbol.

    a. 5% = 0.05   **b.** 135% = 1.35   **c.** 0.8% = 0.008

For part (d), we first write the fraction in decimal form and then proceed as above.

    d. $37\dfrac{1}{2}\% = 37.5\% = 0.375$

_____

Here is the procedure for changing a percent to a decimal fraction.

To Change a Percent to a Decimal Fraction:

1. If the percent involves a common fraction, write it in decimal form.

2. Discard the percent symbol, and move the decimal point two places to the *left*.

Converting Decimal Fractions to Percents

To convert a decimal fraction to percent form, we reverse the process described above.

To Change a Decimal Fraction to a Percent:

1. Move the decimal point two places to the *right*.

2. Write a percent symbol after the number.

EXAMPLE 2

Convert each decimal to a percent.

   a. 0.3   **b.** 0.258   **c.** 3.4

Solution

We move each decimal point two places to the right. We write the percent symbol at the end.

   a. $0.3 = 30\%$ **b.** $0.258 = 25.8\%$ **c.** $3.4 = 340\%$

Converting Percents to Common Fractions

There are several ways to change a percent to a common fraction. The most direct method is to write the percent as the numerator of a fraction with the denominator 100 and then reduce the fraction. For example,

$$60\% = \frac{60}{100} = \frac{3}{5}$$

If the percent itself has decimal places, it is easier to convert the percent to a decimal fraction first. For example,

$$8.4\% = 0.084$$

Now, we can convert the decimal to a common fraction as follows:

$$0.084 = \frac{84}{1000} = \frac{4 \cdot 21}{4 \cdot 250} = \frac{21}{250}$$

If the percent involves a mixed number, we can change the mixed number to an improper fraction and then divide by 100. For example, to convert $12\frac{2}{3}\%$ to a fraction, we first write

$$12\frac{2}{3} = \frac{12}{1} + \frac{2}{3} = \frac{12 \cdot 3}{1 \cdot 3} + \frac{2}{3} = \frac{36 + 2}{3} = \frac{38}{3}$$

Now, we divide by 100 to find

$$\frac{38}{3} \div 100 = \frac{38}{3} \cdot \frac{1}{100} = \frac{38}{300}$$
$$= \frac{2 \cdot 19}{2 \cdot 150} = \frac{19}{150}$$

EXAMPLE 3

Convert each percent to a common fraction.

   a. 71.2%   **b.** $14\frac{1}{6}\%$

Solution

a. First, we change the percent to a decimal by moving the decimal point two places to the left.

$$71.2\% = 0.712$$

Then we convert the decimal fraction to a common fraction, and reduce.

$$0.712 = \frac{712}{1000} = \frac{8 \cdot 89}{8 \cdot 125} = \frac{89}{125}$$

b. First, we write $14\frac{1}{6}$ as an improper fraction.

$$\frac{14}{1} + \frac{1}{6} = \frac{14 \cdot 6}{1 \cdot 6} + \frac{1}{6} = \frac{84 + 1}{6} = \frac{85}{6}$$

Thus $14\frac{1}{6}\% = \frac{85}{6}\%$. Now, we divide by 100:

$$\frac{85}{6} \div 100 = \frac{\overset{17}{\cancel{85}}}{6} \cdot \frac{1}{\underset{20}{\cancel{100}}} = \frac{17}{120}$$

We summarize the procedure as follows:

To Change a Percent to a Common Fraction:

1. Write the percent as a fraction with the denominator 100. Reduce.
2. If the percent has decimal places, change the percent to a decimal fraction first. Then convert the decimal fraction to a common fraction.
3. If the percent involves a mixed number, write the mixed number as an improper fraction first. Then divide by 100.

Converting Common Fractions to Percents

The easiest way to convert a common fraction to percent form is to write the fraction as a decimal fraction first.

To Change a Common Fraction to a Percent:

1. Convert the fraction to a decimal fraction.
2. Move the decimal point two places to the right, and add a percent symbol.

EXAMPLE 4

Convert $\frac{37}{80}$ to a percent.

Solution

First, we change $\dfrac{37}{80}$ to a decimal fraction by dividing the numerator by the denominator:

$$\frac{37}{80} = 37 \div 80 = 0.4625$$

To change 0.4625 to a percent, we move the decimal point two places to the right:

$$0.4625 = 46.25\%$$

HOMEWORK A.9

In Problems 1–12, write each percent as a decimal number.

1. 15%

2. 37%

3. 0.4%

4. 0.7%

5. 6.8%

6. 1.1%

7. 119%

8. 652%

9. $3\dfrac{1}{4}\%$

10. $6\dfrac{3}{8}\%$

11. $\dfrac{2}{5}\%$

12. $\dfrac{3}{2}\%$

In Problems 13–24, write each decimal number as a percent.

13. 0.33

14. 0.54

15. 0.504

16. 0.686

17. 0.787

18. 0.074

19. 0.0201

20. 0.005

21. 0.008

22. 0.0008

23. 5.5

24. 6

In Problems 25–36, write each percent as a fraction.

25. 35%

26. 10%

27. 125%

28. 150%

29. 60%

30. 48%

31. 0.90%

32. 0.11%

33. $37\frac{1}{2}\%$

34. $87\frac{1}{2}\%$

35. $33\frac{1}{3}\%$

36. $66\frac{2}{3}\%$

In Problems 37–48, write each fraction as a percent.

37. $\frac{3}{4}$

38. $\frac{1}{5}$

39. $\frac{3}{8}$

40. $\dfrac{5}{5}$

41. $\dfrac{9}{9}$

42. $\dfrac{3}{20}$

43. $\dfrac{9}{4}$

44. $\dfrac{8}{5}$

45. $\dfrac{12}{5}$

46. $\dfrac{11}{4}$

47. $\dfrac{1}{250}$

48. $\dfrac{1}{500}$

A.10 Laws of Arithmetic

When we add two numbers together, we can compute the sum in either order. For example,

$$3 + 7 = 10 \quad \text{and} \quad 7 + 3 = 10$$

This property is called the **commutative law of addition.**

If a and b are numbers, then

$$a + b = b + a$$

Note that *subtraction is not commutative*. For example, $9 - 5$ is not the same as $5 - 9$.

When we multiply two numbers together, we can compute the product in either order. For example,

$$4 \cdot 5 = 20 \quad \text{and} \quad 5 \cdot 4 = 20$$

This property is called the **commutative law of multiplication.**

If a and b are numbers, then

$$a \cdot b = b \cdot a$$

Note that *division is not commutative*. For example, $12 \div 3$ is not the same as $3 \div 12$.

When we add three or more numbers, we can group the numbers in any way we like. That is, we can add any two of the numbers first and then add the third number to the sum. For example,

$$9 + (6 + 4) = 9 + 10 = 19$$

and

$$(9 + 6) + 4 = 15 + 4 = 19$$

This property is called the **associative law of addition.**

> If *a*, *b*, and *c* are numbers, then
>
> $$a + (b + c) = (a + b) + c$$

Note that *subtraction is not associative*. For example,

$$12 - (6 - 2) = 12 - 4 = 8$$

but

$$(12 - 6) - 2 = 6 - 2 = 4$$

When we multiply three or more numbers together, we may group them in any order. Thus

$$2 \cdot (3 \cdot 5) = 2 \cdot 15 = 30$$

and

$$(2 \cdot 3) \cdot 5 = 6 \cdot 5 = 30$$

This property is called the **associative law of multiplication.**

> If *a*, *b*, and *c* are numbers, then
>
> $$a \cdot (b \cdot c) = (a \cdot b) \cdot c$$

Note that *division is not associative*. For example,

$$12 \div (6 \div 2) = 12 \div 3 = 4$$

but

$$(12 \div 6) \div 2 = 2 \div 2 = 1$$

EXAMPLE 1

Fill in each blank using the indicated law.

   a. Commutative law   **b.** Associative law
     $3 \cdot 15 = 15 \cdot$ _____ $2 + (4 + 6) = (2 +$ _____$) + 6$

Solution

   a. $3 \cdot 15 = 15 \cdot 3$ **b.** $2 + (4 + 6) = (2 + 4) + 6$

EXAMPLE 2

Use the commutative and associative laws to simplify the computations.

    a. $24 + 18 + 6$ **b.** $4 \cdot 27 \cdot 25$

Solution

    a. First, we apply the commutative law of addition.

$$24 + 18 + 6 = 24 + 6 + 18 \quad \text{Now apply the associative law.}$$
$$= (24 + 6) + 18$$
$$= 30 + 18 = 48$$

    b. First, we apply the commutative law of multiplication.

$$4 \cdot 27 \cdot 25 = 4 \cdot 25 \cdot 27 \quad \text{Now apply the associative law.}$$
$$= (4 \cdot 25) \cdot 27$$
$$= (100) \cdot 27 = 2700$$

HOMEWORK A.10

In Problems 1–10, fill in the blanks using the indicated law.

1. Commutative law
    $7 + 10 = 10 + \underline{\hspace{1.5cm}}$

2. Associative law
    $(6 \cdot 4) \cdot 3 = 6 \cdot (4 \cdot \underline{\hspace{1.5cm}})$

3. Associative law
    $(3 + 6) + 9 = \underline{\hspace{1.5cm}} + (6 + 9)$

4. Commutative law
    $8 \cdot 12 = \underline{\hspace{1.5cm}} \cdot 8$

5. Commutative law
    $36 \cdot 147 = \underline{\hspace{1.5cm}} \cdot 36$

6. Commutative law
    $13 + 87 = 87 + \underline{\hspace{1.5cm}}$

7. Associative law

$(17 \cdot 2) \cdot 5 = 17 \cdot (\underline{\hspace{1.5cm}} \cdot \underline{\hspace{1.5cm}})$

8. Associative law

$(44 + 12) + 8 = 44 + (\underline{\hspace{1.5cm}} + \underline{\hspace{1.5cm}})$

9. Commutative law

$(5 + 9) + 4 = (9 + \underline{\hspace{1.5cm}}) + 4$

10. Commutative law

$(8 \cdot 9) \cdot 3 = (9 \cdot \underline{\hspace{1.5cm}}) \cdot 3$

Use the commutative and associative laws to simplify the computations in Problems 11–20.

11. $47 + 28 + 3$

12. $12 + 147 + 8$

13. $26 + 37 + 3 + 4$

14. $55 + 32 + 5 + 8$

15. $2 \cdot 7 \cdot 5$

16. $15 \cdot 6 \cdot 2$

17. $50 \cdot 13 \cdot 2$

18. $4 \cdot 26 \cdot 25$

19. $4 \cdot 6 \cdot 5 \cdot 5$

20. $8 \cdot 8 \cdot 5 \cdot 5$

A.11 Measures of Central Tendency: Mode, Mean, and Median

A first step in analyzing a collection of data is to find an average, or typical, value. Statisticians use three different average values, called the mean, the median, and the mode.

The Mode and the Mean

Consider the following data:
For her sociology class, Alida was given the assignment of investigating how often American families change their place of residence. She asked 15 classmates how many different homes they had lived in before coming to college. Table A.1 shows the results of her survey.

Table A.1

Aaron	2	Hillary	1	Steffi	6
Barbara	2	Juana	6	Tung	2
David	6	Mariel	3	Valerie	4
Elisa	3	Paula	3	Will	3
Geraldo	3	Sean	12	Xavier	4

The value that occurs most frequently is called the **mode** of the data. For the survey data, the mode is 3, because five students lived in three different places.

A second average value is called the **mean**. This is the type of average that most of us are familiar with. This type of average is computed using the formula

$$\text{mean} = \frac{S}{n}$$

where S is the sum of all the data values and n is the number of data values. You should use your calculator to check that the mean of the survey data in Table A.1 is 4.

Which average value is better, the mean or the mode? Both are useful in different situations. The mode is useful if one score occurs more frequently than the others. You can think of it as the most "popular" value in a collection of data. The mean can be strongly affected by the presence of a few outlying data values. In the survey data, most of the values are 6 or below, but there is one much larger value, 12. This value affected the calculation of the mean, causing it to be slightly higher than the typical response from the people surveyed.

The Median

A third type of average value is called the **median.** This is the middle score in a collection of data.

EXAMPLE 1

Compute the mean, the median, and the mode for the following ages of students in a history seminar:

                17   18   19   19   19   20   21   22   22   46

Solution

The mean is

$$\frac{S}{n} = \frac{17 + 18 + 19 + 19 + 19 + 20 + 21 + 22 + 22 + 46}{10}$$

$$= \frac{223}{10} = 22.3$$

The median is the middle age when the ages are arranged in order. Because there is an even number of students in the seminar, we take the age halfway between the middle two ages:

$$\text{median} = \frac{19 + 20}{2} = 19.5$$

Note that the median reflects the typical age of the students more accurately than the mean does. This is because the one atypical age, 46, affects the calculation of the mean more than it affects the median.

The mode is 19, because there are more students (three of them) who are 19 than there are students who are any other age.

_____

The mean, the median, and the mode are called *measures of central tendency* because they usually fall somewhere in the middle of the range of data. Each of these average values is a way of summarizing the data by giving the value of a typical data point.

EXAMPLE 2

The 11 members of the West Hills Cycling Club have the following weights:

$$107 \quad 114 \quad 128 \quad 136 \quad 145 \quad 162 \quad 165 \quad 168 \quad 176 \quad 179 \quad 183$$

Compute the mean, median, and mode of the data.

Solution

The mean of the data is

$$\text{mean} = \frac{S}{n} =$$

$$\frac{107 + 114 + 128 + 136 + 145 + 162 + 165 + 168 + 176 + 179 + 183}{11}$$

$$= \frac{1663}{11} \approx 151.18$$

The median is the middle value, 162. There is no mode for this data set, because no data value occurs more than once.

_____

HOMEWORK A.11

Compute the mean, median, and mode of the data sets in Problems 1–10.

1. Wendy surveyed the students in her geology class to see how many credit hours each is enrolled in this semester. She recorded the following reponses:

   5 6   12 18   12   11   8   15   12   16
   9   12   15   14   9   17   13

2. Dawn asked each of the workers in her office to keep track of how many books they read last year. She recorded the following responses:

   10   3   7   4   6   9   2   4   8   12   6   7
   2   6   3   5   6

3. Professor Jennings asked the students in her philosophy class to fill out a survey about their career goals. She recorded the following responses to the question "How many hours do you work per week?"

   25   20   18   32   6   10   40   12   24   20   35
   16   12   15   30   22   12   15   12   26   18

4. A representative of the Transit District surveyed students in the cafeteria to determine how many minutes they spend commuting to school. Each student agreed to record his or her commuting time each day for a week and report the average time. The representative collected the following results, in minutes.

   10   18   25   8   25   15   45   42   6   22   18
   22   12   16   28   24   33   37   18   32   16

5. Table A.2 gives the average beginning salaries offered to new graduates from various majors. (Data are from the 1994–1995 *American Almanac*.)

Table A.2

Accounting	$27,493
Business	$24,555
Chemical Engineering	$39,482
Chemistry	$28,002
Civil Engineering	$29,211
Computer Engineering	$33,963
Computer Science	$31,329
Electrical Engineering	$34,313
Engineering Technology	$29,236
Humanities	$24,373
Marketing	$24,361
Mathematics	$26,524
Mechanical Engineering	$34,460
Nuclear Engineering	$34,755
Petroleum Engineering	$38,387
Physics	$26,835
Social Sciences	$22,684

6. Table A.3 shows the numbers of Americans, in thousands, who participate in selected sports. (Data are from the 1994–1995 *American Almanac*.)

Table A.3

Aerobic exercise	5771	Golf	3048
Backpacking	1547	Hiking	2644
Baseball	1921	Hunting	2614
Basketball	4896	Racquetball	1980
Bicycling	5547	Running or jogging	4490
Bowling	7024	Skiing	2928
Camping	5297	Soccer	1296
Exercise walking	6310	Softball	2906
Exercising with equipment	6913	Swimming	7690
Fishing	5991	Tennis	3933
Football	2762	Volleyball	5111

7. Participants in an aerobics class were asked to find their heart rates immediately after exercising. They recorded the following results.

136	142	156	148	165	152	168	174
127	148	144	153	135	145	166	151

8. A group of runners were asked how many miles they ran, on average, each week. They gave the following responses.

25	60	24	42	48	30	36	56	35	21
44	32	52	48	45	35				

9. The state police set up a radar speed trap on the highway outside of town and clocked cars traveling at the following speeds, in miles per hour.

68	56	65	72	60	82	67	58	64
76	68	74	80	78	64	66		

10. The gas company conducted a telephone survey to see what room temperature people try to maintain in their homes. The following thermostat readings were reported.

72	70	66	70	68	65	71	74	67
68	75	62	68	61	69	73		

11. There are seven houses for sale in a particular neighborhood. The table gives their listed prices.

House	Price ($)
A	190,000
B	100,000
C	110,000
D	2,500,000
E	100,000
F	170,000
G	120,000

    a. Find the mean, median, and mode of the house prices.

    b. Explain the advantages and disadvantages of each value as an average.

12. There are only seven roles in a new movie. The salaries of the seven actors are shown in the table.

Actor	Salary ($)
A	1,200,000
B	100,000
C	20,000
D	10,000
E	7,000
F	7,000
G	7,000

    a. Find the mean, median, and mode of the salaries.

    b. You are summarizing the costs of the movie to potential producers. Which average salary do you report?

13. A researcher polled 100 families with children in her city in order to estimate the average number of children per family. She obtained the information shown in the following **frequency table**.

Number of children	1	2	3	4	5	6
Number of families with that number of children	28	32	20	12	6	2

Calculate the mean of the data. [*Hint:* Use the formula mean $= \dfrac{S}{n}$. To find S, multiply each data value (number of children) by the frequency of its occurrence (number of families), and add the results.]

14. Grant asked 80 members of his health club how many hours per week they exercised. He recorded the information in a frequency table.

Number of hours	2	3	4	5	6	8	10
Number of people exercising that number of hours	10	18	21	12	8	8	3

Calculate the mean of the data. [*Hint:* Use the formula mean $= \dfrac{S}{n}$. To find S, multiply each data value (number of hours) by the frequency of its occurrence (number of people), and add the results.]

Additional Topics

B.1 Prime Factorization

A **prime number** is a natural number greater than 1 that is exactly divisible only by itself and 1. For example, 2, 3, 5, 7, 11, and 13 are prime numbers. However, 4 is not a prime number, because 4 is divisible by 2, and 15 is not a prime number, because 15 is divisible by 3 and by 5.

EXAMPLE 1

Which of the following are prime numbers?

    a. 19   **b.** 21   **c.** 91

Solution

    a. 19 is prime because it is divisible only by 1 and itself.

    b. 21 is not prime because it is divisible by 3 and 7.

    c. 91 is not prime because it is divisible by 7 and 13.

A natural number greater than 1 that is not a prime number is called a **composite number.** A composite number is always divisible by some natural number other than itself, and so it can be written as a product of smaller numbers. For example, 12 can be factored as $2 \cdot 6$ or as $3 \cdot 4$. However, every natural number can be written in exactly *one* way as a product of primes (if we don't count rearrangements of the factors). The *prime factorization* of 12 is $2 \cdot 2 \cdot 3$. Using exponents, we write

$$12 = 2^2 \cdot 3$$

EXAMPLE 2

Find the prime factorization of 600.

Solution

We begin by finding *any* natural number that divides evenly into 600 and write the resulting factorization in the form of a tree. We might begin with $600 = 6 \cdot 100$, which we write as shown.

Step 1

600
6   100

Now we continue factoring each factor until all the factors are prime numbers. The next steps in the process are

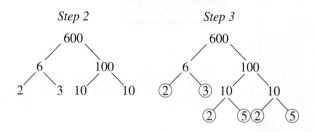

Step 2   *Step 3*

After Step 3, all the factors are prime numbers, so we cannot go any further. The prime factors of 600 are the circled numbers. Thus

$$600 = 2 \cdot 3 \cdot 2 \cdot 5 \cdot 2 \cdot 5$$
$$= 2 \cdot 2 \cdot 2 \cdot 3 \cdot 5 \cdot 5$$

or, using exponents,

$$600 = 2^3 \cdot 3 \cdot 5^2$$

HOMEWORK B.1

1. What is a prime number?

2. What is a composite number?

Which of the numbers in Problems 3–6 are prime numbers?

3. a. 8   **b.** 13   **c.** 14   **d.** 17

4. a. 21   **b.** 23   **c.** 27   **d.** 29

5. a. 37   **b.** 39   **c.** 49   **d.** 51

6. a. 53   **b.** 67   **c.** 87   **d.** 91

Write the numbers in Problems 7–14 in prime-factored form using exponents.

7. 54   **8.** 220   **9.** 180   **10.** 441

11. 210   **12.** 800   **13.** 625   **14.** 286

B.2 Completing the Square

The first two methods you learned in Chapter 4 for solving quadratic equations—factoring and extracting roots—apply only to special cases. In this section, we develop a method that applies to any quadratic equation and can be used to derive the quadratic formula. The method entails writing one side of the quadratic equation as the square of a binomial.

Squares of Binomials

In Section 5.4, we studied squares of binomials and discovered this relationship:

$$(a + b)^2 = a^2 + 2ab + b^2$$

The expanded form on the right side of the equation is called a *perfect-square trinomial*. For example, the following are perfect-square trinomials:

$$(x + 4)^2 = x^2 + 8x + 16$$
$$(x - 5)^2 = x^2 - 10x + 25$$
$$\left(x + \frac{1}{3}\right)^2 = x^2 + \frac{2}{3}x + \frac{1}{9}$$

Note that each example has the form

$$(x + p)^2 = x^2 + 2px + p^2 \quad \text{Equation 1}$$

That is, the constant term of the trinomial is p^2, and the coefficient of the linear term is $2p$. We can use this observation to turn any quadratic expression of the form $x^2 + bx$ into a perfect square by adding an appropriate constant.

Let's start with the expression $x^2 + 6x$. We would like to add a constant term to obtain a perfect-square trinomial. That is, we want to fill in the blanks in this equation:

$$x^2 + 6x + \underline{\hspace{1cm}} = (x + \underline{\hspace{1cm}})^2$$

Comparing this equation with Equation 1, we see that the blanks correspond to p^2 and p. We know the coefficient of x, namely 6, so $2p = 6$. Thus

$$p = \frac{1}{2} \cdot 6 = 3 \quad \text{and} \quad p^2 = 3^2 = 9$$

Filling in the blanks gives

$$x^2 + 6x + 9 = (x + 3)^2$$

We have added the constant term 9 to the expression $x^2 + 6x$ to obtain the square of a binomial. This process is called **completing the square.**

EXAMPLE 1

Complete the square by adding the constant term, and write the result as the square of a binomial.

    a. $x^2 + 18x$ **b.** $x^2 - 5x$

Solution

  a. The coefficient of x is $18 = 2p$. Therefore,

$$p = \frac{1}{2} \cdot 18 = 9 \quad \text{and} \quad p^2 = 9^2 = 81$$

Thus we add 81 to the expression to obtain

$$x^2 + 18x + 81 = (x + 9)^2$$

  b. The coefficient of x is $-5 = 2p$. Therefore,

$$p = \frac{1}{2} \cdot (-5) = \frac{-5}{2} \quad \text{and} \quad p^2 = \left(\frac{-5}{2}\right)^2 = \frac{25}{4}$$

Thus we add $\frac{25}{4}$ to the expression to obtain

$$x^2 - 5x + \frac{25}{4} = \left(x - \frac{5}{2}\right)^2$$

The steps for completing the square can be summarized as follows:

> ## To Complete the Square for the Expression $x^2 + 2px$:
>
> **1.** Multiply the coefficient of x by $\frac{1}{2}$ to get p.
>
> **2.** Square the result to get p^2.
>
> **3.** Add p^2 to the original expression. The new expression is equal to $(x + p)^2$.

Solving Quadratic Equations by Completing the Square

We are now ready to solve quadratic equations by completing the square. Consider the equation

$$x^2 - 6x - 7 = 0$$

We first rewrite the equation so that the constant term is on the right side, leaving a space to complete the square on the left.

$$x^2 - 6x \underline{\hspace{1.5cm}} = 7$$

Now, we complete the square on the left side. We take half the coefficient of x to get p:

$$p = \frac{1}{2}(-6) = -3$$

We square the result to get p^2:

$$p^2 = (-3)^2 = 9$$

We must add 9 to the left side of the original equation to obtain a perfect-square trinomial. However, if we add 9 to the left side of the equation, we must also add 9 to the right side. This gives us

$$x^2 - 6x + 9 = 7 + 9$$

Next, we simplify the right side and write the left side as the square of a binomial:

$$(x - 3)^2 = 16$$

We can now solve the equation by extraction of roots. We take the square root of each side of the equation:

$$x - 3 = \pm\sqrt{16} = \pm 4$$

$$x - 3 = 4 \quad \text{or} \quad x - 3 = -4$$

$$x = 7 \quad \text{or} \quad x = -1$$

The solutions are 7 and -1.

EXAMPLE 2

Solve by completing the square: $x^2 + 3x + 1 = 0$.

Solution

We begin by rewriting the equation so that the constant term is on the right side of the equation.

$$x^2 + 3x \underline{\hspace{2cm}} = -1$$

Now, we complete the square on the left side:

$$p = \frac{1}{2}(3) = \frac{3}{2} \quad \text{and} \quad p^2 = \left(\frac{3}{2}\right)^2 = \frac{9}{4}$$

We add $\frac{9}{4}$ to both sides of the equation:

$$x^2 + 3x + \frac{9}{4} = -1 + \frac{9}{4}$$

We simplify the right side and write the left side as the square of the binomial $(x + p)$:

$$\left(x + \frac{3}{2}\right)^2 = \frac{5}{4} \qquad -1 + \frac{9}{4} = \frac{-4}{4} + \frac{9}{4} = \frac{5}{4}$$

Finally, we solve the equation by extraction of roots. We take the square root of both sides of the equation:

$$x + \frac{3}{2} = \pm\sqrt{\frac{5}{4}}$$ Simplify the radical.

$$x + \frac{3}{2} = \pm\frac{\sqrt{5}}{2}$$ Write as two equations.

$$x + \frac{3}{2} = \frac{\sqrt{5}}{2} \quad \text{or} \quad x + \frac{3}{2} = \frac{-\sqrt{5}}{2}$$ Solve each equation.

$$x = \frac{-3}{2} + \frac{\sqrt{5}}{2} \quad \text{or} \quad x = \frac{-3}{2} - \frac{\sqrt{5}}{2}$$

We can write both solutions together as $\dfrac{-3 \pm \sqrt{5}}{2}$.

Solving quadratic equations by completing the square has one more step that we have not yet discussed. If the coefficient of the quadratic term is not 1, we must divide each term of the equation by that coefficient before we begin.

EXAMPLE 3

Solve by completing the square: $\quad 3x^2 - 2x - 5 = 0$.

Solution

First, we prepare the equation. We divide each term by 3, the coefficient of x^2, and rewrite the equation with the constant term on the right side:

$$x^2 - \frac{2}{3}x \underline{\hspace{1.5cm}} = \frac{5}{3}$$

Next, we complete the square. We multiply the coefficient of x by $\dfrac{1}{2}$:

$$p = \frac{1}{2}\left(\frac{-2}{3}\right) = \frac{-1}{3}$$

We square the result:

$$p^2 = \left(\frac{-1}{3}\right)^2 = \frac{1}{9}$$

We add $\dfrac{1}{9}$ to both sides of the equation to get

$$x^2 - \frac{2}{3}x + \frac{1}{9} = \frac{5}{3} + \frac{1}{9}$$

We write the left side as the square of a binomial, namely $(x + p)^2$, and simplify the right side:

$$\left(x - \frac{1}{3}\right)^2 = \frac{16}{9} \qquad \frac{5}{3} + \frac{1}{9} = \frac{15}{9} + \frac{1}{9} = \frac{16}{9}$$

Finally, we solve the equation by extracting roots. We take the square root of each side of the equation.

$$x - \frac{1}{3} = \pm\sqrt{\frac{16}{9}}$$ Simplify the radical.

$$x - \frac{1}{3} = \pm\frac{4}{3}$$ Write as two equations.

$$x - \frac{1}{3} = \frac{4}{3} \quad \text{or} \quad x - \frac{1}{3} = -\frac{4}{3}$$ Solve each equation.

$$x = \frac{4}{3} + \frac{1}{3} = \frac{5}{3} \quad \text{or} \quad x = -\frac{4}{3} + \frac{1}{3} = -1$$

The solutions are $\frac{5}{3}$ and -1.

Here are the steps for solving an equation by completing the square.

To Solve a Quadratic Equation by Completing the Square:

1. Prepare the equation:
   a. Divide each term by the coefficient of the quadratic term.
   b. Isolate the constant term on the right side of the equation.
2. Complete the square:
   a. Multiply the coefficient of the linear term by $\frac{1}{2}$.
   b. Square the result, and add the square to both sides of the equation.
3. Write the left side as the square of a binomial. Simplify the right side.
4. Solve the equation by extracting roots:
   a. Take the square root of both sides of the equation; this gives two equations.
   b. Solve each equation for the variable to get two solutions.

As noted earlier, any quadratic equation can be solved by completing the square. However, solving an equation by factoring or by extraction of roots is usually faster and easier. You should check to see whether one of these methods will work before completing the square. If the equation has no linear term, you can use extraction of roots. If the right side of the equation has small integer coefficients, it is worth trying to solve by factoring. Of course, if either of these methods works, it will yield the same solutions that you would get by completing the square.

HOMEWORK B.2

1. Explain how to turn the expression $x^2 + bx$ into a perfect square.

2. What two steps are required to prepare the equation $ax^2 + bx + c = 0$ for solution by completing the square?

3. Name three algebraic methods for solving quadratic equations.

4. a. Find the total area of the pieces in part (a) of Figure B.1. Now find the area of the large square in part (b).

Figure B.1

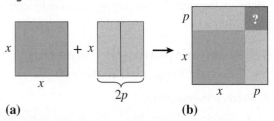

(a)   **(b)**

b. What is the area of the small missing piece in part (b)?

In Problems 5 and 6, isolate the constant term on the right side of the equation.

5. $3w^2 + 6 = w$

6. $5p^2 = 10p - 4$

In Problems 7–10, divide each term by the coefficient of the quadratic term.

7. $6m^2 - 3m = 18$

8. $5q^2 + 3q = 15$

9. $-4r^2 - 6r = 12$

10. $-2s^2 + \dfrac{3}{2}s = 2$

In Problems 11–22, complete the square by adding the constant term, and write the result as the square of a binomial.

11. $x^2 + 2x$

12. $x^2 + 4x$

13. $x^2 - 6x$

14. $x^2 - 8x$

15. $x^2 - 14x$

16. $x^2 + 20x$

17. $x^2 + 3x$

18. $x^2 + 7x$

19. $x^2 + \dfrac{4}{3}x$

20. $x^2 - \dfrac{2}{3}x$

21. $x^2 - \dfrac{3}{2}x$

22. $x^2 + \dfrac{3}{4}x$

Solve Problems 23–34 by completing the square.

23. $x^2 + 4x - 12 = 0$

24. $x^2 - 2x - 14 = 0$

25. $p^2 + 4 = 5p$

26. $r^2 - 10 = 3r$

27. $z^2 + 9z + 20 = 0$ **28.** $u^2 + 5u + 6 = 0$

29. $y^2 - 2y - 1 = 0$ **30.** $y^2 + 4y - 4 = 0$

31. $z^2 = 3z + 3$ **32.** $s^2 + 1 = 3s$

33. $t^2 + 6t = 3$ **34.** $v^2 = -1 - 6v$

Solve Problems 35–46 by completing the square.

35. $4x^2 + 4x = 3$ **36.** $4y^2 - 4y = 3$

37. $6z^2 + 6 = 13z$ **38.** $2a^2 = 2 - 3a$

39. $2t^2 - t = 15$

40. $1 - r = 6r^2$

41. $3b^2 + 2b - 2 = 0$

42. $3w^2 - 4w - 3 = 0$

43. $2p^2 + 2 = 7p$

44. $2q^2 + q = 1$

45. $2k^2 + 8k + 3 = 0$

46. $3n^2 - 6n - 2 = 0$

B.3 Polynomial Division

In Chapter 5, you learned how to reduce algebraic fractions by dividing out common factors from the numerator and denominator. Not all fractions can be reduced. For example, the improper fraction $\dfrac{13}{4}$ cannot be reduced. However, we can write the fraction as a mixed number—that is, as the sum of a whole number and a proper fraction. (See Appendix A.4 for a review of mixed numbers and improper fractions.) To change an improper fraction to a mixed number, we divide the denominator into the numerator.

$$
\begin{array}{r}
3 \\
4{\overline{\smash{\big)}\,13}} \\
\underline{-12} \\
1
\end{array}
$$

The quotient, 3, is the whole number, and the remainder, 1, is the numerator of the proper fraction. Thus

$$\frac{13}{4} = 3\frac{1}{4}$$

$\left(\text{Note that } 3\dfrac{1}{4} \text{ means } \quad 3 + \dfrac{1}{4}, \quad \text{not} \quad 3 \cdot \dfrac{1}{4}.\right)$

We consider an algebraic fraction to be *improper* if the degree of its numerator is greater than or equal to the degree of its denominator. For example, the fraction

$$\frac{x^2 + x - 7}{x + 3}$$

is improper because the degree of the numerator is 2 and the degree of the denominator is 1. In much the same way that we rewrite improper arithmetic fractions, we can write an improper algebraic fraction as the sum of a polynomial and a proper algebraic fraction—one whose numerator is of lesser degree than its denominator. We rewrite improper algebraic fractions by dividing the denominator into the numerator, using a process called *polynomial division*.

Polynomial division is very similar to the process of long division you learned in arithmetic. Compare the steps in the following examples of long division and polynomial division; note that both types of division repeat these steps until the answer is found.

1. Divide

2. Multiply

3. Subtract

4. "Bring down"

$21\overline{)674}$

$x + 3\overline{)x^2 + x - 7}$

Divide 2 into 6; the quotient is 3.

Divide x into x^2; the quotient is x.

$$\begin{array}{r} 3 \\ 21\overline{)674} \end{array}$$

$$\begin{array}{r} x \\ x + 3\overline{)x^2 + x - 7} \end{array}$$

Multiply 3 by 21: $3 \cdot 21 = 63$.

Multiply x by $x + 3$: $x(x + 3) = x^2 + 3x$.

$$\begin{array}{r} 3 \\ 21\overline{)674} \\ 63 \end{array}$$

$$\begin{array}{r} x \\ x + 3\overline{)x^2 + x - 7} \\ x^2 + 3x \end{array}$$

Subtract. (Change the sign and add.)

Subtract. (Change the signs and add.)

$$\begin{array}{r} 3 \\ 21\overline{)674} \\ -63 \\ \hline 4 \end{array}$$

$$\begin{array}{r} x \\ x + 3\overline{)x^2 + x - 7} \\ -x^2 - 3x \\ \hline -2x \end{array}$$

"Bring down" 4.

"Bring down" − 7.

$$\begin{array}{r} 3 \\ 21\overline{)674} \\ -63\downarrow \\ \hline 44 \end{array}$$

$$\begin{array}{r} x \\ x + 3\overline{)x^2 + x - 7} \\ -x^2 - 3x \quad \downarrow \\ \hline -2x - 7 \end{array}$$

Divide 2 into 4; the quotient is 2.

Divide x into $-2x$; the quotient is -2.

$$\begin{array}{r} 32 \\ 21\overline{)674} \\ -63 \\ \hline 44 \end{array}$$

$$\begin{array}{r} x - 2 \\ x + 3\overline{)x^2 + x - 7} \\ -x^2 - 3x \\ \hline -2x - 7 \end{array}$$

Multiply 2 by 21: $2 \cdot 21 = 42$.

$$
\begin{array}{r}
32 \\
21\overline{)674} \\
-63 \\
\hline
44 \\
42 \leftarrow
\end{array}
$$

Multiply -2 by $x + 3$: $-2(x + 3) = -2x - 6$.

$$
\begin{array}{r}
x - 2 \\
x + 3\overline{)\,x^2 + x - 7} \\
\underline{-x^2 - 3x} \\
-2x - 7 \\
-2x + 6 \leftarrow
\end{array}
$$

Subtract. (Change the sign and add.)

$$
\begin{array}{r}
32 \\
21\overline{)674} \\
-63 \\
\hline
44 \\
-42 \leftarrow\\
\hline
2 \leftarrow
\end{array}
$$

Subtract. (Change the signs and add.)

$$
\begin{array}{r}
x - 2 \\
x + 3\overline{)\,x^2 + x - 7} \\
\underline{-x^2 - 3x} \\
-2x - 7 \\
2x - 6 \leftarrow\\
\hline
-1 \leftarrow
\end{array}
$$

The remainder is 2.

$$
\begin{array}{r}
32\tfrac{2}{21} \leftarrow\\
21\overline{)674} \\
-63 \\
\hline
44 \\
-42 \\
\hline
2
\end{array}
$$

The answer is $32\dfrac{2}{21}$.

The remainder is -1.

$$
\begin{array}{r}
x - 2 + \dfrac{-1}{x + 3} \leftarrow\\
x + 3\overline{)\,x^2 + x - 7} \\
\underline{-x^2 - 3x} \\
-2x - 7 \\
2x + 6 \\
\hline
-1
\end{array}
$$

The answer is $x - 2 + \dfrac{-1}{x + 3}$.

We can now use polynomial division to rewrite improper algebraic fractions.

EXAMPLE 1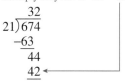

Write $\dfrac{2x^2 + 3x - 3}{2x - 1}$ as the sum of a polynomial and a proper algebraic fraction.

Solution

We divide the denominator into the numerator, using polynomial division:

$$2x - 1\overline{)2x^2 + 3x - 3}$$

We begin by dividing the *first* term of the denominator, $2x$, into the *first* term of the numerator, $2x^2$:

$$\frac{2x^2}{2x} = x$$

We write the quotient, x, above the division bar. Next we multiply *each* term of the divisor, $2x - 1$, by x and write the products below the numerator. We make sure that terms of like degree are aligned vertically.

$$
\begin{array}{r}
x \\
2x - 1\overline{)2x^2 + 3x - 3} \\
\underline{2x^2 - x}
\end{array}
$$
  Multiply: $x(2x - 1) = 2x^2 - x$.

Finally, we subtract the product from the numerator and bring down the next term of the numerator.

$$\begin{array}{r} x \\ 2x-1\overline{)2x^2+3x-3} \\ \underline{-2x^2+x} \\ 4x-3 \end{array}$$
Change signs and add; bring down -3.

Now, we begin the process again. We divide the *first* term of the divisor, $2x-1$, into the *first* term of $4x-3$. The quotient is $\dfrac{4x}{2x}=2$. We write the quotient above the division bar and multiply each term of the divisor by 2. Write the product under $4x-3$ and subtract.

$$\begin{array}{r} x+2 \\ 2x-1\overline{)2x^2+3x-3} \\ \underline{-2x^2+x} \\ 4x-3 \\ \underline{4x-2} \\ -1 \end{array}$$

Multiply: $2(2x-1)=4x-2$.

Change signs and add.

The quotient, $x+2$, is the polynomial, and the remainder, -1, is the numerator of the proper fraction, with denominator $2x-1$. Thus

$$\frac{2x^2+3x-3}{2x-1}=x+2+\frac{-1}{2x-1} \qquad \text{or} \qquad x+2-\frac{1}{2x-1}$$

In order to use polynomial division, you should write the numerator and the denominator of the fraction in descending powers of the variable. Also, it is helpful to insert a term with zero coefficient for all powers of the variable that are missing between the highest-degree term and the lowest-degree term.

EXAMPLE 2

Write $\dfrac{3x-1+4x^3}{2x+1}$ as the sum of a polynomial and a proper algebraic fraction.

Solution

First, we write the numerator in descending powers of x. Because there is no x^2 term, we insert the term $0x^2$ between $4x^3$ and $3x$:

$$4x^3+0x^2+3x-1$$

Now, we use polynomial division.

Divide: $\dfrac{4x^3}{2x}=2x^2$

Divide: $\dfrac{-2x^2}{2x}=-x$

Divide: $\dfrac{4x}{2x}=2$

$$\begin{array}{r} 2x^2-x+2 \\ 2x+1\overline{)4x^3+0x^2+3x-1} \\ \underline{4x^3+2x^2} \\ -2x^2+3x \\ \underline{-2x^2-x} \\ 4x-1 \\ \underline{4x+2} \\ -3 \end{array}$$

Multiply: $2x^2(2x+1)=4x^3+2x^2$

Subtract: Change signs and add.

Multiply: $-x(2x+1)=-2x^2-x$

Subtract: Change signs and add.

Multiply: $2(2x+1)=4x+2$

Subtract: Change signs and add.

Thus

$$\frac{3x - 1 + 4x^3}{2x + 1} = 2x^2 - x + 2 - \frac{3}{2x + 1}$$

Zero Remainders

If the remainder in a polynomial division is zero, the original improper fraction is equivalent to a polynomial. A zero remainder also means that the denominator is a factor of the numerator, so that the same result can be achieved by factoring the numerator and denominator, as we did in Section 5.6. For example, consider this division:

$$
\begin{array}{r}
2x - 3 \\
x + 4 \overline{)2x^2 + 5x - 12} \\
\underline{2x^2 + 8x} \\
-3x - 12 \\
\underline{-3x - 12} \\
0
\end{array}
$$

Because the remainder is zero, we have

$$\frac{2x^2 + 5x - 12}{x + 4} = 2x - 3$$

We can obtain the same result by reducing the original fraction:

$$\frac{2x^2 + 5x - 12}{x + 4} = \frac{(x + 4)(2x - 3)}{x + 4} = 2x - 3$$

HOMEWORK B.3

1. What is an improper algebraic fraction?

2. Is the fraction $\dfrac{x - 1}{x + 1}$ improper?

3. What four steps are repeated in the polynomial division process?

4. How is the remainder in a polynomial division written?

5. a. Use long division to simplify $\dfrac{312}{24}$. What is the remainder?

6. a. Use polynomial division to simplify $\dfrac{a^2 - 7a + 12}{a - 4}$. What is the remainder?

b. Reduce the fraction in part (a) by factoring numerator and denominator and then canceling any common factors.

b. Reduce the fraction in part (a) by factoring numerator and denominator and then canceling any common factors.

In Problems 7–10, write each polynomial in descending powers of the variable. Insert terms with zero coefficients if necessary, to include all powers of the variable from the highest-degree term to the constant term.

7. $b^3 - 1$

8. $1 + 2a^3 - a^2$

9. $2c - 3c^4$

10. $2x^4 - 3x^2$

In Problems 11–22, write each fraction as the sum of a polynomial and a proper algebraic fraction.

11. $\dfrac{x^2 + 3x + 1}{x + 2}$

12. $\dfrac{x^2 - x + 3}{x + 1}$

13. $\dfrac{x^2 + 3x - 9}{x + 5}$

14. $\dfrac{x^2 - 2x - 2}{x - 3}$

15. $\dfrac{2x^2 + x - 2}{x + 1}$

16. $\dfrac{3x^2 - 8x - 1}{x - 3}$

17. $\dfrac{2x^2 + 5x - 3}{2x - 1}$

18. $\dfrac{2x^2 - 9x - 5}{2x + 1}$

19. $\dfrac{4x^2 - 4x - 5}{2x + 1}$

20. $\dfrac{6x^2 + x + 2}{3x + 2}$

21. $\dfrac{2x^3 + 3x^2 - x + 2}{x + 2}$

22. $\dfrac{3x^3 - x^2 - 4x + 2}{x - 1}$

In Problems 23–40, write each fraction as the sum of a polynomial and a proper algebraic fraction.

23. $\dfrac{x^2 + 1}{x + 1}$

24. $\dfrac{x^2 + 4}{x + 2}$

25. $\dfrac{-7 + x^2}{x + 6}$

26. $\dfrac{-10 + x^2}{x - 7}$

27. $\dfrac{1 + 2x^3 - x^2}{x - 1}$

28. $\dfrac{2x + 4x^3 - 3}{x + 2}$

29. $\dfrac{x^3 - 8}{x - 2}$

30. $\dfrac{x^3 + 1}{x + 1}$

31. $\dfrac{x^2 + 4x + 3x^3}{3x - 2}$

32. $\dfrac{4x^3 - 3 - 2x^2}{2x - 3}$

33. $\dfrac{15x^2 - 12 - 8x + 6x^3}{2x + 5}$

34. $\dfrac{3 - 9x + 8x^3 - 6x^2}{4x + 3}$

35. $\dfrac{3x^4 + 2x}{x - 2}$

36. $\dfrac{2x^4 - 3x^2}{x + 3}$

37. $\dfrac{x^4 - 8}{x^2 - 4}$

38. $\dfrac{x^6 - 6}{x^3 + 3}$

39. $\dfrac{x + 4 + 6x^2 + 2x^3}{2x^2 + 1}$

40. $\dfrac{4 - 6x - 4x^2 + 3x^3}{x^2 - 2}$

Recall that you can check a division problem as follows: Multiply the quotient by the divisor and then add the remainder. This should give the dividend in the original division problem. For example:

$$\begin{array}{r} 17 \text{ R. } 9 \\ 26\overline{)451} \\ -26 \\ \hline 191 \\ -182 \\ \hline 9 \end{array}$$

Check

$$17 \cdot 26 + 9 = 442 + 9$$

$$\text{quotient} \cdot \text{divisor} + \text{remainder}$$

$$= 451$$

$$\text{dividend}$$

41. Check your answer to Problem 15.

42. Check your answer to Problem 16.

43. Check your answer to Problem 31.

44. Check your answer to Problem 32.

45. a. Divide $(x^2 - 27x + 180) \div (x - 12)$.

46. a. Divide $(x^2 + 15x - 216) \div (x - 9)$.

b. Write $x^2 - 27x + 180$ in factored form.

b. Write $x^2 + 15x - 216$ in factored form.

47. a. Divide $(x^3 + 3x^2 - 4x - 12) \div (x - 2)$.

48. a. Divide $(x^3 - 8x^2 + 11x + 20) \div (x - 4)$.

b. Write $x^3 + 3x^2 - 4x - 12$ in completely factored form.

b. Write $x^3 - 8x^2 + 11x + 20$ in completely factored form.

49. Reduce $\dfrac{a^3 + 1}{a + 1}$.
(*Hint:* Use polynomial division.)

50. Reduce $\dfrac{w^5 - 1}{w - 1}$.
(*Hint:* Use polynomial division.)

B.4 Inequalities in Two Variables

Figure B.2

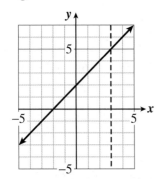

Recall that the graph of an equation in two variables is a picture of its solutions. Consider the graph of

$$y = x + 2$$

shown in Figure B.2. Choose any x-value, say $x = 3$,   and consider *all* points whose x-coordinate is 3. These points lie on the vertical line $x = 3$,   which is shown as a dashed line in the figure. Now, among all the points whose x-coordinate is 3, which one lies on the graph in Figure B.2? It is the point whose y-coordinate is given by the equation when we substitute 3 for x:

$$y = x + 2$$
$$= 3 + 2 = 5$$

Thus, for $x = 3$,   the point on the graph has a y-value of 5.

Where are all the points with x-coordinates equal to 3 but with y-coordinates *greater than* 5? These points lie on the dashed line $x = 3$,   but *above* the graph of $y = x + 2$.

The observations we have made are true for any value of x, not just $x = 3$. For any particular choice of x, the point whose y-coordinate is *equal* to $x + 2$ lies on the graph of $y = x + 2$.   All points that lie *above* the graph have y-coordinates *greater* than $x + 2$.   In fact, any point that lies *above* the graph is a solution of the *inequality*

$$y > x + 2$$

A solution of an inequality in two variables is an ordered pair that makes the inequality true.

EXAMPLE 1

Decide whether the given points are solutions of the inequality

$$y > -x - 1$$

Verify your answers by comparing the points with the graph of the equation $y = -x - 1$ shown in Figure B.3.

   a. (1, 3)   **b.** (−2, −1)   **c.** (3, −4)

Figure B.3

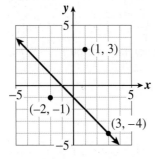

Solution

   a. We substitute 1 for x and 3 for y into the inequality:

$$3 > -1 - 1$$
$$3 > -2$$

Because this is a true statement, (1, 3) is a solution of the inequality. In Figure B.3, note that (1, 3) lies above the graph of $y = -x - 1$.

   b. We substitute −2 for x and −1 for y into the inequality:

$$-1 > -(-2) - 1$$
$$-1 > 1$$

This is a false statement, so (−2, −1) is not a solution of the inequality. Note that (−2, −1) does not lie above the graph of $y = -x - 1$.

c. We substitute 3 for x and -4 for y into the inequality:

$$-4 > -3 - 1$$
$$-4 > -4$$

This is a false statement, so $(3, -4)$ is not a solution of the inequality. Note that $(3, -4)$ does not lie above the graph of $y = -x - 1$.

Graphing the Solutions of Inequalities

As we observed earlier, the solutions of the inequality

$$y > x + 2$$

consist of *all* points that lie above the graph of the equation $y = x + 2$. We indicate the solutions by shading the half-plane above the line, as shown in Figure B.4(a). In this graph, the graph of the equation $y = x + 2$ is shown as a dashed line. This is because the points *on* the line are *not* solutions of the inequality $y > x + 2$. They are not part of the graph.

In Figure B.4(b), we show the graph of the solutions of the inequality

$$y \geq x + 2$$

These solutions consist of all points whose y-coordinates are greater than *or equal to* $x + 2$. In other words, the points on the line $y = x + 2$ *are* solutions of the inequality $y \geq x + 2$. We indicate this by using a solid line to graph the equation $y = x + 2$. These points *are* part of the graph of $y \geq x + 2$.

EXAMPLE 2

Graph the solutions of the inequalities.

a. $y < -\dfrac{1}{2}x + 4$ **b.** $y \geq 2x$

Solution

a. We begin by graphing the line $y = -\dfrac{1}{2}x + 4$. This will be the edge, or boundary, of the shaded region. The y-intercept of the line is $(0, 4)$ and its slope is $-\dfrac{1}{2}$. We use a dashed line to show the graph, because points on the line are not included among the solutions of the inequality.

For this inequality, solution points have y-coordinates *less than* $-\dfrac{1}{2}x + 4$, so we shade the half-plane *below* the dashed line. The completed graph is shown in Figure B.5(a).

b. We begin by graphing the line $y = 2x$. This line has y-intercept $(0, 0)$ and slope 2. Because the inequality symbol in this example is $\geq$, greater than *or equal to*, the solutions of the equation $y = 2x$ are also solutions of the inequality $y \geq 2x$. This means that the points on the line are also included in the graph of the inequality. We indicate this by a solid line instead of a dashed line.

Because the solution points for this inequality have y-coordinates *greater than* $2x$, we shade the half-plane *above* the line. The completed graph is shown in Figure B.5(b).

Figure B.4

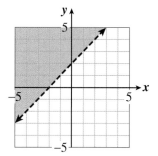

(a)

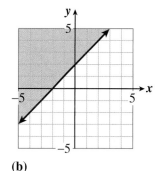

(b)

Figure B.5

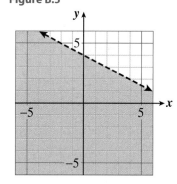

(a)

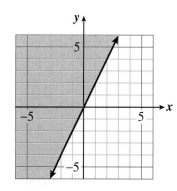

(b)

Using a Test Point

When graphing an inequality, it is not always easy to determine which side of the boundary line to shade. For example, it is not obvious whether the solutions to the inequality

$$3x - 2y \le 6$$

lie above the line $3x - 2y = 6$ or below the line. However, we know that all the points on one side of the line are solutions and that all the points on the other side of the line are not solutions. Therefore, we need only decide *which* side of the boundary line contains the solutions. We can do this by testing a single point.

Choose a point that does not lie on the boundary line itself, say the point $(0, 0)$. We call this point the *test point*. The test point happens to lie above the line $3x - 2y = 6$. [See Figure B.6(a).] To determine whether the test point is a solution of the inequality, we substitute 0 for x and 0 for y into the inequality:

$$3x - 2y \le 6$$
$$3(0) - 2(0) \le 6$$
$$0 - 0 \le 6$$

Because $0 \le 6$ is a true statement, the point $(0, 0)$ is a solution. Recall that *all* the solutions lie on one side of the boundary line or the other. Therefore, if one of the solutions lies above the boundary line, all of them do. We shade the half-plane that contains the test point $(0, 0)$, and we use a solid line for the graph of $3x + 2y = 6$ because the points of the boundary line are included among the solutions. The completed graph is shown in Figure B.6(b).

We can summarize this strategy for graphing the solutions to a linear inequality as follows:

Figure B.6

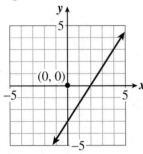

(a)

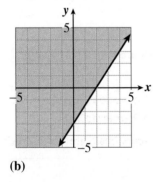

(b)

To Graph a Linear Inequality:

1. Graph the corresponding linear equation. This line is the boundary for the shaded region.

2. If the inequality symbol is $>$ or $<$, use a dashed line for the graph of the equation. If the inequality symbol is $\ge$ or $\le$, use a solid line.

3. Choose a test point that does not lie on the boundary line. Determine whether the test point is a solution of the inequality.

   a. If the test point *is* a solution, shade the half-plane that contains the test point.

   b. If the test point is *not* a solution, shade the half-plane that does *not* contain the test point.

The origin, $(0, 0)$, is usually an easy test point to work with. However, if the boundary line passes through the point $(0, 0)$, then you must choose some other test point that is not on the line.

EXAMPLE 3

Graph the solutions of the inequality $x - 2y > 0$.

Solution

Step 1 First, we graph the boundary line $x - 2y = 0$.   We write the equation in slope-intercept form by solving for y: $y = \frac{1}{2}x$.   The line has slope $\frac{1}{2}$ and y-intercept $(0, 0)$.

Step 2 We use a dashed line for the graph, because the symbol in the inequality is $>$.

Step 3 Now, we choose a test point that does not lie on the line, say $(0, 3)$. We determine whether the test point is a solution of the original inequality by substituting 0 for x and 3 for y:

$$x - 2y > 0$$
$$0 - 2(3) > 0$$
$$-6 > 0$$

This statement is false, so the point $(0, 3)$ is *not* a solution of the inequality. Because $(0, 3)$ lies above the boundary line, the solutions must lie below the boundary line. We shade the half-plane below the dashed line. The graph is shown in Figure B.7.

Figure B.7

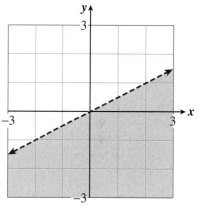

HOMEWORK B.4

Make a sketch on a Cartesian coordinate system illustrating each set of points in Problems 1–6.

1. All points with x-coordinate less than 4

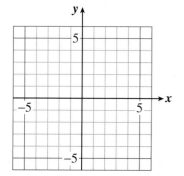

2. All points with y-coordinate greater than -2

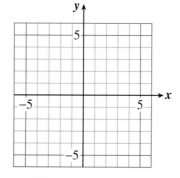

3. All points with *x*-coordinate greater than −3 and *y*-coordinate less than or equal to 5

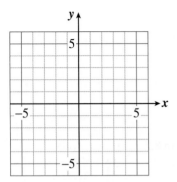

4. All points with *x*-coordinate less than or equal to −1 and *y*-coordinate less than 2

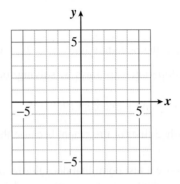

5. All points with *x*-coordinate equal to 1 and *y*-coordinate less than 6

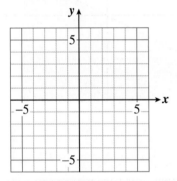

6. All points with *y*-coordinate equal to −3 and *x*-coordinate greater than 2

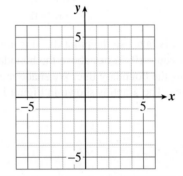

7. a. Graph the line $y = 2x - 4$.

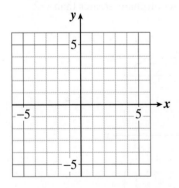

b. On the same graph, plot the points $(0, -2)$, $(2, 1)$, $(3, -1)$, and $(1, -3)$.

c. From your graph, determine which of the given points are solutions of the inequality $y < 2x - 4$.

8. a. Graph the line $y = 2 - x$.

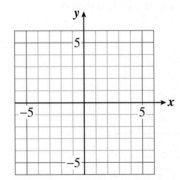

b. On the same graph, plot the points $(-1, 2)$, $(2, 1)$, $(3, -1)$, and $(4, -1)$.

c. From your graph, determine which of the given points are solutions of the inequality $y \geq 2 - x$.

9. Explain how to use a test point when graphing a linear inequality.

10. Explain when the boundary line should be a dashed line and when it should be a solid line.

In Problems 11–16, determine whether the given point is a solution of the inequality.

11. $y > x$, $(1, 3)$

12. $y \geq x - 5$, $(3, 2)$

13. $x + y < -1$, $(-1, 3)$

14. $2x - y \leq 3$, $(-2, -4)$

15. $x < 3y + 5$, $(2, 0)$

16. $2 > x + 2y$, $(0, -2)$

Graph each inequality in Problems 17–40.

17. $y > 3x + 2$

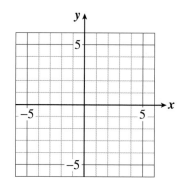

18. $y > -2x + 1$

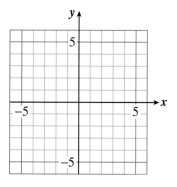

19. $y < -x - 3$

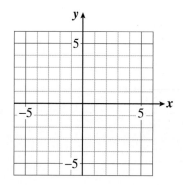

20. $y < x - 3$

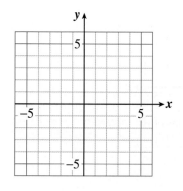

21. $y \leq 2x$

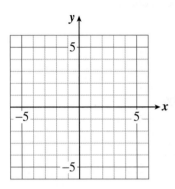

22. $y \geq -2x$

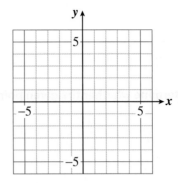

23. $y > -4x$

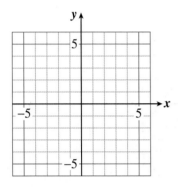

24. $y < 3x$

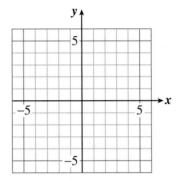

25. $y - 3x \leq 5$

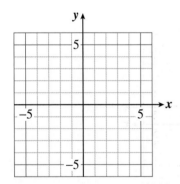

26. $y + 3x \leq 6$

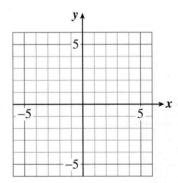

27. $3x - 4y \geq 12$

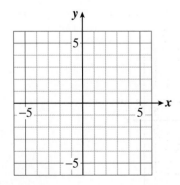

28. $2x + 5y \geq 10$

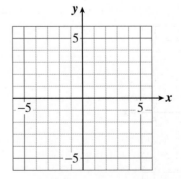

29. $y + 4x < 8$

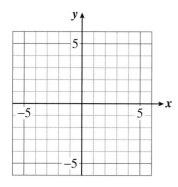

30. $2x - y < 4$

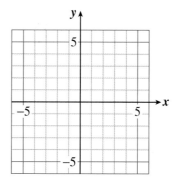

31. $6x + 4y \leq 12$

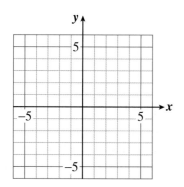

32. $3x + 2y \leq 12$

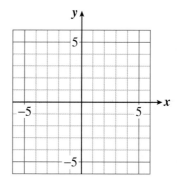

33. $y \leq 2$

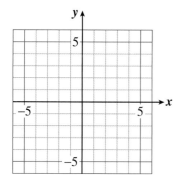

34. $y \leq -1$

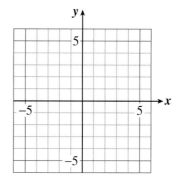

35. $x > -3$

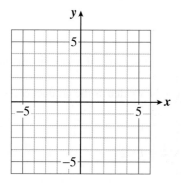

36. $x > 2$

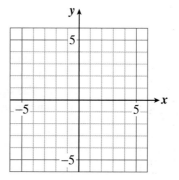

37. $3x > 2y$

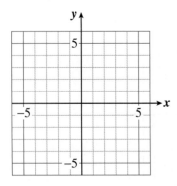

38. $5y \geq 4x$

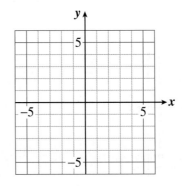

39. $2x + 3y \leq 0$

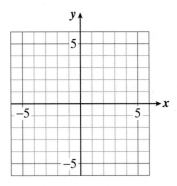

40. $3x - 4y > 0$

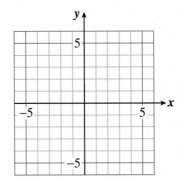

B.5   Complex Numbers

In Chapter 4, we saw that the square root of a negative number, such as $\sqrt{-4}$, is not a real number. We said that such a radical was "undefined." However, square roots of negative numbers are so important for some applications that a new kind of number was invented to make it possible to work with these radicals.

Imaginary Numbers

We begin with the square root of -1, $\sqrt{-1}$. We give this number a name, i. That is,

$$i = \sqrt{-1}$$

The letter i used in this way is not a variable. It is the name of a specific number, and hence it is a constant.

The square root of any negative number can be written as the product of a real number and i. For example,

$$\sqrt{-4} = \sqrt{-1 \cdot 4}$$
$$= \sqrt{-1}\,\sqrt{4} = i \cdot 2$$

or $\sqrt{-4} = 2i.$ Any number that is the product of i and a real number is called an **imaginary number**. Examples of imaginary numbers are

$$3i, \quad \frac{7}{8}i, \quad -38i, \quad \sqrt{5}i$$

Complex Numbers

Consider the quadratic equation

$$x^2 - 2x + 5 = 0$$

We can use the quadratic formula to solve this equation, as follows:

$$x = \frac{-(-2) \pm \sqrt{(-2)^2 - 4(1)(5)}}{2(1)} \qquad \text{Substitute } a = 1, b = -2, c = 5.$$

$$= \frac{2 \pm \sqrt{-16}}{2}$$

If we now replace $\sqrt{-16}$ by $4i$, we have

$$x = \frac{2 \pm 4i}{2} = 1 \pm 2i$$

The two solutions of the equation are $1 + 2i$ and $1 - 2i.$ These numbers are examples of **complex numbers**.

> A complex number is a number of the form $a + bi,$ where a and b are real numbers.

Thus, these are all complex numbers:

$$3 - 5i, \quad 2 + \sqrt{7}i, \quad \frac{4 - i}{3}, \quad 6i, \quad -9$$

In the complex number $a + bi,$ a is called the *real part*, and b is called the *imaginary part*. Note that all real numbers are also complex numbers (with imaginary part equal to zero) that and all imaginary numbers are complex numbers with real part equal to zero.

Sums and Differences of Complex Numbers

We can add and subtract complex numbers by combining their real and imaginary parts separately. For example,

$$(4 + 5i) + (2 - 3i) = (4 + 2) + (5 - 3)i$$
$$= 6 + 2i$$

In general, we have

> $$(a + bi) + (c + di) = (a + c) + (b + d)i$$
> $$(a + bi) - (c + di) = (a - c) + (b - d)i$$

EXAMPLE 1

Subtract $(8 - 6i) - (5 + 2i)$.

Solution

We combine the real and imaginary parts.

$$(8 - 6i) - (5 + 2i) = (8 - 5) + (-6 - 2)i$$
$$= 3 + (-8)i$$
$$= 3 - 8i$$

Products of Complex Numbers

Because we have defined $i = \sqrt{-1}$,   it must also be true that

$$i^2 = -1$$

We use this fact to find the product of two imaginary numbers. For example,

$$(3i) \cdot (4i) = 3 \cdot 4i^2$$
$$= 12(-1) = -12$$

To find the product of two complex numbers, we use the FOIL method, as though the numbers were binomials. For example,

$$(2 + 3i)(3 - 5i) = 6 - 10i + 9i - 15i^2$$

Because $i^2 = -1$,   the last term, $-15i^2$, can be replaced by $(-15)(-1)$,   or 15, to obtain

$$6 - 10i + 9i + 15$$

Finally, we combine the real parts and imaginary parts to obtain

$$(2 + 3i)(3 - 5i) = 21 - i$$

EXAMPLE 2

Multiply $(7 - 4i)(-2 - i)$.

Solution

$$(7 - 4i)(-2 - i) = -14 - 7i + 8i + 4i^2 \quad \text{Replace } i^2 \text{ by } -1.$$
$$= -14 - 7i + 8i - 4 \quad \text{Combine real parts and imaginary parts.}$$
$$= -18 + i$$

Quotients of Complex Numbers

In order to find the quotient of two complex numbers, we must use the technique of rationalizing the denominator that we discussed in Section 6.5. For example, consider the quotient

$$\frac{3 + 4i}{2i}$$

Because i is really a radical (remember that $i = \sqrt{-1}$), we multiply the numerator and the denominator of the quotient by i.

$$\frac{(3 + 4i) \cdot i}{2i \cdot i} = \frac{3i + 4i^2}{2i^2}$$
   Apply the distributive law to the numerator. Recall that $i^2 = -1$.

$$= \frac{3i - 4}{-2}$$

Finally, we can divide -2 into each term of the numerator to get

$$\frac{3i}{-2} - \frac{4}{-2} = -\frac{3}{2}i + 2$$

EXAMPLE 3

Divide $\dfrac{10 - 15i}{5i}$.

Solution

We multiply numerator and denominator by i.

$$\frac{10 - 15i}{5i} = \frac{(10 - 15i) \cdot i}{5i \cdot i}$$

$$= \frac{10i - 15i^2}{5i^2} = \frac{10i + 15}{-5}$$
   Replace i^2 by -1.

$$= \frac{10i}{-5} + \frac{15}{-5} = -2i - 3$$
   Divide -5 into each term of numerator.

If the divisor has both a real and an imaginary part, we must multiply numerator and denominator by the conjugate of the denominator. The **conjugate** of a complex number $a + bi$ is $a - bi$.

EXAMPLE 4

Divide $\dfrac{2 + 3i}{4 - 2i}$.

Solution

We multiply numerator and denominator by the conjugate of the denominator, $4 + 2i$.

$$\frac{2 + 3i}{4 - 2i} = \frac{(2 + 3i)(4 + 2i)}{(4 - 2i)(4 + 2i)}$$
   Expand numerator and denominator.

$$= \frac{8 + 4i + 12i + 6i^2}{16 + 8i - 8i - 4i^2}$$
   Replace i^2 by -1.

$$= \frac{8 + 16i - 6}{16 - (-4)}$$
   Combine like terms.

$$= \frac{2 + 16i}{20}$$
   Divide 20 into each term of numerator.

$$= \frac{2}{20} + \frac{16i}{20} = \frac{1}{10} + \frac{4}{5}i$$

Quadratic Equations

As we noted earlier in this section, a quadratic equation with real coefficients may have solutions that are complex numbers.

EXAMPLE 5

Solve $x^2 + x + 1 = 0$.

Solution

We apply the quadratic formula with $a = 1$, $b = 1$, and $c = 1$.

$$x = \frac{-b \pm \sqrt{b^2 - 4ac}}{2a}$$

$$= \frac{-1 \pm \sqrt{1^2 - 4(1)(1)}}{2(1)}$$

$$= \frac{-1 \pm \sqrt{-3}}{2}$$

$$= \frac{-1 \pm i\sqrt{3}}{2}$$

The two solutions of the equation are $-\dfrac{1}{2} + \dfrac{\sqrt{3}}{2}\, i$ and $-\dfrac{1}{2} - \dfrac{\sqrt{3}}{2}\, i$.

In Example 5, the solutions are complex because the expression under the radical, $b^2 - 4ac$, is a negative number, -3. For any quadratic equation with real coefficients, if one of the solutions is complex, both solutions are complex. In fact, they are complex conjugates; that is, they differ only in the sign of their imaginary parts.

If the expression under the radical is positive or zero, then the solutions of the quadratic equation are real numbers. Thus we can determine whether the solutions are real or complex numbers by finding the sign of the expression $b^2 - 4ac$. Because it helps us discriminate between real and complex solutions to a quadratic equation, the expression $b^2 - 4ac$ is called the **discriminant** of the quadratic equation. We have the following result.

The solutions of the quadratic equation $ax^2 + bx + c = 0$ are described by the discriminant, $D = b^2 - 4ac$, as follows. (The coefficients a, b, and c are real numbers, $a \neq 0$.)

1. **If $D > 0$** then the equation has two distinct real solutions.
2. **If $D = 0$** then the equation has two identical real solutions.
3. **If $D < 0$** then the equation has two complex-conjugate solutions.

EXAMPLE 6

Use the discriminant to determine the nature of the solutions to each equation.

a. $4x^2 - 12x + 9 = 0$ **b.** $2x^2 - 4x + 5 = 0$

Solution

a. We compute the discriminant with $a = 4$, $b = -12$, and $c = 9$.

$$D = b^2 - 4ac = (-12)^2 - 4(4)(9)$$
$$= 144 - 144 = 0$$

Because the discriminant is zero, the equation has two identical real solutions.

b. We compute the discriminant with $a = 2$, $b = -4$, and $c = 5$.

$$D = b^2 - 4ac = (-4)^2 - 4(2)(5)$$
$$= 16 - 40 = -24$$

Because the discriminant is negative, the equation has two complex-conjugate solutions.

HOMEWORK B.5

1. What does i stand for?

2. What real number is i^2 equal to?

3. What is an imaginary number? Give an example.

4. What is a complex number? Give an example.

5. Explain how to add two complex numbers.

6. Explain how to multiply two complex numbers.

7. When we divide two complex numbers, we multiply numerator and denominator by _____.

8. State three possibilities for the nature of the roots of a quadratic equation.

9. What is the discriminant of the quadratic equation $ax^2 + bx + c = 0$?

10. Explain how to use the discriminant to determine the nature of the roots of a quadratic equation.

11. Simplify $\sqrt{-9}$. Explain why $\sqrt{-9}$ is not a real number.

12. Can a quadratic equation with real coefficients have as its solutions $2 + 3i$ and $3 - 4i$? Why or why not?

Add or subtract the expressions in Problems 13–22.

13. $(6 - 3i) + (-2 - 5i)$

14. $(-7 + 2i) + (4 + 3i)$

15. $(3i - 7) - (6 - 4i)$

16. $(i - 1) - (2i - 6)$

17. $\left(\dfrac{2}{3} - \dfrac{1}{3}i\right) - \left(\dfrac{2}{3}i + \dfrac{5}{3}\right)$

18. $\left(\dfrac{1}{5}i - \dfrac{2}{5}\right) - \left(\dfrac{4}{5} - \dfrac{3}{5}i\right)$

19. $\left(\dfrac{7}{2} - \dfrac{4}{5}i\right) + \left(\dfrac{2}{3} + \dfrac{3}{10}i\right)$

20. $\left(\dfrac{3}{8} - \dfrac{5}{2}i\right) + \left(\dfrac{2}{3} - \dfrac{5}{7}i\right)$

21. $(8.5 + 2.6i) + (-5.3i - 2.9)$

22. $(14.7 - 11.1i) - (8.5 - 0.4i)$

Multiply the expressions in Problems 23–32.

23. $3i(7 - 4i)$

24. $-2i(-7 + i)$

25. $(2 - 3i)(4 - i)$

26. $(-6 + 7i)(2 - 3i)$

27. $(3 - 4i)(3 - 4i)$

28. $(2 + 5i)(2 + 5i)$

29. $(4 + i \sqrt{2})^2$

30. $(1 - i \sqrt{7})^2$

31. $(4 + i \sqrt{2})(4 - i \sqrt{2})$

32. $(1 - i \sqrt{7})(1 + i \sqrt{7})$

Divide the expressions in Problems 33–50.

33. $\dfrac{8 - 6i}{2i}$

34. $\dfrac{6 + 9i}{-3i}$

35. $\dfrac{35 - 63i}{7i}$

36. $\dfrac{30 - 48i}{-6i}$

37. $\dfrac{4 + 6i}{8i}$

38. $\dfrac{3 - 5i}{15i}$

39. $\dfrac{5 + 10i}{2 + i}$

40. $\dfrac{15 - 5i}{1 - 2i}$

41. $\dfrac{6 + 8i}{1 - i}$

42. $\dfrac{12 - 2i}{1 + i}$

43. $\dfrac{3 + 5i}{3 - 5i}$

44. $\dfrac{7 - 2i}{7 + 2i}$

45. $\dfrac{-5i}{1 + 3i}$

46. $\dfrac{-3i}{3 - 2i}$

47. $\dfrac{\sqrt{3}}{\sqrt{3} - i}$

48. $\dfrac{2\sqrt{2}}{1 - i\sqrt{2}}$

49. $\dfrac{1 + i\sqrt{5}}{1 - i\sqrt{5}}$

50. $\dfrac{\sqrt{2} - i}{\sqrt{2} + i}$

In Problems 51–60, use the quadratic formula to solve each equation.

51. $2t^2 + 3t + 2 = 0$

52. $r^2 + r + 2 = 0$

53. $4n^2 - 4n + 3 = 0$

54. $3m^2 - 6m + 1 = 0$

55. $b^2 + 1 = \dfrac{4}{3}b$

56. $\dfrac{a^2}{2} + \dfrac{5}{4} = a$

57. $0.4z^2 = 0.16z - 0.2$

58. $0.04q^2 = -0.1q - 0.07$

59. $3x^2 - 8x + 2 = 0$

60. $x^2 + 16x - 9 = 0$

In Problems 61–70, use the discriminant to determine the nature of the solutions to each equation. You do not need to solve the equation.

61. $2u^2 - 3u + 7 = 0$

62. $4p^2 - 4p + 1 = 0$

63. $v^2 + 6v + 9 = 0$

64. $3w^2 - w + 6 = 0$

65. $5d^2 - 12d + 7 = 0$

66. $8g^2 + 13g + 5 = 0$

67. $27x^2 + 35x - 74 = 0$

68. $31z^2 - 62z - 19 = 0$

69. $81x^2 - 36x + 4 = 0$

70. $15x^2 + 27x + 50 = 0$

In Problems 71–80, use the discriminant to determine the nature of the solutions. Do not use pencil, paper, or calculator.

71. $4x^2 - 5x + 2 = 0$

72. $2x^2 - 3x + 1 = 0$

73. $3x^2 + 5 = 0$

74. $11x^2 + 2 = 0$

75. $7x^2 - 2 = 0$

76. $13x^2 - 11 = 0$

77. $8x^2 - 7x - 1 = 0$

78. $5x^2 + 3x - 2 = 0$

79. $-3x^2 + 2x + 8 = 0$

80. $-12x^2 - 10x + 11 = 0$

In Problems 81–86, perform the indicated operations.

81. $(3 - 6i) - (4 + 11i)$

82. $(14 + 5i) + (-8 - 3i)$

83. $(5 - 8i)(-2 + 6i)$

84. $(2 - 7i)^2$

85. $\dfrac{8i}{2 - 5i}$

86. $\dfrac{3i + 2}{2i - 3}$

Glossary

absolute value, *n.* the distance on a number line between a number and 0; for example, the absolute value of -7 is 7, or $\ |-7| = 7$

algebraic expression, *n.* a meaningful combination of numbers, variables, and operation symbols; also called an *expression*

algebraic fraction, *n.* a fraction whose numerator and denominator are polynomials; also called a *rational expression*

altitude, *n.* (1) distance above the ground or above sea level; (2) the vertical distance between the vertex and the base of a geometric object or the distance between parallel sides of a parallelogram, trapezoid, or rectangle; also called *height*

area, *n.* a measure of the two-dimensional space enclosed by a polygon or curve, typically expressed in square units, such as square meters or square feet

associative law of addition: If a, b, and c are any numbers, then

$$(a + b) + c = a + (b + c)$$

associative law of multiplication: If a, b, and c are any numbers, then

$$(a \cdot b) \cdot c = a \cdot (b \cdot c)$$

average, *n.* a typical or middle value for a set of data

axis (*plural* **axes**), *n.* a line used as a reference for position and/or orientation

axis of symmetry, *n.* a line that cuts a plane figure into two parts, each a mirror image of the other

bar graph, *n.* a picture of numerical information in which the lengths or heights of bars represent the values of variables

base, *n.* (1) a number or algebraic expression that is used as a factor repeated the number of times indicated by the exponent—for example, in 3^5, the base is 3; (2) the bottom side of a polygon; (3) the bottom face of a solid

binomial, *n.* a polynomial consisting of exactly two terms

build a fraction, *v.* to find an equivalent fraction by multiplying numerator and denominator by the same nonzero expression

building factor, *n.* an expression by which both numerator and denominator of a given fraction are multiplied (in order to *build the fraction*)

Cartesian coordinate system, *n.* a grid that associates points in the coordinate plane with ordered pairs of numbers

Cartesian plane, *n.* a plane with a pair of coordinate axes; also called a *coordinate plane*

circle, *n.* the set of all points in a plane at a fixed distance (the *radius*) from the center

circumference, *n.* the distance around a circle

coefficient, *n.* the numerical factor in a term; for example, in the expression $32a + 7b$, the coefficient of a is 32 and the coefficient of b is 7

combine like terms, *v.* to simplify an expression by adding or subtracting like terms as indicated

common factor, *n.* a quantity that divides evenly into each of two or more given expressions

commutative law of addition: If a and b are any numbers, then $a + b = b + a$

commutative law of multiplication: If a and b are any numbers, then $a \cdot b = b \cdot a$

complementary angles, *n.* two angles whose measures add up to 90°

completing the square, *n.* adding a constant to a quadratic polynomial so that the result is the square of a binomial; a technique used to solve quadratic equations

complex fraction, *n.* a fraction that contains one or more fractions in its numerator and/or in its denominator

complex conjugates, *n.* a pair of complex numbers such as $1 + 2i$ and $1 - 2i$ whose real parts are equal and whose imaginary parts are opposites

complex number, *n.* a number that can be written in the form $a + bi$, where a and b are real numbers and $i^2 = -1$

composite number, *n.* a natural number with a whole-number factor other than itself and 1

compound inequality, *n.* a mathematical statement involving two order symbols. For example, the compound inequality $1 < x < 2$ says that "1 is less than x, and x is less than 2."

cone, *n.* a three-dimensional object consisting of a circular base, a vertex, and the points on the line segments joining the circle to the vertex.

conjugate, *n.* see *complex conjugates*

consistent system of equations, *n.* a system of equations having at least one solution

constant, *adj.* unchanging, not variable; for example, the product of two variables is *constant* if the product is always the same number, for any values of the variables

constant, *n.* a number (as opposed to a *variable*)

constant of proportionality, *n.* the quotient of two directly proportional variables, or the product of two inversely proportional variables; also called the *constant of variation*

constant of variation, see *constant of proportionality*

continuous, *adj.* without holes or gaps; for example, a curve is continuous if it can be drawn without lifting the pencil from the page

coordinate, *n.* a number used with a number line or an axis to designate position

coordinate axis, *n.* one of the two perpendicular number lines used to define the coordinates of points in the plane

coordinate plane, *n.* a plane containing a pair of coordinate axes; also called the *Cartesian plane*, or *xy-plane*

corresponding angles, *n.* as shown in the following figure, the four pairs of angles formed when two lines ℓ_1 and ℓ_2 are both intersected by a third line ℓ_3; angles A and A', B and B', C and C', and D and D' are the four pairs of corresponding angles formed

costs, *n.* money that an individual or group must pay out; for example, a company's costs might include payments for wages, supplies, and rent

counting number, *n.* one of the numbers 1, 2, 3, 4, . . . ; also called a *natural number*

cross-multiply, *v.* to rewrite a proportion as an equivalent equation without fractions by equating the two products obtained by multiplying a numerator of one side of the proportion by the denominator of the other side

cube, *n.* (1) a three-dimensional box whose six faces all consist of squares; (2) an expression raised to the power 3

cube, *v.* to raise an expression to the power 3; for example, to cube 2 means to form the product of three 2's, or $2^3 = 2 \times 2 \times 2 = 8$

cube root, *n.* a number that when raised to the power 3 gives a desired value; for example, 2 is the cube root of 8 because $2^3 = 8$

cylinder, *n.* a three-dimensional figure in the shape of a tin can

decimal fraction, *n.* a decimal number (usually less than 1 in absolute value)

decimal number, *n.* a number such as 2.718 in which the fractional part is indicated by the digits to the right of the decimal point

decimal place, *n.* the relative position of a digit on the right of the decimal point; for example, in the number 3.14159, the digit 4 is in the second decimal place, or hundredths place

decimal point, *n.* the period or dot between the whole-number part and the fractional part of a decimal number; for example, the decimal form of $1\frac{3}{10}$ is 1.3

degree, *n.* a measure of angle equal to $\frac{1}{360}$ of a complete revolution

degree of a polynomial (in one variable), *n.* the largest exponent that appears on the variable in the polynomial

denominator, *n.* the expression below the fraction bar in a fraction

dependent system of equations, *n.* a system of equations having infinitely many solutions

dependent variable, *n.* a variable whose value is determined by specifying the value of the *independent variable*; on a coordinate plane, the values of the dependent variable are displayed on the vertical axis

descending powers, *n.* terms in a polynomial arranged with the term of highest degree written first, then the term with the second-highest degree, and so on

diameter, *n.* the distance across a circle through its center, equal to twice the radius

difference, *n.* the result of a subtraction, for example, the expression $a - b$ represents the difference of a and b

difference of squares, *n.* an expression of the form $a^2 - b^2$

direct variation, *n.* a relation between two variables such that one is a constant multiple of the other (so that the ratio between the two variables is the constant)

directed distance, *n.* the difference between the coordinates of the ending and starting points of a segment of a number line; the directed distance is negative if the ending value is smaller than the starting value; for example, the directed distance from 5 to 2 is $2 - 5 = -3$

directly proportional, *adj.* describing variables related by direct variation

discriminant, *n.* the quantity $b^2 - 4ac$, where a, b, and c are the coefficients of the quadratic equation $ax^2 + bx + c = 0$

distributive law: For any numbers a, b, and c, $a(b + c) = ab + ac$

dividend, *n.* a quantity that is divided by another quantity; for example, in the expression $a \div b$, the dividend is a

divisor, *n.* a quantity that is divided into another quantity; for example, in the expression $a \div b$, the divisor is b

elimination method, *n.* a method for solving a system of equations that involves adding together corresponding terms of the equations or multiples of the equations

equation, *n.* a mathematical statement that two expressions are equal

equilateral triangle, *n.* a triangle with three sides of equal length

equivalent equations, *n.* equations that have the same solutions

equivalent expressions, *n.* expressions that represent the same value for any value(s) of the variable(s)

evaluate, *v.* to compute the value of an expression when the variable in the expression is replaced by a number

exponent, *n.* the number or expression that indicates how many times a *base* is used as a factor; for example, in 3^5, the exponent is 5, and $3^5 = 3 \times 3 \times 3 \times 3 \times 3$

expression, see *algebraic expression*

extraction of roots, *n.* a method used to solve quadratic equations

extraneous solution, *n.* a value that is not a solution to a given equation but is a solution to an equation derived from the original

extrapolation, *n.* an estimate of the value of a variable that is beyond the range of the given data

factor, *n.* an expression that divides evenly into another expression; for example, 2 is a factor of 6

factor, *v.* to write as a product; for example, to factor 6 we write $6 = 2 \times 3$

FOIL, *n.* an acronym for a method of computing the product of two binomials: multiply the **F**irst, **O**uter, **I**nner, and **L**ast terms

formula, *n.* an equation involving two or more variables

fraction, *n.* a number in the form $\dfrac{a}{b}$, where a and b are numbers and $b \neq 0$

fraction bar, *n.* the line segment separating the numerator and denominator of a fraction; it indicates division

frequency table, *n.* a table that summarizes a data set by indicating how often each data value occurs

fundamental principle of fractions: If a is any number and b and c are nonzero numbers, then $\dfrac{a \cdot c}{b \cdot c} = \dfrac{a}{b}$

geometrically similar, *adj.* having the same shape (but possibly different sizes)

graph, *v.* to draw a graph

graph of an equation (or **inequality**), *n.* a picture of the solutions of an equation (or inequality) using a number line or coordinate plane

greatest common factor (GCF), *n.* the largest factor that divides evenly into two or more expressions

height, see *altitude*

hemisphere, *n.* half a sphere

histogram, *n.* a type of bar graph in which the height of each bar gives the frequency of a particular outcome or event

horizontal axis, *n.* the horizontal coordinate axis; often called the *x-axis*

horizontal intercept, *n.* the point where a graph meets the horizontal axis; often called the *x-intercept*

hypotenuse, *n.* the longest side of a right triangle (always the side opposite the right angle)

imaginary number, *n.* a complex number of the form *bi*, where *b* is a real number and $i^2 = -1$

improper fraction, *n.* a fraction in the form $\dfrac{a}{b}$, where *a* and *b* are natural numbers and $a > b$

inconsistent system of equations, *n.* a system of equations having no solution

independent system of equations, *n.* a system of equations having at most one solution

independent variable, *n.* a variable whose value determines the value of the *dependent variable;* on a coordinate plane, the values of the independent variable are displayed on the horizontal axis

inequality, *n.* a mathematical statement that two quantities are not equal; it takes the form $a < b$, $a \le b$, $a > b$, $a \ge b$, or $a \ne b$

integer, *n.* a whole number or the negative of a whole number

intercept, *n.* a point where a graph meets a coordinate axis

intercept method, *n.* a method for graphing a line by finding its horizontal and vertical intercepts

interest, *n.* money paid for the use of money; for example, after borrowing money, the borrower must pay the lender not only the original amount borrowed (known as the *principal*) but also the interest, which is a certain percentage of the principal

interest rate, *n.* the fraction of the principal that is paid as interest for 1 year; for example, an interest rate of 10% means that interest for 1 year is 10% of the principal

interpolation, *n.* estimating the value of a variable on the basis of data that include both larger and smaller values of the variable

inverse variation, *n.* a relation between two variables in which one is a constant divided by the other (so that the product of the two variables is the constant)

inversely proportional, *adj.* describing variables related by inverse variation

irrational number, *n.* a number that cannot be expressed as a quotient of two integers but does correspond to a point on the number line

isosceles triangle, *n.* a triangle with two sides of equal length

leg, *n.* one of the two shorter sides of a right triangle (one side of the right angle)

like fractions, *n.* fractions with equivalent denominators

like radicals, *n.* square roots with equivalent radicands (or cube roots with equivalent radicands)

like terms, *n.* terms with equivalent variable parts

line graph, *n.* a type of graph in which the points are connected with line segments

linear equation, *n.* an equation such as $2x + 3y = 4$ or $x - 3y = 7$ in which each term has degree 0 or 1

linear regression, *n.* the process of using a line to predict values of a (dependent) variable

linear term, *n.* a term that consists of a constant times a variable

lowest (or least) common denominator (LCD), *n.* the smallest denominator that is a multiple of the denominators of two or more fractions

lowest (or least) common multiple (LCM), *n.* the smallest natural number that two or more natural numbers divide into evenly

mean, *n.* the average of a set of numbers, computed by adding the numbers and dividing by how many are in the set. For example, the mean of 5, 2, and 11 is $\dfrac{5 + 2 + 11}{3} = 6.$

median, *n.* the middle number in a set of numbers written in order; for example, the median of 5, 2, and 11 is 5; if the set has two numbers in the middle, then the median is the mean of those two: The median of 6, 1, 9 and 27 is $\dfrac{6 + 9}{2} = 7.5.$

mixed number, *n.* a number such as $33\dfrac{1}{3}$ written in the form $N\dfrac{a}{b}$ to mean $N + \dfrac{a}{b}$, where N is a natural number and $\dfrac{a}{b}$ is a positive proper fraction

mode, *n.* the number that occurs most frequently in a set of numbers. For example, the mode of 1, 1, 2, and 3 is 1.

model, *n.* a mathematical expression used to represent a situation in the world or a situation described in words; for example, the equation $P = R - C$ is a model for the relationship among the variables profit, revenue, and cost

model, *v.* to create a model

monomial, *n.* an algebraic expression with only one term

natural number, *n.* a counting number

negative number, *n.* a number that is less than 0

negative of, *n.* the opposite of

number line, *n.* a line with coordinates marked on it; see the figure below

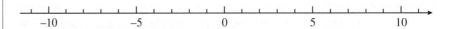

numerator, *n.* the expression above the fraction bar in a fraction

numerical coefficient, *n.* see *coefficient*

operation, *n.* addition, subtraction, multiplication, division, raising to a power, or taking a root

opposite of a number, *n.* the number with the same absolute value as a given number but the opposite sign; for example, the opposite of -5 is 5

order of operations, *n.* rules that prescribe the order in which to carry out the operations in an expression

order symbol, *n.* one of the four symbols $<$, $\le$, $>$, and $\ge$

ordered pair, *n.* a pair of numbers enclosed in parentheses: (x, y); often used to specify a point or a location on the coordinate plane

origin, *n.* the point where the coordinate axes meet, having coordinates $(0, 0)$

parabola, *n.* a curve with the shape of the graph of $y = ax^2$, where $a \ne 0$

parallel lines, *n.* lines that lie in the same plane but do not intersect, even if extended indefinitely

parallelogram, *n.* a four-sided figure in the plane with two pairs of parallel sides

percent, *n.* a fraction with (an understood) denominator of 100; for example, the fraction $\dfrac{51}{100}$ is equal to 51 percent, or 51%

perfect square, *n.* the square of an integer; for example, 9 is a perfect square because $9 = 3^2$

perimeter, *n.* the distance around the edge or boundary of a two-dimensional figure

perpendicular lines, *n.* lines that form a right angle

point-slope formula, *n.* an equation for a line in terms of its slope and one point on the line:

$$y - y_1 = m(x - x_1) \qquad \text{or} \qquad \frac{y - y_1}{x - x_1} = m$$

polygon, *n.* a closed geometric figure in the plane consisting of line segments that meet only at their endpoints; for example, triangles are polygons with three sides

polynomial, *n.* a sum of terms, each of which is either a constant or a constant times a power of a variable with the exponent a positive integer

positive number, *n.* a number greater than 0

power, *n.* an expression that consists of a base and an exponent

prime number, *n.* an integer greater than 1 whose only whole-number factors are itself and 1

principal, *n.* the original amount of money deposited in an account or borrowed from a lender; see also *interest*

principal square root, *n.* the positive square root

product, *n.* the result of a multiplication, for example, the expression $a \cdot b$ represents the product of a and b

profit, *n.* money left after subtracting the costs from the revenue

proper fraction, *n.* a fraction in the form $\dfrac{a}{b}$,   where a and b are positive integers and $a < b$

proportion, *n.* an equation in which each side is a ratio

proportional, see *directly proportional* and *inversely proportional*

pyramid, *n.* a three-dimensional object with a polygonal base and triangular faces that meet in a vertex

Pythagorean theorem: If the legs of a right triangle are a and b and the hypotenuse is c, then $a^2 + b^2 = c^2$

quadrant, *n.* one of the four regions into which the coordinate axes divide the plane: the *first quadrant* consists of the points whose two coordinates are positive; the *second quadrant* of points whose first coordinate is negative and second coordinate is positive; the *third quadrant* of points whose two coordinates are negative; and the *fourth quadrant* of points whose first coordinate is positive and second coordinate is negative

quadratic, *adj.* relating to the square of a variable or an expression

quadratic formula, *n.* the formula that gives the solutions of a quadratic equation $ax^2 + bx + c = 0$,   namely,

$$x = \frac{-b \pm \sqrt{b^2 - 4ac}}{2a}$$

quadratic polynomial, *n.* a polynomial whose degree is 2

quadratic term, *n.* a term whose degree is 2

quotient, *n.* the result of a division; for example, the expression $a \div b$ represents the quotient of a and b

radical, *n.* a square root or a cube root

radical equation, *n.* an equation in which the variable appears under a radical

radical sign, *n.* the symbol $\sqrt{}$ used to indicate the principal square root, or the symbol $\sqrt[3]{}$ used to indicate cube root

radicand, *n.* an expression under a radical sign

radius, *n.* the distance from any point on a circle or sphere to its center

raise to a power, *v.* use as a repeated factor; for example, to raise x to the power 2 is the same as multiplying $x \cdot x$

rate, *n.* a ratio that compares two quantities with different units

ratio, *n.* a comparison of two quantities that is expressed as a fraction

rational, *adj.* having to do with ratios

rational expression, *n.* a ratio of two polynomials; also called an *algebraic fraction*

rational number, *n.* a number that can be expressed as the ratio of two integers $\frac{a}{b}$, with $b \neq 0$

rationalize the denominator, *v.* to find an equivalent fraction that contains no radical in the denominator; for example, rationalizing $\frac{1}{\sqrt{2}}$ gives $\frac{\sqrt{2}}{2}$

real line, see *number line*

real number, *n.* a number that corresponds to a point on a number line

reciprocal, *n.* the result of dividing 1 by a given number; for example, the reciprocal of 2 is $\frac{1}{2}$

rectangle, *n.* a four-sided figure (in the plane) with four right angles and opposite sides that are equal in length

reduce a fraction, *v.* to find an equivalent fraction whose numerator and denominator share no common factors (other than 1)

regression line, *n.* the line used for linear regression

repeating decimal, *n.* a decimal number with a digit or block of digits that repeats itself endlessly; for example, the repeating decimal 0.16666 . . . (with repeating 6's) is equivalent to $\frac{1}{6}$, and the repeating decimal 0.8414141 . . . (with repeating 41's) is equivalent to $\frac{833}{990}$

revenue, *n.* money that an individual or group receives; for example, a person might have revenues from both a salary and investments

right angle, *n.* an angle of 90°

right triangle, *n.* a triangle that includes one right angle

root, see *cube root, principal square root,* and *square root*

round, *v.* to give an approximate value of a number by choosing the nearest number of a specified accuracy; for example, rounding 3.14159 to two decimal places gives 3.14

satisfy, *v.* to make an equation or inequality true (said of a value that is substituted for the variable); for example, the number 5 satisfies the equation $x - 2 = 3$

scale, *v.* to multiply (measurements) by a fixed number (the *scale factor*)

scatterplot, *n.* a type of graph used to represent pairs of data values, where each pair provides the coordinates for one point on the scatterplot

scientific notation, *n.* a standard method for writing very large or very small numbers using powers of 10; for example, the scientific notation for 12,000 is 1.2×10^4

semicircle, *n.* half a circle

signed number, *n.* a positive or negative number

similar, see *geometrically similar*

simplify, *v.* to write in an equivalent but simpler or more convenient form; for example, the expression $\sqrt{16}$ can be simplified to 4

slope, *n.* a measure of the steepness of a line or of the rate of change of one variable with respect to another, denoted by *m*

slope-intercept form, *n.* a standard form for the equation of a nonvertical line: $y = mx + b$

slope-intercept method, *n.* a method for graphing a line that uses the slope and the *y*-intercept

solution, *n.* a value for a variable that makes an equation or an inequality true; a solution of an equation in two variables is an ordered pair that satisfies the equation

solution of a system, *n.* an ordered pair that satisfies each equation of the system

solve, *v.* (1) to find any and all solutions of an equation, inequality, or system; (2) to write an equation for one variable in terms of any other variables; for example, solving $5x + y = 3$ for *y* gives $y = -5x + 3$

sphere, *n.* a three-dimensional object in the shape of a ball and consisting of all the points in space at a fixed distance (the radius) from the center of the sphere

square, *n.* (1) the product of any number or expression and itself; (2) a rectangle whose sides are all the same length

square, *v.* to multiply a number or expression by itself—that is, to raise to the power 2

square root, *n.* a number that when squared gives a desired value; for example, 7 is a square root of 49 because $7^2 = 49$

subscript, *n.* a small number written below and to the right of a variable to identify it; for example, in the equation $x_1 = 3$, the variable *x* has the subscript 1

substitution method, *n.* a method for solving a system of equations that begins by expressing one variable in terms of the other

sum, *n.* the result of an addition; for example, the expression $a + b$ represents the sum of *a* and *b*

surface area, *n.* the total area of the faces or surfaces of a three-dimensional object

supplementary angles, *n.* two angles whose measures add up to 180°

system of (linear) equations, *n.* a pair of (linear) equations involving the same two variables

term, *n.* a quantity that is added to another; for example, in the expression $x - 4$, both *x* and -4 are terms

trapezoid, *n.* a four-sided figure in the plane with one pair of parallel sides

triangle, *n.* a three-sided figure in the plane

trinomial, *n.* a polynomial with exactly three terms

two-point formula for slope, *n.* a formula that shows how to compute the slope of the line passing through the points (x_1, y_1) and (x_2, y_2):

$$m = \frac{y_2 - y_1}{x_2 - x_1}$$

unlike fractions, *n.* fractions whose denominators are not equivalent

unlike terms, *n.* terms with variable parts that are not equivalent

variable, *adj.* not constant, subject to change

variable, *n.* a numerical quantity that changes over time or in different situations

variation, see *direct variation* and *inverse variation*

vertex, *n.* (1) a point where two sides of a polygon meet; (2) a corner or extreme point of a geometric object; (3) the highest or lowest point on a parabola

vertical angles, *n.* as shown in the figure, the two pairs of equal angles formed when two lines intersect; angles *A* and *D* are one pair of vertical angles, and angles *B* and *C* are another pair of vertical angles

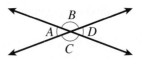

vertical axis, *n.* the vertical coordinate axis; often called the *y-axis*

vertical intercept, *n.* the point where the graph meets the vertical axis; also called the *y-intercept*

volume, *n.* a measure of the three-dimensional space enclosed by a three-dimensional object, typically expressed in cubic units, such as cubic meters or cubic feet

whole number, *n.* one of the numbers 0, 1, 2, 3, . . .

***x*-axis,** see *horizontal axis*
***x*-intercept,** see *horizontal intercept*
***xy*-plane,** see *coordinate plane*

***y*-axis,** see *vertical axis*
***y*-intercept,** see *vertical intercept*

zero factor principle: If $ab = 0$, then either $a = 0$ or $b = 0$

Answers to Odd-Numbered Problems

Section 1.1   Activity 1

Table 1.1

Airlines	Complaints	Passengers (in thousands)	Calculations and results
Alaska	53	10,084	0.005
American	497	79,511	0.006
American Eagle	35	11,900	0.003
Continental	368	35,013	0.011
Delta	504	86,909	0.006
Hawaiian Airlines	31	4776	0.006
Markair	263	990	0.266
Nations Air Express	36	82	0.439
Northwest	257	49,313	0.005
Southwest	107	50,039	0.002
Sun Jet International	142	486	0.292
TWA	291	21,551	0.014
United	597	78,664	0.008
USAir	379	55,674	0.007
ValuJet	83	5145	0.016

Source: Los Angeles Times, June 2, 1996.

1. United; Hawaiian   **3.** Nations Air Express, American Eagle had far more passengers than Nations Air Express but about the same number of complaints.   **5.** Nations Air Express

Section 1.1   Activity 2

1. 59; 72   **3.** From 1920 to 1930 the life expectancy at age 50 decreased.   **5.** Probably not

Section 1.1   Activity 3

1. Yes   **3.** 1970   **5.** More slowly. It supports that answer, because it looks like life expectancies will level off.

Section 1.1   Activity 4

1. 10

Section 1.1 Activity 5

1.

Monthly rent	$0–$200	$200–$400	$400–$600
Number of freshmen	12	9	13

Monthly rent	$600–$800	$800–$1000	$1000–$1200
Number of freshmen	3	0	3

3. $600–$800; it has the taller bar.

Homework 1.1

1a. Yes. High degrees resulted in higher salaries for all groups.
b. African-Americans have higher salaries on average than Asians in all educational levels except master's degree.

c. No high school diploma: $\dfrac{16,487}{26,115} \approx 0.63 = 63\%$

High school diploma: $\dfrac{21,121}{27,376} \approx 0.77 = 77\%$

Bachelor's degree: $\dfrac{33,817}{44,426} \approx 0.76 = 76\%$

Master's degree: $\dfrac{41,431}{52,787} \approx 0.78 = 78\%$

Doctorate: $\dfrac{46,873}{59,348} \approx 0.79 = 79\%$

Professional degree: $\dfrac{41,029}{77,877} \approx 0.53 = 53\%$

d. Anglo: $\dfrac{44,426 - 27,376}{27,376} = \dfrac{17,050}{27,376} \approx 0.623 = 62.3\%$

Latino: $\dfrac{33,817 - 21,121}{21,121} = \dfrac{12,696}{21,121} \approx 0.601 = 60.1\%$

African-American: $\dfrac{34,290 - 22,040}{22,040} = \dfrac{12,250}{22,040} \approx 0.556 = 55.6\%$

Asian: $\dfrac{33,758 - 21,608}{21,608} = \dfrac{12,150}{21,608} \approx 0.562 = 56.2\%$

3a.

Nationality	Number in United States in 1980 (in thousands)	Increase from 1980 to 1990	Increase from 1980 to 1990	Number in United States in 1990 (in thousands)
Chinese	806	104.1%	1.041 × 806 ≈ 839	806 + 839 + 1645
Filipino	775	81.6%		0.816 × 775 + 775 ≈ 1407.4
Japanese	701	20.9%		0.209 × 701 + 701 ≈ 847.5
Indian	361	125.6%		1.256 × 361 + 361 ≈ 814.4
Korean	353	126.3%		1.263 × 353 + 353 ≈ 798.8
Vietnamese	262	134.8%		1.348 × 262 + 262 ≈ 615.2

Source: Christian Science Monitor, July 27, 1993.

b.

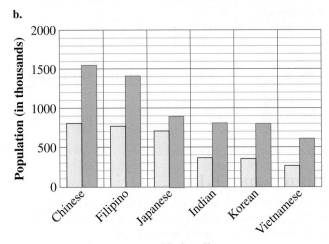

5a. 9 items   **b.** 150
c.

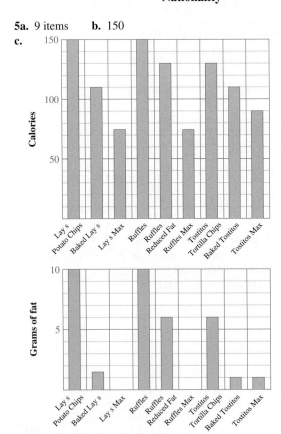

d. 10
7a.

Quiz score	0	1	2	3	4	5	6	7	8	9	10
Number of students	1	2	1	4	7	9	10	8	3	2	3

b. Mode = 6   **c.** 50 students   **d.** Median = 6
e. Total = 0 × 1 + 1 × 2 + 2 × 1 + 3 × 4 + 4 × 7 + 5 × 9 + 6 × 10 + 7 × 8 + 8 × 3 + 9 × 2 + 10 × 3 = 0 + 2 + 2 + 12 + 28 + 45 + 60 + 56 + 24 + 18 + 30 = 277
f. Mean = $\frac{277}{50}$ = 5.54

Section 1.2   Activity 1

1. 24   **3.** $1.47; $2.22; 23   **7.** The price of natural gas is highest in the winter and lowest in the summer.

Section 1.2   Activity 2

1.

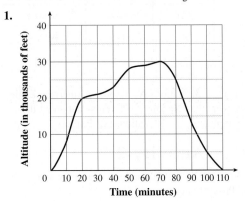

3.

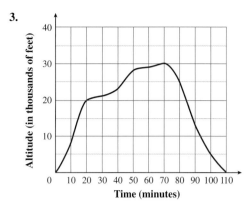

5. About 43 minutes into the flight; also at 80 minutes

Homework 1.2

1a. The year; the number of U.S. bicycle commuters in millions
d. The graph is a smooth curve because the number of bicycle commuters changes smoothly.
3a. 12 tick marks   **b.** The high is 80°F and the low is 40°F.
c.

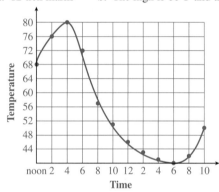

d. The graph is a smooth curve because the temperature changes smoothly.
e. The temperature is about 72°F at 1 P.M. and about 45°F at 9 A.M.
f. About 12:30 P.M. and again at about 6:15 P.M.
g. 6 P.M. to 8 P.M. had the greatest temperature change. We can see that this is the greatest change in a 2-hour period by noting that it corresponds to the steepest section of the graph—that is, the 2-hour segment with the largest vertical change.
5a. The graph is a smooth curve because the life expectancy changes smoothly.

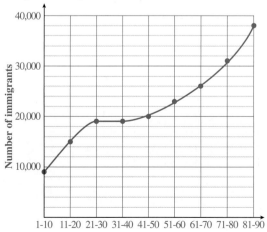

b. Life expectancy dropped from 1979 to 1980.   **c.** The increase was greatest from 1977 to 1978 and from 1980 to 1981.
d. Life expectancy remained constant from 1978 to 1979 and from 1984 to 1985. The graph is horizontal between these intervals.
7a. Number of years with account; amount in account   **b.** About $3000   **c.** About 13.7 yr   **d.** About $2550 - 2350 = \$200$
e. About $5600 - 5200 = \$400$ **9a.** Number of minutes after being poured; temperature of soup   **b.** About 3.7 min
c. About 70°F   **d.** About $180 - 125 = 55°F$
e. About 60°F
11a.

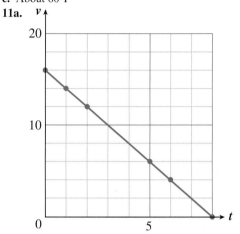

b. This is a straight line.
13a.

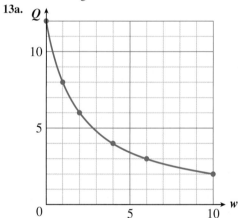

b. This is not a straight line   **15.** Graph (c) shows speed increasing for a time and then instantly going to zero.
17a. i. Graph (c)   ii. Graph (d)   iii. Graph (a)
b. Graph (b); Delbert walks toward school, but before he gets there he returns home where he stays for a time. He then walks to school and stays there.

Section 1.3   Activity 1

1. Subtract $20 from his paycheck.   **3.** $k = p - 20$

Section 1.3   Activity 2

1. Multiply Liz's hours by 6.   **3.** $w = 6 \times h$

Section 1.3   Activity 3
1. Subtract Ralph's weight from 320.   **3.** $W = 320 - R$

Section 1.3   Activity 4
1. Multiply the number of calories by 0.30.   **3.** $F = C \times 0.30$

Section 1.3   Homework 1.3
1a.

Temperature in Ridgecrest	70	75	82	86	90	R
Calculation	70 + 12	75 + 12	82 + 12	86 + 12	90 + 12	R + 12
Temperature in Sunnyvale	82	87	94	98	102	R + 12

b. Add 12 to Ridgecrest's temperature.   **c.** temperature in Ridgecrest + 12   **d.** $S = R + 12$

e.

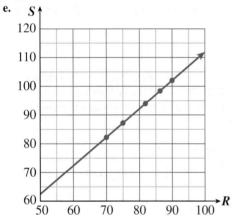

3a.

Miles driven	40	60	95	120
Calculation	200 − 40	200 − 60	200 − 95	200 − 120
Miles remaining	160	140	105	80

Miles driven	145	170	d
Calculation	200 − 145	200 −170	200 − d
Miles remaining	55	30	200 − d

b. Subtract the miles driven from 200.   **c.** 200 − miles driven
d. $r = 200 - d$

e.

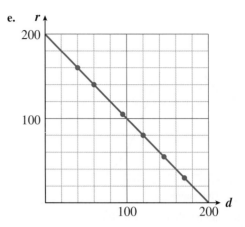

5a.

Total bill	24	30	33	45	54	81	b
Calculation	$\frac{24}{3}$	$\frac{30}{3}$	$\frac{33}{3}$	$\frac{45}{3}$	$\frac{54}{3}$	$\frac{81}{3}$	$\frac{b}{3}$
Milton's share	8	10	11	15	18	27	$\frac{b}{3}$

b. Divide the total bill by 3.   **c.** $\frac{\text{total bill}}{3}$ **d.** $s = \frac{b}{3}$

e.

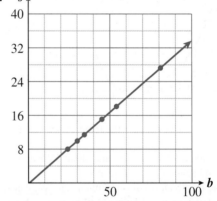

7.

m	2	3	5	10	12	16	18	m
g	5	6	8	13	15	19	21	m + 3

$m + 3$

9.

t	0	2	4	5	6	10	12	t
w	20	18	16	15	14	10	8	20 − t

$20 - t$

11.

b	0	2	4	5	6	8	9	b
x	0	1	2	2.5	3	4	4.5	$\frac{b}{2}$

$\frac{b}{2}$

13.

z	3	6	8	12	15	18	20	z
r	2	4	$\dfrac{16}{3}$	8	10	12	$\dfrac{40}{3}$	$\dfrac{2}{3}z$

$\dfrac{2}{3}z$

15.

n	0	5	10	15	20
W	0	6	12	18	24

17.

x	0	2	4	6	8
M	0	3	6	9	12

19a.

x	0	10	30	40	60	70
y	70	60	40	30	10	0

b. $y = 70 - x$

Homework 1.4

1. sum of, increased by, more than, exceeded by, total **3.** times, twice, (fraction) of, product of **5.** $4y$ **7.** $2b$ **9.** $1.15g$

11. $t - 5$ **13.** $\dfrac{7}{w}$ **15.** Cost of light bulb: b; $3b$

17. Savings account balance: s; $\dfrac{3}{5}s$ **19.** Price of a pizza: p; $\dfrac{p}{6}$

21. Weight of copper: w; $\dfrac{w}{16}$ **23.** Rebate: r; Sale price: p; $p - r$

25. Base: b; Height: h; bh
27. Height of the tree: t; Heights of the roof: r; $t - r$
29a. $35

b.

w	12	16	20	30	36	40
p	15	20	25	37.50	45	50

31a. $d = 180t$ **b.** 360 mi; 630 mi; 2160 mi
33a. $P = 6000 - C$ **b.** \$5200; \$5000; \$3500

35a. $A = \dfrac{540}{n}$ **b.** 27; 21.6; 18 **37.** $2x$ **39.** $4b$

41. $2v$ **43.** (Algebraic) expression **45.** product; factors
47. difference **49.** To <u>evaluate</u> an expression is to substitute a specific value for each of the variables.
51.

x	0	5	15	20	25	30
y	15	20	30	35	40	45

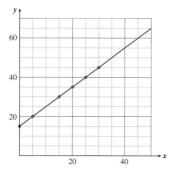

b. $y = x + 15$ **c.** For $x = 40$: $y = 55$; Yes; For $x = 50$: $y = 65$
53a.

x	0	500	1000	2000	2500	3000
y	0	15	30	60	75	90

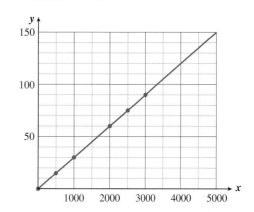

b. $y = 0.03x$; **c.** For $x = 4500$: $y = 135$; Yes; For $x = 5000$: $y = 150$

Homework 1.5

1. graph **3.** horizontal **5.** ordered pair **7.** solution
9. To graph an equation: *Step 1* Make a table of values. *Step 2* Choose scales for the axes. *Step 3* Plot the points and connect them with a smooth curve.
11a. $D = T - 5$

b.

T	5	10	15	20	50
D	0	5	10	15	45

(other answers possible)

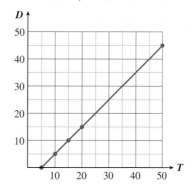

13a. $r = \dfrac{w}{2}$

b.

w	1	2	4	6	10
r	0.5	1	2	3	5

(other answers possible)

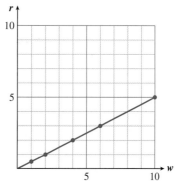

15a. $d = 1000 - l$

b.

l	0	300	500	800	1000
d	1000	700	500	200	0

(other answers possible)

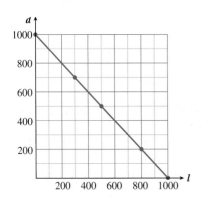

17a. $S = \dfrac{3600}{m}$

b.

m	10	20	30	40	100
S	360	180	120	90	36

(other answers possible)

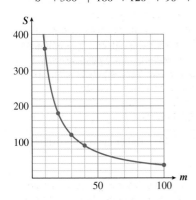

19a. yes   **b.** no   **c.** no   **d.** yes   **21a.** no   **b.** no
c. yes   **d.** yes   **23a.** No   **b.** Yes   **c.** No   **d.** No
25a. Yes   **b.** Yes   **c.** No   **d.** No

27.

x	0	2	5	6	10	12	16
y	16	14	11	10	6	4	0

a. The first variable is x, and the second variable is y.
b. (other last two columns possible)   **c.** $y = 16 - x$

29.

h	3	6	7.5	9	10.5	12	15
d	1	2	2.5	3	3.5	4	5

a. The first variable is h, and the second variable is d.
b. (other last three columns possible)   **c.** $d = \dfrac{h}{3}$

31. Equation (f)

x	0	5	10	15
y	15	20	25	30

33. Equations (c) and (e)

x	0	1	2	3
y	0	0.2	0.4	0.6

35a.–c.

x	y
4	0.5
8	1
10	1.25
16	2

All the tables are the same because all the equations are equivalent.

Midchapter Review

1. The heights of the bars represent the frequencies of each range.
3. An algebraic expression is any meaningful combination of numbers, variables, and operation symbols. An equation is a statement that two expressions are equal.   **5a.** $w - p$
b. $56 - 5.6 = 50.4$ **7a.** rt **b.** $R - C$ **c.** Prt
d. rW **e.** $\dfrac{S}{n}$ **9.** The first, or independent, variable is displayed on the horizontal axis; the second, or dependent, variable is displayed on the vertical axis.
11a. Great Britain; Israel   **b.** Sweden; Israel
c.

Country	Total number of seats	Number of seats held by women	Percent of seats held by women
Canada	295	63	$\dfrac{63}{295} \approx 0.214 = 21.4\%$
France	577	63	$\dfrac{63}{577} \approx 0.109 = 10.9\%$
Great Britain	659	120	$\dfrac{120}{659} \approx 0.182 = 18.2\%$
Israel	120	9	$\dfrac{9}{120} = 0.075 = 7.5\%$
Japan	500	23	$\dfrac{23}{500} = 0.046 = 4.6\%$
Mexico	500	71	$\dfrac{71}{500} = 0.142 = 14.2\%$
South Africa	400	100	$\dfrac{100}{400} = 0.250 = 25.0\%$
Sweden	349	141	$\dfrac{141}{349} \approx 0.404 = 40.4\%$
United States	435	51	$\dfrac{51}{435} \approx 0.117 = 11.7\%$

Source: Time, June 16, 1997.

d. Sweden; Japan   **13a.** 9 states   **b.** 13%
c. Mode = 14%; 10 states   **d.** 3 states   **e.** 27 states
f. Median = 14%; Mean = 13.5%   **15a.** 13,000 yr
b. 0.3, or 30%   **c.** About 5700 yr
17a. About $15,000; about $11,000
b. 1986 to 1987; 1980 to 1981   **c.** $3700; $9000; $12,500

d.

Year	Average earnings		Percent
	High school graduates	College graduates	
1970	9567	13,264	72.1%
1975	13,542	17,477	77.5%
1980	19,469	24,311	80.1%
1985	23,853	32,822	72.5%
1990	26,653	39,238	67.9%

Source: U.S. Bureau of the Census, Digest of Educational Statistics, 1994.

19. $t + 6$ **21.** $0.08t$ **23a.** $d = 3r$ **b.** 18 mi; 60 mi
25a. $n = 2s$

b.

s	5	10	20	50
n	10	20	40	100

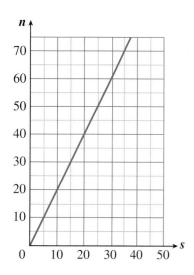

Homework 1.6

1.

t	q
2	11
4	13
6	15
9	18
21	30
30	39

To find a value for q, add 9 to the value of t. To find a value for t, subtract 9 from the value of q.

3.

n	p
0	0
2	10
4	20
5	25
7	35
11	55

To find a value for p, multiply the value of n by 5. To find a value for n, divide the value of p by 5.
5. It is a solution.   **7.** It is not a solution.   **9.** It is a solution.
11. $x = 14$ **13.** $y = 2.8$ **15.** $y = 36$ **17.** $b = 12$
19. $a = 3.9$ **21.** $x = 4$ **23.** $x = \dfrac{106}{17}$ **25.** $z = \dfrac{10}{3}$
27. $k = 0$ **29.** $75 = 0.03P$; $2500 = P$; Clive loaned his
brother \$2500.   **31.** $38.25 = \dfrac{S}{8}$; $306 = S$; Andy had 306
total points   **33.** $234 = 13 \cdot t$; $18 = t$; It takes 18 hr.
35. $400 = l \cdot 16$; $25 = l$; The roll is 25 ft long.   **37.** An equation is a statement where the two quantities are equal.
39. A given number is a solution if, when the given number is substituted for the variable, the equation becomes a true statement.
41. A *trial-and-error method* for solving an equation involves substituting different values for the variable. When a value gives a true statement, that value is a solution to the equation.
43. $x + 80 + 35 = 180°$; $x = 65°$ **45.** $x + 90 + 37 = 180°$; $x = 53°$ **47.** $x + 135 = 180°$; $x = 45°$
49. $x + 18 = 180°$; $x = 162°$

Homework 1.7

1. Evaluating an expression involves substituting a value for the variable and performing the indicated operations. In $B = 1.08P$, if $P = 100$, we can substitute 100 for P and evaluate the expression $1.08(100)$ to find the value of B. Solving an equation involves finding the value(s) for a variable to make the equation true. In $B = 1.08P$, if $B = 216$, we can solve the equation $216 = 1.08P$ to find the value of P.   **3.** *Step 2* Find a quantity that can be described in two different ways, and write an equation using the variable to model the problem; *Step 3* Solve the equation and answer the question.   **5a.** Price of used car: u; Price of new car: n; $u = n - 3400$ **b.** A used car costs \$11,100.
c. A new car costs \$12,600.   **7a.** Number of games won: w; Number of games played: p; $w = 0.60p$ **b.** They won 72 games.   **c.** They played 160 games.   **9a.** Profit: P; Selling price: S; $P = 0.18S$ **b.** The profit is \$10.80.   **c.** The selling price is \$40.   **11.** $89 = p - 26$; Her mother paid \$115.
13. $360 = 0.40I$; She makes \$900 per month.
15a. $200 = r \cdot 19.32$; About 10.35 meters per second
b. $100 = r \cdot 9.84$; About 10.16 meters per second   **c.** Johnson
was faster.   **17.** $x + 7 = 26$ **19.** $\dfrac{x}{7} = 26$ **21.** $\dfrac{x}{26} = 7$
23a. $p = 110$ **b.** $p = 60$ **25a.** $g \approx 13.9$ **b.** $g \approx 11.5$
27a. $(50, 4)$ **b.** When the price is \$50, the tax is \$4.
29a. $(4, 16)$ **b.** When the height is 4 inches, the weight is
16 ounces.   **31.** 3   **33.** 38 cm.   **35.** The bus made six
stops.

Homework 1.8

1. If a, b, and c are any numbers, then $(a + b) + c = a + (b + c)$ and $(a \cdot b) \cdot c = a \cdot (b \cdot c)$ **3.** Parentheses and brackets are used as grouping symbols to show which part of an expression to simplify first.   **5.** False. Multiplication and division are done left to right.   **7a.** $(20 - 2) \cdot 8 + 1$ **b.** $20 - 2 \cdot (8 + 1)$

c. $20 - 2 \cdot 8 + 1$ (no change needed)   **9.** $\dfrac{(5 + 7) \cdot 4}{10 - 8}$

11. First multiply 32 times 12, and divide that result by 4. Subtract that amount from 825, and then add 2.

13. Incorrect to subtract 9 minus 3.

$(5 + 4) - 3(8 - 3 \cdot 2)$	Add inside parentheses.
$= 9 - 3(8 - 3 \cdot 2)$	Multiply inside parentheses.
$= 9 - 3(8 - 6)$	Subtract inside parentheses.
$= 9 - 3(2)$	Multiply before subtracting.
$= 9 - 6$	
$= 3$	

15. 14   **17.** 3   **19.** 72   **21.** 18   **23.** $\dfrac{3}{2}$ **25.** 23

27. 0   **29.** 9   **31.** 4   **33.** 9   **35.** 3   **37.** 0.859

39. 42.705   **41.** 2204.533

43.

x	2	2.5	3	3.5	4	4.5	5
y	1.9	3.15	4.4	3.65	3.9		

$x = 2.8$

45a. 18   **b.** 50   **47a.** 3   **b.** 18   **49a.** 2   **b.** 8
51a. 42   **b.** 12   **53a.** 2   **b.** 18
55. $(2.3 + 5.7)6 - (1.2 + 3.3)2 = 39$ sq cm

Homework 1.9

1a. $4400; $2600   **b.** Multiply the number of weeks by 200; then subtract that from 5000.

c.

Number of weeks	2	4	5
Calculation	5000 − 200(2)	5000 − 200(4)	5000 − 200(5)
Savings left	4600	4200	4000

Number of weeks6	6	10	15	20
Calculation	5000 − 200(6)	5000 − 200(10)	5000 − 200(15)	5000 − 200(20)
Savings left	3800	3000	2000	1000

d. $5000 - 200w$ **e.** $S = 5000 - 200w$ **3a.** $720; $960
b. Subtract 2000 from her income; then multiply the result by 0.12.

c.

Income	Calculation	State tax
5000	0.12 (**5000** − 2000)	**360**
7000	0.12(7000 − 2000)	600
12,000	0.12(12,000 − 2000)	1200
15,000	0.12(15,000 − 2000)	1560
20,000	0.12(20,000 − 2000)	2160
24,000	0.12(24,000 − 2000)	2640
30,000	0.12(30,000 − 2000)	3360

d. $0.12(I - 2000)$ **e.** $T = 0.12(I - 2000)$

5.

z	$5z$	$5z - 3$
2	10	7
4	20	17
5	25	22

7.

Q	$12 + Q$	$2(12 + Q)$
0	12	24
4	16	32
8	20	40

9a. $2w - 3$ **b.** 23 in.   **11a.** $20 + 0.40P$ **b.** $220
13a. $\dfrac{1}{3}(c + t)$ **b.** 4 vehicles   **15a.** $\dfrac{1}{2}w + \dfrac{2}{3}m$
b. 17   **17a.** $0.80(v - m)$ **b.** 20   **19.** 26   **21.** 53
23. $\dfrac{7}{12}$ **25.** $\dfrac{19}{8}$ **27.** 7.5   **29.** 29.8 m   **31.** 24 sq cm
33. $5400   **35.** 37°C

37.

d	0	2	5	10	15
P	100	88	70	40	10

$P = 100 - 6d$

39.

U	20	40	80	100	200
M	204	208	216	220	240

$M = 200 + \dfrac{U}{5}$

41.

S	70	100	112	130	160
D	5	10	12	15	20

$D = S - \dfrac{40}{6}$

43. $\dfrac{n}{n+4}$

n	0	1	2	3	4	5
$\dfrac{n}{n+4}$	0	$\dfrac{1}{5}$	$\dfrac{1}{3}$	$\dfrac{3}{7}$	$\dfrac{1}{2}$	$\dfrac{5}{9}$

45. $(a+5)(a-5)$

a	5	6	7	8	9	10
$(a+5)(a-5)$	0	11	24	39	56	75

47. $W = I + \dfrac{1}{2}(A - 25)$ **49.** $T = 0.08(I - D) + 300$

51. $10 + 2x$ **53.** $x + 6$

Homework 1.10

1.

x	$2x$	$2x+4$
3	6	10
6	12	16
5	10	14
8	16	20

3.

q	$q-3$	$5(q-3)$
3	0	0
5	2	10
4	1	5
7	4	20

5a. Multiply the value of x by 2 (to get 6); then add 4 to the result (to get 10).   **b.** Subtract 4 from the value of y (to get 10); then divide the result by 2 (to get 5).   **7a.** Subtract 3 from the value of q (to get 0); then multiply the result by 5 (to get 0).
b. Divide the value of R by 5 (to get 2); then add 3 to the result (to get 5).
9. Add 6 to 29 (to get 35); then divide 35 by 5 (to get 7). The number is 7.   **11.** $x = 3$ **13.** $a = 20$ **15.** $x = 4$
17. $x = 0$ **19.** $p = 13$ **21.** $z = 0$ **23.** $k = 41$
25. $x = 18$ **27.** $b = 6$ **29.** $w = 13.4$ **31a.** \$540
b. \$30   **c.** \$360   **d.** 18 payments   **e.** $B = 540 - 30m$
f. $0 = 540 - 30m; m = 18$
33a. About 8.5 lb   **b.** About 37 wk   **c.** $W = 3.8 + 0.6t$
d. $26 = 3.8 + 0.6t; 37 = t$
35a. $s = 20 + 4x$
b.

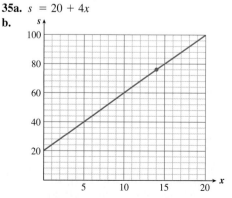

c. $76 = 20 + 4x; x = 14$

37a. $P = 600 + 3r$

b.

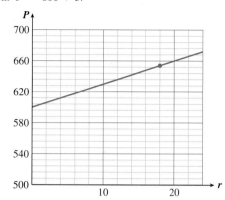

c. $654 + 600 + 3r; r = 18$
39. $1200 + 36a = 10{,}200; a = \$250$ **41.** $5(d + 0.5) = 22.5$;
$d = 4$ mi   **43.** $0.06p - 38 = 112; p = 250$ seeds
45. 175 yd $= l$ **47.** 6 m $= h$ **49.** $x = 7$ **51.** $x = 17$

Chapter 1 Summary and Review

1. The <u>commutative law of addition</u> says that two numbers may be added in either order to give the same answer. If a and b are any numbers, then $a + b = b + a$. The <u>commutative law of multiplication</u> says that two numbers may be multiplied in either order to give the same answer. If a and b are any numbers, then $a \cdot b = b \cdot a$

3.
$$d = rt \begin{cases} \text{Distance: } d \\ \text{Rate: } r \\ \text{Time: } t \end{cases} \qquad P = rW \begin{cases} \text{Part: } P \\ \text{Percentage rate: } r \\ \text{Whole: } W \end{cases}$$

$$P = R - C \begin{cases} \text{Profit: } P \\ \text{Revenue: } R \\ \text{Cost: } C \end{cases} \qquad A = \dfrac{S}{n} \begin{cases} \text{Average: } A \\ \text{Sum of scores: } S \\ \text{Number of scores: } n \end{cases}$$

$$I = Prt \begin{cases} \text{Interest: } I \\ \text{Principal: } P \\ \text{Interest rate: } r \\ \text{Time: } t \end{cases} \qquad A = lw \begin{cases} \text{Area: } A \\ \text{Length: } l \\ \text{Width: } w \end{cases}$$

5. To write an algebraic expression:
Step 1 Identify the unknown quantity and write a short phrase to describe it.
Step 2 Choose a variable to represent the unknown quantity.
Step 3 Use mathematical symbols to represent the relationship described.

7. To solve an equation algebraically:
Step 1 Ask yourself what operation has been performed on the variable.
Step 2 Perform the opposite operation on both sides of the equation to isolate the variable.
If two or more operations have been performed on the variable, perform the opposite operations in reverse order to isolate the variable.

9a. The year   **b.** Percent of 18- to 21-year-olds that registered and that voted   **c.** 47%; 38%   **d.** 1980   **e.** Participation is declining.

11a.

Year	Number of degrees per 100 students
1960	13
1965	18
1970	22
1975	23
1980	22
1985	18
1990	22
1994	26

b. 1975–1985   **c.** 6500 degrees

13.

Gallons of gas	3	5	8	10	11	12
Calculation	22 × 3	22 × 5	22 × 8	22 × 10	22 × 11	22 × 12
Miles	66	110	176	220	242	264

a. Multiply 22 by the number of gallons.
b. 22 × gallons   **c.** $m = 22g$
d.

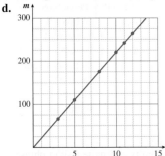

15.

x	0.5	1.0	1.5	2.0	4.0	6.0	7.5
y	0.125	0.25	0.375	0.5	1.0	1.5	1.875

$y = \dfrac{x}{4}$

17. $z + 5$ **19.** $f - 60$ **21a.** $d = 88t$ **b.** 440 ft; 44 ft;
2640 ft   **23a.** $n = b - 60$
b.

b	100	120	150	180	200
n	40	60	90	120	140

25. $y = \dfrac{5}{2}x$ **a.** yes   **b.** no   **c.** yes   **d.** yes

27a. Yes   **b.** No   **29.** $y = 9$ **31.** $a = 0.4$
33. $w = 0$ **35.** $171 = 0.095P$; She deposited $1800.
37. $106{,}000 = 0.53 \cdot W$; 200,000 people voted.

39. $\dfrac{w}{85} = 0.7$; Puppy should weigh 59.5 lb   **41a.** $x = 32$

b. $x = 24$ **43.** 3   **45.** 6   **47.** $\dfrac{1}{2}d + 4$
49. $3(l + 5.6)$ **51.** 4   **53.** 5   **55.** $1300
57.

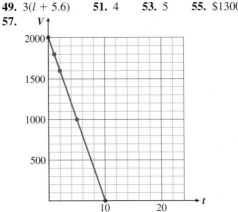

$V = 2000 - 200t$
59. $x = \dfrac{5}{3}$ **61.** $v = 8$ **63.** $5.75 + 4.50b = 19.25$;
$b = 3$ bushels

Homework 2.1

1.

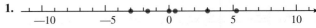

3. 6   **5.** −9   **7.** 8.5   **9.** 18   **11.** 0   **13.** $0 > -4$
15. $-5 > -9$ **17.** $13.6 < 13.66$ **19.** $-2 < |-2|$
21. $-|-4.1| > -5.8$ **23a.** −14   **b.** 8   **c.** 21
d. −17   **25a.** 9   **b.** −4   **c.** −13   **d.** 15
27a. p is positive means $p > 0$ **b.** n is negative means $n < 0$
29. The opposite of a number can be positive or negative. For
example, the opposite side of 3 is −3, but the opposite of −3 is 3.
The absolute value of a number is never negative. For example,
$|-3| = 3$ and also $|3| = 3$.   **31a.** $-2.9, -2, -1$ (other answers
possible)
b.

33a. 3.9, 3, 2 (other answers possible)
b.

35. Note: If $-2 > x$, then $x < -2$.
a. $-2.1, -3, -4$ (other answers possible)
b.

37. 2   **39.** −8   **41.** 6   **43.** −25   **45.** 4.1
47. $-\dfrac{1}{6}$ **49.** −3   **51.** −22   **53.** −11   **55.** −20
57. $500   **59.** 40 ft   **61.** down $3\dfrac{1}{2}$ **63.** up $1\dfrac{3}{4}$
65a. 3%; −3%   **b.** Kellogg's Corn Flakes and Crispix
c. Lucky Charms and Frosted Flakes; −21%
d. Their original prices were different.

Homework 2.2

1. −4   **3.** 12   **5.** −2   **7.** −11   **9.** 6   **11.** 10
13. −10   **15.** 2   **17.** −11   **19.** −9   **21.** 0
23. −8   **25.** −6°   **27.** 14,776 ft   **29.** −$35.40

31a. 10   **b.** 10   **c.** 10   **33a.** −8   **b.** −8   **c.** −8
35. 2   **37.** 12   **39.** 20   **41.** 2   **43.** 17   **45.** 1
47. −18   **49.** 2   **51.** −15   **53.** −15   **55.** −5
57. −9   **59.** 5   **61.** 9   **63.** 5   **65.** 13   **67.** −12
69. $x = 3.01; x = 4; x = 5$ **71.** $x = 1; x = 0; x = -1$
73. $x = -3.1; x = -4; x = -5$ **75a.** 7.4

b.

Score, x	Deviation from the mean, $x - \bar{x}$
4	$4 - \bar{x} = -3.4$
5	−2.4
6	−1.4
7	−0.4
7	−0.4
8	0.6
8	0.6
9	1.6
10	2.6
10	2.6

c. 0; The total amount by which some of the scores exceed the mean is balanced by the total amount by which the other scores are less than the mean.   **77a.** −9%   **b.** First quarter of 1997
c. +6%   **d.** −8%   **e.** +7%

Homework 2.3

1. 32   **3.** −3   **5.** 4   **7.** 18   **9.** Undefined   **11.** 0
13. −48 m   **15.** −28.75°   **17.** −$1000   **19a.** −24
b. 24   **c.** −24   **d.** 48   **e.** −48   **f.** 48   **21a.** −16
b. 48   **c.** −8   **d.** −16   **23a.** (a) and (d)   **b.** (b)
25a. −6   **b.** 9   **c.** 0   **d.** −6   **e.** 1
27a. 3 and −3 because $3 + (-3) = 0$ (other answers possible)
b. 3 and 3 because $3 - 3 = 0$ (other answers possible)
c. 3 and 0 because $3(0) = 0$ (other others possible)
d. 0 and 3 because $0 \div 3 = 0$ (other answers possible)
29. 2; Multiply; Subtract   **31.** −2; Multiply before subtracting
33. 7; Simplify inside parentheses; Multiply   **35.** −22;
Multiply first; Add and subtract in order from left to right
37. −6; Simplify above and below fraction bar; Divide
39. $\dfrac{-13}{5}$; Multiply above and below fraction bar; Simplify above
and below fraction bar; Reduce   **41a.** −60; Multiply
b. 7; Multiply before subtracting   **c.** 27; Add inside parentheses; Multiply   **d.** 6; Add inside parentheses
e. −23; Multiply before subtracting   **f.** 35; Add inside parentheses; Multiply   **43a.** −0.4   **b.** No   **45.** −3.75; All the
same   **47.** Not the same   **49.** −1   **51.** −18   **53.** $\dfrac{1}{13}$
55. −25   **57.** $-\dfrac{1}{9}$ or $\dfrac{-1}{9}$ **59.** 1.75   **61.** $x = 0; x = 1;$
$x = 1.5$; (other answers possible)   **63.** $x = -4.1; x = -5;$
$x = -6$; (other answers possible)   **65.** $x = 2.1; x = 3; x = 4;$
(other answers possible)

Section 2.4   Activity 1

a.

h	T
−3	−18
−2	−16
−1	−14
0	−12
1	−10
2	−8
3	−6

b. $T = -12 + 2h$

c.

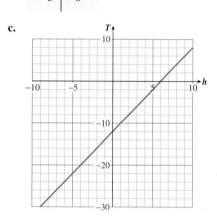

d. −28°   **e.** 4 P.M., 8 P.M.   **f.** 8°F

Section 2.4   Activity 2

a.

x	y
−3	12
−1	8
0	6
2	2
4	−2

c.

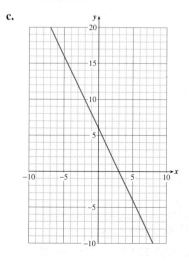

Homework 2.4

1a.

d	−15	−5	0	5	15	20	25
B	3500	2500	2000	1500	500	0	−500

b. $B = 2000 - 100d$

c.

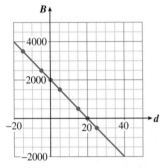

d. $3500   **e.** 15 days from now
f. $2500 - (-2000) = 2500 + 2000 = \$4500$

3a.

F	−5	−2	0	1	2	4	6
E	−45	−18	0	9	18	36	54

b. $E = 9F$

c.

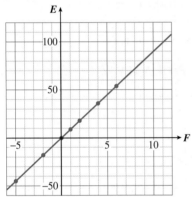

d. −27 ft   **e.** Floor 8   **f.** $90 - (-45) = 90 + 45 = 135$ ft

5.

x	−4	−2	0	2	4
y	−1	1	3	5	7

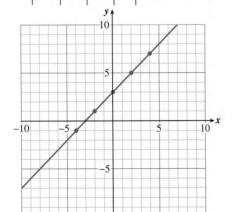

7.

x	−2	−1	0	1	2
y	−3	−1	1	3	5

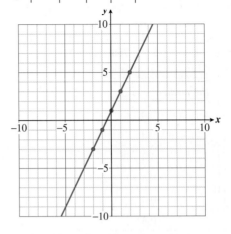

9.

x	−10	−2	0	2	10
y	0	−4	−5	−6	−10

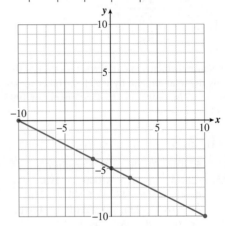

11.

x	−4	−2	0	4	8
y	−9	−6.5	−4	1	6

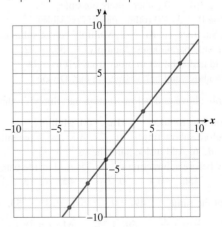

13.

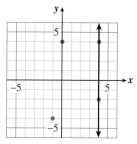

a. No   **b.** Yes   **c.** No   **d.** Yes

15.

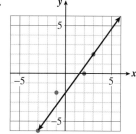

a. Yes   **b.** Yes   **c.** No   **d.** No   **17.** Graph (d)
19. Graph (b)

21.

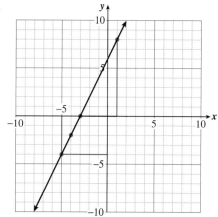

a. When $x = -5$, $y = -4$.   **b.** When $y = -4$, $x = -5$.
c. $y = 2x + 6$; For $(-5, -4)$; $-4 \stackrel{?}{=} 2(-5) + 6$; $-4 \stackrel{?}{=} -10 + 6$;
$-4 = -4$ True   **d.** When $y = 8$, $x = 1$.   **e.** Any point
above $(-5, -4)$ on the graph will have $y > -4$. Therefore,
$x = -4.9$ or -4. (other answers possible)

23.

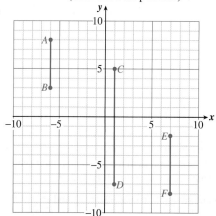

a. Distance = 5   **b.** Distance = 12   **c.** Distance = 6
25a. $|8 - 3| = |5| = 5$

b. $|5 - (-7)| = |5 + 7| = |12| = 12$
c. $|-2 - (-8)| = |-2 + 8| = |6| = 6$ **27a.** On a vertical line
the x-coordinates are the same.   **b.** To find the distance between two points on a vertical line, subtract the y-coordinates and
take the absolute value of the result.

Homework 2.5

1. $x = 5$ **3.** $z = \dfrac{-4}{3}$ **5.** $a = -32$ **7.** $x = 6$
9. $c = -2$ **11.** $t = 1$ **13.** $b = 18$ **15.** $y = -10$
17. $x = -4$ **19.** $x = 3$ **21.** $x = 6$ **23.** $x = -3$
25.

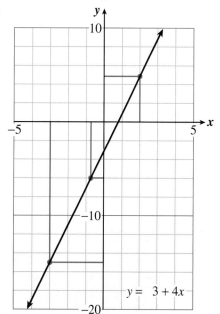

a. $x = -3$ **b.** $x = \dfrac{-1}{2}$ **c.** $x = 3$

27.

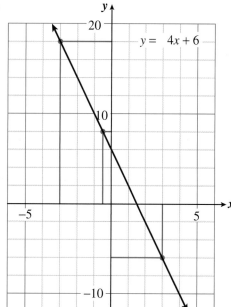

a. $x = 2$ **b.** $x = \dfrac{-3}{4}$ **c.** $x = -3$

29. $t = 1.5$

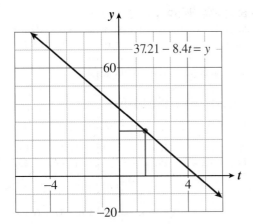

$37.21 - 8.4t = y$

31. $x = -2.5$

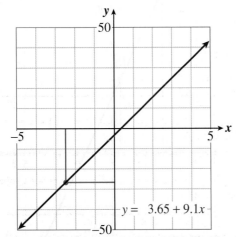

$y = 3.65 + 9.1x$

33a. $s = 100 - 5x$ **b.**
c. $65 = 100 - 5x$
   7 wrong answers

x	2	5	6	12
s	90	75	70	40

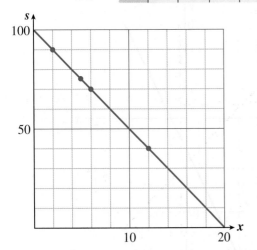

35a. $D = 10 - \dfrac{1}{4}w$ **b.**

c. $3.5 = 10 - \dfrac{1}{4}w$
   It will take 26 weeks.

w	2	8	10	28
D	9.5	8	7.5	3

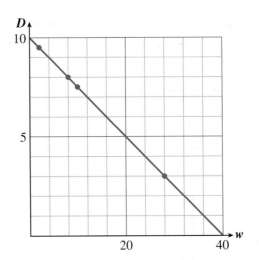

37. 8°F   **39.** 8.5 wk   **41.** $645,000   **43.** $135

45.

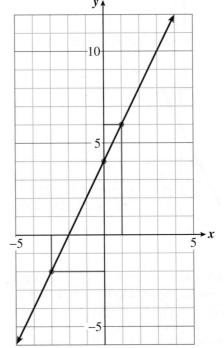

a. When $y = 6$, $x = 1$.   **b.** When $y = 4$, $x = 0$. Because $mx + k < 4$, any x-values below $(0, 4)$ on the graph will satisfy the inequality. Therefore, $x = -0.1$, -1, or -2. (other answers possible)   **c.** When $y = -2$, $x = -3$. Because $mx + k > -2$, any x-values above $(-3, -2)$ on the graph will satisfy the inequality. Therefore, $x = -2.9$, -2, or -1. (other answers possible)

Midchapter Review

1. The <u>natural numbers</u> are the counting numbers: 1, 2, 3, 4, . . . ; The <u>whole numbers</u> are the natural numbers and zero: 0, 1, 2, 3, 4, . . .; The <u>integers</u> are the natural numbers, zero, and the negatives of the natural numbers: . . . , -3, -2, -1, 0, 1, 2, 3, ...

3a. Multiplication and division: $(-6)(-2) = 12$ and $\dfrac{-6}{-2} = 3$

b. Addition: $-6 + (-2) = -8$ **c.** Subtraction:
$-6 - (-2) = -6 + 2 = -4$ and $-2 - (-6) = -2 + 6 = 4$

5a. 14   **b.** −2   **c.** 8   **d.** 4   **7.** −90   **9.** 16

11. −2   **13.** $\dfrac{-3}{5}$ **15.** 60   **17.** −10.48   **19.** −$28.26

21. $x = -9.1$; $x = -10$; $x = -11$ (other answers possible)

23a.

d	−3	0	4	8	12
W	−28	−22	−14	−6	2

b. $W = 2d - 22$

c.

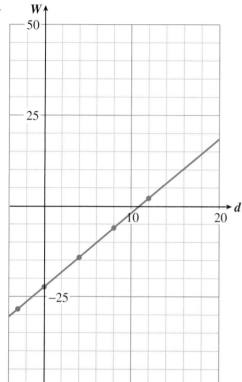

d. −8 ft   **e.** In 11 days   **f.** On the 6th day, the water level is −10 ft; On the 14th day, the water level is 6 ft; The change = 6 − (−10) = 6 + 10 = 16 ft   **25.** 26

27.

x	0	1	2	3	4
y	8	6	4	2	0

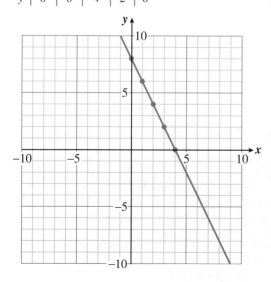

29. $c = 2$ **31.** $y = 18$

33a.

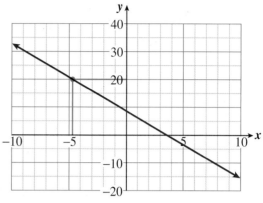

$x \approx -4.7$ **b.** $x \approx -4.7$

35. $-5 + 7t = 16$;   3°F

Homework 2.6

1a. $x \le -15$

b.

c. $x = -15.1$ is a solution, and $x = -14.9$ is not a solution.

3a. $y > -5$

b.

c. $y = -4.9$ is a solution, and $y = -5$ is not a solution.

5a. $x \le 12$

b.

c. $x = 12$ is a solution, and $x = 12.1$ is not a solution.

7a. $t \le 7.5$

b.

c. $t = 7$ is a solution, and $t = 8$ is not a solution.

9a. $x \le -4$

b.

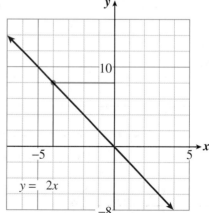

c. $x \le -4$

11a. $x \leq -3$

b.

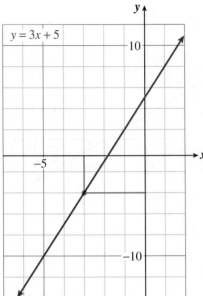

$y = 3x + 5$

c. $x \leq -3$

13a. $x < 9$

b.

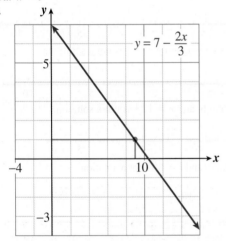

$y = 7 - \dfrac{2x}{3}$

c. $x < 9$

15. $x > 2$

17. $x \geq -3$

19. $x < -6$

21. $T = 56 - 4d$

a. On day 6

b. On day 18 and after

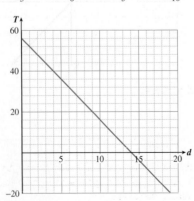

23. $h = -200 + 15m$

a. For the first 12 minutes

b. After $13\frac{1}{3}$ minutes

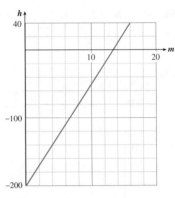

25. $-1 \leq x \leq 4$

27. $-7 < b \leq -2$

29. $-7 \leq w < -3$

31. x **33.** y, n, p, and $53°$ **35.** $n = 53°$ **37.** $z = 127°$

39. $m = 127°$

Homework 2.7

1a. $(0, 2)$; $(4, 0)$

b.

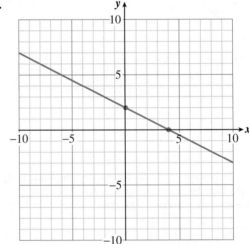

3a. $(0, -5)$; $(-10, 0)$

b.

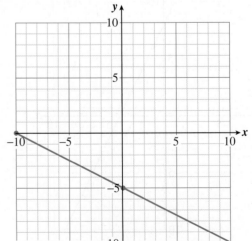

5a. $(7, 0)$; $(0, -2)$

b.

7a. $(0, 8)$; $(2, 0)$

b.

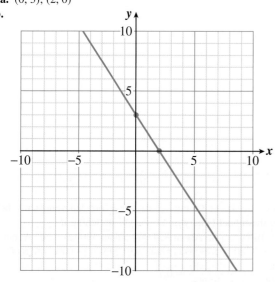

9a. $(0, 3)$; $(2, 0)$

b.

11a. $(0, -60)$; $(40, 0)$

b.

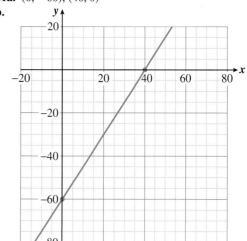

13a. 360 mi   **b.** 24 hr   **15a.** $(0, 200)$; $\left(13\frac{1}{3}, 0\right)$

b.

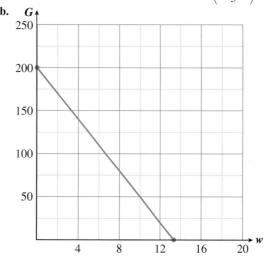

c. The G-intercept at $G = 200$ shows that there were 200 gallons in the tank when they turned on the furnace. The w-intercept at $w = 13\frac{1}{3}$ shows that the fuel will run out after $13\frac{1}{3}$ weeks.

17a. $(0, 225)$; $(-9, 0)$

b.

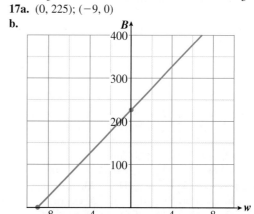

c. The B-intercept at $B = 225$ shows that Dana has \$225 this week. The w-intercept at $w = -9$ shows that she had a zero balance 9 weeks ago.

19a. $(0, -600)$; $(15, 0)$
b.

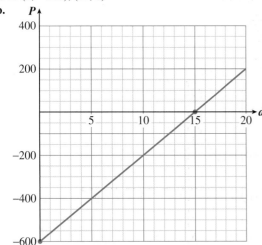

c. The P-intercept at $P = -600$ shows that he spent $600 on equipment. The d-intercept at $d = 15$ shows that he must groom 15 dogs to break even.
21. Graph (d)   **23.** Graph (c)   **25.** Graph (a)
27a. $(12.5, 0)$ **b.** $(0, -5)$ **29.** Increasing
31a. $x - -3$ **b.** $x - \dfrac{-k}{2}$ **33a.** $x = 3$ **b.** $x = \dfrac{k - 15}{-4}$
35a. $x = \dfrac{-10}{3}$ **b.** $x = \dfrac{-10}{k}$ **37a.** $61°$ **b.** $56°$
c. $63°$ **d.** $61°$ **39a.** $50°$ **b.** $35°$ **c.** $35°$ **d.** $50°$
e. $145°$ **41a.** $41°$ **b.** $43°$

Homework 2.8

1. For $x = 0$: $2 + 7x = 2 + 7(0) = 2 + 0 = 2$; $9x = 9(0) = 0$
3. For $a = 0$: $-(a - 3) = -(0 - 3) = -(-3) = 3$; $-a - 3 = -0 - 3 = 0 - 3 = -3$ **5.** For $x = 0$: $5(x + 3) = 5(0 + 3) = 5(3) = 15$; $5x + 3 = 5(0) + 3 = 0 + 3 = 3$ **7.** $-4x$
9. $-12.8a$ **11.** t **13.** $7ab$ **15.** $12y - 4$
17. $-8st + 9s - 2$ **19a.** -21 **b.** $-6y - 6$ **c.** -21
21. $9x - 2$ **23.** $6y - 10$ **25.** $2a$ **27a.** $3x$
b. $x + 3x = 4x$ **29a.** $10t - 0.2t = 9.8t$ **b.** 98 ft
c. After 15 sec   **31a.** $847m - (251m + 1355) = 847m - 251m - 1355 = 596m - 1355$ **b.** 30 stereos
33a. $0.08x$ **b.** $x + 0.08x$ **c.** $860   **35.** $m = 4$
37. $t = \dfrac{-9}{2}$ **39.** $s = 0$ **41.** $x > -2$ **43.** $g = \dfrac{8}{3}$
45. $y \le -4$ **47.** $3x + 4y + 3x + 4y = 6x + 8y$
49. $5a + 3a + 5a + 4 + 3 = 13a + 7$ **51a.** $32°$ **b.** $124°$

Homework 2.9

1a. $32c$ **b.** $32 + 8c$; The distributive law is used in part (b).
3a. $-16 - 2t$ **b.** $-16t$; The distributive law is used in part (a).   **5.** $10y - 15$ **7.** $-8x - 16$ **9.** $-5b + 3$
11. $36 - 12t$ **13.** $-4x - 6$ **15.** $23 + x$
17. $-15 - 12z$ **19a.** 48 **b.** $-12x + 12$ **c.** 48
21a. 54 **b.** 24 **c.** -25 **d.** -54 **23.** (d)
25a. $260 - a$ **b.** $2a + 3(260 - a)$ **c.** 120 calories
27a. $2(2w + 6) + 2w$ **b.** width = 5 yd, length = 16 yd
29a. $47 - x$ **b.** $10x$ **c.** $6(47 - x) = 282 - 6x$
d. $10x + 282 - 6x = 4x + 282$ **31.** 12   **33.** 28 mph

35a. 1 mph   **b.** 40 mi   **37.** $y = -2$ **39.** $w = 6$
41. $c = 0$ **43.** $t < 2$ **45.** $x = 9$ **47.** $5(2x + 3)$;
$10x + 15$ **49.** $8x(3y + 2)$; $24xy + 16x$

Homework 2.10

1a. M **b.** A **c.** D
3. 15 oz

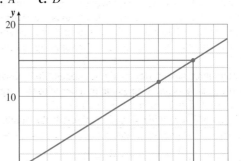

5.

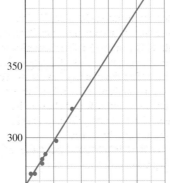

a. ≈ 395 mi
b. ≈ 10.4 gal

7.

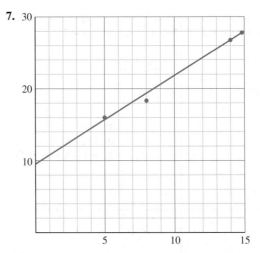

a. ≈ 24 oz   **b.** ≈ 8.5 cm   **c.** The y-intercept represents the length of the rubber band with no weight stretching it.
9a. 3 yr; 10 yr   **b.** About $24,000; about $9000   **c.** About $28,000 - 8000 = $20,000; About $39,000 - 13,000 = $26,000
d. $\dfrac{8000}{13,000} \approx 0.6$; $\dfrac{28,000}{39,000} \approx 0.7$

Chapter 2 Summary and Review

1. If we add or subtract the same quantity on both sides of an inequality, the direction of the inequality is unchanged. If we multiply or divide both sides of an inequality by the same positive quantity, the direction of the inequality is unchanged. If we multiply or divide both sides of an inequality by the same negative quantity, the direction of the inequality must be reversed.

3. To add signed numbers: If the numbers have the same sign, add their absolute values. The sum has the same sign as the numbers. $2 + 5 = 7$ and $-2 + (-5) = -7$
If the numbers have different signs, subtract their absolute values. The sum has the same sign as the number with the larger absolute value. $-2 + 5 = 3$ and $2 + (-5) = -3$

5. To multiply or divide signed numbers: If the numbers have the same sign, their product or quotient is positive. $6(2) = 12$ and $-6(-2) = 12; \dfrac{6}{2} = 3$ and $\dfrac{-6}{-2} = 3$
If the numbers have different signs, their product or quotient is negative. $6(-2) = -12$ and $-6(2) = -12$
$\dfrac{6}{-2} = -3$ and $\dfrac{-6}{2} = -3$

7. Quadrant I: Both coordinates are positive; Quadrant II: The x-coordinate is negative, and the y-coordinate is positive; Quadrant III: Both coordinates are negative; Quadrant IV: The x-coordinate is positive, and the y-coordinate is negative.   **9.** To graph a line using the intercept method: *Step 1* Find the x- and y-intercepts. To find the y-intercept, substitute 0 for x and solve for y. To find the x-intercept, substitute 0 for y and solve for x. *Step 2* Draw a line through the two intercepts. *Step 3* Find a third point as a check. Choose any convenient value for x and solve for y.

11. A regression line is a "line of best fit" obtained from a set of data. It is used to estimate or to predict other values of the variables.

13a. 3   **b.** 3   **c.** -3 **d.** -3 **15.** $-|-2| > -|-3|$
17. $2 - (-5) > -7$ **19.** 20   **21.** 9   **23.** 31
25. $\dfrac{-13}{16}$ **27.** -3 **29.** -1 **31.** $-4 + 10 = 6°F$
33. $280{,}000 - (-180{,}000) = \$460{,}000$ **35.** $3(-4) = -12$ yd
37. $x = 6.9; x = 6; x = 5$ **39.** $x = 3; x = 2; x = 1$
41a.

h	-4	-2	0	1	3	5	8
T	30	24	18	15	9	3	-6

b. $T = 18 - 3h$
c.
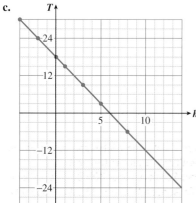
d. $24°$
e. 11 P.M.
f. $-9 - 9 = -18°F$

43a. 5   **b.** $= 10$
45.

x	y
0	7
1	5
2	3

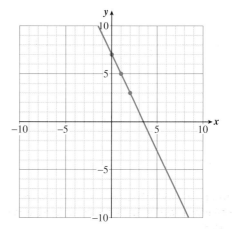

47a. $x = -5$ **b.** $x \le -2$
49a. $30m$ **b.** $-30 + 5m$;
The distributive law is used in part (b).
51. $2m + 7n$ **53.** $-15w + 26$
55. $z = -1$ **57.** $w = -3$
59. $h = -6$ **61.** $p = 25$
63a. $S = 7800 - 600m$

b.

m	S
0	7800
13	0

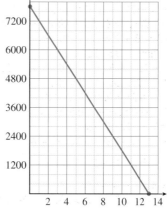

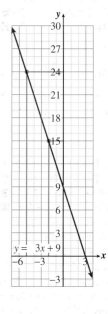

c. The S-intercept at $S = 7800$ show that she started with $7800. The m-intercept at $m = 13$ shows that she will use all the money in 13 mo.
d.
$$\begin{array}{r} 2400 = -7800 - 600m \\ \underline{-7800 \qquad -7800} \\ -5400 = \qquad -600m \\ \dfrac{-5400}{-600} = \dfrac{-600m}{-600} \\ 9 \text{ mo} = m \end{array}$$

65.

x	y
0	−3
2	0

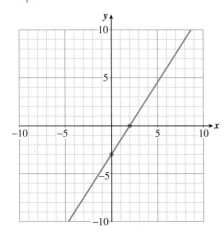

67a. 3x   **b.** 2(3x) + 2(x) = 6x + 2x = 8x
c. 6 cm by 18 cm   **69a.** 30 − x   **b.** 1200x
c. 800(30 − x)   **d.** 1200x + 800(30 − x)
= 1200x + 24,000 − 800x = 400x + 24,000
e. 400x + 24,000 = 28,800

$$\frac{-24,000 \quad -24,000}{400x \qquad = \quad 4,800}$$

$$\frac{400x}{400} = \frac{4800}{400}$$

x = 12 computers with speakers and
30 − x = 18 computers without

71. z ≥ 3

73. k > −6

75. 7 > n ≥ 3

77.

a. 120 chirps/min   **b.** 65°F

Homework 3.1

1. A <u>ratio</u> is a comparison of two quantities. Written as a fraction, the ratio of *a* to *b* is $\frac{a}{b}$. A <u>proportion</u> is a statement that two ratios are equal, such as $\frac{a}{b} = \frac{c}{d}$.   **3.** $\frac{125}{175} = \frac{5}{7}$

5. $\frac{\$56.68}{6.5 \text{ hr}} = \$8.72/\text{hr}$ **7.** $\frac{11}{6}$ **9.** $\frac{30}{7}$ or ≈ 4.3

11. x = 6   **13.** w = 7.5   **15.** a = 4   **17.** 3.25 = b
19. Yes   **21.** No   **23.** No   **25.** 40 lb   **27.** 100 ℓ
29. 80 km   **31.** 5753 in favor   **33.** 13.78 in.

35. $13,714,286   **37a.** 3 to 2; $416\frac{2}{3}$ servings; $7\frac{1}{2}$ cans

b. 2.4 oz; 15.5 lb   **c.** 24.75 cans; 90 lb

Homework 3.2

1. Similar; Ratio $= \frac{2.5}{4} = \frac{2.5(10)}{4(10)} = \frac{25}{40} = \frac{5}{8}$
3. Not similar because corresponding ratios are not equal: $\frac{7}{12} \neq \frac{7}{8}$
5. Similar; Ratio $= \frac{5}{4}$ **7.** Not similar because corresponding
ratios are not equal: $\frac{9}{6} \neq \frac{10}{7}$ **9.** (b), (c), and (d)
11. A = B = 37°   **13.** A = B = C = 45°; d = 5
15. A = 60°; w = 3.5   **17.** Yes, their corresponding sides are
proportional. $\frac{3}{6} = \frac{4}{8} = \frac{6}{12}$ **19.** Yes, their corresponding
angles are equal.   **21.** h = 12   **23.** 10 = q   **25.** h = 30
27a. 0.88 = 88%   **b.** m = 154 ft   **29.** h ≈ 555 ft

31. x = 1 mi   **33.** y = 2x   **35.** $y = \frac{12x}{17}$ **37.** h = 7.5
39. c = 15   **41.** 6 = s   **43.** $y = \frac{3x}{5}$ **45.** $y = \frac{3x + 20}{4}$

Section 3.3 Activity

1. 1.50, 1.50, 1.50, 1.50, 1.50
5a. $\frac{D}{h} = 60$ mi/h
b. & d.
Table 3.5

Units	Distance	Distance / Hour
3	180	60
5	300	60
8	480	60
10	600	60
12	960	60

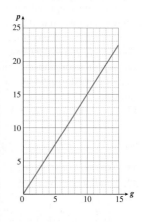

c.

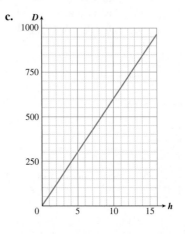

Homework 3.3

1. If two variables are proportional, their ratios are always the same.   **3.** Figures 3.12 and 3.15 illustrate direct variation.
5. If S varies directly with w, the graph of S verses w is a straight line through the origin.   **7.** No, the ratios are not constant.

$\dfrac{76.90}{43} \approx 1.79$; $\dfrac{156.51}{77} \approx 2.03$; $\dfrac{220.17}{101} \approx 2.18$

9a. $\dfrac{16}{2} = \dfrac{24}{3} = 8 = k$ **b.** $d = 8t$

c.

t	d
2	16
3	24

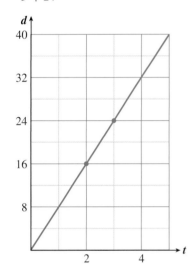

11a. $\dfrac{0.72}{12} = \dfrac{0.90}{15} = 0.06 = k$ **b.** $t = 0.06p$

c.

p	t
12	0.72
15	0.90

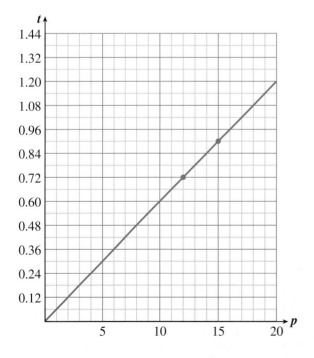

13a.

Time (hours)	Distance (miles)	Time (hours)	Distance (miles)
2	16	4	32
3	24	6	48
5	40	10	80

b. The distance is doubled when the time is doubled.

15a.

Price	Sales tax	Price	Sales tax
12	0.72	6	0.36
20	1.20	10	0.60
30	1.80	15	0.90

b. The sales tax is halved when the price is halved.
17. If y varies directly with x, and we multiply the value of x by a constant n, then the value of y is also multiplied by the same constant n.   **19a.** No

b.

Units	Tuition	Units	Tuition
3	620	6	740
5	700	10	900
8	820	16	1140

c. No   **21.** No   **23.** Yes   **25.** Yes   **27.** No
29. No   **31.** No

33.

x	y	z	w
-4	-8	-12	-6
-2	-4	-6	-3
0	0	0	0
2	4	6	3
4	8	12	6

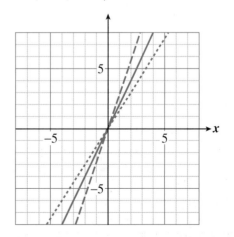

35.

x	y	z	w
−6	2	4	8
−3	1	2	4
0	0	0	0
3	−1	−2	−4
6	−2	−4	−8

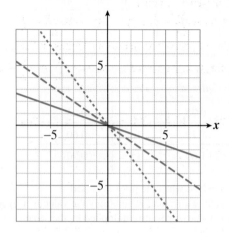

37. The origin $(0, 0)$

Section 3.4 Activity

1. 1.5 **3.** 30 **5a.** Yes **b.** Yes **7.** Figure 3.12 a. Dollars per gallon; b. The slope gives the price per gallon. Figure 3.13 a. People per year; b. The slope gives the rate of growth of the suburb. Figure 3.14 a. Dollars per unit; b. The slope gives the cost of tuition per unit. It is the rate of change of tuition when the number of units taken increases; Figure 3.15 a. Miles per hour; b. The slope gives the speed at which the train is traveling.

Homework 3.4

1a. $\dfrac{\Delta y}{\Delta x} = \dfrac{6}{4} = \dfrac{3}{2}$ **b.** $\dfrac{\Delta y}{\Delta x} = \dfrac{-8}{4} = -2$

3a. $\dfrac{\Delta y}{\Delta x} = \dfrac{-2}{8} = \dfrac{-1}{4}$ **b.** $\dfrac{\Delta y}{\Delta x} = \dfrac{9}{3} = 3$

5.

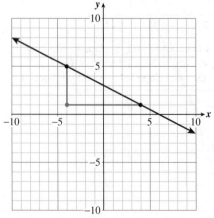

$$\frac{\Delta y}{\Delta x} = \frac{-4}{8} = \frac{-1}{2}$$

7.

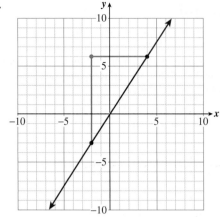

$$\frac{\Delta y}{\Delta x} = \frac{9}{6} = \frac{3}{2}$$

9a.

x	y
0	4
6	0

b.

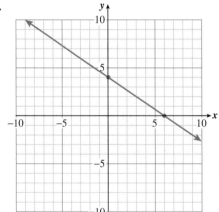

c. $\dfrac{\Delta y}{\Delta x} = \dfrac{-4}{6} = \dfrac{-2}{3}$ **d.** $\dfrac{\Delta y}{\Delta x} = \dfrac{-4}{6} = \dfrac{-2}{3}$

11a.

x	y
0	−5
2	0

b.

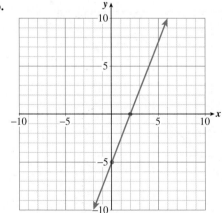

c. $\dfrac{\Delta y}{\Delta x} = \dfrac{5}{2}$ **d.** $\dfrac{\Delta y}{\Delta x} = \dfrac{15}{6} = \dfrac{5}{2}$

© 2003 Brooks/Cole

13a.

x	y
0	5
5	0

b.

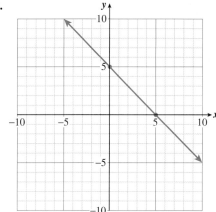

c. $\dfrac{\Delta y}{\Delta x} = \dfrac{-5}{5} = -1$ **d.** $\dfrac{\Delta y}{\Delta x} = \dfrac{-11}{11} = -1$

15a.

x	y
0	-2
4	0

b.

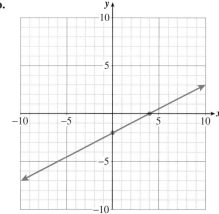

c. $\dfrac{\Delta y}{\Delta x} = \dfrac{2}{4} = \dfrac{1}{2}$ **d.** $\dfrac{\Delta y}{\Delta x} = \dfrac{-5}{-10} = \dfrac{1}{2}$ **17.** $\dfrac{5}{3}$ **19.** -2

21. $\Delta y = \dfrac{35}{2}$ **23.** $\dfrac{3}{2} = \Delta x$ **25.** $\Delta y = 950.4$ ft

27. 16 ft $= \Delta x$ **29.** $m = \dfrac{\Delta y}{\Delta x} = \dfrac{30}{2} = 15$

31.

x	-2	0	3	4
y	56	32	-4	-16

$m = \dfrac{\Delta y}{\Delta x} = \dfrac{-24}{2} = -12$

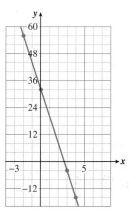

33. $m = \dfrac{\Delta y}{\Delta x} = \dfrac{2}{3}$ **35.** $m = \dfrac{\Delta y}{\Delta x} = \dfrac{50}{-30} = \dfrac{-5}{3}$

37a. $d = 25g$ **b.** $m = \dfrac{\Delta d}{\Delta g} = \dfrac{75 \text{ mi}}{3 \text{ gal}} = \dfrac{25 \text{ mi}}{1 \text{ gal}}$

c. The car gets 25 miles per gallon. **39a.** $T = .04p$

b. $m = \dfrac{\Delta T}{\Delta p} = \dfrac{20 \text{ cents}}{5 \text{ dollars}} = \dfrac{4 \text{ cents}}{1 \text{ dollar}}$ **c.** The sales tax is 4 cents

per dollar. **41.** $m = \dfrac{\Delta y}{\Delta x} = \dfrac{0}{1} = 0$ **43.** $m = \dfrac{\Delta y}{\Delta x} = \dfrac{1}{0}$

undefined **45a.** All the y-coordinates on a horizontal line are

the same, so the change in y's is zero. **b.** $m = \dfrac{\Delta y}{\Delta x} = \dfrac{0}{\Delta x} = 0$

Section 3.5 Activity

1a. $W = 400 + 30u$ **b.** $X = 200 + 30u$ **c.** $Y = 30u$

u	W	X	Y
3	490	290	90
5	550	350	150
8	640	440	240
10	700	500	300
12	760	560	360

d.

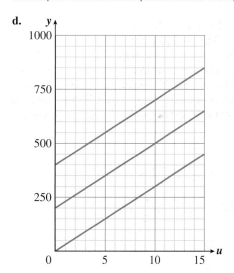

e. W: slope $= 30$; y-intercept $= 400$; X: slope $= 30$,
y-intercept $= 200$; Y: slope $= 30$, y-intercept $= 0$
f. The graphs have the same slope but different y-intercepts.

Homework 3.5

1a. **i.**

x	-1	0	1	2	3
y	-8	-6	-4	-2	0

ii.

x	-1	0	1	2	3
y	-1	1	3	5	7

iii.

x	-1	0	1	2	3
y	1	3	5	7	9

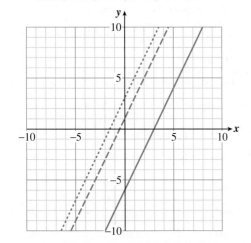

b. i. $m = 2$ ii. $m = 2$ iii. $m = 2$

3a. **i.**

x	-6	-4	-2	0	2
y	5	2	-1	-4	-7

ii.

x	-6	-4	-2	0	2
y	11	8	5	2	-1

iii.

x	-6	-4	-2	0	2
y	15	12	9	6	3

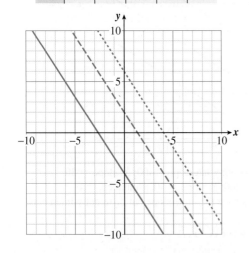

b. i. $m = \dfrac{-3}{2}$ ii. $m = \dfrac{-3}{2}$ iii. $m = \dfrac{-3}{2}$

5a. **i.**

x	-4	-2	0	2	4
y	1	$\frac{3}{2}$	2	$\frac{5}{2}$	3

ii.

x	-4	-2	0	2	4
y	0	1	2	3	4

iii.

x	-4	-2	0	2	4
y	-2	0	2	4	6

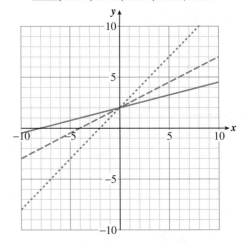

b. i. $m = \dfrac{1}{4}$ ii. $m = \dfrac{1}{2}$ iii. $m = 1$

7a. **i.**

x	-6	-3	0	3	6
y	16	7	-2	-11	-20

ii.

x	-6	-3	0	3	6
y	10	4	-2	-8	-14

iii.

x	-6	-3	0	3	6
y	8	3	-2	-7	-12

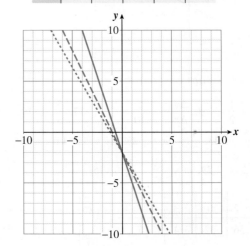

b. i. $m = -3$ ii. $m = -2$ iii. $m = \dfrac{-5}{3}$

9. The slope is 3 and the y-intercept is $(0, 4)$. **11.** The slope is -2 and the y-intercept is $\left(0, \dfrac{5}{3}\right)$. **13.** The slope is $\dfrac{2}{3}$ and the y-intercept is $(0, -2)$. **15.** The slope is $\dfrac{5}{4}$ and the y-intercept is $(0, 0)$. **17a.** Slope $= \dfrac{6}{2} = 3$; y-intercept $= (0, -4)$ **b.** $y = 3x - 4$ **19a.** Slope $= \dfrac{4}{6} = \dfrac{2}{3}$; y-intercept $= (0, -3)$

b. $y = \dfrac{2}{3}x - 3$ **21a.** Slope $= \dfrac{-8}{5}$; y-intercept $= (0, 1)$

b. $y = \dfrac{-8}{5}x + 1$ **23a.**

x	y
0	3
4	0

b.

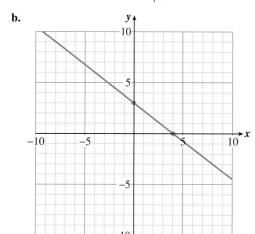

c. $\dfrac{\Delta y}{\Delta x} = \dfrac{-3}{4}$ **d.** $y = \dfrac{-3}{4}x + 3$

25a.

x	y
0	8
$\frac{8}{3}$	0

b.

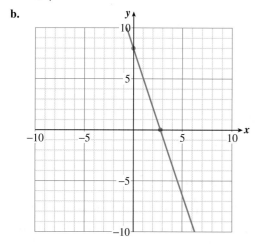

c. $\dfrac{\Delta y}{\Delta x} = \dfrac{-8}{\frac{8}{3}} = -8 \div \dfrac{8}{3} = \dfrac{-8}{1} \cdot \dfrac{3}{8} = -3$ **d.** $y = -3x + 8$

27a. $3x - 5y = 0$
$$-5y = -3x$$
$$\frac{-5y}{-5} = \frac{-3x}{-5}$$
$$y = \frac{3}{5}x$$

b. y-intercept $= (0, 0)$; slope $= \dfrac{3}{5}$ **c.** $(5, 3)$; $(10, 6)$; $(-5, -3)$; $(-10, -6)$

d.

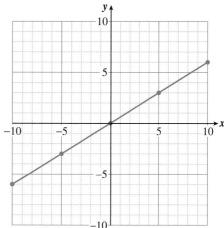

29a. $5x + 4y = 0$
$$4y = -5x$$
$$\frac{4y}{4} = \frac{-5x}{4}$$
$$y = \frac{-5}{4}x$$

b. y-intercept $= (0, 0)$; slope $= \dfrac{-5}{4}$ **c.** $(4, -5)$; $(8, -10)$; $(-4, 5)$; $(-8, 10)$

d.

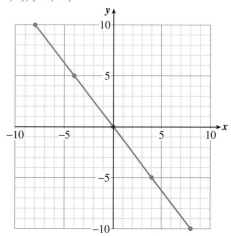

31.

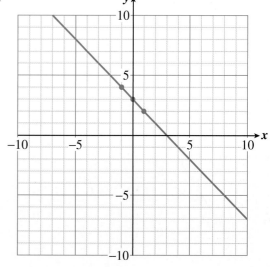

33.

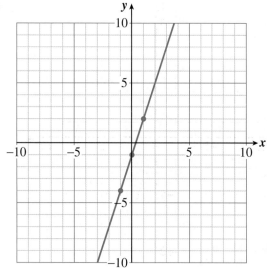

35.

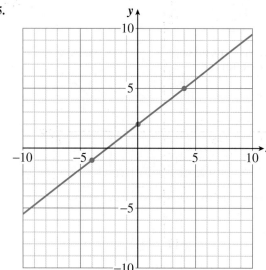

37.

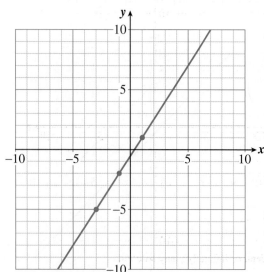

39.

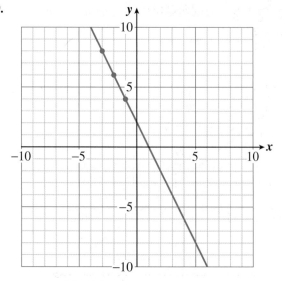

41.

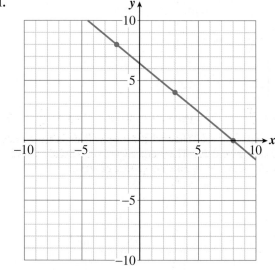

Midchapter Review

1. A <u>ratio</u> is a comparison of two quantities written as a fraction.
The ratio of a to b is $\dfrac{a}{b}$. A <u>proportion</u> is a statement that two ratios
are equal, such as $\dfrac{a}{b} = \dfrac{c}{d}$.   **3.** Equal; proportional

5. $y = kx$ **7.** $y = mx + b$; For example, if $y = \dfrac{1}{2}x + 3$, the
slope is $\dfrac{1}{2}$ and the y-intercept is $(0, 3)$.   **9.** Slope; zero

11. 14 men
     $35 - 14 = 21$ women
     $\dfrac{14}{21} = \dfrac{2}{3}$

13. $x = 180$ **15.** $z = 14$ **17.** No   **19.** $x = 825$ bricks

21a. 10 mi; 15 mi   **b.** 50 mi; 30 cm   **c.** $\dfrac{50 \text{ mi}}{30 \text{ cm}} = \dfrac{5 \text{ mi}}{3 \text{ cm}}$

d. 150 sq mi; 54 sq cm   **e.** $\dfrac{150 \text{ sq mi}}{54 \text{ sq cm}} = \dfrac{25 \text{ sq mi}}{9 \text{ sq cm}}$

23. $\dfrac{x}{x + 6} = \dfrac{2}{6}$ **25.** 8515 ft

$6x = 2x + 12$

$4x = 12$

$x = 3$

27a.

x	y
0	−5
4	0

b. $m = \dfrac{5}{4}$ **c.** $y = \dfrac{5}{4}x - 5$

d.

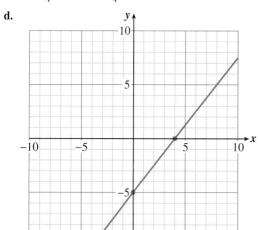

29a. $m = \dfrac{\Delta y}{\Delta x} = \dfrac{-3}{5}$; y-intercept $= (0, 3)$ **b.** $y = \dfrac{-3}{5}x + 3$

31. The slope of -5000 shows that the water is draining at a rate of 5000 gallons per hour. The y-intercept of $(0, 500,000)$ shows that the pool started with 500,000 gallons of water.

33.

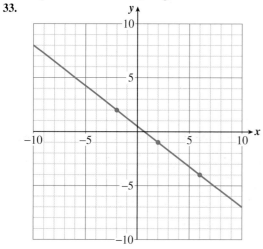

Homework 3.6

1. False   **3.** True   **5.** False   **7.** True   **9.** No
11. Yes   **13.** Intercept method because the equation is in general form, $Ax + By = C$.   **15.** Slope-intercept method because the equation is in slope-intercept form, $y = mx + b$
17. Slope-intercept method because both intercepts are the same point $(0, 0)$.

19.   (1, −3)

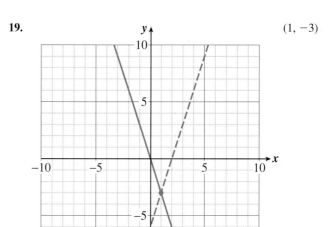

21.   (−4, −5)

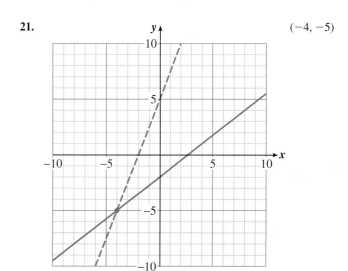

23.

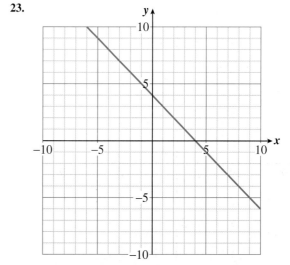

Dependent, or infinitely many solutions

25.

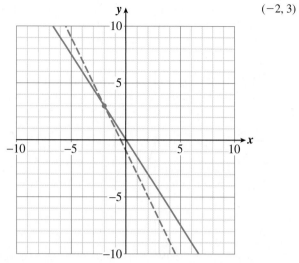

(−2, 3)

27. From their slope-intercept forms, both lines have slope 3, but the *y*-intercepts are different, so their graphs are parallel lines.

29a. Plan A: $y = 0.03x + 20,000$; Plan B: $y = 0.05x + 15,000$

b.

x	Earnings under Plan A	Earnings under Plan B
0	20,000	15,000
50	21,500	17,500
100	23,000	20,000
150	24,500	22,500
200	26,000	25,000
250	27,500	27,500
300	29,000	30,000
350	30,500	32,500
400	32,000	35,000

c.

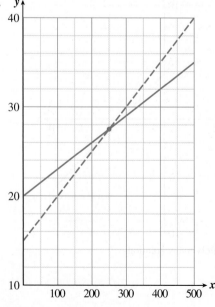

d. $250,000   **e.** He should prefer plan A if he expects less than $250,000 in sales.   **f.** More than $300,000; more than $333,333
g. $0.05x + 15,000 > 30,000$; $x > 300,000$; $0.03x + 20,000 > 30,000$; $x > 333,333$ **31a.** Cost $y = 60x + 6000$; Revenue $y = 80x$

b.

x	Cost	Revenue
0	6,000	0
50	9,000	4,000
100	12,000	8,000
150	15,000	12,000
200	18,000	16,000
250	21,000	20,000
300	24,000	24,000
350	27,000	28,000
400	30,000	32,000

c.

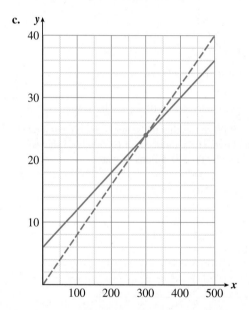

d. More than 250 clarinets   **e.** 300 clarinets
f. $4000 profit; −$2000 or a loss of $2000
33.

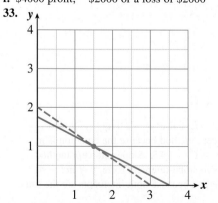

A rose costs $1.50, and a carnation costs $1.00.

35.

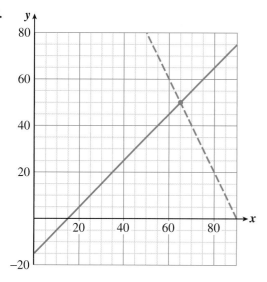

The base angles are 65°, and the vertex angle is 50°.

Homework 3.7

1. $(2, 4)$ **3.** $(-1, 2)$ **5.** $(5, 2)$ **7.** $(22, -28)$

9. $\left(\dfrac{2}{3}, \dfrac{-4}{3}\right)$ **11.** $\left(\dfrac{1}{3}, -2\right)$ **13.** Hamburger: 650;
Shake: 380   **15.** He can buy 12 tables and 48 chairs.
17. $(3, 2)$ **19.** $(1, 1)$ **21.** $(1, 2)$ **23.** $(-3, -2)$

25. $(1, -1)$ **27.** $\left(\dfrac{3}{4}, \dfrac{-1}{4}\right)$ **29.** The rectangle is 17 m
by 4 m.   **31.** Bacon costs \$2.20 per lb, and coffee costs
\$5.60 per lb.   **33.** Inconsistent   **35.** Inconsistent
37. Independent   **39.** Elimination   **41.** Elimination

Homework 3.8

1a. Amount in bonds: x; Amount in mutual fund: y;
$x + y = 150{,}000$ **b.** $0.065x$; $0.118y$
c. $0.065x + 0.118y = 12{,}930$ **d.** They invested \$90,000 in
bonds and \$60,000 in mutual funds.
3a. Amount of first loan: x Amount of second loan: y

	Principle	Rate	Interest
First Loan	x	0.12	$0.12x$
Second Loan	y	0.15	$0.15y$

b. $x + y = 30{,}000$; $0.12x + 0.15y = 3750$ **c.** He borrowed
\$25,000 at 12% and \$5000 at 15%   **5.** He borrowed \$10,000 on
his car loan and \$5000 on his student loan.   **7a.** 28 women;
9 women   **b.** 140 students; 37 women   **c.** $\approx 26.4\%$
9a. 15 mℓ; 1.8 mℓ   **b.** 16.8 mℓ; 42 mℓ   **c.** 40%
d.

	Number of milliliters (W)	Percent acid (r)	Amount of acid (P)
50% solution	30	0.50	15
15% solution	12	0.15	1.8
Mixture	42	0.40	16.8

11a.

	Number of Liters	Percent Salt	Amount of Salt
12% Solution	x	0.12	$0.12x$
30% Solution	y	0.30	$0.30y$
Mixture	45	0.24	$0.24(45)$

b. $x + y = 45$; $0.12x + 0.30y = 0.24(45)$ **c.** He needs
15 liters of the 12% solution and 30 liters of the 30% solution.
13. 160 women were polled.
15a.

	Rate (R)	Time (T)	Distance (D)
Delbert	x	6	$6x$
Francine	y	6	$6y$

Cedar Rapids
b. Delbert ⟷ ● ⟷ Francine
    $6x$ $6y$
c. $y = x - 5$; $6x + 6y = 570$ **d.** Delbert's speed is 50 mph,
and Francine's speed is 45 mph.

Bonnie
17a. Dallas
Clyde
b.

	Rate	Time	Distance
Bonnie	40	t	d
Clyde	70	$t - 3$	d

c. $40t = d$; $70(t - 3) = d$ **d.** Bonnie drove for 7 hr and
traveled 280 mi   **19.** 10 hr; 400 mi   **21.** The tour rode
40 mi   **23a.** $g = 0.40t + 0.60(72)$; $g = 0.40t + 43.2$
b. 69.2; 75.2   **c.** 92   **d.** $g = 0.40(92) + 43.2 = 36.8 + 43.2 = 80$

Section 3.9 Activity

1.

3. $(3, 5)$ **7a.** Yes   **b.** No   **c.** No

Homework 3.9

1. $\dfrac{-4}{3}$ **3.** 0 **5.** -4 **7.** $\dfrac{5}{3}$ **9.** -1 **11.** $\dfrac{1}{9}$

13a. $(0, -3)$ **b.** $(-2, -7)$

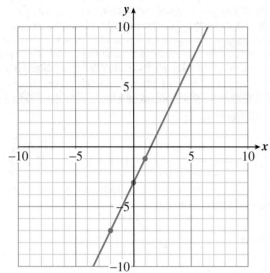

15a. $(6, -8)$ **b.** $-4, -3)$

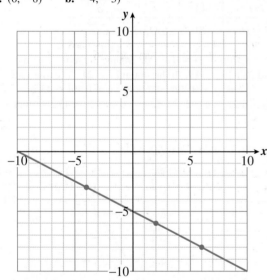

17. $y = -2x - 2$ **19.** $y = \dfrac{1}{2}x - 5$ **21.** $y = \dfrac{-2}{3}x - 2$

23. $m = \dfrac{3}{5}$; $(0, -7)$ **25.** $m = 3$; $(-5, 2)$ **27.** $m = \dfrac{4}{5}$; $(0, 0)$

29.

x	y
0	7
5	0

$\dfrac{-7}{5}$

31.

x	y
0	1.6
-2.4	0

$\dfrac{2}{3}$

33.

x	y
0	$\frac{7}{2}$
$\frac{1}{3}$	0

$\dfrac{-21}{2}$

Homework 3.10

1. Parallel **3.** Neither **5.** Perpendicular **7.** $y = -5$
9. $x = 2$ **11.** $x = -8$ **13.** $y = 0$ **15a.** -3 **b.** -3
c. $y = -3x + 6$ **17a.** -2 **b.** $\dfrac{1}{2}$ **c.** $y = \dfrac{1}{2}x$

19a.

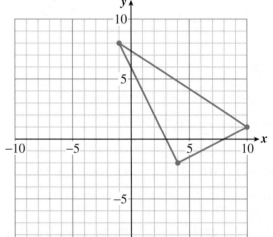

b. Because $m_{BC}m_{CD} = -1$, the lines are perpendicular.

21a.

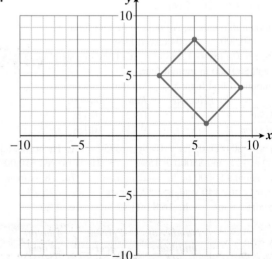

b. From the graph $Z(5, 8)$. **c.** Because $m_{WX} = m_{ZY}$, the lines are parallel; Because $m_{XY} = m_{WZ}$, the lines are parallel.

23. $y = x + 6$ **25.** $y = \dfrac{5}{3}x$ **27.** $y = \dfrac{-1}{8}x + \dfrac{19}{4}$

29a. $y = -0.0025x + 85$ **b.** $m = \dfrac{-1 \text{ degree}}{400 \text{ feet}}$; The temperture is dropping 1°F every 400 ft. **c.** 60°F **d.** 85°F

31a. $y = 200x + 25{,}000$ **b.** $m = 200$ dollars per dryer; It costs \$200 to produce one dryer.   **c.** 375   **d.** \$25,000

33. $y = \dfrac{-5}{7}x + \dfrac{11}{7}$ **35.** $y = \dfrac{1}{4}x - \dfrac{1}{4}$

37.

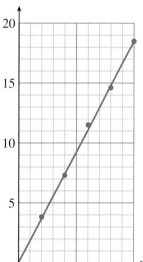

$y = 0.37x$

a. ≈ 6.4 in.   **b.** ≈ 227 in.

39.

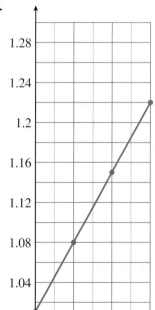

$y = 0.0035x + 0.94$

a. ≈ 1.07 atm   **b.** $\approx 160°C$ **c.** The y-intercept represents the pressure when the temperature is 0°C.   **d.** The x-intercept represents the temperature when the pressure is 0 atm.

41a. Yes   **b.** 1.4 sec   **c.** 1.1 sec   **d.** 6; 5
e. 0.96 sec; 1.26 sec   **f.** $y \approx -0.0094x + 10.53$;
$y \approx -0.0173x + 11.94$ **g.** The women's   **h.** ≈ 9.82 sec
i. ≈ 10.63 sec   **j.** The intersection point is approximately $(177, 8.8)$; The intersection point predicts that the men's and the women's times will be the same, approximately 8.8 sec, in the year 2105.   **k.** The x-intercept would represent a running time of zero seconds.

Chapter Summary and Review

1. If $\dfrac{a}{b} = \dfrac{c}{d}$, then $ad = bc$ **3.** $m = \dfrac{\Delta y}{\Delta x}$ **5.** $m = \dfrac{y_2 - y_1}{x_2 - x_1}$
7. If two lines are parallel, then $m_1 = m_2$.   **9.** $x = a$ is the equation of a vertical line with undefined slope.   **11.** Two variables are proportional if their ratios are always the same.
13. Solve for the equation for y. The coefficient of x is the slope of the line.   **15.** Substitution and elimination   **17.** Principal Rate Interest   **19.** Rate Time Distance   **21a.** They are not proportional   **b.** They are proportional   **23.** $x = 11$
25. 320 g protein; 2880 g carbohydrates   **27.** 25   **29.** 14 ft

31a. $B = \dfrac{3}{4}h$ **b.** 52 hr

c.

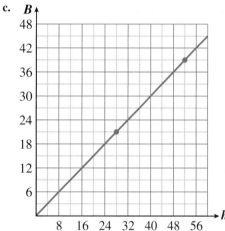

h	B
28	21
52	39

33. 15

35a.

x	y
0	-6
3	0

b.

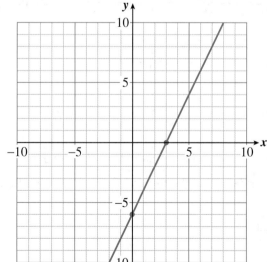

c. $\dfrac{\Delta y}{\Delta x} = \dfrac{6}{3} = 2$

37. $\dfrac{\Delta y}{\Delta x} = \dfrac{-12}{8} = \dfrac{-3}{2}$ **39.** 288 ft

41.

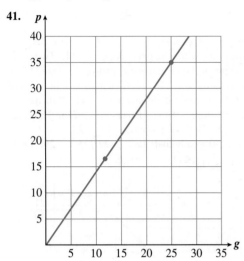

a. $m = \dfrac{\Delta p}{\Delta g} = \dfrac{35 - 16.8}{25 - 12} = \dfrac{\$18.20}{13 \text{ gal}} = \$1.40/\text{gal}$

b. The slope indicates the price per gallon.

43a. The slope is $\dfrac{-5}{2}$ and the y-intercept is $(0, -5)$

b.

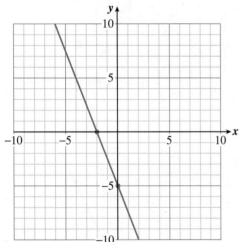

45. Slope: 0; y-int: $(0, 4)$ **47.** Not a solution
49. $(3, 2)$

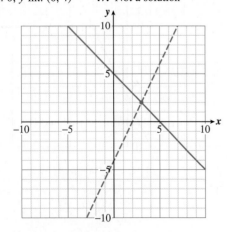

51a. Inconsistent   **b.** Dependent   **53.** $(-3, -5)$
55. $(1, -3)$ **57.** He needs 12 lb of cereal and 18 lb of dried fruit.   **59.** She invested $500 in the first account and $700 in the second account.   **61.** He needs 2 lb of the 60% alloy and 6 lb of the 20% alloy.   **63.** Alida is traveling at 62 mph, and Steve is traveling at 31 mph.   **65.** $\dfrac{-5}{8}$ **67.** $y = \dfrac{1}{5}x - 1$
69. $x = -4$ **71.** $y = -x - 2$
73.

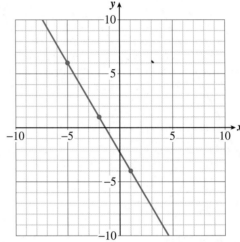

Homework 4.1

1a. 64   **b.** 125   **c.** 625   **3a.** $\dfrac{16}{81}$ **b.** $\dfrac{64}{125}$ **c.** $\dfrac{121}{81}$
5a. ≈ 29.79 **b.** ≈ 45.70 **c.** ≈ 0.41 **7a.** -25
b. -125 **c.** 25   **d.** -125 **9a.** -4 **b.** 8
c. -12 **d.** -36 **11.** 49   **13.** ≈ 17.17 **15.** 54
17a. -40 **b.** 20   **c.** 1   **d.** 13   **19a.** 192   **b.** 61
c. 361   **d.** -84 **21a.** $3x$ **b.** x^3 **23a.** $4a^2$ **b.** $4a$
25a. $-3q$ **b.** $-q^3$ **27a.** $-6m$ **b.** $9m^2$
29a. For $z = 2$
$\left. \begin{array}{l} 3z^2 = 3(2)^2 \cdot 4 = 12 \\ (3z)^2 = (3 \cdot 2)^2 = 6^2 = 36 \end{array} \right\}$ Not the same
b. $(3z)^2 = (3z)(3z) = 9z^2$
31a. $x = 0; x = 2$ **b.** $x = 1; x = 3; x = 4; x = 5$
c. No   **33.** $16w^2 - 36t^2$ **35.** $4\pi w^2 - 25$ **37.** $\dfrac{32\pi h^3}{3}$
39. $42c^3$ **41a.** $\dfrac{128{,}000\pi}{3}$ cu ft   **b.** 3200π sq ft
c.

43a. 210π sq cm   **b.** 90 ft $\approx h$ **c.**

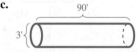

45a. 16   **b.** $-3, 3$ **c.** ≈ 6.3 **d.** ≈ 12.3 **e.** ≈ -4.5, ≈ 4.5 **f.** There is no solution since the graph never has a y-coordinate of -5.

47. $\dfrac{v}{lh} = w$ **49.** $\dfrac{2E}{v^2} = m$ **51.** $\dfrac{2A}{b + c} = h$
53. $\dfrac{5}{9}(F - 32) = C$ **55.** $\dfrac{A - 2\pi r^2}{\pi r} = h$ **57.** 2400
59. 30   **61.** 234

Homework 4.2

1a. p^2 **b.** $\pm\sqrt{k}$ **3a.** The square root of a negative number is undefined because the square of any number is positive or zero (never negative). **b.** $\sqrt{-x}$ is the square root of a negative number, which is undefined. $-\sqrt{x}$ indicates the negative of the square root of the positive number x, which is defined.

5a. -12 **b.** -2 **7a.** 9 **b.** 3 **9a.** $\dfrac{19}{2}$ **b.** $\dfrac{-2}{3}$

11. In a linear equation, such as $2x + 3 = 0$, the variable cannot have an exponent other than 1. In a quadratic equation, such as $x^2 + 2x + 3 = 0$, the variable must have an exponent of 2.

13. ±11 **15.** ±7 **17.** $\approx \pm2.236$ **19.** $\approx \pm36.742$
21. ±4 **23.** 8.660 **25.** -3.055 **27.** 1.899
29. -15.544 **31.** 1.293 **33.** -1.512 **35.** 5.463 cm
37. 1405 ft

39.

x	$x + x$	$2x$	x^2
3	6	6	9
5	10	10	25
-4	-8	-8	16
-1	-2	-2	1

$x + x = 2x$

41.

x	$x^2 + x^2$	x^4	$2x^2$
2	8	16	8
3	18	81	18
-2	8	16	8
-1	2	1	2

$x^2 + x^2 = 2x^2$

43.

x	$x + x^2$	$x \cdot x^2$	x^3
1	2	1	1
4	20	64	64
-3	6	-27	-27
-1	0	-1	-1

$x \cdot x^2 = x^3$

45. $-2a^2$ **47.** Cannot be simplified **49.** $-2m^2$
51. $12k^2$ **53.** Cannot be simplified **55.** $7k^2$ **57.** 5
59. x **61.** $\sqrt{6}$ **63.** $22a^2$ **65.** $4\sqrt{3}m^2 + 4\sqrt{3}m + 6m$
67. ±4.243 **69.** ±7.246 **71.** ±15 **73.** No solution

75. $\pm\sqrt{\dfrac{A}{4\pi}}$ **77.** $\pm\sqrt{\dfrac{3V}{\pi h}}$

Section 4.3 Activity

1a. \$2.25 **b.** 8 in. **c.** 12 in.

r	C
0	0
1	0.25
2	1
4	4
6	9
9	10.25
10	25
14	49

3a. $3.14r^2$
b.

r	A
0	0
0.5	0.8
1	3.1
1.5	7.1
2	12.6
2.5	19.6
3	28.3
3.5	38.5
4	50.2

c. 15.9 in^2 **d.** 3.1 in.

5a.

h	d
2000	54.6
5000	86.3
10,000	122
15,000	149.4
20,000	172.5
25,000	192.9
30,000	211.3

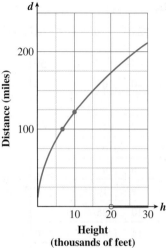

b. 122 mi   **c.** 6719 ft   **d.** 20,576 ft

Homework 4.3

1. A <u>rational number</u> can be written as a fraction $\frac{a}{b}$, where a and b are both integers and $b \neq 0$. For example, $2 = \frac{2}{1}$ is rational. An <u>irrational number</u> cannot be written as a fraction $\frac{a}{b}$, where a and b are both integers and $b \neq 0$. For example, $\sqrt{2}$ is irrational.
3. Rational numbers: $\sqrt{1}$, $\sqrt{4}$, $\sqrt{9}$; Irrational numbers: $\sqrt{2}$, $\sqrt{3}$, $\sqrt{5}$ **5.** True   **7.** False   **9.** False   **11.** Irrational
13. Rational   **15.** Rational   **17.** Irrational   **19.** Rational
21. Irrational
23.

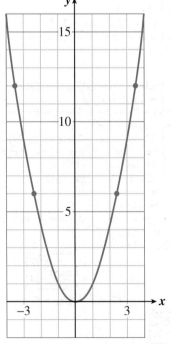

a. ± 3.5 **b.** $-2.45 \leq x \leq 2.45$

25.

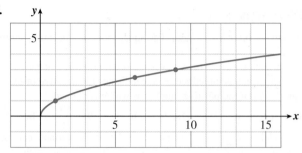

a. 6.3   **b.** $1 < x \leq 9$
27.

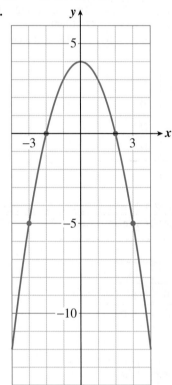

a. ± 3 **b.** $x > 2$ or $x < -2$
29.

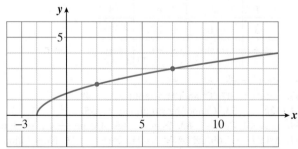

a. 7   **b.** $x > 2$

31b. $A = (2w)w = 2w^2$

c.

w	0	1	2	3	4	5
A	0	2	8	18	32	50

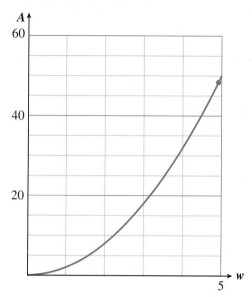

d. 4.9 in.

33a.

r	0	100	200	300	400	500	600	700
v	0	15.8	22.4	27.4	31.6	35.4	38.7	41.8

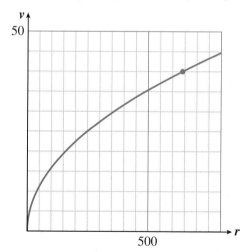

b. 640 ft   **35.** (b)   **37.** (c)   **39.** (b)
41a. For $x = 3$: $\sqrt{3^2} = \sqrt{9} = 3$
For $x = 5$: $\sqrt{5^2} = \sqrt{25} = 5$
For $x = 8$: $\sqrt{8^2} = \sqrt{64} = 8$
For $x = 12$: $\sqrt{12^2} = \sqrt{144} = 12$
b. $\sqrt{a^2} = a$ **43.** 16   **45.** 7   **47.** 5   **49.** 9
51. $\sqrt{3}$ **53.** $-\sqrt{11}$ **55.** $2b$ **57.** $2\sqrt{m}$
59. x when $x \geq 0$ **61.** 64 because 64 is 8^2, and 64 is 4^3.

Homework 4.4

1. 8   **3.** 37   **5.** $\sqrt{68}$ **7.** $\sqrt{15}$ **9.** $k = 5\sqrt{2}$;
$3k = 15\sqrt{2}$ **11.** No   **13.** Yes   **15.** No   **17.** Yes
19. 44.72 ft   **21.** 39.05 ft   **23.** 2.65 mi

25.

a	b	$a + b$	$(a + b)^2$	a^2	b^2	$a^2 + b^2$
2	3	5	25	4	9	13
3	4	7	49	9	16	25
1	5	6	36	1	25	26
-2	6	4	16	4	36	40

No, $(a + b)^2 \neq a^2 + b^2$

27.

a	b	$a + b$	$\sqrt{a + b}$	$\sqrt{a}$	$\sqrt{b}$	$\sqrt{a} + \sqrt{b}$
2	7	9	3	$\sqrt{2}$	$\sqrt{7}$	4.1
4	9	13	3.6	2	3	5
1	5	6	2.4	1	$\sqrt{5}$	3.2
9	16	25	5	3	4	7

No, $\sqrt{a + b} \neq \sqrt{a} + \sqrt{b}$.
29. 2.236, 1, 0   **31.** $-1, -0.646, -0.149$
33. 30, -1.657, $\dfrac{-7}{8}$ **35.** $\dfrac{5}{4}$, 2, 2.3, $\sqrt{8}$
37. $\sqrt{6}$, 3, $2\sqrt{3}$, $\dfrac{23}{6}$ **39.** $t = \pm 100$ **41.** $u = \pm\dfrac{1}{7}$
43. $n = \pm 0.5$ **45a.** $\sqrt{27}$ ft   **b.** $3\sqrt{27}$ sq ft
47a. $\sqrt{8}$ cm   **b.** $\sqrt{14}$ **c.** $\dfrac{4\sqrt{14}}{3}$ cu cm   **49.** 18
51. True; $\dfrac{15}{\sqrt{5}} = \dfrac{3 \cdot 5}{\sqrt{5}} = \dfrac{3 \cdot \sqrt{5}\sqrt{5}}{\sqrt{5}} = 3\sqrt{5}$
53. True; $(\sqrt{13})^3 = \sqrt{13}\sqrt{13}\sqrt{13} = 13\sqrt{13}$
55. False. We can add only like terms that have the same variable with the same exponent; $5x + 3x^2 \neq 8x^3$.
57. False; $(\sqrt{2}m)(\sqrt{2}m) = (\sqrt{2}\sqrt{2})(m \cdot m) = 2m^2 \neq 2m$
59. True; $(\sqrt[3]{b})(2\sqrt[3]{b})(3\sqrt[3]{b}) = (2 \cdot 3)(\sqrt[3]{b}\sqrt[3]{b}\sqrt[3]{b}) = 6b$

Homework 4.5

1a. $A = 10a^2 + 6a$ **b.** $2a(5a + 3) = 2a(5a) + 2a(3) = 10a^2 + 6a$ **3a.** $A = 6x^2y - 15xy + 3xy^2$
b. $3xy(2x - 5 + y) = 3xy(2x) + 3xy(-5) + 3xy(y) = 6x^2y - 15xy + 3xy^2$ **5.** $-2b(6b - 2) = -12b^2 + 4b$
7. $-4x^2(2x + 3y) = -8x^3 - 12x^2y$ **9.** $(y^3 + 3y - 2)(2y) = 2y^4 + 6y^2 - 4y$ **11.** $-xy(2x^2 - xy + 3y^2) = -2x^3y + x^2y^2 - 3xy^3$ **13.** $-ax + 15a$ **15.** $4x + 2$ **17.** -3
19. $-3m$ **21.** $4; -3h$ **23.** 1512   **25.** 1312
27a. $x^2 + 4x + 3x + 12$ **b.** $(x + 3)(x + 4)$
29a. $4x^2 - 4x - 2x + 2$ **b.** $(4x - 2)(x - 1)$
31. $a^2 - 8a + 15$ **33.** $3y^2 + y - 2$ **35.** $20x^2 + 7x - 6$
37. $x^2 + xy - 2y^2$ **39.** $6s^2 + 11st + 3t^2$
41. $2x^2 - 7ax + 3a^2$ **43a.** $-3x$ **b.**

	z	-9
x	x^2	$-9x$
6	$6x$	-54

45a. $3x$ **b.**

	x	4
$2x$	$2x^2$	$8x$
-5	$-5x$	-20

47. $6x^2 - 20x + 6$

49. $-3x^2 - 9x + 12$ **51.** $-4x^2 + 5x + 6$ **53.** $x^2 - 9$
55. $x^2 - 4a^2$ **57.** $9x^2 - 1$ **59.** $w^2 + 8w + 16$
61. $z^2 - 12z + 36$ **63.** $9a^2 - 12ac + 4c^2$

65.

a	b	$a - b$	$(a - b)^2$	a^2	b^2	$a^2 - b^2$
5	3	2	4	25	9	16
2	6	-4	16	4	36	-32
-4	-3	-1	1	16	9	7

No, $(a - b)^2 \ne a^2 - b^2$.
67a. $x^2 + 7x + 7x + 49$ **b.** $(x + 7)^2$ **69a.** $x^2 - 5x - 5x$
$+ 25$ **b.** $(x - 5)^2$ **71.** No; $(x + 4)^2$ will have a linear
(middle) term because both the inner and the outer terms are $4x$,
and their sum will be $8x$. For $x = 1$: $(x + 4)^2 = (1 + 4)^2 = 5^2 =$
$25; x^2 + 4^2 = 1^2 + 4^2 = 1 + 16 = 17$ **73.** $x^2 - 4x + 4$
75. $4x^2 + 4x + 1$ **77.** $9x^2 - 24xy + 16y^2$
79. $h^2 + (\sqrt{7})^2 = (h + 1)^2$ **81.** $12^2 + (x - 1)^2 = (2x + 1)^2$

Midchapter Review

1. $-3^2 = -(3 \cdot 3) = -9; (-3)^2 = (-3)(-3) = 9$ **3.** Cube
root   **5.** $x^2 = 9$ is true for $x = \pm 3$, but $x = \sqrt{9}$ is true only for
$x = 3$.   **7.** The Pythagorean theorem applies only to right triangles, and c must be the hypotenuse.   **9.** $-\sqrt{2}$ (other answers
possible)   **11a.** 147   **b.** -441 **13a.** 12   **b.** 8
15a. 1   **b.** $\dfrac{7}{5}$ **17a.** -2 **b.** -8 **19a.** 4   **b.** 4
21a. Irrational   **b.** Undefined   **c.** Rational   **d.** Irrational
23a. -4 **b.** -4 **25a.** -4 **b.** 4   **27a.** 12   **b.** 18
29a. 6   **b.** 12   **31.** 126; $30\sqrt{7} + 84$ **33.** $\pm\dfrac{2}{3}$
35. ± 2

37.

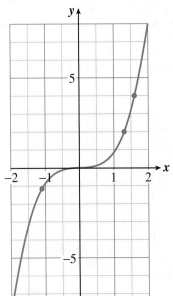

a. 1.3
b. -1.1
c. 1.6

39. $\dfrac{s}{t} = \dfrac{at}{2}$ **41.** $\pm\sqrt{\dfrac{3V}{h}}$ **43a.** 3 sec   **b.** 36 ft
45. $V = 16b^3$ **47a.** 17   **b.** 40   **c.** 60   **49a.** 1
b. $3 + \sqrt{3}$ **c.** $\dfrac{\sqrt{3}}{2}$ **51.** 13.9 yd   **53.** $-20x + 15x^2$
55. $9a^2 - 3a$ **57.** 1   **59a.** $5x^2 - 2x + 5x - 2$
b. $(x + 1)(5x - 2)$ **61.** $x^2 + 2x - 8$ **63.** $25a^2 - 1$
65. $4q^2 + 20q + 25$

Section 4.6 Activity 1

a.

Price per umbrella p	Number of umbrellas sold $150 - 5p$	Revenue $R =$
11	95	1045
13	85	1105
15	75	1125
17	65	1105
18	60	1080
20	50	1000
22	40	980
24	30	720
28	10	280
30	0	0

b., c.

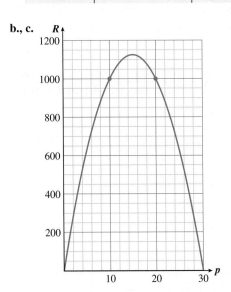

d. $20;
50 umbrellas
e. $1125
f. $15

Section 4.6 Activity 2

a.

t	h
0	360
0.5	374
1	380
1.5	378
2	368
2.5	350
3	324
4	248
5	140
6	0

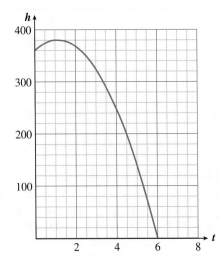

b. 380 ft; 324 ft   **c.** 4.5 sec   **d.** 6 sec
e. $-16t^2 + 36t + 360 = 0$ **f.** $-16(6)^2 + 36(6) + 360 = 0$

5. $(0, 1)$; $(0, -3)$; $(0, -1)$; $(0, 2)$; $(0, -2)$; $(0, 5)$; $(0, 3)$; $(0, 9)$

Homework 4.6

1.

x	y		x	y
-3	10		-3	6
-2	5		-2	1
-1	2		-1	-2
0	1		0	-3
1	2		1	-2
2	5		2	1
3	10		3	6

7.

x	y		x	y
-4	4		-3	16
-3	1		-2	9
-2	0		-1	4
-1	1		0	1
0	4		1	0
1	9		2	1
2	16		3	4

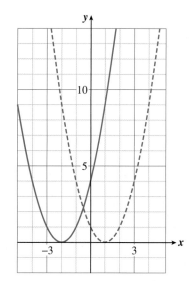

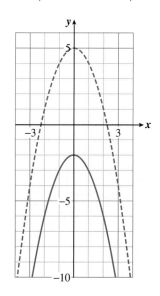

3.

x	y		x	y
-3	-11		-3	-4
-2	-6		-2	1
-1	-3		-1	4
0	-2		0	5
1	-3		1	4
2	-6		2	1
3	-11		3	-4

9.

x	y		x	y
-3	-4		0	-16
-2	-1		1	-9
-1	0		2	-4
0	-1		3	-1
1	-4		4	0
2	-9		5	-1
3	-16		6	-4

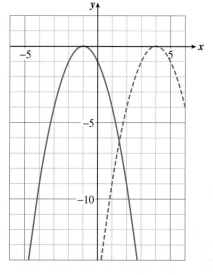

11. $(-2, 0)$; $(1, 0)$; $(2, 0)$; $(-3, 0)$; $(-1, 0)$; $(4, 0)$; $(3, 0)$; $(-4, 0)$

13a.

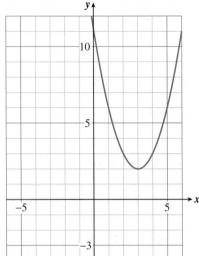

x	y
0	11
1	6
2	3
3	2
4	3
5	6
6	11

b. $(3, 2)$ **c.** $y = x^2 - 6x + 11$

15a.

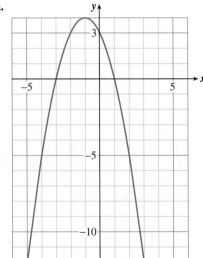

x	y
-3	0
-2	3
-1	4
0	3
1	0
2	-5
3	-12

b. $(-1, 4)$ **c.** $y = -x^2 - 2x + 3$

17.

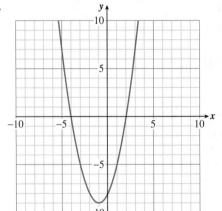

x	y
-5	7
-4	0
-3	-5
-2	-8
-1	-9
0	-8
1	-5
2	0
3	7

a. $x = -4, x = 2$ **c.** $x^2 + 2x - 8$ **d.** $0; 0$

19.

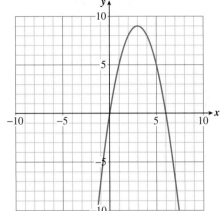

x	y
-1	-7
0	0
1	5
2	8
3	9
4	8
5	5
6	0
7	-7

a. $x = 0, x = 6$ **c.** $-x(x - 6) = -x^2 + 6x$ **d.** $0; 0$

21.

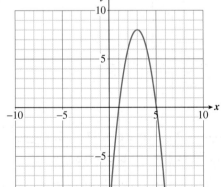

x	y
-1	16
0	9
1	4
2	1
3	0
4	1
5	4
6	9
7	16

a. $x = 3$ **c.** $x^2 - 6x + 9$ **d.** 0

23.

x	y
-1	-24
0	-10
1	0
2	6
3	8
4	6
5	0
6	-10
7	-24

a. $x = 1$ or $x = 5$ **c.** $-2x^2 + 12x - 10$ **d.** $0; 0$

25. To find the solutions to a quadratic equation, set each factor equal to zero and solve for x.

27a.

t	0	0.5	1	1.5	2	2.5	3
h	48	60	64	60	48	28	0

b.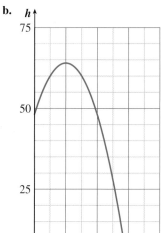

c. 55 ft
d. 2.2 sec
e. $40 = 48 + 32t - 16t^2$
f. 1 sec
g. 2 sec

b.

n	R
0	0
5	350
10	600
20	800
30	600
40	0

c. $800; 20 bracelets

33a. $-16t(t-5)$

b.

t	h
0	0
1	64
2	96
2.5	100
3	96
4	64
5	0

c. 100 ft; 2.5 sec

35. $(3, 0)$ and $\left(\dfrac{-5}{2}, 0\right)$

37. $(0, 0)$ and $(3, 0)$ **39.** $(0, 0)$ and $\left(\dfrac{8}{3}, 0\right)$

41a. $y^2 + 6y + 8$

b.

	y	2
y	y^2	$2y$
4	$4y$	8

43a. $w^2 - 3w - 18$

b.

	w	3
w	w^2	$3w$
-6	$-6w$	-18

45a. $2t^2 + 13t + 15$

b.

	2t	3
t	$2t^2$	$3t$
5	$10t$	15

29a.

x	10	20	30	40	50	60	70	80	90	100
C	330	280	250	240	250	280	330	400	490	600

b.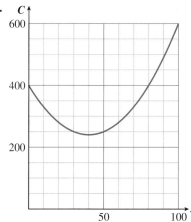

c. $260
d. 10 or 70 toasters
e. $330 = 0.1x^2 - 8x + 400$
f. 40 toasters
g. 90 or fewer

Homework 4.7

1. $x = -1$ or $x = 4$ **3.** $p = 0$ or $p = -7$ **5.** $v = \dfrac{-3}{2}$
or $v = \dfrac{1}{4}$ **7a.** $x^2 - 6x - 27$ **b.** $x = 9$ or $x = -3$

c. $0 = 0; 0 = 0$ **9a.** $-3x(x - 16) = -3x^2 + 48x$
b. $x = 0$ or $x = 16$ **c.** $0 = 0; 0 = 0$ **11.** $2a(3a - 4)$
13. $6v(-3v - 1)$ or $-6v(3v + 1)$ **15.** $h(4 - 9h)$
17. $k(1 - at)$ **19.** $r(2h - \pi r)$ **21.** $h(\pi r^2 - \dfrac{1}{3}s^2)$

23. $0, \dfrac{3}{2}$ **25.** $0, \dfrac{1}{20}$ **27.** $0, -48$

29. The result is not a value for x that is a solution to the original
equation; it gives only another (equivalent) equation with x's on
both sides.   **31a.** $-2n(n - 40)$

Homework 4.8

1a. 6; 5 **b.** $(x + 5)(x + 6) = x^2 + 6x + 5x + 30 = x^2 + 11x + 30$ **3a.** −9; −9x **b.** $(x − 9)(x − 3) = x^2 − 3x − 9x + 27 = x^2 − 12x + 27$ **5a.** 2; x; 8; 8x
b. $(x + 8)(x + 2) = x^2 + 2x + 8x + 16 = x^2 + 10x + 16$
7. $(n + 2)(n + 8)$ **9.** $(h + 2)(h + 24)$ **11.** $(a − 2)(a − 6)$
13. $(t − 3)(t − 12)$ **15.** $(x + 2)(x − 5)$
17. $(a − 2)(a + 10)$ **19.** $(x − 2)(x − 15)$ **21.** cannot be factored **23.** $(y + 1)(y − 45)$ **25.** cannot be factored
27. $(q + 1)(q − 6)$ **29.** $(n − 2)(n − 3)$
31. $(x + 3)(x − 3)$ **33.** $(2 + w)(2 − w)$ **35.** $b^2 − 121 = (b + 11)(b − 11)$ **37.** $x = 2; x = −5$ **39.** $t = 6; t = −7$
41. $x = −1; x = 6$ **43.** $n = 7$ **45.** $q = 0; q = 2$
47. $x = −3; x = 7$ **49.** $x = −2; x = 3$
51a. x-intercepts: $(0, 0)$ and $(−2, 0)$; y-intercept: $(0, 0)$
b. vertex: $(−1, −1)$
c.

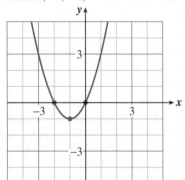

53a. x-intercepts: $(−1, 0)$ and $(3, 0)$; y-intercept: $(0, −3)$
b. vertex: $(1, −4)$
c.

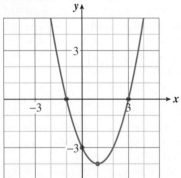

55a. x-intercepts: $(−3, 0)$ and $(3, 0)$; y-intercept: $(0, 9)$
b. vertex: $(0, 9)$
c.

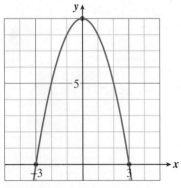

57b. $A = w(w + 3)$ **c.** $w(w + 3) = 54$
d. The width $w = 6$ yd **59.** It takes 6 hr.

Homework 4.9

1. $(2x + 1)(x + 5)$ **3.** $(5t + 2)(t + 1)$ **5.** $(3x − 5)(x − 1)$
7. $(2x − 9)(x − 2)$ **9.** $(x + 4)(5x − 4)$
11. $(3h + 2)(2h + 1)$ **13.** $(n − 1)(9n + 1)$
15. $(2t − 5)(3t + 5)$ **17.** $(x − 4)(5x + 6)$
19. $(3x − 8)(3x + 1)$ **21.** $(8z − 5)(2z + 1)$

23. $(4a − 1)(a + 6)$ **25.** $n = \dfrac{4}{3}; n = −1$

27. $t = \dfrac{1}{3}; t = \dfrac{3}{2}$ **29.** $y = \dfrac{1}{2}$ **31.** $z = \dfrac{−5}{2}; z = \dfrac{1}{3}$

33. $y = \dfrac{2}{3}; y = −2$ **35.** $x = \dfrac{3}{2}; x = 1$ **37.** $1\dfrac{1}{2}$ sec

39. The width is $w = 6$ cm, and the length is $2w − 2 = 2(6) − 2 = 10$ cm. **41.** The altitude is $a = 5$ in., and the base is $a + 11 = 5 + 11 = 16$ in. **43.** The width is $w = 5$ ft, and the length is $2w + 2 = 2(5) + 2 = 12$ ft. **45a.** x-intercepts: $(5, 0)$ and $(−3, 0)$; y-intercept: $(0, −15)$; x-intercepts: $(−5, 0)$ and $(3, 0)$; y-intercept: $(0, −15)$ **b.** vertex: $(1, −16)$; vertex: $(−1, −16)$
c.

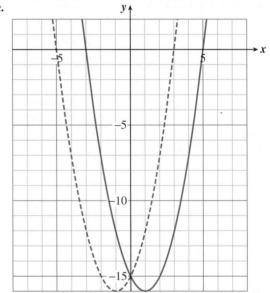

47a. x-intercepts: $(2, 0)$ and $(−1, 0)$; y-intercept: $(0, −2)$; x-intercepts $(2, 0)$ and $(−1, 0)$; y-intercept: $(0, 2)$

b. vertex: $\left(\dfrac{1}{2}, \dfrac{−9}{4}\right)$; vertex: $\left(\dfrac{1}{2}, \dfrac{9}{4}\right)$

c.

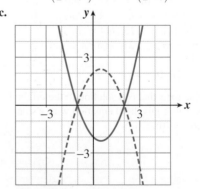

49a. *x*-intercept: (1, 0); The other factor is the same and will give the same intercept. *y*-intercept: (0, 3); *x*-intercept: (−1, 0); *y*-intercept: (0, 3)   **b.** vertex: (1, 0); vertex: (−1, 0)

c.

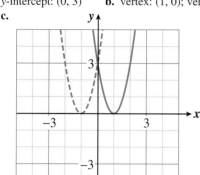

51. $p = 5; p = -4$ **53.** $m = \dfrac{1}{4}$ **55.** $(x + 7)(x + 8)$

57. $(b - 12)(b + 20)$ **59.** $(n - 100)(n + 3)$

61. $(u - 6)(3u + 1)$ **63.** $(t - 6)(2t - 9)$

Homework 4.10

1a. $\dfrac{8 \pm \sqrt{48}}{2}$ **b.** 7.46; 0.54   **3a.** $\dfrac{-2 \pm \sqrt{28}}{6}$

b. 0.55; −1.22   **5a.** $\dfrac{1 \pm \sqrt{5}}{2}$ **b.** 1.62; −0.62

7a. $\dfrac{-2 \pm \sqrt{20}}{-8}$ **b.** −0.31; 0.81   **9a.** ±1.29   **b.** ±1.29

11. $0, \dfrac{1}{3}$ **13.** $\dfrac{3}{2}, 2$ **15.** $\dfrac{-6 \pm \sqrt{52}}{2}$ **17.** $\dfrac{7}{8}$ or 1

19. $\dfrac{9 \pm \sqrt{73}}{4}$ **21.** 15 or 3

23a. *x*-intercepts: (2.73, 0) and (−0.73, 0)

b.

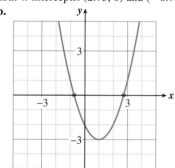

x	y
−1	1
0	−2
1	−3
2	−2
3	1

25a. No *x*-intercepts

b.

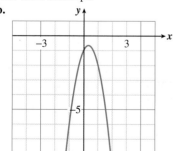

x	y
−1	−6
0	−1
1	−2
2	−9

27a. *x*-intercepts: (4.24, 0) and (−0.24, 0)

b.

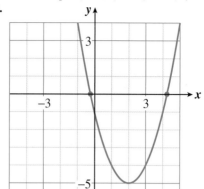

x	y
−1	4
0	−1
1	−4
2	−5
3	−4
4	−1
5	4

29a. $2\dfrac{1}{2}$ sec   **b.** 5.13 sec   **31.** width: 15 in., length: 40 in., height: 20 in.   **33.** 9 in. by 12 in.   **35.** The sides are $x = 8$ and $2x - 1 = 15$.   **37.** The sides are $k = 6$ and $2k - 5 = 7$.

Chapter 4 Summary and Review

1. Pythagorean theorem: $a^2 + b^2 = c^2$ where c is the hypotenuse of a right triangle and a and b are the two legs. This theorem gives us a formula for finding the third side of a right triangle if we know the other two sides.   **3.** Property of binomial products: When we represent the product of two binomials by the area of a rectangle, the products of the entries on the two diagonals are equal.   **5.** Order of operations: 1. Perform any operations inside parentheses, under a radical, or above or below a fraction bar. 2. Compute all powers and roots. 3. Perform all multiplications and divisions in order from left to right. 4. Perform all additions and subtractions in order from left to right.   **7.** To solve a quadratic equation by extraction of roots: 1. Isolate the square of the variable. 2. Take the square root of both sides. 3. Solve for the variable. Note: Because every positive number has two square roots use ±.   **9.** To multiply two binomials, use FOIL. Multiply each term of the first binomial by each term of the second binomial.   **11.** To find the *x*-coordinate of the vertex, find the average of the *x*-intercepts. To find the *y*-coordinate of the vertex, substitute the *x*-coordinate into the equation for the parabola.

13. −14   **15.** −2   **17a.** −13   **b.** 48

19. The coefficient indicates how many times the power is added together. For example, $2x^3$ means $x^3 + x^3$ and $3x^2$ means $x^2 + x^2 + x^2$. The exponent indicates how many times the base is multiplied together. For example, x^3 means $x \cdot x \cdot x$ and x^2 means $x \cdot x$.   **21a.** $-2t^2$ **b.** cannot be simplified   **c.** $-48t^3$

23. −20.247   **25.** $150p^3; 190p^2$ **27.** $\pm\dfrac{2}{3}$ **29a.** 2.5 ft

b. 78.54 sq ft   **31.** $\dfrac{2S}{n} - f$ **33.** $\pm\sqrt{\dfrac{C}{br}}$

35. Irrational; 17.32   **37.** Irrational; 7.47

39a.

h	d
0	0
1000	110
2000	155
3000	190
4000	219
5000	245

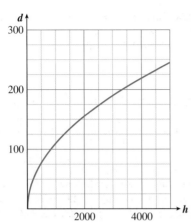

b. 225 mi   **c.** 223.13 mi   **d.** 800 m   **41.** $3x$ **43.** $25a$
45. 11.40 m   **47.** 10 m   **49.** $6x^2y - 12xy - 3xy^2$
51. $-2a^2 - 9a + 8$ **53.** $u^2 - 7u + 10$
55. $a^2 + 12a + 36$ **57.** $6a^2 - 11ac - 10c^2$
59a.

t	0	0.5	1	1.25	1.5	1.75	2	2.25	2.5
h	100	96	84	75	64	51	36	19	0

b.

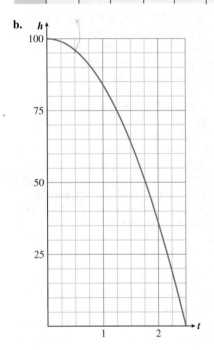

c. 36 ft; $h = 100 - 16t^2 = 100 - 16(2)^2 = 100 - 16(4)$
$= 100 - 64 = 36$

d. 1.25 sec;

$$h = 100 - 16t^2$$
$$75 = 100 - 16t^2$$
$$-25 = -16t^2$$
$$\frac{-25}{-16} = t^2$$
$$\sqrt{\frac{25}{16}} = t$$
$$\frac{5}{4} = t$$

61.

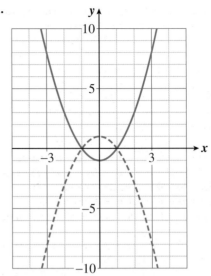

x	y
-2	3
-1	0
0	-1
1	0
2	3

x	y
-2	-3
-1	0
0	1
1	0
2	-3

63. $6x(4x - 3)$ **65.** $(a - 15)(a - 3)$ **67.** $(2y + 3)(2y + 5)$
69. $(3w - 1)(w + 5)$ **71.** $(z + 11)(z - 11)$
73. $3(2x + 1)(x + 3)$ **75.** -5 **77.** 0, 4   **79.** 3, 1
81a. x-intercepts: $(0, 0)$ and $(-6, 0)$; y-intercept: $(0, 0)$
b. vertex: $(-3, -9)$
c.

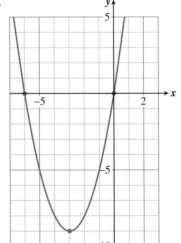

83. $\dfrac{-6 \pm \sqrt{12}}{4}$ **85.** 3.81 sec
87a. $R = p(160 - 2p) = 160p - 2p^2$ **b.** $-2p^2 + 136p - 80$
c. They charged 30 cents/lb or 38 cents/lb   **89.** The sides are
$m = 3.9$, $m + 5 = 8.9$, and 8.

Homework 5.1

1. Allowed   **3.** Not allowed   **5.** A polynomial of degree 3
has one term of degree three and no terms with a higher degree.

Examples: x^3 or $2x^3 + x^2 - x + 3$; A trinomial has exactly three terms and can have any degree higher than 1. Examples: $x^2 - x + 3$ or $2x^{11} + 8x^4 - 3x$ **7.** In 7–10, other examples possible. $2x^4$ **9.** $x^2 + 5x - 6$ **11a.** Yes   **b.** No, variable is in the denominator   **c.** No, variable is in the denominator   **d.** Yes   **13.** 2   **15.** 7   **17.** 4
19. $-1.9x^3 + x + 6.4$ **21.** $-2x^2 + 6xy + 2y^3$
23. 14   **25.** 15.9872   **27.** 24   **29.** ≈$1225.04
31. 130 ft; No   **33.** 567   **35.** 1001
37. $x^3 + 3x^2 + 4x + 1$ **39.** $x^4 - 7x^3 - 5x^2 + 8x - 3$
41a. 29. When evaluating mentally, it is easier before expanding.
b. 6.632   **43.** $12b^3$ **45.** $6x + 6xy - 9y$ **47.** The exponent should not change when like terms are combined. $14w^3$
49. Unlike terms cannot be combined. $6 + 3x^2$ cannot be simplified.   **51.** The sign of each term in parentheses must be changed. $t^2 + 12$ **53.** $2y^3 + 2y^2 + y + 1$
55. $3x^3 + 3x^2 + 11$ **57.** $6x^2 + 2x - 3$ **59.** $-7x^2 + 7x - 1$
61a. $-0.3x^2 + 9x - 50$ **b.** $10; $17.50; $10
63a. $28t^2 - 56t + 28$ **b.** 28 ft, 7 ft   **c.** They will be at the same height in 1 sec.   65a. $0.30h$ **b.** $50 - h$
c. $0.25(50 - h)$ **d.** $0.05h + 12.5$ **67a.** $2w$ **b.** $w + 30$
c. $2(w + 30)$ **d.** $4w + 80$ mi
69a.

n	S_n	$\frac{1}{2}n^2 + \frac{1}{2}n$
1	1	$\frac{1}{2}(1)^2 + \frac{1}{2}(1) = 1$
2	3	$\frac{1}{2}(2)^2 + \frac{1}{2}(2) = 3$
3	6	$\frac{1}{2}(3)^2 + \frac{1}{2}(3) = 6$
4	10	$\frac{1}{2}(4)^2 + \frac{1}{2}(4) = 10$
5	15	$\frac{1}{2}(5)^2 + \frac{1}{2}(5) = 15$
6	21	$\frac{1}{2}(6)^2 + \frac{1}{2}(6) = 21$
7	28	$\frac{1}{2}(7)^2 + \frac{1}{2}(7) = 28$
8	36	$\frac{1}{2}(8)^2 + \frac{1}{2}(8) = 36$
9	45	$\frac{1}{2}(9)^2 + \frac{1}{2}(9) = 45$
10	55	$\frac{1}{2}(10)^2 + \frac{1}{2}(10) = 55$

c. They are the same.   **71.** -7.2

Homework 5.2

1. x^9 **3.** 5^{14} **5.** b^9 **7.** 5   **9.** 4   **11.** 2
13a. $-6x^8$ **b.** $2x^6$ **c.** $-x^4y^3$ **d.** $-9x^5y^5$
15a. $-x^4$ **b.** cannot be simplified   **c.** cannot be simplified
d. cannot be simplified   **17.** $24a^2, 11a$ **19.** $-10b^6, 3b^3$
21. $-28p^4q^2, -3p^2q$ **23a.** $50m^2$ **b.** $144m^4$
25a. $48w^5$ **b.** $144w^{10}$ **27a.** $2x^2$ **b.** x^4 **c.** 0
d. $-x^4$ **29a.** Cannot be simplified   **b.** x^3 **c.** Cannot be simplified   **d.** $-x^3$ **31.** $-8w^7$ **33.** $-10s^7t$ **35.** $-36x^3y^6$ **37a.** $4x$ **b.** $-4x^2$ **c.** $-2x$ **d.** $-3x - 2$
39a. $7b$ **b.** $12b^2$ **c.** Cannot be simplified   **d.** b^7
41. $-x^3y - x^2y^2 - xy^3$ **43.** $-9s^4t^4 + 3s^3t^3 - 18s^2t^2$
45. $-3ax$ **47.** $3a^2b^2 + 2ab^2$ **49.** $4a^3 + 16a^2 - 20a$
51. $-4x^3 + 4x^2 - x$ **53.** $6s^4t^2 + s^3t^3 - s^2t^4$
55. $x^3 - 5x^2 + 8x - 4$ **57.** $27x^3 - 18x^2 + 6x - 1$
59a. $x^3 + 6x^2 + 11x + 6$ **b.** 0, 0, 0

61a. $8x^3 - 12x^2 + 6x - 1$ **b.** $\frac{1}{2}$

63a. $16 - n$ **b.** $16n - n^2$ **65a.** Length: $w + 3$,
Height: $w - 2$ **b.** $w^3 + w^2 - 6w$ **c.** $6w^2 + 4w - 12$
67. 80,000,000   **69.** 60,000   **71.** 66,000

Homework 5.3

1. a^3 **3.** 3^5 **5.** $\frac{1}{z^3}$ **7.** For $t = 2$, $t^2 \cdot t^2 = 32$; $t^6 = 64$

9. For $v = 2$, $\frac{v^8}{v^2} = 64$; $v^4 = 16$ **11.** For $n = 2$, $\frac{n^3}{n^5} = \frac{1}{4}$;

$n^2 = 4$ **13.** $\frac{1}{4xy^4}$ **15.** $\frac{-3x^2}{2}$ **17.** $\frac{5}{y^2}$ **19.** 2
21. $4ab$ **23.** $3x^2(3 - 4x^3 + x)$ **25.** $7xy(2x^2 - 5xy + 3y^2)$
27. $-b(b + c + a)$ **29.** $-2k(2k^3 - 2k + 1)$
31. $(x + 6)(2x - 3)$ **33.** $(2x + 3)(3x^2 - 1)$
35. $-1(x + 7)(x - 3)$ **37.** $-1(a - 27)(a + 3)$
39. $2(x + 2)(x + 3)$ **41.** $b(2a + 7)(2a - 1)$
43. $2z(2z + 3)(z + 1)$ **45.** $9b(2a - 3)(a + 1)$
47. $(x - 2y)(x - 3y)$ **49.** $(x + 11a)(x - 7a)$
51. $4x(x + 2y)(x + y)$ **53.** $9ab(a + 2b)(a - b)$
55. $(2t + s)(t - 3s)$ **57.** $(4by + 1)(by + 1)$
59. $3a(4b + a)(b + a)$

Homework 5.4

1. $64t^8$ **3.** $144a^4$ **5.** $100h^4k^2$ **7.** $9a^2 + 6ab + b^2$
9. $49b^6 - 84b^3 + 36$ **11.** $4h^2 + 20hk^4 + 25k^8$
13. $9p^2 - 16$ **15.** $4x^4 - 1$ **17.** $h^4 - 49t^2$
19. $(y + 3)^2$ **21.** $(m - 15)^2$ **23.** $(x + 2y)^2$
25. $(x + 3)(x - 3)$ **27.** $(6 + ab)(6 - ab)$
29. $(8y + 7x)(8y - 7x)$ **31.** $(a^2 + 5)^2$
33. $(6y^4 + 7)(6y^4 - 7)$ **35.** $(4x^3 + 3y^2)(4x^3 - 3y^2)$
37. $3(a + 5)(a - 5)$ **39.** $2a(a - 3)^2$
41. $9x^3(x^2 + 3)(x^2 - 3)$ **43.** $3(4h^2 + k^6)$
45. $(9x^4 + y^2)(3x^2 + y)(3x^2 - y)$
47. $2a^4(9b^4 + a^2)(3b^2 + a)(3b^2 - a)$ **49.** The area of the large rectangle is $(a + b)^2$. The sum of the areas of the four smaller rectangles is $a^2 + 2ab + b^2$.   **51a.** $a^3 - 3a^2b + 3ab^2 - b^3$
b. $8x^3 - 36x^2 + 54x - 27$ **c.** $(a - b)^3 = 27$; $a^3 - b^3 = 117$
53a. $a^3 + b^3$ **b.** $(a + b)(a^2 - ab + b^2)$
c. $(x + 2)(x^2 - 2x + 4)$

Section 5.5 Activity

1a.

w	l
2	30
3	20
4	15
5	12
6	10
10	6
12	5
15	4
20	3

b. $l = \dfrac{60}{w}$

c.

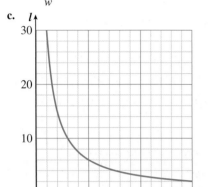

d. l decreases toward 0 as w increases.

Homework 5.5

1a.

n	s
5	240
6	200
10	120
12	100
20	60
25	48
30	40
40	30

b. $s = \dfrac{1200}{n}$

c.

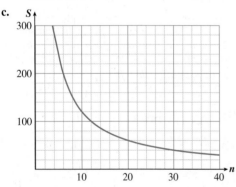

d. $\dfrac{3}{2}$ **3a.** $c = \dfrac{360}{p}$ **b.**

p	c
6	60
9	40
10	36
12	30
15	24
18	20
20	18

c.

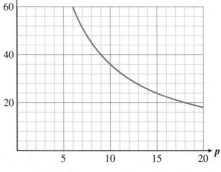

d. $\dfrac{2}{3}$

5.

w	t
4.5	8
7.2	5
7.5	4.8

$t = \dfrac{36}{w}$

7.

R	C
20	200
16	250
125	32

$C = \dfrac{4000}{R}$

9. $y = \dfrac{16}{x}$ **11.** $y = \dfrac{120}{x}$ **13.** Yes; the product of the variables is constant.   **15.** No; the product of the variables is not constant.

17.

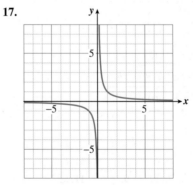

a. As x gets large and is positive, y gets close to zero and is positive.   **b.** As x gets close to zero and is positive, y gets large and is positive.   **c.** As x gets large and is negative, y gets close to zero and is negative.   **d.** As x gets close to zero and is negative, y gets large and is negative.

19.

a. As x gets large and is positive, y gets close to zero and is negative.   **b.** As x gets close to zero and is positive, y gets large and is negative.   **c.** As x gets large and is negative, y gets close to zero and is positive.   **d.** As x gets close to zero and is negative, y gets large and is positive.   **21.** graph (c)   **23.** graph (e)
25. graph (b)

Midchapter Review

1. $\frac{x}{2} + 3$ is equivalent to $\frac{1}{2}x + 3$. $\frac{2}{x} + 3$ has a variable in the denominator.   **3.** In 3–9, other examples possible. Degree 5; $2x^5 + x^4 - 3x^2 + 1$ **5.** $(x + 1) + (x - 1) = 2x$ (one term); $(x + 1) + (x + 1) = 2x + 2$ (two terms); $(x^2 + 1) + (x + 1) = x^2 + x + 2$ (three terms); $(x^3 + x^2) + (x + 1) = x^3 + x^2 + x + 1$ (four terms)
7. Square of monomial: $(3x)^2 = 9x^2$ (one term); Square of binomial: $(x + 3)^2 = x^2 + 6x + 9$ (three terms)   **9.** Sum of two squares: $a^2 + b^2$; Square of a binomial: $(a + b)^2 = a^2 + 2ab + b^2$
11. $-\frac{x^6}{720} + \frac{x^4}{24} - \frac{x^2}{2} + 1$; degree 6   **13.** 26   **15.** $\frac{1}{3}$
17. 0   **19a.** 4845   **b.** 0   **c.** 1; There is only one way to choose 4 items from a list of 4; choose all 4 items.

21. $2a^2 + a - 9$ **23.** $\frac{1}{2c}$ **25.** $56q^2 - 49q^4 - 7q^6$
27. $81v^2 - 25w^2$ **29.** $-3p^4 + 9p^3 + 30p^2$ **31.** $8x^3 - 27$
33a. $10a^4$ **b.** $6a^8$ **35a.** $8m + 2n$
b. $3m^2 + 5mn - 2n^2$ **37.** -32.6 **39.** Equation: $x = \frac{-3}{2}, 1$
41. Polynomial: $(2x + 3)(x - 1)$ **43.** Equation: $x = \frac{-1}{2}, 0$
45. Equation: $x = \frac{-3}{2}$ **47.** Polynomial: $2x(x + 1)(x - 1)$
49. Equation: $p = -1, 0, 1$ **51.** $4x^3(3x^2 - 2x + 5)$
53. $-5d^2(2d^2 - 4d + 1)$ **55.** $(q - 3)(7q - 1)$
57. $3z(z + 2)(z - 2)$ **59.** $(a^2 + 5)^2$ **61.** $2a^2b^2(b^4 + 16a^4)$
63. $-2a^2b(a + 5b)(a - 3b)$

65. $t = \frac{18}{v}$

v	t
2.4	7.5
3.6	5
15	1.2

67. $y = \frac{15}{x}$ **69.** No; the product of the variables is not constant.

Homework 5.6

1a. $5; \frac{3}{7}$ **b.** 3   **3a.** $\frac{-3}{10}; \frac{-3}{2}$ **b.** None

5. (b), (c), and (d)   **7a.** $\frac{18}{x - 4}$ dollars per gallon
b. 1.8; If the tank holds 14 gal, then gas cost \$1.80/gal.
9a. $\frac{2800}{h - 10}$ miles per hour   **b.** If he drove 50 hr, then his
average speed was 70 mph.   **11a.** $\frac{200}{2x - 1}$ square centimeters
b. 8; If $x = 13$, then the area of cross-section is 8 cm^2.
13. $3x^3 - 2 - \frac{2}{3x^2}$ **15.** $n - 1 + \frac{2}{3n} - \frac{1}{n^2}$ **17.** $x - 2 + \frac{3}{y}$
19. $\frac{3}{4}$ **21.** $\frac{-b^2}{5a}$ **23.** $\frac{x + 4}{2}$ **25.** $\frac{a}{a - 3}$ **27.** $\frac{3}{4}$
29. $\frac{1}{x + 1}$ **31.** (b)   **33.** (a)   **35.** (a)
37. Cannot be reduced   **39.** Cannot be reduced
41a. 30 min   **b.** 50 min   **c.** 50 min   **d.** 8 mph
e. As the current increases, the time increases. If the current is
10 mph, the team will not be able to row upstream.   **43.** $\frac{-1}{2}$
45. $\frac{1}{a + b}$ **47.** $\frac{3x + 2y}{2y - 3x}$ **49.** $\frac{-a}{a + 1}$ **51.** (b), (c), (d)
53. (a), (d)   **55.** (b), (c)   **57.** The 3 in the numerator is not a factor; no cancellation is possible.   **59.** The $5z$ in the numerator is not a factor; no cancellation is possible.

Homework 5.7

1. $\frac{3}{2x}$ **3.** $\frac{8b}{3(b + 1)}$ **5.** $\frac{3}{4}$ **7.** 5   **9.** $\frac{2x}{3}$
11. $\frac{3a - 3b}{4}$ **13.** $-8t + \frac{1}{4} - \frac{3}{t}$ **15.** $\frac{8v^2}{9} - \frac{8}{v} - 1$
17. $\frac{4LR}{D^2}$ **19.** $\frac{2L}{c} + \frac{2LV^2}{c^3}$ **21.** $\frac{3q}{8\pi R} - \frac{a^2q}{8\pi R^3}$
23. $\frac{9}{14}$ **25.** $\frac{b^2}{6a}$ **27.** $\frac{2y}{x}$ **29.** $\frac{3a}{2}$ **31.** $\frac{x^2}{y - 1}$
33. $\frac{-5}{2a}$ **35.** $\frac{(c - 1)(c + 1)}{(c + 5)(c - 3)}$ or $\frac{c^2 - 1}{c^2 + 2c - 15}$ **37.** $\frac{4}{a}$
39. $\frac{8}{x - 1}$ **41.** $\frac{2x + y}{3x}$ **43.** $m - 1$ **45.** $z - 3$
47. $\frac{4}{b - 1}$ **49a.** $\frac{4 - 2m}{m(m + 1)}$ **b.** $\frac{-1}{6}$ **51.** $-3x + 2$
53. $\frac{y^2 + 4y - 3}{(y - 1)(y + 1)}$ **55.** $\frac{5}{8}$ **57a.** $\frac{x}{2}$ **b.** $2x$ **c.** $\frac{1}{2x}$
59a. $\frac{1}{a + b}$ **b.** $\frac{3}{4(a + b)}$ **c.** $\frac{4}{3(a + b)}$ **61.** $x + 3$
63. $6n + 9$ **65.** $-18w + 3$ or $3 - 18w$

Homework 5.8

1. $\frac{3y - 4x}{xy}$ **3.** $\frac{a - 3}{a}$ **5.** $\frac{yz + x^2z + x^2y}{xyz}$
7. $\frac{uv - 4v + 18}{3v}$ **9.** $\frac{x + 3}{x(x + 1)}$ **11.** $\frac{4n^2 + 15n - 9}{(n + 3)(n - 3)}$
13a. $\frac{144}{x(x - 2)}$ sq ft   **b.** $\frac{48x - 48}{x(x - 2)}$ ft   **15a.** $\frac{100}{x - 10}$
b. $\frac{100}{x + 10}$ **c.** $\frac{200x}{(x - 10)(x + 10)}$ **d.** $\frac{5}{7}, \frac{5}{8}, \frac{75}{56}$

17a. $\dfrac{1}{d}$ **b.** $\dfrac{1}{d-5}$ **c.** $\dfrac{2d-5}{d(d-5)}$ of bag per day

d. $\dfrac{100}{9}$, or $11\dfrac{1}{9}$, days   **19.** $\dfrac{h^2+2h-3}{h+2}$ **21.** $\dfrac{v^2+v-1}{v(v-1)}$

23. $\dfrac{-x^2+6x-4}{x(x-2)}$ **25.** LCD $= 4x^2$; $\dfrac{10x+3}{4x^2}$

27. LCD $= 24z$; $\dfrac{10z-17}{24z}$ **29.** LCD $= 4\cdot(2a-b)$; $\dfrac{13}{4(2a-b)}$

31. LCD $= x(d-x)$; $\dfrac{d}{x(d-x)}$ **33.** LCD $= 2a$; $\dfrac{a^2+N}{2a}$

35. LCD $= 2at$; $\dfrac{a^2-2mt}{2at}$ **37.** LCD $= 4\pi r^2$; $\dfrac{qr+2aq}{4\pi r^2}$

39. LCD $= (c-V)(c+V)$; $\dfrac{2Lc}{(c-V)(c+V)}$

41. LCD $= r(r-a)(r+a)$; $\dfrac{2a^2q}{r(r-a)(r+a)}$ **43.** $6y$

45. $\dfrac{5}{16}$ **47.** $\dfrac{nq}{p+q}$ **49.** $\dfrac{2x+1}{x}$ **51.** $\dfrac{b^2+d^2}{2d}$

53. $\dfrac{y^2+x^2}{xy}$ **55.** $\dfrac{-22}{17}$

57a.

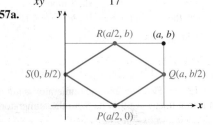

$R(a/2, b)$ (a, b)

$S(0, b/2)$ $Q(a, b/2)$

$P(a/2, 0)$

b. $\dfrac{b}{a}$ $\dfrac{-b}{a}$ $\dfrac{b}{a}$ $\dfrac{-b}{a}$ **59.** $\dfrac{n^2-k^2}{n^2}$ **61.** $\dfrac{2cd}{c^2-u^2}$

63. $\dfrac{LK}{LN-FN}$ **65.** $\dfrac{8}{x-1}$ **67.** $\dfrac{19}{6(p-2)}$

69. $\dfrac{-m+6}{m(m+3)(m-3)}$ **71.** $\dfrac{5k+1}{k(k-3)(k+1)}$

73. $\dfrac{2x^2+2x-3}{x(x+3)^2}$ **75.** $\dfrac{AK_1K_2(T_2-T_1)}{K_2L_1+K_1L_2}$ **77.** $\dfrac{1}{m+2h}$

Homework 5.9

1. 1   **3.** 12   **5.** 4   **7.** $\dfrac{2}{3}$ $\dfrac{-1}{2}$ **9.** 1   **11.** 2

13. $-6, 5$ **15.** No solution.   **17.** $\dfrac{1}{2}, 2$ **19.** 50%

21a. .207   **b.** $\dfrac{x+34}{x+164}$ **c.** 36 hits

23a. $AE = 1$; $ED = x - 1$; $CD = 1$ **b.** $\dfrac{x}{1} = \dfrac{1}{x-1}$ (other

answers possible).   **c.** $\dfrac{1+\sqrt{5}}{2}$ **25.** $\dfrac{PV}{h}$

27. $\dfrac{F-aM}{a}$ **29.** $\dfrac{y-k+hm}{m}$ **31.** $\dfrac{BR}{B-R}$

33. $\dfrac{Pw}{w+0.622}$ **35.** $\dfrac{En-Ir}{In}$ **37.** $\dfrac{r-dc}{cr}$

39. freight train 40 mph; express train 60 mph

41. Sam 40 mph; Reginald 60 mph   **43.** 5 hr   **45.** $34\dfrac{2}{7}$ min

47a. $\dfrac{17x-5}{20}$ **b.** $\dfrac{25}{17}$ **49.** Because $x = 1$, dividing by

$x - 1$ in the fourth step is dividing by zero.

51. The left side of the equation has been multiplied by 6, but the right side has not; $a = 6$

53. In the third equation, dividing by m is division by zero because $m = 0$ is one of the solutions; $m = 6, 0$ **55.** 2 days

Chapter Summary and Review

1. To multiply two powers with the same base, add the exponents and leave the base unchanged; $a^m \cdot a^n = a^{m+n}$ $x^5 \cdot x^2 = x^{5+2} = x^7$ **3.** $(a+b)^2 = a^2 + 2ab + b^2$; $(a-b)^2 = a^2 - 2ab + b^2$ **5.** $y = \dfrac{k}{x}$, where k is the constant of variation.   **7.** To add two polynomials, remove the parentheses and combine like terms. To subtract one polynomial from another, change the sign of each term in the polynomial being subtracted and combine like terms.   **9.** To divide one monomial by another, divide the coefficients, and use the second law of exponents to find the quotients of the variables.   **11.** $(a+b)(a-b)$; cannot be factored   **13.** 1. Factor the numerator and denominator completely. 2. Divide numerator and denominator by any common factors.   **15.** Multiply the first fraction by the reciprocal of the second fraction.   **17.** 1. Find the least common denominator (LCD) for the fractions. 2. Build each fraction to an equivalent one with the same denominator. 3. Add or subtract the numerators and keep the same denominator. 4. Reduce if necessary.

19. Multiply both sides of the equation by the LCD   **21.** 8

23. (a) and (b)   **25.** $9p^2q - 5pq^2$ **27.** $5b^3 + 2b^2 + 6b + 4$

29a. $r+1$ **b.** $2(r+1) = 2r + 2$ **c.** $3r$ **d.** $r-2$

31a. $4x^2$ **b.** $3x^4$ **33a.** Cannot be simplified   **b.** $7a^{10}$

35a. $\dfrac{b^6}{3}$ **b.** $\dfrac{3}{b^6}$ **37.** $-30m^5n^3$ **39.** $-7x^3$

41. $-2y^3 + 2y^2 + 24y$ **43.** $d^3 - 6d^2 + 12d - 8$

45. $144x^5y - 216x^4y^2 + 81x^3y^3$ **47.** $w^3 + 8w^2 - 20w$

49. $6w^4(5w^5 - 7 + 9w^4)$ **51.** $-vw^2(w^3 + w^2 - 1)$

53. $2q^2(q+8)(q-5)$ **55.** $(x-2)(x-y)$

57. $3b(a+3b)(a+b)$ **59.** $(3xy+2)(3xy-1)$

61. $(h-12)^2$ **63.** cannot be factored   **65.** $(w^4+6)^2$

67. $3(s^2+4)(s+2)(s-2)$ **69.** (c); $y = \dfrac{48}{x}$

71a. $I = \dfrac{120}{R}$ **b.** 30 amps   **73a.** $\dfrac{-1}{2}$ **b.** -5 2

75. Cannot be reduced   **77.** $\dfrac{5+y}{y}$ **79.** $\dfrac{1}{v+2}$

81. $\dfrac{-1}{2x+3}$ **83a.** $\dfrac{19}{24}$ **b.** $\dfrac{5}{32}$ **85a.** $\dfrac{10x^2+2x+1}{6x(x+1)}$

b. $\dfrac{1}{4}$ **87a.** $\dfrac{2x+1}{x}$ **b.** $\dfrac{2}{x}$ **89.** $\dfrac{2c}{9d}$ **91.** $m-1$

93. $\dfrac{8q^2-16q+21}{q(q-3)}$ **95a.** $\dfrac{5}{x-2}$ **b.** $\dfrac{5}{x+2}$

c. $\dfrac{10x}{(x+2)(x-2)}$ **97.** $\dfrac{2ab-a^2}{2ab-b^2}$ **99.** $8, -2$

101. $\dfrac{-1}{2}, 4$ **103.** $\dfrac{yz}{y-z}$ **105.** 1.53 mph

Homework 6.1

1a. t^{15} **b.** b^8 **c.** w^{144} **3a.** $125x^3$ **b.** $81w^4z^4$

c. $-a^5b^5$ **5a.** $\dfrac{w^6}{64}$ **b.** $\dfrac{625}{v^4}$ **c.** $\dfrac{-m^3}{p^3}$ **7a.** x^9

b. x^{18} **c.** $\dfrac{1}{x^3}$ **d.** x^3

9. The first law of exponents simplifies a product of powers with the same base by adding the exponents. For Example: $x^5 \cdot x^2 = x^{5+2} = x^7$. The third law of exponents simplifies a power raised to a power by multiplying the exponents. Example: $(x^5)^2 = x^{5 \cdot 2} = x^{10}$. **11.** The 3 should be raised to the second power before multiplying. $2 \cdot 3^2 = 2 \cdot 9 = 18$
13. The exponent does not apply to the negative sign. $-10^2 = -10 \cdot 10 = -100$ **15.** When like bases are multiplied, the exponents should be *added*, not multiplied.
$a^4 \cdot a^3 = a^{4+3} = a^7$ **17.** $32p^{15}$ **19.** $\dfrac{-243}{q^{20}}$ **21.** $\dfrac{16h^8}{m^{12}}$
23. x^{13} **25.** $4x^{10}y^{14}$ **27.** $a^{21}b^{15}$ **29.** $-a^4$
31. $-x^3y^4$ **33.** $-4p^5q^{10}$ **35.** $-52y^7$ **37.** $2a^9 + 2a^8$
39. $-6v^5 + 3v^4 + 16v^3$ **41a.** $4x^2 + 2x^4$ **b.** $8x^6$
43a. $-x^7$ **b.** x^{16} **45a.** $324x^{16}$ **b.** $77x^8$
47a. $6x^3 - 3x^6$ **b.** $-18x^9$ **49a.** $6x^3 - 3x^6$ **b.** $3x^6$
51. 6 **53.** 6 **55.** 6 **57.** $x^2 + x^4$
59. $1 - m^4 + 2m^{12}$ **61.** $<$ **63.** $=$ **65.** $<$ **67.** $=$
69. $>$ **71.** $=$

Homework 6.2

1. $\dfrac{1}{25}$ **3.** $\dfrac{1}{x^6}$ **5.** 1 **7.** $\dfrac{64}{27}$ **9.** $\dfrac{81}{b^4}$ **11.** $\dfrac{1}{32q^5}$
13. $\dfrac{3}{64}$ **15.** $\dfrac{4}{x^2}$ **17.** 2^{-3} **19.** $3 \cdot 5^{-2}$ **21.** 3^{-3}
23. $5^{-4}x$ **25.** $2z^{-2}$ **27.** $10^{-5}z^5$ **29.** Any nonzero number rasied to the zero power is 1. $x^0 = 1$ **31.** A negative exponent is the reciprocal of the corresponding positive power (not a negative number). $w^{-3} = \dfrac{1}{w^3}$ **33.** The exponent applies only to the base x. $2x^{-4} = 2 \cdot x^{-4} = 2 \cdot \dfrac{1}{x^4} = \dfrac{2}{x^4}$ **35.** x^5
37. 5^{-7} **39.** $15b^{-3}$ **41.** c^{-3} **43.** $2b^4$ **45.** 6^8
47. 216 **49.** 192 **51.** $8x^3y^5$ **53.** 8^{-10} **55.** w^{18}
57. d^{-24} **59.** $p^{-5}q^{-5}$ **61.** $3^{-2}x^{-2}$ **63.** $5 \cdot 2^{-3}r^{-3}$
65. $\dfrac{a^{12}}{c^6}$ **67.** $\dfrac{1}{8u^{14}}$ **69.** $\dfrac{5}{6k^{10}}$ **71.** $\dfrac{p^{10}}{4}$ **73.** $\dfrac{1 + 2a}{1 - 2a}$
75. $\dfrac{1 - c}{c}$ **77.** $\dfrac{y + x}{xy}$ **79.** $\dfrac{w}{v}$ **81.** 10^5 **83.** 10^{-2}
85. 10^2 **87.** 10^1 **89.** 10^{-8}

Homework 6.3

1. 43,000 **3.** 0.000 008 **5.** 0.000 02 **7.** 10^2
9. 10^{-1} **11.** 10^6 **13.** 4.834×10^3 **15.** 7.2×10^{-2}
17. 7×10^{-6} **19.** 6.85×10^8 **21.** 5.674×10^5
23. 3.85×10^{-1} **25.** Positive **27.** Negative
29. Positive **31.** Negative **33.** 2.9141×10^4 ft
35. 7.6×10^{-5} cm **37.** 0.000 005 g **39.** 5,000,000,000 yr
41. 140 **43.** 0.04 **45.** 2400 **47.** $\approx 5.88 \times 10^{12}$ mi
49a. $\approx 1.30 \times 10^9$ ft **b.** $\approx 2.46 \times 10^5$ mi
51a. 1.12×10^{18} insects **b.** 3.36×10^{14} g
53. $\approx 1.14 \times 10^8$ gal

Midchapter Review

1. $b^2 \cdot b^4$ is a product of powers with the same base, therefore, the exponents are added; $b^2 \cdot b^4 = b^{2+4} = b^6$; $(b^2)^4$ is a power raised to a power, therefore, the exponents are multiplied; $(b^2)^4 = b^{2 \cdot 4} = b^8$

3. A negative exponent is the reciprocal of the corresponding positive power. Example: $2^{-3} = \dfrac{1}{2^3} = \dfrac{1}{8}$. **5.** There should be exactly one nonzero digit to the left of the decimal point.
7a. The power of a sum is not equal to the sum of the powers.
b. In a quotient of powers with the same base, subtract the exponents. $\dfrac{m^{-8}}{m^{-4}} = m^{-8-(-4)} = m^{-4}$ **9.** $-8x^6$ **11.** $576b^{11}$
13. $324a^6$ **15.** $2t^6(1 - 6t^6)$ **17.** $\dfrac{1}{16x^4}$ **19.** $\dfrac{4}{3x^4}$
21. $\dfrac{b^{16}}{3}$ **23.** $\dfrac{1}{t}$ **25.** 0.000 234 **27.** 10^3
29. 6.3×10^{-3} **31.** 10^{26} **33.** 400,000 lb

Homework 6.4

1. True **3.** False **5.** False **7.** y^4 **9.** n^{18}
11. $\pm 4x^2$ **13.** $-11ab^3$ **15.** $3(x + y)$ **17.** $\dfrac{-8}{b^3}$
19. $2\sqrt{2}$ **21.** $-2\sqrt{5}$ **23.** $5\sqrt{5}$ **25.** $x\sqrt{x}$
27. $-b^5\sqrt{b}$ **29.** $p^{12}\sqrt{p}$ **31.** $2a\sqrt{2a}$ **33.** $\pm 6m^4\sqrt{2m}$
35. $\dfrac{x^4}{3\sqrt{3}}$ **37.** $4c^3\sqrt{3d}$ **39.** $\dfrac{-3bd\sqrt{5d}}{2}$ **41.** $\dfrac{3w\sqrt{w}}{2\sqrt{7z}}$
43. $6x\sqrt{x}$ **45.** $-10a^2b\sqrt{2a}$ **47.** $-2bk\sqrt{bk}$ **49.** $3\sqrt{3}$
51. Cannot be simplified **53.** $-7\sqrt{6}$ **55.** $-3\sqrt{5}$
57. $-3\sqrt{3} - 3\sqrt{2}$ **59.** $-2\sqrt{2a}$ **61.** $-3x\sqrt{5x} - 6x\sqrt{5}$
63. The square root of a sum does not equal the sum of the square roots. Addition under the radical should be done first. $\sqrt{36 + 64} = \sqrt{100} = 10$ **65.** When adding like radicals, add their coefficients and leave the radicand unchanged. $\sqrt{3} + \sqrt{3} = 2\sqrt{3}$
67. The square root of a sum does not equal the sum of the square roots. Because 9 and x^2 are not like terms, the addition cannot be done; therefore, $\sqrt{9 + x^2}$ cannot be simplified. **69.** (c)
71. (b) **73.** (a) **75.** $\sqrt{20}$ **77.** $\sqrt{12}$ **79.** $\sqrt{54}$
81. Valid **83.** Invalid; $r^2 - s^2 \neq (r - s)^2$

Homework 6.5

1. 4 **3.** $x\sqrt{6}$ **5.** $36a^2$ **7.** $3\sqrt{2} + \sqrt{6}$
9. $4\sqrt{5} + 10\sqrt{3}$ **11.** $2p\sqrt{2} - 2p\sqrt{2p}$ **13.** $1 - 2\sqrt{2}$
15. $16 - a$ **17.** $7 + 4\sqrt{3}$ **19.** $2w - 3\sqrt{5w} - 10$
21. $3 - \sqrt{5}$ **23.** $\dfrac{-4 + \sqrt{2}}{2}$ **25.** $\dfrac{2a - \sqrt{2}}{2a}$
27. $x = \dfrac{-1 \pm 2\sqrt{2}}{2}$ **29.** $4 \pm \sqrt{5}$ **31.** $x = -3 \pm 3\sqrt{2}$
33. $x = \pm\sqrt{b^2 - a^2}$ **35.** $x = \pm 2\sqrt{1 + y^2}$ **37.** $\dfrac{5 + 6\sqrt{2}}{4}$
39. $\dfrac{9 + \sqrt{3}}{6a}$ **41.** $\dfrac{3\sqrt{3} + 6}{2}$ **43.** $\dfrac{3 - 8\sqrt{y}}{4}$
45. Only like radicals can be added or subtracted. Since the radicands are different, $8\sqrt{7} - 2\sqrt{5}$ cannot be subtracted.
47. The product rule for radicals states that two numbers can be multiplied if both are under the radical.
$6\sqrt{8} = \sqrt{36}\sqrt{8} = \sqrt{36 \cdot 8} = \sqrt{288}$ **49.** Since 5 is a term of the numerator, not a factor, it cannot be canceled. If we factor the numerator, then $\dfrac{5 - 10\sqrt{3}}{5} = \dfrac{5(1 - 2\sqrt{3})}{5} = 1 - 2\sqrt{3}$.
51. $0 = 0$ **53.** $8 + 4\sqrt{3} = 8 + 4\sqrt{3}$ **55.** $0 = 0$
57. $\sqrt{9} = 3$ **59.** $\sqrt{25x^2} = 5x$ **61.** $\dfrac{4}{3}$ **63.** $\dfrac{5\sqrt{2}}{2}$

65. $\sqrt{2a}$ **67.** $\sqrt{7b}$ **69.** $\dfrac{\sqrt{21x}}{6}$ **71.** $\dfrac{\sqrt{15}}{3}$

73. $\dfrac{\sqrt{6}-2}{\sqrt{2}}$ **75.** $\dfrac{2\sqrt{2}-2}{3\sqrt{6}}$ **77a.** $h = \dfrac{w\sqrt{3}}{2}$

b. $\dfrac{w^2\sqrt{3}}{4}$ **79a.** $h = \dfrac{k\sqrt{14}}{2}$ **b.** $\dfrac{k^3\sqrt{14}}{6}$

Homework 6.6

1. $x = 21$; $5 = 5$ **3.** $x = 81$; $5 = 5$ **5.** $x = 4$; $4 \neq 8$.
Therefore, NO solution.   **7.** $x = 5$; $8 = 8$ **9.** $x = 5$; $5 = 5$
11a. $x = 16$ **b.** $x = 6.25$ **c.** No solution   **d.** $x = 28.09$
13a.

x	4	5	6	10	16	19	24
y	0	1	1.4	2.4	3.5	3.9	4.5

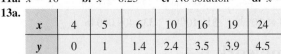

b. $x = 13$

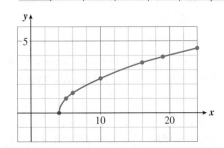

15a.

x	-3	-2	0	1	4	8	16
y	4	3	2.3	2	1.4	0.7	-0.4

b. $x = 6$

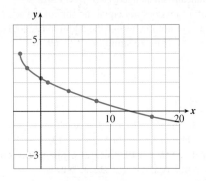

17. $x = 1$ is the only solution.   **19.** $x = 5$ is the only solution.
21. $x = -3$ is the only solution.   **23.** $x = \dfrac{-3}{4}$ is the only
solution.   **25.** $x = 8$ is the only solution.   **27a.** ≈ 144.5 km
b. It is at least $161\frac{1}{3}$ m.   **29.** It is at least 24,867 ft.
31a. $r \approx 3$ in.   **b.** ≈ 2091 in.^{3} **33.** $\sqrt{5}$ or $-\sqrt{5}$
35. -125 **37.** 16   **39.** $\sqrt[3]{7}$

Chapter Summary and Review

1. Third law of exponents: $(a^m)^n = a^{m \cdot n}$; To raise a power to a
power, multiply the exponents; First law of exponents:
$a^m \cdot a^n = a^{m+n}$; To multiply powers with the same base, add the
exponents.   **3.** Negative expression: $a^{-n} = \dfrac{1}{a^n}$;

Example: $5^{-2} = \dfrac{1}{5^2} = \dfrac{1}{25}$ **5.** $1 = \dfrac{2^5}{2^5} = 2^{5-5} = 2^0$

7. Product rule for radicals: $\sqrt{ab} = \sqrt{a}\,\sqrt{b}$; Quotient rule for
radicals: $\sqrt{\dfrac{a}{b}} = \dfrac{\sqrt{a}}{\sqrt{b}}$ **9.** To simplify a square root: 1.
Factor any perfect squares from the radicand. 2. Use the product
rule to write the radical as a product of two roots. 3. Simplify the
square root of the perfect squares.   **11.** To rationalize the denominator, multiply the numerator and denominator by any radicals
in the denominator that cannot be simplified.   **13a.** a^{10}
b. a^{24} **15a.** $8a^6$ **b.** $2a^8$ **17a.** $32x^7$
19a. $-32t^8$ **21a.** $\dfrac{3}{x^2}$ **b.** $\dfrac{1}{9x^2}$
23a. $\dfrac{z^2}{25}$ **b.** $5z^2$ **25.** $\dfrac{6}{p^7}$ **27.** $\dfrac{8}{k}$ **29.** 5
31. $\dfrac{2xy}{y+x}$ **33.** 5.86×10^5 **35.** 7×10^{-4}
37. 4.83×10^5 **39.** $18{,}240{,}000{,}000{,}000{,}000$
41. $0.000\,025$ **43.** $\approx 9.99 \times 10^{-4}$ kg   **45a.** $2x^3$
b. Cannot be simplified   **c.** $(4 + x)^3$ **47.** $-3m^2\sqrt{3m}$
49. $\dfrac{a\sqrt{ac}}{4}$ **51.** $\dfrac{4b^2\sqrt{3b}}{3}$ **53.** $\sqrt{6} + 6\sqrt{2}$
55. $w^3\sqrt{6}$ **57.** $2 - 2\sqrt{3}$ **59.** $\dfrac{4 - 3\sqrt{3}}{6}$
61. $a = \dfrac{2 \pm 2\sqrt{6}}{3}$ **63.** $b = \dfrac{\pm\sqrt{c^2 - 2a^2}}{2}$
65. $\sqrt{6} - 3\sqrt{2}$ **67.** $4 - d$ **69.** $16 + 6\sqrt{7}$
71. $\dfrac{2\sqrt{x}}{x}$ **73.** $\dfrac{\sqrt{10}}{2}$ **75.** $\dfrac{5\sqrt{7}}{7}$ **77.** $0 = 0$
79. $2\sqrt{10a} = c$ **81.** $x = 7$ is the solution.
83. $x = 12$ is the only solution.   **85.** $x = 41$ **87.** ≈ 90 ft

Homework A.1

1. $\dfrac{3}{5}$ **3.** $\dfrac{4}{25}$ **5.** $\dfrac{1}{3}$ **7.** $\dfrac{3}{4}$ **9.** $\dfrac{3}{5}$ **11.** 12
13. $\dfrac{6}{9}$ **15.** $\dfrac{12}{8}$ **17.** $\dfrac{40}{100}$ **19.** $\dfrac{24}{72}$ **21.** $\dfrac{90}{144}$
23. $\dfrac{0}{6}$

Homework A.2

1. $\dfrac{10}{21}$ **3.** $\dfrac{4}{5}$ **5.** $\dfrac{1}{2}$ **7.** $\dfrac{1}{3}$ **9.** $\dfrac{1}{2}$ **11.** $\dfrac{3}{5}$
13 $\dfrac{6}{5}$ **15.** $\dfrac{5}{12}$ **17.** $\dfrac{2}{15}$ **19.** 18   **21.** $\dfrac{22}{3}$ **23.** 5

Homework A.3

1. $\dfrac{4}{11}$ **3.** $\dfrac{7}{15}$ **5.** $\dfrac{2}{3}$ **7.** $\dfrac{1}{2}$ **9.** 1   **11.** $\dfrac{2}{5}$
13. $\dfrac{1}{4}$ **15.** $\dfrac{3}{5}$ **17.** 6   **19.** 12   **21.** 21   **23.** 60
25. 30   **27.** 90   **29.** 36   **31.** 72   **33.** $\dfrac{5}{6}$ **35.** $\dfrac{1}{12}$
37. $\dfrac{17}{24}$ **39.** $\dfrac{7}{30}$ **41.** $\dfrac{37}{30}$ **43.** $\dfrac{73}{120}$ **45.** $\dfrac{41}{100}$
47. $\dfrac{479}{840}$

Homework A.4

1. $\dfrac{11}{3}$ **3.** $\dfrac{25}{2}$ **5.** $\dfrac{104}{5}$ **7.** $\dfrac{221}{50}$ **9.** $\dfrac{39}{8}$

11. $\dfrac{149}{12}$ **13.** $\dfrac{55}{8}$ **15.** $\dfrac{139}{24}$ **17.** $\dfrac{33}{2}$ **19.** $\dfrac{50}{7}$

21. $\dfrac{5}{12}$ **23.** 2 **25.** $3\dfrac{2}{3}$ **27.** $5\dfrac{3}{8}$ **29.** $6\dfrac{11}{16}$

31. $9\dfrac{29}{32}$

Homework A.5

1. $\dfrac{17}{100}$ **3.** $\dfrac{7}{100}$ **5.** $\dfrac{23}{1000}$ **7.** $\dfrac{3}{5}$ **9.** $\dfrac{13}{50}$ **11.** $\dfrac{3}{8}$

13. $\dfrac{9}{4}$ **15.** $\dfrac{18}{5}$ **17.** 0.84 **19.** 0.46 **21.** 0.31

23. 0.031 **25.** 0.3125 **27.** 0.375 **29.** $0.8\overline{3}$

31. $0.\overline{27}$

Homework A.6

1. 10; 14.8; 14.77; 14.774 **3.** 80; 76.3; 76.28; 76.283
5. 170; 169.9; 169.90; 169.899 **7.** 5550; 5545.9; 5545.91;
5545.910 **9.** 700; 701.0; 700.96; 700.960 **11.** 20; 20.0;
19.95; 19.951 **13.** 1.9; 1.91; 1.907 **15.** 0.9; 0.92; 9.920
17. 0.1; 0.10; 0.099 **19.** 6.2; 6.17; 6.170

Homework A.7

1. 4.73 **3.** 4 **5.** 1.66 **7.** 13.26 **9.** 6.227
11. 34.45 **13.** 3.47 **15.** 5.74 **17.** 5.07 **19.** 9.74
21. 362.15 **23.** 12.9994 **25.** 3.71 **27.** 542.046
29. 577.08

Homework A.8

1. 4.08 **3.** 0.128 **5.** 0.0408 **7.** 0.12036 **9.** 3.25
11. 10.368 **13.** 1.7304 **15.** 0.00384 **17.** 4.51
19. 4.21875 **21.** 13.7 **23.** 15.3 **25.** 605 **27.** 230.4
29. $4.\overline{54}$ **31.** 800

Homework A.9

1. 0.15 **3.** 0.004 **5.** 0.068 **7.** 1.19 **9.** 0.0325
11. 0.004 **13.** 33% **15.** 50.4% **17.** 78.7%

19. 2.01% **21.** 0.8% **23.** 550% **25.** $\dfrac{7}{20}$ **27.** $\dfrac{5}{4}$

29. $\dfrac{3}{5}$ **31.** $\dfrac{9}{1000}$ **33.** $\dfrac{3}{8}$ **35.** $\dfrac{1}{3}$ **37.** 75%

39. 37.5% **41.** 100% **43.** 225% **45.** 240%
47. 0.4%

Homework A.10

1. 7 **3.** 3 **5.** 147 **7.** 2; 5 **9.** 5 **11.** 50 + 28
13. 30 + 40 **15.** 10 · 7 **17.** 100 · 13 **19.** 20 · 30

Homework A.11

1. Mean: 12; median: 12; mode: 12 **3.** Mean: 20; median: 18;
mode: 12 **5.** Mean: $29,998; median: $29,211; mode: none

7. Mean: 150.625; median: 149.5; mode: 148 **9.** Mean:
68.625; median: 67.5; mode: none **11a.** Mean: $470,000;
median: $120,000; mode: $100,000 **b.** Knowing the mean
price and the number of houses allows you to compute the total
price of all the houses. However, the highest-priced house is so
much more expensive than the rest that it causes the mean to be
higher than all but one price. The median tells a middle price (half
the prices will be less than and half greater than the median) but
gives you no way to estimate what the highest price is. The mode
is the most common price but tells nothing about whether most of
the prices are higher or lower. **13.** Mean: 2.42

Homework B.1

1. See page 789. **3.** b, d **5.** a **7.** $2 \cdot 3^3$
9. $2^2 \cdot 5 \cdot 3^2$ **11.** $2 \cdot 3 \cdot 5 \cdot 7$ **13.** 5^4

Homwork B.2

1. Add $\left(\dfrac{b}{2}\right)^2$ **3.** Factoring, extracting roots, completing the
square. (The quadratic formula gives still another method.)

5. $3w^2 - w = -6$ **7.** $m^2 - \dfrac{1}{2}m = 3$ **9.** $r^2 + \dfrac{3}{2}r = -3$

11. $x^2 + 2x + 1 = (x + 1)^2$ **13.** $x^2 - 6x + 9 = (x - 3)^2$

15. $x^2 - 14x + 49 = (x - 7)^2$ **17.** $x^2 + 3x + \dfrac{9}{4} = \left(x + \dfrac{3}{2}\right)^2$

19. $x^2 + \dfrac{4}{3}x + \dfrac{4}{9} = \left(x + \dfrac{2}{3}\right)^2$ **21.** $x^2 - \dfrac{3}{2}x + \dfrac{9}{16} = \left(x - \dfrac{3}{4}\right)^2$

23. $x = -6, x = 2$ **25.** $p = 1, p = 4$ **27.** $z = -5, z = -4$

29. $y = 1 \pm \sqrt{2}$ **31.** $z = \dfrac{3}{2} \pm \dfrac{\sqrt{21}}{2}$ **33.** $r = -3 \pm 2\sqrt{3}$

35. $x = \dfrac{1}{2}, x = -\dfrac{3}{2}$ **37.** $z = \dfrac{2}{3}, z = \dfrac{3}{2}$ **39.** $t = 3, t = -\dfrac{5}{2}$

41. $b = -\dfrac{1}{3} \pm \dfrac{\sqrt{7}}{3}$ **43.** $p = \dfrac{7}{4} \pm \dfrac{\sqrt{33}}{4}$

45. $k = -2 \pm \dfrac{\sqrt{10}}{2}$

Homework B.3

1. In an improper algebraic fraction, the degree of the numerator is
greater than or equal to the degree of the denominator.
I3. Division, multiplication, subtracting, "bringing down"

5a. $\dfrac{312}{24} = 13, R = 0$ **b.** $\dfrac{312}{24} = \dfrac{13 \cdot \cancel{4} \cdot \cancel{3} \cdot \cancel{2}}{\cancel{4} \cdot \cancel{3} \cdot \cancel{2}} = 13$

7. $b^3 + 0b^2 + 0b - 1$ **9.** $-3c^4 + 0c^3 + 0c^2 + 2c + 0$

11. $x + 1 - \dfrac{1}{x + 2}$ **13.** $x - 2 + \dfrac{1}{x + 5}$

15. $2x - 1 - \dfrac{1}{x + 1}$ **17.** $x + 3$ **19.** $2x - 3 - \dfrac{2}{2x + 1}$

21. $2x^2 - x + 1$ **23.** $x - 1 + \dfrac{2}{x + 1}$ **25.** $x - 6 + \dfrac{29}{x + 6}$

27. $2x^2 + x + 1 + \dfrac{2}{x - 1}$ **29.** $x^2 + 2x + 4$

31. $x^2 + x + 2 + \dfrac{4}{3x - 2}$ **33.** $3x^2 - 4 + \dfrac{8}{2x + 5}$

35. $3x^3 + 6x^2 + 12x + 26 + \dfrac{52}{x - 2}$ **37.** $x^2 + 4 + \dfrac{8}{x^2 - 4}$

39. $x + 3 + \dfrac{1}{2x^2 + 1}$ **41.** $(2x - 1)(x + 1) - 1 = 2x^2 + x - 2$

43. $(x^2 + x + 2)(3x - 2) + 4 = x^2 + 4x + 3x^3$ **45a.** $x - 15$
b. $(x - 12)(x - 15)$ **47a.** $x^2 + 5x + 6$
b. $(x - 2)(x + 2)(x + 3)$ **49.** $a^2 - a + 1$

Homework B.4

1.

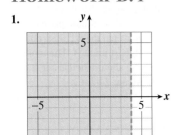

3.

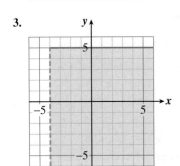

5.

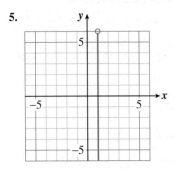

7.

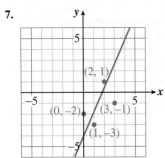

c. $(3, -1)$ and $(1, -3)$ **9.** See page 810.   **11.** Yes
13. No   **15.** Yes

17.

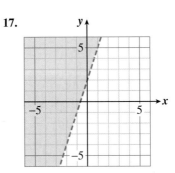

19.

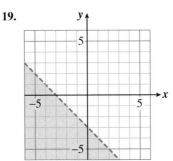

21.

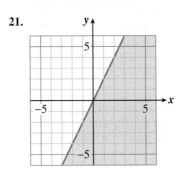

23.

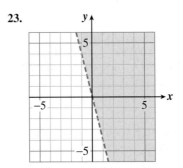

25.

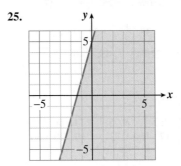

27.

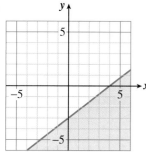

29.

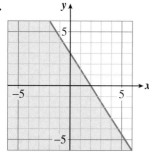

31.

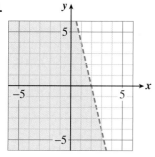

33.

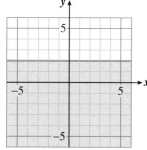

35.

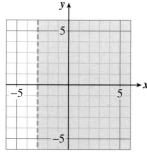

37.

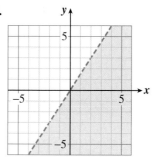

39.

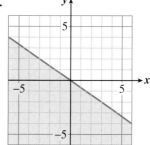

Homework B.5

1. $\sqrt{-1}$ **3.** See page 817.   **5.** See page 817.
7. The conjugate of the denominator   **9.** $b^2 - 4ac$
11. $3i$; The square of a real number cannot be negative.

13. $4 - 8i$ **15.** $-13 + 7i$ **17.** $-1 - i$ **19.** $\dfrac{25}{6} - \dfrac{1}{2}i$

21. $5.6 - 2.7i$ **23.** $12 + 21i$ **25.** $5 - 14i$
27. $-7 - 24i$ **29.** $14 + 8\sqrt{2}i$ **31.** 18 **33.** $-3 - 4i$

35. $-9 - 5i$ **37.** $\dfrac{3}{4} - \dfrac{1}{2}i$ **39.** $4 + 3i$ **41.** $-1 + 7i$

43. $\dfrac{-8}{17} + \dfrac{15}{17}i$ **45.** $-\dfrac{3}{2} - \dfrac{1}{2}i$ **47.** $\dfrac{3}{4} - \dfrac{\sqrt{3}}{4}i$

49. $\dfrac{-2}{3} + \dfrac{\sqrt{5}}{3}i$ **51.** $\dfrac{-3}{4} \pm \dfrac{\sqrt{7}}{4}i$ **53.** $\dfrac{1}{2} \pm \dfrac{\sqrt{2}}{2}i$

55. $\dfrac{2}{3} \pm \dfrac{\sqrt{5}}{3}i$ **57.** $\dfrac{1}{5} \pm \dfrac{\sqrt{46}}{10}i$ **59.** $\dfrac{4}{3} \pm \dfrac{\sqrt{10}}{3}$

61. $D = -47$; two complex-conjugate solutions   **63.** $D = 0$;
two identical real solutions   **65.** $D = 4$; two distinct real
solutions   **67.** $D = 9217$; two distinct real solutions
69. $D = 0$; two identical real solutions   **71.** $D < 0$; two complex-conjugate solutions   **73.** $D < 0$; two complex-conjugate
solutions   **75.** $D > 0$; two distinct real solutions
77. $D > 0$; two distinct real solutions   **79.** $D > 0$; two distinct
real solutions   **81.** $-1 - 17i$ **83.** $38 + 46i$

85. $\dfrac{-40}{29} + \dfrac{16}{29}i$

Index